Smart Sensors, Measurement and Instrumentation

Volume 51

The Smart Sensors, Measurement and Instrumentation series (SSMI) publishes new developments and advancements in the fields of Sensors, Instrumentation and Measurement technologies. The series focuses on all aspects of design, development, implementation, operation and applications of intelligent and smart sensors, sensor network, instrumentation and measurement methodologies. The intent is to cover all the technical contents, applications, and multidisciplinary aspects of the field, embedded in the areas of Electrical and Electronic Engineering, Robotics, Control, Mechatronics, Mechanical Engineering, Computer Science, and Life Sciences, as well as the methodologies behind them. Within the scope of the series are monographs, lecture notes, selected contributions from specialized conferences and workshops, special contribution from international experts, as well as selected PhD theses.

Indexed by SCOPUS and Google Scholar.

Gianluca Rossi

Mechanical and Thermal Measurements by Images and Waves

Principles and Applications

Springer

Gianluca Rossi
Dipartimento di Ingegneria
Università di Perugia
Perugia, Italy

ISSN 2194-8402 ISSN 2194-8410 (electronic)
Smart Sensors, Measurement and Instrumentation
ISBN 978-3-032-03463-2 ISBN 978-3-032-03464-9 (eBook)
https://doi.org/10.1007/978-3-032-03464-9

This Springer imprint is published by the registered company Springer Nature Switzerland AG
The registered company address is: Gewerbestrasse 11, 6330 Cham, Switzerland

"The sands of time were eroded by the river of constant change"
Genesis, Firth of Fifth, 1973

Preface

This book originated from a course I organized during the COVID-19 lockdown in 2020, held from my home for Mechanical Engineering students at the University of Perugia, and in part for my Ph.D. students at the University of Padova, both in Italy. The course aimed to outline the theoretical foundations common to various measurement techniques of shapes, motion, deformations, velocity and other mechanical and thermal fields, involving image analysis and wave beam scanning. It also presented the operating principles behind these techniques, many of which were developed and applied over 36 years of my research activity: first at the University of Ancona, now Politecnica delle Marche, (from 1988 to 1997) and later at the University of Perugia (from 1998 to 2025). These methods have been widely used in the field of experimental mechanics and measurement science, with applications spanning engineering, medicine, design, diagnostics, art analysis and digitalization, digital twin development, machine vision, and industrial quality control.

All these measurement techniques rely on the processing of images and wave phenomena to carry out mechanical and thermal measurements useful to take decisions, qualifications, or improvement. As illustrated in the following picture, our sun emits visible light along with other forms of electromagnetic radiation, which illuminates the surface of a lake, where waves transmit both energy and information useful to us.

So, it is a pleasure for me to acknowledge all the people working with me in all my scientific career, started in 1988 with my scientific father Enrico Primo Tomasini at the University of Ancona where, together with my first colleague and friend Nicola Paone, I start my first laboratory experiences in Laser Dopper Vibrometry and Anemometry, Particle Image Velocimetry, acoustic field measurements, and many other techniques developed together for vibration measurements on mechanical components, turbomachines, single- and two-phase flows, to solve industrial problems, to make experimental studies, etc. Many of these research works are illustrated in the book as applications of the measurement techniques explained.

Since 1998, at the University of Perugia in the research group I managed, where the director in these years, I wish to thanks, Prof. Raffale Balli, not yet with us, allow me to work in a very good way. In Perugia, I work mainly on the application and development of measurement techniques of temperature fields by thermography, stress field by thermoelectricity, shape measurements by image analysis, deformation, motion, and vibration measurement by digital image correlation, optical flow, and markers. Many of these experiences are illustrated in the book as applications, after the description of the measurement techniques itself.

I would like to extend my sincere thanks to all my collaborators at the University of Perugia since 1998, beginning with Roberto Marsili, who started as a fellow researcher and has been working alongside me ever since.

I am deeply grateful to all the Ph.D. students, postdoctoral researchers, and research fellows who have worked with me over the years. It all began with the remarkable group I fondly refer to as the "magnificent four" during the early period of 1998–2002: Nicola Borgarelli, Giuseppe Brustenga, Claudio Rondini, and Paolo Maggiorana. A few years later, I had the pleasure of working with Matteo Becchetti,

Michele Moretti, and Jacopo Pirisinu. Between 2003 and 2020, many more talented individuals joined the team, including Gloria Allevi, Riccardo Bartolucci, Maurizio Bosi, Giacomo Buitoni, Luigi Casacanditella, Marco Cibeca, Matteo Cruciani, Antonio Di Renzo, Roberta Flori, Ilario Lori, Emiliano Mingardi, Emanuele Moriconi, Francesco Panerai, Alessandra Salviuolo, Erik Tagliaferri, Roberto Tomassini, Tommaso Tocci, and Iva Xhimitiku.

I would also like to acknowledge my recent team at the University of Perugia, whose contributions have been invaluable in recent years and continue to this day. Once again, special thanks go to Roberto Marsili, Chiara Valenti, and Sara Fabri for remaking some figures, Lorenzo Capponi, who also assisted in refining Chap. 8. I am equally grateful to Giulio Tribbiani for his support in improving Chap. 6, Tiberio Truffarelli for his contributions to Chap. 7, and Federico Paolucci for his work on Chap. 5. These chapters reflect the core areas of their respective scientific research and Ph.D. work. I would also like to thank my colleague and friend Giuseppe Saccomandi for helping me improve the derivation of the d'Alembert equation from the Navier-Stokes equations in Chap. 1, and Luigi Montalto for reviewing the part on photoelasticity.

I would also like to thank Stephanie De Matos for her highly professional work in redesigning many of the figures included in this book. Finally, my heartfelt gratitude goes to my family, not for scientific or technical support, but for something even more essential: their unwavering presence and encouragement. To my father Bruno, who is no longer with us, my mother Neda, my partner Susanna, my daughter Giulia, and my son Emanuele. Emanuele has supported me not only emotionally but also practically, helping me to process and organize many of the formulas and tables in this book using innovative methods based on artificial intelligence, techniques he has been patiently teaching me.

Perugia, Italy Gianluca Rossi

Competing Interests The author has no competing interests to declare that are relevant to the content of this manuscript.

Contents

Chapter 1
Fields and Waves Fundamentals

Abstract This chapter provides a concise introduction to the physical concept of a *field*, with a focus on fluid dynamics and electromagnetic fields. It explores how the fundamental laws and properties of fields give rise to wave phenomena. The classical wave equation formulated by d'Alembert, whose solutions describe wave propagation, is derived from both Maxwell's equations for the electromagnetic field and the Navier–Stokes equations for fluids. Additional examples are provided to demonstrate how the same wave equation emerges in the context of elastic continua, such as vibrating strings and metal bars. Key wave characteristics and phenomena relevant to measurement applications are discussed, including different types of waves, propagation behaviours, and interactions such as interference. The chapter also covers the behaviour of waves at interfaces between different media, explaining phenomena such as reflection and refraction. Snell's law is introduced and interpreted through Huygens' principle. Finally, the chapter explores advanced topics critical to modern measurement systems: the Doppler effect, wave polarization, the quantum description of waves via photon flux, and fundamental concepts related to LASER light sources. These elements are essential for measurement systems applications involving the generation and processing of images and waves.

1.1 Field

In physics, a field F is a quantity, typically a single number, or more numbers like a vector or a matrix (a tensor), defined on every point P in space and time. In the simplest way it is a function that give us a value, named F(x, y, z, t), where x, y, z is the position of the point P at time instant t. In many cases it describes how forces are distributed in space, affecting any object that comes within its influence. The concept of force field in physics is an evolution of the initial idea of direct interaction (between masses, electric charges, etc.), i.e. the presence of something that changes the space around it.

For instance, in the case of a gravitational field, the classical physic concept describes how an object with mass m_2 creates an invisible field F around it that affects

G. Rossi, *Mechanical and Thermal Measurements by Images and Waves*, Smart Sensors, Measurement and Instrumentation 51, https://doi.org/10.1007/978-3-032-03464-9_1

other objects with mass m_1. This force is what we experience as the gravitational attraction ($f = G(m_1 \cdot m_2)/r^2$).

So, we can say that exist the gravitational field $F = G\, m_2/r^2$ (force per unit of mass), generated by the mass m_2 that multiplied by the mass m_1 gives us the force f.

Similarly, an electric field describes how electric charges create a force field around them that affects other charges. This influence determines how other charges move and interact within the field.

An example of scalar fields defined by a single numerical value on each point could be:

- A temperature field $T(x, y, z, t)$: temperature in a room at point P(x, y, z) and time t;
- A pressure field $p(x, y, z, t)$: for example, the ground atmospheric pressure in a position P(x, y, z) and time t illustrated in Fig. 1.1.

Example of vector fields are:

The wind velocity field $\mathbf{V}(x, y, z, t)$ in a region, illustrated in Fig. 1.2; The gravitational field $\boldsymbol{F}(x, y, z)$: gravitational force in a certain point P, normally not changing with time t.

Fields F(x, y, z) that don't depend on time are called static. Fields F(t) that do not depend on position are named uniform.

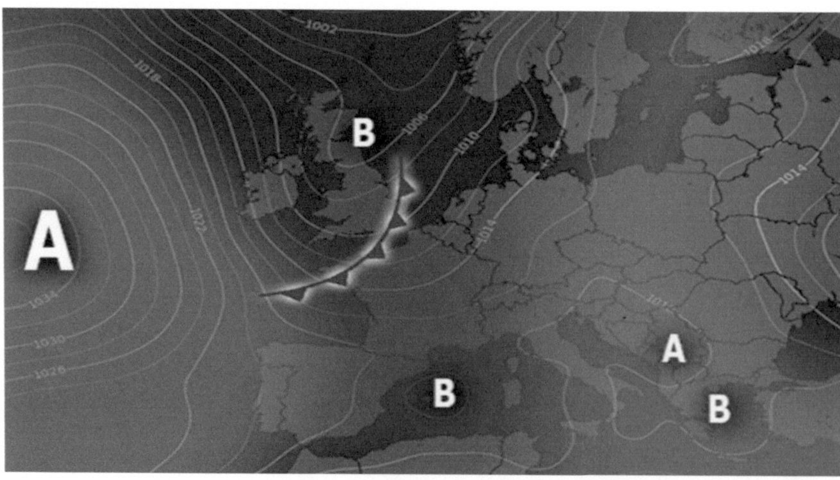

Fig. 1.1 A barometric pressure field at ground level influencing whether

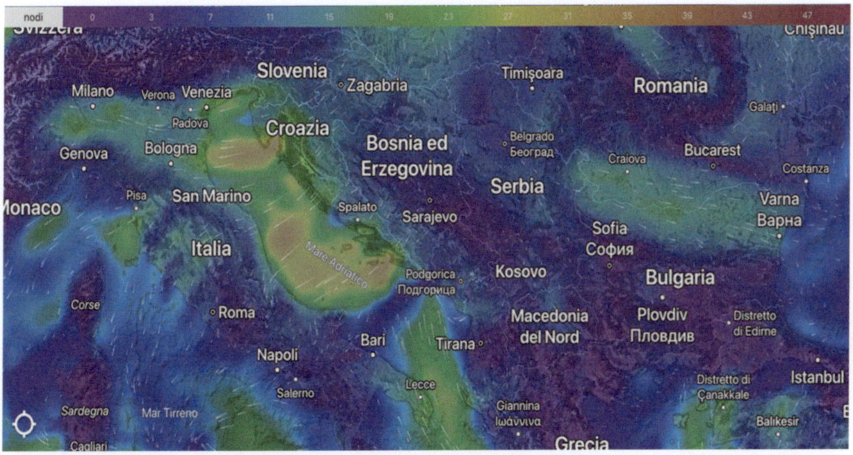

Fig. 1.2 A velocity field of the wind in a region at ground level

1.2 Field Differential and Integral Operators

To analyse how a field change in space and time is used the concept of partial derivatives. A partial derivative is the speed of change of a field in a particular direction (x, y or z) or in time t, so the ratio between field change ΔF to the corresponding displacement Δx, Δy, Δz or to the time evolution Δt. For infinitesimal changes of 3D fields we have these four speeds of change:

$$\frac{\partial F}{\partial x} \;;\; \frac{\partial F}{\partial y} \;;\; \frac{\partial F}{\partial z} \;;\; \frac{\partial F}{\partial t} \tag{1.1}$$

For example, the shape of a mountain summit like in Fig. 1.3 have two possible partial derivatives with respect to the variables x and y that are the slope of the surface moving along the two directions x and y, that could be Nord and East.

By partial derivatives can be defined operators that transform a field into another, the most useful operators are [1]:

- *Gradient;*
- *Divergence;*
- *Curl;*
- *Laplacian.*

Gradient is defined as a new vector field obtained by making the sum of the three partial derivatives of the original field, each multiplied with the tree versors **i**, **j**, **k** of the axis:

$$\nabla = \boldsymbol{i}\frac{\partial}{\partial x} + \boldsymbol{j}\frac{\partial}{\partial y} + \boldsymbol{k}\frac{\partial}{\partial z} \tag{1.2}$$

Fig. 1.3 Partial derivatives of a surface: slope of the two lines on xz and yz planes. A picture by Fulvio Zuanni of the author near Grunhorn summit in the Alps

If applied to a scalar-valued differentiable field θ, *gradient* is given by:

$$\nabla\theta = \boldsymbol{i}\frac{\partial\theta}{\partial x} + \boldsymbol{j}\frac{\partial\theta}{\partial y} + \boldsymbol{k}\frac{\partial\theta}{\partial z} \tag{1.3}$$

So gradient operator returns a vector field whose value at a point P is the direction and rate of fastest change. Naturally it could be applied to the three k = 1, 2, 3 components of a vector field giving the three vector fields:

$$\nabla A_k = \boldsymbol{i}\frac{\partial A_k}{\partial x} + \boldsymbol{j}\frac{\partial A_k}{\partial y} + \boldsymbol{k}\frac{\partial A_k}{\partial z} \quad k = 1, 2, 3 \tag{1.4}$$

Figure 1.4 gives an example of the gradient field vectors of an U shaped surface.

Divergence is defined as:

$$\nabla \cdot \mathbf{A} = \frac{\partial A_k}{\partial x} + \frac{\partial A_k}{\partial y} + \frac{\partial A_k}{\partial z} \tag{1.5}$$

Mathematically is the scalar product of gradient operator and vectorial field, remembering from linear algebra that the scalar product of two vector $\mathbf{v_1}$ and $\mathbf{v_2}$ is the number $v_1\ v_2\cos\ \alpha$, were α is the angle between vectors $\mathbf{v_1}$ and $\mathbf{v_2}$. If the two vectors have the cartesian components (a, b, c) and (d, e, g) respectively the scalar product is the number given by:

$$(\mathrm{a},\ \mathrm{b},\ \mathrm{c}) \cdot (\mathrm{d},\ \mathrm{e},\ \mathrm{g}) = (\mathrm{ad} + \mathrm{be} + \mathrm{cg}). \tag{1.6}$$

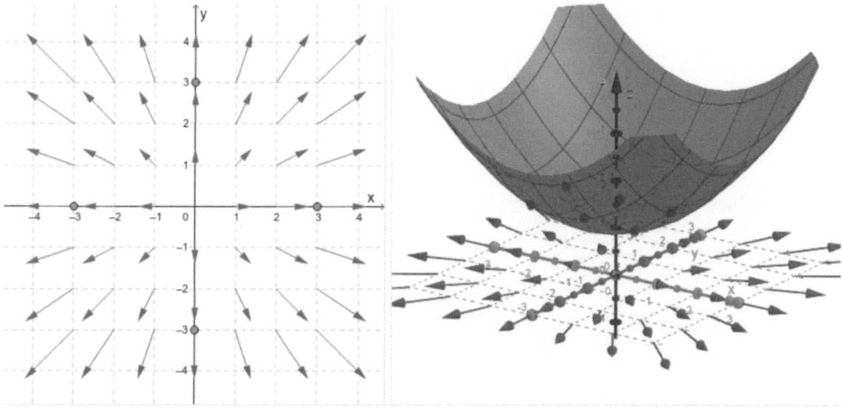

Fig. 1.4 Gradient field of a U-shaped surface

So:

$$\nabla \cdot \mathbf{A} = \left(\boldsymbol{i}\frac{\partial}{\partial x} + \boldsymbol{j}\frac{\partial}{\partial y} + \boldsymbol{k}\frac{\partial}{\partial z} \right) \cdot \left(A_x\boldsymbol{i} + A_y\boldsymbol{j} + A_z\boldsymbol{k} \right) \tag{1.7}$$

Can be seen as an operator that applied to a vector field produces a scalar field. The *curl* operator is defined as:

$$\frac{\mathrm{dx}}{\mathrm{dy}}\nabla \times \mathbf{A} = \boldsymbol{i}\left(\frac{\partial A_z}{\partial y} - \frac{\partial A_y}{\partial z} \right) + \boldsymbol{j}\left(\frac{\partial A_x}{\partial z} - \frac{\partial A_z}{\partial x} \right) + \boldsymbol{k}\left(\frac{\partial A_y}{\partial x} - \frac{\partial A_x}{\partial y} \right) \tag{1.8}$$

Remembering from linear algebra that cross (vector) product of two vector $\mathbf{v_1}$ and $\mathbf{v_2}$ of Cartesian component (a, b, c) and (d, e, f) is the vector given by:

$$(\mathrm{a, b, c}) \times (\mathrm{d, e, f}) = (\mathrm{bf} - \mathrm{ce, cd} - \mathrm{af, ae} - \mathrm{bd}) \tag{1.9}$$

That is a vector perpendicular to $\mathbf{v_1}$ and $\mathbf{v_2}$ plane, versus defined by right hand rule and of modulus equal to $v_1\ v_2\ \sin(\alpha)$, where α is the angle between $\mathbf{v_1}$ and $\mathbf{v_2}$. So is the result of cross product of gradient operator and a vector field, it gives a vector field that describes for each point magnitude and axis of maximum circulation.

Laplacian or Laplace operator $\Delta\Phi$ of a scalar field Φ is defined [1] as:

$$\Delta\Phi = \frac{\partial^2\Phi}{\partial x^2} + \frac{\partial^2\Phi}{\partial y^2} + \frac{\partial^2\Phi}{\partial z^2}. \tag{1.10}$$

It is applied on a scalar field and returns a scalar field. It could be seen also as the divergence of the gradient of a scalar field:

$$\Delta\Phi = \vec{\nabla} \cdot \vec{\nabla}\Phi = \nabla^2\Phi \tag{1.11}$$

since that:

$$\Delta = \nabla \cdot \nabla = \left(\boldsymbol{i}\frac{\partial}{\partial x} + \boldsymbol{j}\frac{\partial}{\partial y} + \boldsymbol{k}\frac{\partial}{\partial z} \right) \cdot \left(\boldsymbol{i}\frac{\partial}{\partial x} + \boldsymbol{j}\frac{\partial}{\partial y} + \boldsymbol{k}\frac{\partial}{\partial z} \right) = \frac{\partial^2}{\partial x^2} + \frac{\partial^2}{\partial y^2} + \frac{\partial^2}{\partial z^2} \tag{1.12}$$

Naturally it could be applied on each of the A_k components of a vector field obtaining:

$$\Delta A_k = \frac{\partial^2 A_k}{\partial x^2} + \frac{\partial^2 A_k}{\partial y^2} + \frac{\partial^2 A_k}{\partial z^2} \quad k = 1, 2, 3 \tag{1.13}$$

To analyse global effects on fields it's possible to integrate (i.e. infinitesimal summing) fields, generally in the three main ways that are [1]:

Line integral it's a one-dimensional integral defined as:

$$\oint_C \mathbf{A} \cdot d\vec{l} \tag{1.14}$$

where **A** is the vector field, integrated along the line C. It is the sum of infinitesimal dot products of the vector value of the field **A** in each point of the line C with the infinitesimal vector element $d\vec{l}$ of the line C. It can be approximated by the sum:

$$\oint_C \vec{A} \cdot d\overrightarrow{l} \simeq \mathbf{A}_1 \cdot \overrightarrow{dl_1} + \mathbf{A}_2 \cdot \overrightarrow{dl_2} + \mathbf{A}_3 \cdot \overrightarrow{dl_3} + \cdots$$

i.e. the sum of the projections of field vector along the infinitesimal line element, as illustrated in Fig. 1.5a.

If the vector field is a force field the dot product of a force with an infinitesimal displacement is a work, so the line integral is the total work along the line path of the force field.

Flux of a vector field on a surface S is defined as:

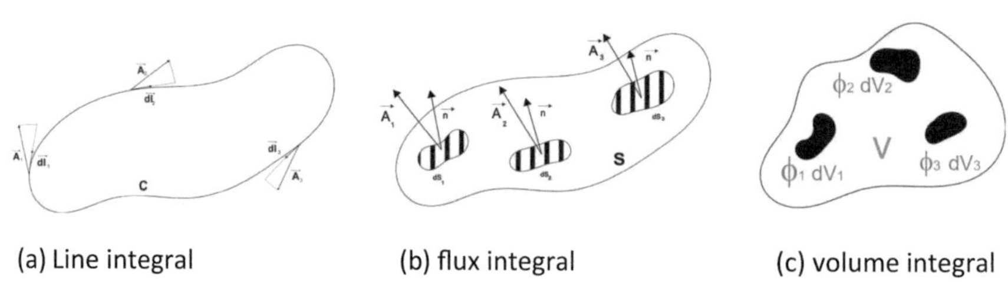

Fig. 1.5 Field integrals

$$\iint_S \mathbf{A} \cdot \vec{n} \cdot da = \iint_S \mathbf{A} \cdot d\vec{a} \tag{1.15}$$

Flux integral, as illustrated in Fig. 1.5b, can be approximated by the sum:

$$\iint_S \mathbf{A} \cdot \hat{n} da \simeq \mathbf{A}_1 \cdot \overrightarrow{da_1} + \mathbf{A}_2 \cdot \overrightarrow{da_2} + \mathbf{A}_3 \cdot \overrightarrow{da_3} + \cdots$$

If $\vec{A}$ is a velocity field, this integral gives us the volume flow rate across the surface S.

Volume integral of a scalar field Φ is defined as:

$$\iiint_V \Phi dV \tag{1.16}$$

Volume integral, as illustrated in Fig. 1.5c, can be approximated by the sum:

$$\iiint_V \Phi dV \simeq \Phi_1 dV_1 + \Phi_2 dV_2 + \Phi_3 dV_3 \ldots.$$

For example, if Φ is the density field of a solid, or of a fluid, this integral gives the value of the total mass of the volume V.

1.3 Electromagnetic Field Equations

Electromagnetic theory is based on the four Maxwell's equations that describe how electric and magnetic fields are related and change in space and time [2], they are:

1. Gauss's Law describing the relationship between static electric field and electric charge, in the integral form it states that the flux of the electric field out of an arbitrary closed surface is proportional to the electric charge enclosed by the surface:

$$\Phi_E = \iint_S \mathbf{E} \cdot d\vec{A} = \frac{q_{int}}{\varepsilon_0} \tag{1.17}$$

2. Gauss law for magnetism, states that the magnetic field has divergence equal to 0, so in integral form is:

$$\Phi_B = \iint_S \boldsymbol{B} \cdot d\vec{A} = 0 \tag{1.18}$$

3. Faraday's law of induction describing how a time-varying magnetic field correspond to the curl of an electric field. In the integral form it states that the work per unit charge required to move a charge around a closed loop equals the rate of change of the magnetic flux through the enclosed surface, so:

$$\oint_{\gamma} \mathbf{E} \cdot d\vec{l} = -\frac{d\Phi_{\mathrm{B}}}{dt} \tag{1.19}$$

4. Ampere-Maxwell's law that states that magnetic fields relate to electric current and changing electric fields. The integral form states that electric and displacement current (changing electric fields) are associated with a proportional magnetic field along any enclosing curve in this way:

$$\oint_{\gamma} \mathbf{B} \cdot d\vec{l} = \mu_0 \cdot i + \varepsilon_0 \cdot \mu_0 \frac{d\Phi_{\mathrm{E}}}{dt} \tag{1.20}$$

Maxwell equations can be written also in differential form as follows.

Gauss's law for electric field in general $\nabla \cdot \boldsymbol{E} = \frac{\rho}{\varepsilon_0}$; for free space, $\rho = 0$, so it becomes:

$$\nabla \cdot \boldsymbol{E} = 0 \tag{1.21}$$

Gauss's law for magnetism:

$$\nabla \cdot \boldsymbol{B} = 0 \tag{1.22}$$

Faraday's law of induction:

$$\nabla \times \boldsymbol{E} = -\frac{\partial \boldsymbol{B}}{\partial t} \tag{1.23}$$

Ampère's law (with Maxwell's correction):

$$\nabla \times \boldsymbol{B} = \mu_0 \varepsilon_0 \frac{\partial \boldsymbol{E}}{\partial t} \tag{1.24}$$

1.4 Fluidodynamic Fields Equations

In fluid mechanics, the physical state of a fluid is described by several key field variables: the velocity field $\mathbf{u}(\mathrm{x, y, z}, \ t)$, the pressure field $p(\mathrm{x, y, z}, t)$, the density field $\rho(\mathrm{x, y, z}, \ t)$, and the viscosity μ, which is generally considered constant for a given fluid. These fields obey the fundamental laws of mass, momentum, and energy conservation.

The laws governing fluid dynamics are expressed through the Navier-Stokes equations. These are partial differential equations that relate the changes in momentum and mass to external forces and internal stresses. The general form for a Newtonian, compressible fluid is:

$$\text{Mass conservation}: \frac{\partial \rho}{\partial t} + \nabla \cdot (\rho \mathbf{u}) = 0, \tag{1.25}$$

$$\text{Momentum conservation}: \rho\left(\frac{\partial \mathbf{u}}{\partial t} + (\mathbf{u} \cdot \nabla)\mathbf{u}\right) = -\nabla p + \mu \nabla^2 \mathbf{u} + \rho \mathbf{g}, \tag{1.26}$$

$$\text{Energy conservation}: \rho\left(\frac{\partial e}{\partial t} + (\mathbf{u} \cdot \nabla)e\right) = -p\nabla \cdot \mathbf{u} + \Phi + \nabla \cdot (k\nabla T). \tag{1.27}$$

Here, e denotes the specific internal energy, T the temperature, k the thermal conductivity, and Φ represents viscous dissipation, for a Newtonian fluid typically given by

$$\Phi = \mu\left(\nabla \mathbf{u} + (\nabla \mathbf{u})^T - \frac{2}{3}(\nabla \cdot \mathbf{u})\mathbf{I}\right) : \nabla \mathbf{u}$$

These equations must be supplemented by constitutive relations and an equation of state, such as:

$$p = p(\rho, T),$$
$$e = e(\rho, T),$$

to close the system. In many practical applications, especially when dealing with air and water at moderate speeds, fluids can be treated as incompressible. In this case, the mass conservation law simplifies to:

$$\nabla \cdot \mathbf{u} = 0. \tag{1.28}$$

The others governing equations simplify to:
Momentum conservation (Navier–Stokes equation):

$$\rho\big(\tfrac{\partial \boldsymbol{u}}{\partial t} + (\boldsymbol{u} \cdot \nabla)\boldsymbol{u}\big) = -\nabla p + \mu \nabla^2 \boldsymbol{u} + \rho \boldsymbol{g} \tag{1.29}$$

Energy conservation (temperature equation). Assuming constant specific heat c_p and thermal conductivity k, the temperature evolution is governed by:

$$\rho c_p\big(\tfrac{\partial T}{\partial t} + \boldsymbol{u} \cdot \nabla T\big) = k\nabla^2 T + \Phi \tag{1.30}$$

where the viscous dissipation term Φ is given by:

$$\Phi = 2\mu \boldsymbol{D} : \boldsymbol{D}, \text{ with } \boldsymbol{D} = \tfrac{1}{2}\left(\nabla \boldsymbol{u} + (\nabla \boldsymbol{u})^T\right) \tag{1.31}$$

We remember that the assumptions to deduce these equations are

- Constant density: $\rho = \text{const.}$
- Constant dynamic viscosity μ and thermal conductivity k
- No compressibility effects: $\nabla \cdot \boldsymbol{u} = 0$
- No pressure work (i.e., $-p\nabla \cdot \boldsymbol{u} = 0$)

1.5 From Fields to Waves

Any little disturbance of a field (be it mechanical motion, or a pressure fluctuation in a fluid, a stress fluctuation in a solid, an electric or magnetic fluctuation, etc.) propagates in the field and produces a wave. Field values oscillate around a mean value on each point in time and propagates in other points in field space with a velocity c. This phenomenon is described by the well-known d'Alembert equation, an equation that relates second order spatial derivatives of a field parameter with second order time derivative of same parameter.

If the generic field parameter is u d'Alembert equation have this form:

$$\frac{\partial^2 u}{\partial t^2} = c^2\left(\frac{\partial^2 u}{\partial x^2} + \frac{\partial^2 u}{\partial y^2} + \frac{\partial^2 u}{\partial z^2}\right) \quad \text{or in compact form :} \quad \frac{\partial^2 u}{\partial t^2} = c^2\nabla^2 u \tag{1.32}$$

where u is the variable that describe the field (physical quantity) and c is velocity with which perturbation moves (phase velocity).

Taking (for example) a field u changing in time and only in x direction as $u = u_a \sin(kx - \omega t)$ and deriving it two times in space and in time, substituting these two derivatives in the d'Alembert equation we obtain $c^2 = \frac{\omega^2}{k^2}$. So, d'Alembert equation tell us that this simple sinusoidal function is a possible solution, describing wave evolution in space and time and propagating with a velocity c equal to ω/k.

1.5.1 From Electromagnetic Fields to Electromagnetic Waves

Deriving the wave equation of an electromagnetic field is possible starting from Faraday's law. Taking the curl of Faraday's law (Eq. 1.23), we have:

$$\nabla \times (\nabla \times \boldsymbol{E}) = \nabla \times \left(-\frac{\partial \boldsymbol{B}}{\partial t}\right)$$

And using the vector identity:

$$\nabla \times (\nabla \times \boldsymbol{A}) = \nabla(\nabla \cdot \boldsymbol{A}) - \nabla^2 \boldsymbol{A} \quad \textit{and} \quad \nabla \cdot \boldsymbol{E} = 0,$$

we get:

$$\nabla \times (\nabla \times \boldsymbol{E}) = -\nabla^2 \boldsymbol{E}$$

Thus,

$$-\nabla^2 \boldsymbol{E} = \nabla \times \left(-\frac{\partial \boldsymbol{B}}{\partial t}\right)$$

Simplifying:

$$\nabla^2 \boldsymbol{E} = \frac{\partial}{\partial t}(\nabla \times \boldsymbol{B})$$

Substitute Ampère's law into the above equation we get:

$$\nabla^2 \boldsymbol{E} = \frac{\partial}{\partial t}\left(\mu_0 \varepsilon_0 \frac{\partial \boldsymbol{E}}{\partial t}\right)$$

But because $\mu_0 \varepsilon_0$ are constant above equation becomes:

$$\nabla^2 \boldsymbol{E} = \mu_0 \varepsilon_0 \frac{\partial^2 \boldsymbol{E}}{\partial t^2}$$

Rearrange to get the wave equation for magnetic field, once again we obtain a D'Alembert equation of the form:

$$\frac{\partial^2 \boldsymbol{E}}{\partial t^2} = \frac{1}{\mu_0 \varepsilon_0} \nabla^2 \boldsymbol{E}$$

Whose solution is an oscillation of electric field **E** propagating with the velocity (in empty space) equal to $c = \sqrt{\frac{1}{\mu_0 \varepsilon_0}}$.

Same can be done for the magnetic field **B**. Taking the curl of Ampère's law we have:

$$\nabla \times (\nabla \times \boldsymbol{B}) = \nabla \times \left(\mu_0 \varepsilon_0 \frac{\partial \boldsymbol{E}}{\partial t}\right)$$

And using the vector identity and $\nabla \cdot \boldsymbol{B} = 0$, we get:

$$\nabla \times (\nabla \times \boldsymbol{B}) = -\nabla^2 \boldsymbol{B}$$

Thus,

$$-\nabla^2 \boldsymbol{B} = \mu_0 \varepsilon_0 \vec{\nabla} \times \frac{\partial \boldsymbol{E}}{\partial t}$$

Substitute Faraday's law into the above equation we have:

$$\nabla^2 \boldsymbol{B} = \mu_0 \varepsilon_0 \frac{\partial}{\partial t} (\boldsymbol{\nabla} \times \boldsymbol{E})$$

Using $\boldsymbol{\nabla} \times \boldsymbol{E} = -\frac{\partial \boldsymbol{B}}{\partial t}$ we get:

$$\nabla^2 \boldsymbol{B} = \mu_0 \varepsilon_0 \frac{\partial}{\partial t} \left(-\frac{\partial \boldsymbol{B}}{\partial t} \right)$$

Simplifying and rearranging we finally have:

$$\frac{\partial^2 \boldsymbol{B}}{\partial t^2} = \frac{1}{\mu_0 \varepsilon_0} \nabla^2 \boldsymbol{B}$$

So, once again, a d'Alembert equation for the magnetic fields who's solution is an oscillation of magnetic field **B** propagating with the velocity (in empty space) equal to $c = \sqrt{\frac{1}{\mu_0 \varepsilon_0}}$. Electric wave and magnetic wave are orthogonal. Two specific Maxwell equations explain the orthogonality: Faraday's Law, Eq. (1.23) and Ampère's Law, Eq. (1.24). These curl equations show how a changing electric field induces a circulating magnetic field and vice versa. The versus follow the right-hand rule, meaning the magnetic field is perpendicular to the electric field and both to the direction of wave propagation. By substituting in the above equation (written only in one dimension x) the functions

$$E = E_{max} \cos(kx - \omega t) \quad \text{and} \quad B = B_{max} \cos(kx - \omega t)$$

and making the derivatives in space and time is easy to verify that these functions are possible solutions of the above equations. Electromagnetic waves are therefore two orthogonal fluctuating vectors, both orthogonal to the direction of propagation of the wave and both propagating in empty space with the same velocity:

$$c = \frac{1}{\sqrt{\mu_0 \varepsilon_0}} \tag{1.33}$$

This velocity c is called the speed of light, which is a fundamantal phisical constant. According to Einstein's special theory of relativity, this represents the maximum possible speed in the universe for any form of energy or information transfer.

Considering from Faraday's law, Eq. (1.23), in only one dimension x, that:

$$\frac{\partial E}{\partial x} = -\frac{\partial B}{\partial t}$$

And taking once again the two possible solutions of d'Alembert equation in one dimension x

$$E = E_{\max}\cos(kx - \omega t) \quad \text{and} \quad B = B_{\max}\cos(kx - \omega t)$$

We have for the partial derivatives of the above expressions:

$$\frac{\partial E}{\partial x} = -kE_{\max}\sin(kx - \omega t) \quad \text{and} \quad \frac{\partial B}{\partial t} = \omega B_{\max}\sin(kx - \omega t).$$

And so, for Faraday's law in one dimension x:

$$-kE_{\max}\sin(kx - \omega t) = -\omega B_{\max}\sin(kx - \omega t)$$

Therefore:

$$kE_{\max} = \omega B_{\max} \quad \text{and also} \quad \frac{E_{\max}}{B_{\max}} = \frac{E}{B} = \frac{\omega}{k} = c \tag{1.34}$$

That give us another fundamental relation between wave velocity and magnetic and electric fields amplitudes, valid for light and any other kind of electromagnetic wave.

For an electromagnetic wave we can define the so-called Poynting Vector as:

$$\mathbf{S} = \frac{1}{\mu_0}\mathbf{E} \times \mathbf{B} \tag{1.35}$$

Since **E** and **B** are perpendicular, their cross product gives a vector **S** perpendicular to both, of amplitude (E B), which aligns with the direction in which the wave propagates and so the direction of wave energy propagation.

To understand why **S** represents the power per unit area, i.e. the wave intensity let us to consider the rate of energy transfer. The total energy density u (energy per unit volume) of an electromagnetic wave is the sum of the electric field u_E and magnetic field u_B energy, given in the simplest condition by [3]:

$$u = u_E + u_B = \frac{1}{2}\varepsilon_0 E^2 + \frac{1}{2\mu_0}B^2$$

Substituting in the above Eq. (1.34), from where we can write B = E/c and E = Bc and using for c the expression given by Eq. (1.33) we obtain:

$$u = \varepsilon_0 E^2 \quad \text{and also} \quad u = \frac{1}{\mu_0}B^2 \tag{1.36}$$

So, the energy stored in the wave is proportional to the squared field amplitude. Wave energy moves at the speed of light c, meaning that in a small time interval dt,

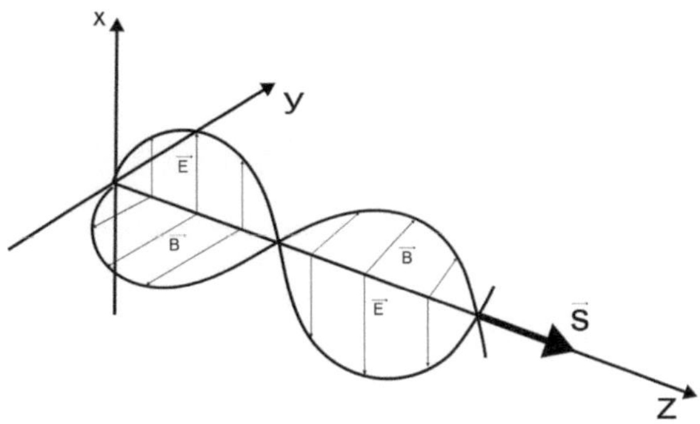

Fig. 1.6 Electromagnetic wave and its Poynting vector **S**

energy flows through a volume of thickness cdt. The energy passing through an area dA is:

$$\text{Energy in volume} = (\text{energy per unit of volume}) \times (\text{Volume}) = \text{u(cdt)dA}$$

Dividing by dt, the power crossing the surface per unit area is:

$$\frac{uc\ dA\ dt}{dt} = uc = \varepsilon_0\ c\ E^2$$

But because S = (E B) = E/μ_0 c, taking the expression of c given by Eq. (1.33) we obtain:

$$\text{S} = \text{u c}$$

which confirms that S represents the instantaneous power per unit area carried by the wave, i.e. its intensity (Fig. 1.6).

An electric field wave can be generated by moving electric charges, as in Fig. 1.7, but a changing electric field **E** generates also a changing magnetic field **B**, as in Fig. 1.8.

If we put a wire where arrive an electromagnetic wave, charges (electrons) on the wire will be forced to move accordingly to the wave electric field fluctuations, this

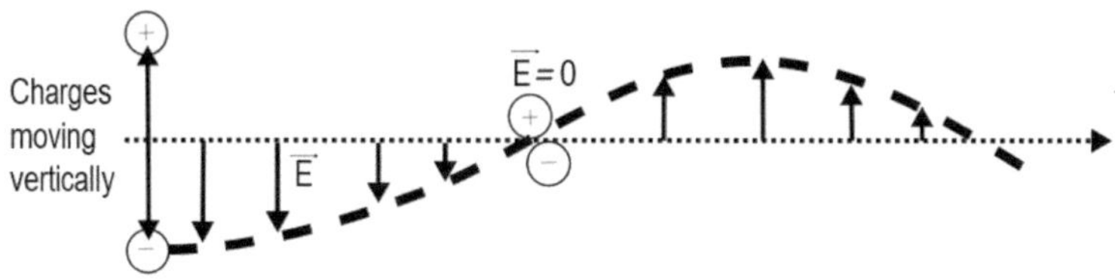

Fig. 1.7 Moving charges and electric field waves

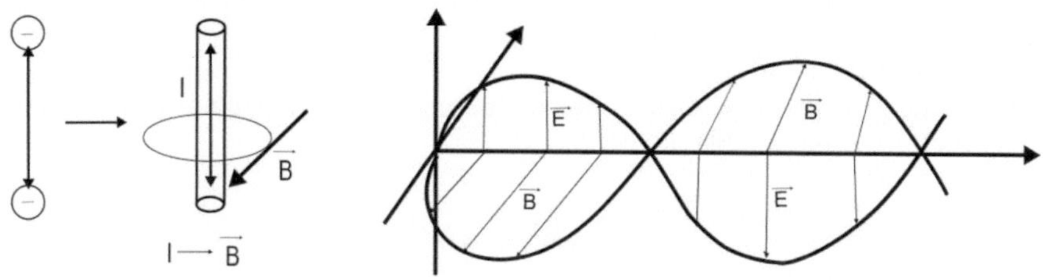

Fig. 1.8 Moving electric charges generating also a perpendicular magnetic field and propagation of both generating the electromagnetic wave

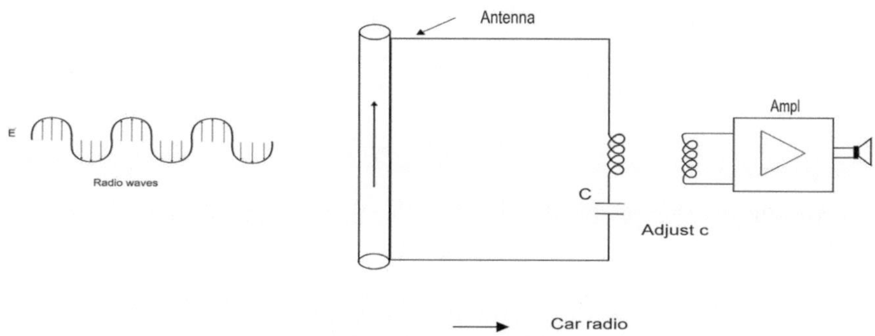

Fig. 1.9 Detection of an electromagnetic wave

wire is an antenna where an alternating current is induced on it and this effect can be improved by a resonant circuit tuned with a proper capacity, than amplified and used to detect and convert the wave into an electrical current signal, as illustrated in Fig. 1.9.

The working principle of a car radio.

1.5.2 From Navier–Stokes Equations to Waves in Fluids

A disturbance in any of the field variables of a fluid (e.g., pressure or velocity) related by the Navier–Stokes equations (1.25)–(1.27) may propagate through the fluid, resulting in a wave. When the flow is assumed to be irrotational ($\nabla \times \mathbf{u} = 0$), the velocity field can be represented as the gradient of a scalar potential ϕ, i.e. by the Laplace equation:

$$\mathbf{u} = \nabla\phi.$$

which governs many steady or quasi-steady flow scenarios. When compressibility is taken into account, and under linearized assumptions (small disturbances, inviscid and irrotational flow), the governing equation for the velocity potential becomes the classical wave equation.

We begin with the linearized forms of the mass and momentum conservation equations for a compressible, inviscid fluid:

$$\frac{\partial \rho'}{\partial t} + \rho_0 \nabla \cdot \mathbf{u} = 0$$

$$\rho_0 \frac{\partial \mathbf{u}}{\partial t} = -\nabla p'$$

where ρ_0 is the equilibrium density, and ρ' and p' are the small perturbations in density and pressure, respectively.

Assuming a linear relation between pressure and density perturbations, $p' = v^2 \rho'$, and substituting into the momentum equation, we get:

$$\rho_0 \frac{\partial \mathbf{u}}{\partial t} = -\nabla\left(v^2 \rho'\right)$$

Taking the divergence of both sides and using the linearized continuity equation, we obtain:

$$\rho_0 \frac{\partial}{\partial t}(\nabla \cdot \mathbf{u}) = -v^2 \nabla^2 \rho'$$

$$-\rho_0 \frac{\partial^2 \rho'}{\partial t^2} = -v^2 \nabla^2 \rho'$$

This simplifies to the wave equation for ρ':

$$\frac{\partial^2 \rho'}{\partial t^2} = v^2 \nabla^2 \rho'$$

Since $\mathbf{u} = \nabla \phi$ and ρ' is related to ϕ through the continuity equation, we can similarly derive the wave equation for the velocity potential:

$$\frac{\partial^2 \phi}{\partial t^2} = v^2 \nabla^2 \phi \tag{1.37}$$

which is the classical wave equation describing the propagation of waves in a compressible, inviscid, and irrotational fluid. Once again a D'Alembert equation where

$$v = \sqrt{\frac{\rho_0}{\rho}} \tag{1.38}$$

is the speed of the wave propagation inside the fluid.

In one spatial dimension, the wave equation reduces to:

$$\frac{\partial^2 \phi}{\partial t^2} = v^2 \frac{\partial^2 \phi}{\partial x^2}$$

The general solution to this equation is given by

$$\phi(x, t) = f(x - vt) + g(x + vt)$$

The d'Alembert's formula, given:

$$\phi(x, t) = \tfrac{1}{2}[\phi_0(x - vt) + \phi_0(x + vt)] + \tfrac{1}{2v}\int_{x-vt}^{x+vt} \psi_0(s)ds$$

where $\phi_0(x)$ is the initial displacement and $\psi_0(x)$ is the initial velocity, is the exact solution of the Cauchy problem on the whole real line for the 1-D wave equation.

Consider the initial value problem:

$$\phi_0(x) \equiv \phi(x, 0) = \sin(kx)$$
$$\psi_0(x) \equiv \frac{\partial \phi}{\partial t}(x, 0) = 0$$

This means that at $t = 0$ we have a sinusoidal wave with wavenumber k at rest. Substituting into the d'Alembert formula we obtain:

$$\phi(x, t) = \frac{1}{2}[\sin(k(x - vt)) + \sin(k(x + vt))].$$

The solution consists of two traveling waves:

- A left-moving wave $\frac{1}{2}\sin(k(x + vt))$ moving with velocity $-v$.
- A right-moving wave $\frac{1}{2}\sin(k(x - vt))$ moving with velocity $+v$.

These waves propagate without distortion along the characteristics $x \pm ct =$ constant. The initial sine profile splits symmetrically into two waves traveling in opposite directions. Due to the zero initial velocity, the waves have equal amplitudes (no asymmetry). Therefore at $t = 0$, the wave coincide to form $\sin(kx)$ and as time progress the left and right moving waves separates.

In higher dimensions, solutions can be more complex and depend heavily on the geometry of the domain and boundary conditions. In two or three dimensions, wave fronts expand as circles or spheres, and the solution techniques typically involve Fourier transforms, separation of variables, or Green's functions. Nevertheless, the essential nature of wave propagation transport of energy and information through the medium-remains the same.

Solution of d'Alembert equation can be obtained also by discretization techniques and numerical methods like FEM and BEM models using many software available for mechanical and electromagnetic field waves simulation like ANSYS, Sysnoise, etc.

1.5.3 Mechanical Waves on Elastic Media: String Vibrations

It is also possible to derive a d'Alembert-type equation for elastic solid materials, demonstrating that elastic waves can propagate through such media. Consider, for instance, an infinitesimal element of a stretched string of length *dl*, subjected to a uniform tension *T*. If this element is perturbed by a small transverse displacement *dy*, as illustrated in Fig. 1.10, the tension in the string gives rise to force components along both the x and y directions. These forces can then be analysed to establish the dynamic equilibrium and derive the corresponding wave equation. Considering that:

$$F_x = T\left(\cos\alpha' - \cos\alpha\right) \text{ and } F_y = T\left(\sin\alpha' - \sin\alpha\right);$$

For small algles (small *dy*) we can use the small angle approximations: $\cos\alpha' \sim 1,\ \cos\alpha \sim 1,\ \sin\alpha \sim \alpha \sim tg\,\alpha,\ \sin\alpha' \sim \alpha' \sim tg\,\alpha'$ so:
$F_x = 0$ and $F_y = \frac{\partial}{\partial x}(tg\alpha)dx$ and with $tg\alpha \sim \frac{dy}{dx}$ we'll have:

$$F_y = \frac{\partial^2 y}{\partial x^2} dx. \tag{1.39}$$

Appling Newton second's low, i.e. force=mass x acceleration of this string element we can write:

$$F_y = (dm)a_y = \rho_l dx \frac{\partial^2 y}{\partial t^2} \tag{1.40}$$

where dm is the mass of the string element and ρ_l is its linear density, i.e. the string's total mass divided by its length. From Eq. (1.39) and Eq. (1.40) we get:

$$\frac{\partial^2 y}{\partial t^2} = \frac{T}{\rho_l}\frac{\partial^2 y}{\partial x^2} \text{ and defining } v = \sqrt{\frac{T}{\rho_l}} \tag{1.41}$$

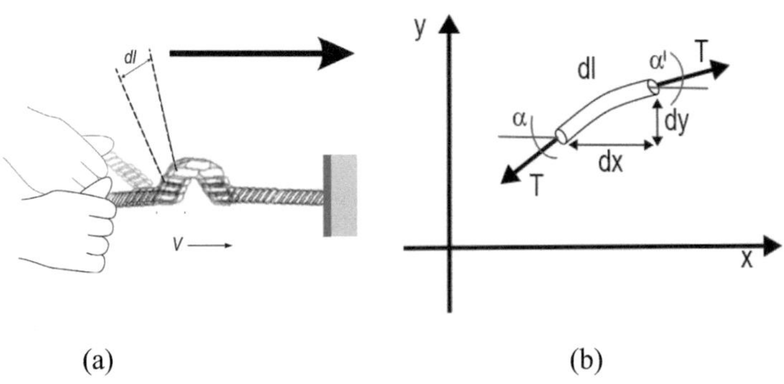

Fig. 1.10 Waves on a string

We obtain once again a one dimensional d'Alembert equation of the form:

$$\frac{\partial^2 y}{\partial t^2} - v^2 \frac{\partial^2 y}{\partial x^2} = 0 \tag{1.42}$$

where wave propagation velocity increase with tension T and decrease with mass per unit length ρ_l of the rope.

1.5.4 *Example of Mechanical Waves on Elastic Media: Acoustic Waves on Gas, Liquid or Solids*

Solutions of d'Alembert wave equations on fluids are travelling waves on gases (acoustic waves on air) but same kind of waves generates also on solid elastic materials. These waves are made of density, pressure variations and displacements, like in Fig. 1.11.

As illustrated for a string, but generalizing for any other liner elastic medium like a metal bar or a tube of gas or liquid, etc. we can write, in one dimension:

$$\frac{\partial^2 \xi}{\partial t^2} - v^2 \frac{\partial^2 \xi}{\partial x^2} = 0 \tag{1.43}$$

where $\xi(x, t)$ can be the displacement (or gas pressure or level, density, or ...) value of the phisical quantity at point x and time t. By substituting it on the d'Alembert equation it can be easily proved that the general solution of this equation has this form:

$$\xi_1(x, t) = \xi_1(x - vt) \text{ propagation to the right} \tag{1.44}$$

$$\xi_2(x, t) = \xi_2(x + vt) \text{ propagation to the left} \tag{1.45}$$

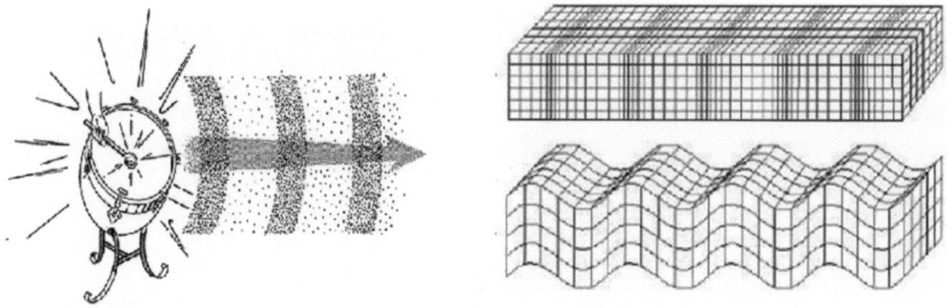

Fig. 1.11 Acoustic waves on gases, air or on elastic media

Propagation velocity is in general a square ratio of an elastic and an inertial property of the medium. For a string we have previously seen that propagation velocity is $v = \sqrt{\frac{T}{\rho_l}}$ where tension T is the elastic property and ρ_l is the inertial property; for a metal bar is $v = \sqrt{\frac{E_m}{\rho}}$ were E_m, the Young modulus of the materials, is the elastic property and ρ, the density, is the inertial property; for a gas $v = \sqrt{\frac{\beta}{\rho}}$ where β, the compression modulus, is the elastic property and ρ, the density, is the inertial property.

1.6 Waves Description and Parameters

As we know from Fourier theory every periodic waveform having a general shape can be approximated by a weighted sum of sinusoidal functions, so a basic class of waves, from which we can derive by superimposition (summing) all the others.

Fundamental harmonic waves, characterized by a sinusoidal shape we can express in a general form as:

$$\xi(x, t) = A \sin k(x - vt) \tag{1.46}$$

Using the parameter k Eq. (1.46) is plotted in space and time in Fig. 1.12. It is possible to identify two periods, one in space λ, named wavelength, and the other in time T, named period.

Harmonic waves (sine or cosine functions) illustrated in Fig. 1.12 oscillate from +A to –A amplitude value, so wave amplitude determines maximum displacement from the equilibrium position or maximum pressure fluctuations for a mechanical wave, maximum fluctuation of electric and magnetic field for electromagnetic waves, etc.

Sinusoidal wave functions can be expressed also as:

$$\xi(x, t) = A \sin(kx - \omega t) \tag{1.47}$$

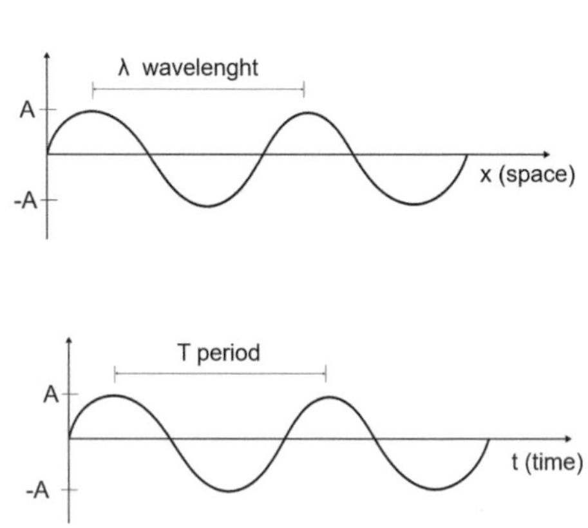

Fig. 1.12 Parameters of a harmonic wave

Using the parameter $\omega = kv$ named *angular frequency*. Or also, expressing its evolution only in time, as:

$$\xi(x, t) = A \sin(\omega t + \phi) \tag{1.48}$$

Using the parameter $\phi = kx$ named *phase*, considering it constant if we stay in a point looking at wave evolution in time.

Expression (1.47) is periodic in both space and time. Period of a sine function is, as well known, given by 2π so space period, named λ (wavelenght) is given by:

$$\xi(x, t) = \xi(x + \lambda, t) \text{ when } k\lambda = 2\pi \text{ and so } \lambda = \frac{2\pi}{k} \tag{1.49}$$

Time period T is given by

$$\xi(x, t) = \xi(x, t + T) \text{ when } \omega T = 2\pi$$

And so:

$$\omega = \frac{2\pi}{T} = 2\pi f, \quad \text{where } f = \frac{1}{T} \text{ is the wave frequency} \tag{1.50}$$

So, from the previous result, $\omega = 2\pi f$, and we can say that frequency f and angular frequency ω are proportional, i.e. are the same, with a scale factor equal to 2π.

Wave velocity v, space/time, can obviously be calculated by taking wavelength as space and wave period as corresponding time, so:

$$v = \frac{\lambda}{T} = \lambda f \tag{1.51}$$

From this relation it is possible to get the important conclusion that product between wavelength and frequency is the wave velocity, constant in general, so lower frequency waves have larger wavelengths and vice versa. In Fig. 1.13a acoustic wavelength range is illustrated. Acoustic wave giving sound that our ear can listen are pressure fluctuations in air ranging from 20 to 20.000 Hz in frequency. Propagation velocity of sound in air is typically a bit more 300 m/s, so a sound of 100 Hz is "long" about 3 m and a sound of 10000 Hz is "long" about 3 cm.

In Fig. 1.13b the electromagnetic spectrum is also illustrated that start from radio waves of 100 Hz frequency and 1000 Km of wavelength up to x and gamma rays of wavelength below nanometers. In the middle there is the visible light that our eyes can see of wavelength from 0.4 micrometers (red color) to 0.7 micrometers (violet). Propagation velocity of electromagnetic waves in void is a bit more than 300.000 Km/s, the value of the product between frequency and wavelength of an electromagnetic wave, accordingly to Eq. (1.51).

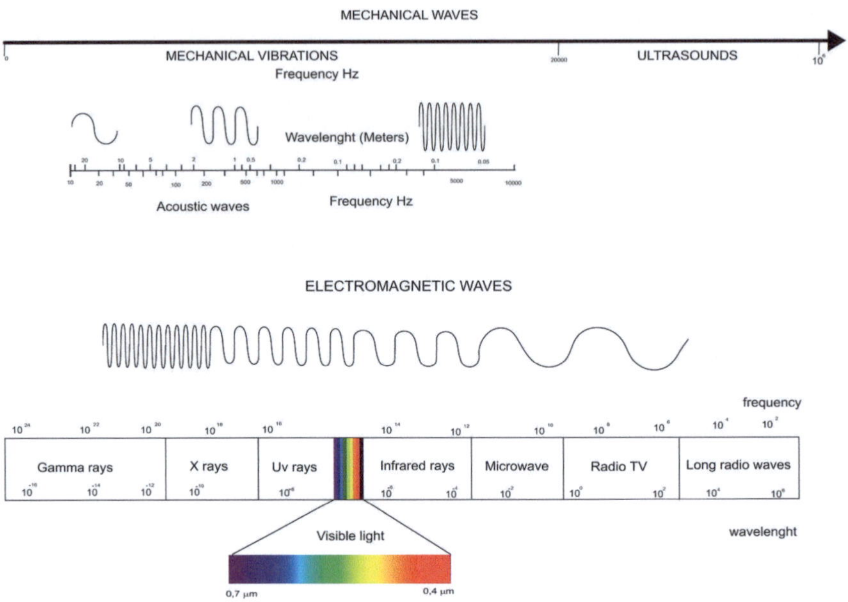

Fig. 1.13 Mechanical and electromagnetic wave spectrum, wavelength and frequencies

1.6.1 *Waveform in Time and Frequency*

A waveform $\xi(t)$ in a point can be recorded in time like in Fig. 1.14a, and, as previously said, its shape can be a sine function or another one. Fourier theory gives us the way to reconstruct the waveform by summing weighted sine or cosine functions, so the wave can be seen from another point of view: instead of looking to its time evolution, looking to its "content" in frequency, i.e. the number of sine or cosine and their amplitude (Fig. 1.14b) and phase we need to reconstruct it. The mathematic expression of this principle, from [3], is:

$$\xi(t) = \sum_{-\infty}^{\infty} X_n e^{i\omega_n t} \xi(t) = a_0 + \sum_{n=1}^{\infty} (a_n \cdot \cos(\omega_n t) + b_n \cdot \sin(\omega_n t)) \tag{1.52}$$

where $\omega_n = 2\pi n/T; f_n = \omega_n/2\pi$ and coefficients a_0, a_n e X_n may be calculated by the integrals:

$$a_0 = \frac{1}{T}\int_0^T x(t) \cdot dt \quad a_n = \frac{2}{T}\int_0^T x(t)\cos \omega_n t \cdot dt \quad b_n = \frac{2}{T}\int_0^T x(t)\sin \omega_n t \cdot dt$$

$$X_n = \frac{1}{T}\int_0^T x(t) \cdot e^{-i\omega_n t} \cdot dt \tag{1.53}$$

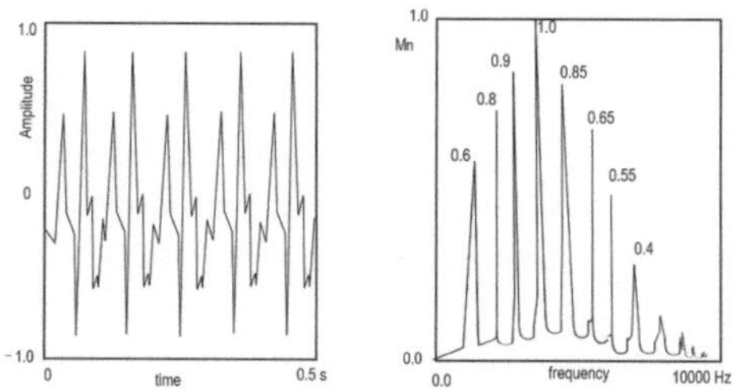

(a) (b)

f_n	M_n	φ_n
800	0.6	0.234
1000	0.8	1.432
1500	0.9	0.841
3000	1.0	0.131
3500	0.85	0.761
5000	0.65	0.516
6000	0.55	0.021
7000	0.4	0.341

(c)

Fig. 1.14 Frequency analysis of a periodic wave

Fourier series can be seen as a table (a vector) of complex numbers $[a_n, b_n]$, each one for the frequency of the nth sine or cosine function. Each of this complex number can also be expressed by its modulus $M_n = \left(a_n^2 + b_n^2\right)^{1/2}$ and its phase $\varphi_n = \text{arctg}(b_n/a_n)$, like the list of numbers in Fig. 1.14c.

So the time evolution of a waveform can be seen in frequency domain, simply from another point of view, like in Fig. 1.15. Looking from this point of view is called looking at wave spectrum. If there are many frequencies in the spectrum it can be seen as a continuum function of frequency f, or the proportional angular frequency ω or, typically for light, of wavelength λ.

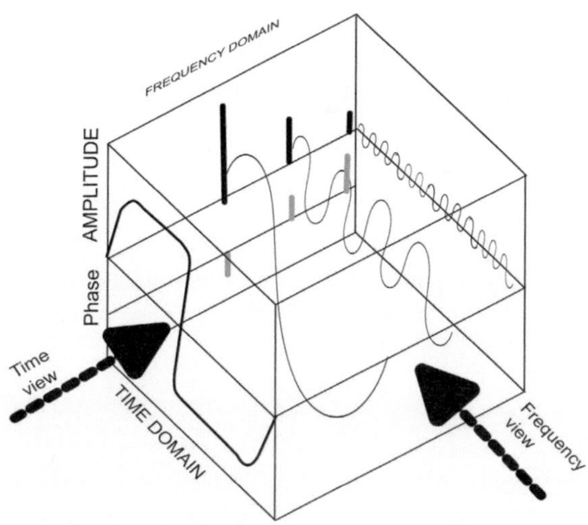

Fig. 1.15 Time and frequency domain: two different view directions for looking at same thing

1.6.2 Energy, Power and Intensity of Mechanical Waves

Waves transports energy, not matter; in other words, physical quantities and particles in elastic media oscillates around an equilibrium position, but do not move from there, so not particles but only their motion is transmitted. Energy of a mechanical wave described in its displacement by a harmonic function $s = A\ \cos(\omega t + \phi)$ can be expressed by its kinetic energy $E = \frac{1}{2}mv^2$. Deriving displacement s to obtain velocity we have $\mathrm{v} = -A\,\omega \sin(\omega t + \phi)$. Energy E, proportional to velocity squared is proportional to square amplitude of velocity $A\omega$, so to the quantity $\omega^2 A^2$. The wave power P_m, i.e. the energy on a wave period, is proportional to $\omega^2 A^2/T$, but because $f = 1/T$ and $\omega = 2\pi f$ we get:

$$P_m \propto \omega^2 A^2 f \propto \omega^3 A^2 \propto f^3 A^2 \tag{1.54}$$

So the power of a wave increase with the square of its amplitude and with the cube of its frequency.

Intensity of a wave is defined as the power carried by the wave through the unit of area crossed by the wave along the direction of propagation, along the normal to the surface crossed by the wave, as illustrated in Fig. 1.16a. If S is the area crossed by the power P_m its modulus is therefore:

$$I = \frac{(\Delta E/\Delta t)_{\text{average}}}{\mathrm{S}} = \frac{P_m}{\mathrm{S}} \tag{1.55}$$

If we have a wave point source emitting a power P undisturbed, uniformly on all space directions we will have at a distance r on the points of the sphere of Fig. 1.17a the intensity I:

$$I = \frac{P_m}{S} \quad \Rightarrow \quad \mathrm{S} = 4\pi r^2 \quad \Rightarrow \quad I \propto \frac{1}{r^2} \tag{1.56}$$

So wave intensity decreases as the square of the distance from the power source, if no other reflections or interference take places. A condition named free space

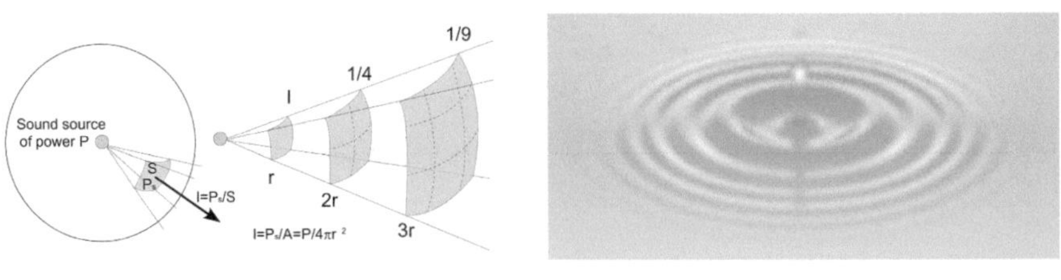

(a) (b)

Fig. 1.16 Amplitude, power and intensity distribution of a wave propagating in free space

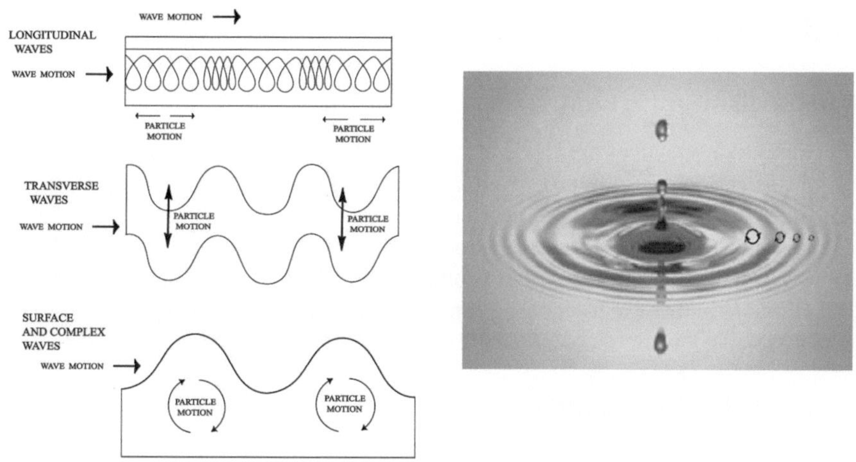

Fig. 1.17 Different kind of waves

propagation, as illustrated in Fig. 1.16b for a fixed amount of energy of a falling drop, generating a wave on a liquid surface, quickly reducing in its amplitude because of intensity reduction as $1/r^2$ and because of power diffusion as described by Eq. 1.54.

Wave intensity vector **I** can be also described, for mechanical waves inside fluids, as the product of fluctuation of fluid pressure (Force/Area) with propagation wave velocity, because:

$$\begin{aligned} \mathbf{I} &= \text{Power/Area} = \text{Force} \times \text{Velocity/Area} \\ &= (\text{Force/Area})\,\text{Velocity} = \text{Pressure} \times \text{Velocity} \end{aligned}$$

Therefore we can write for the mechanical wave instantaneous intensity vector:

$$\boldsymbol{I}(t) = p(t)\boldsymbol{v}(t) \tag{1.57}$$

We can therefore estimate mechanical average wave intensity flow in a point of a fluid by the integral:

$$\boldsymbol{I} = \frac{1}{T}\int_0^T p(t)\boldsymbol{v}(t)dt \tag{1.58}$$

where p(t) is the fluctuating acoustic pressure in a point and **v**(t) is the velocity of the wave particle motion, i.e. the first derivative of the displacement generated by the wave.

So intensity of a mechanical wave is a vector, like the Poynting vector for electromagnetic waves, having direction and versus of wave propagation velocity and modulus equal to the time integral of the Pressure-Velocity product, so to its average value in time gives us the energy flux across a unitary surface perpendicular to intensity vector.

1.6.3 Kind of Waves

Waves, as illustrated in Fig. 1.17, can be classified in:

(1) Longitudinal waves are those in which the particle oscillations are parallel to the direction the wave propagation, waves on a coil spring for example.
(2) Transverse waves are those in which the particle oscillations are perpendicular to the direction the wave propagation, waves on a guitar string for example.
(3) Surface and complex waves are combination of the first two giving circular or elliptical particle motion; waves on sea surface are of this kind.

1.6.4 Waves Visualization

The pressure oscillations of an acoustic wave (i.e., sound) can be detected at any point in space in front of a loudspeaker using a microphone. By connecting the microphone to an amplifier and subsequently to a light source, the amplitude of the pressure fluctuations can be converted into variations in light intensity. By moving this microphone–light source setup in front of a photo camera, it is possible to create a visual representation of the acoustic wave's intensity, essentially an image of the propagating sound wave, as shown in Fig. 1.18a. The alternating light and dark regions in the image correspond to areas of varying pressure, and their distances are the wavelength of the acoustic wave, typically ranging from approximately 3 cm to 3 m, as previously discussed.

In a similar way, an electromagnetic wave can be visualized by connecting a small antenna to an amplifier, and then linking the amplifier's output to a light source, as illustrated in Fig. 1.18b. This setup produces a visual representation of the electromagnetic wave, provided that its wavelength falls within a range detectable by the camera's field of view.

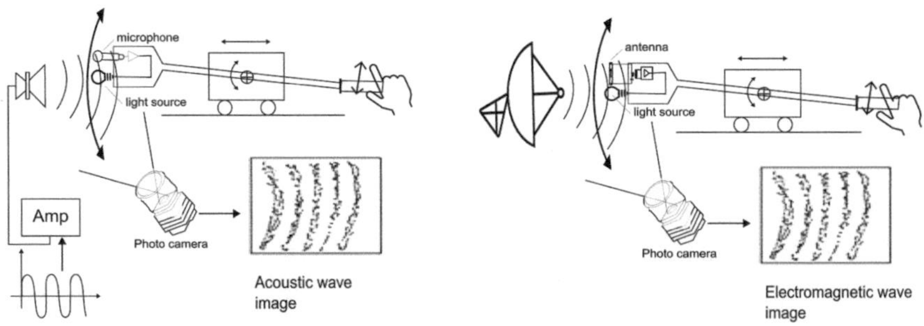

(a) acoustic waves imaging (b) Electromagnetic waves imaging

Fig. 1.18 Visualization of waves

1.7 Waves Interactions

As previously discussed, the general one-dimensional solutions of the d'Alembert equation take the following form:

$$\xi_1(x, t) = \xi_1(x - vt) \text{ propagation to the right}$$

$$\xi_2(x, t) = \xi_2(x + vt) \text{ propagation to the left}$$

For the superposition principle in general we have:

$$\xi(x, t) = \xi_1(x - vt) + \xi_2(x + vt)$$

As illustrated in Fig. 1.19.

1.7.1 Stationary Waves

A stationary wave can be generated for example on a string with one fixed end: if we moves the free end with a dispacement like $s_I = A \ \sin\ (kx - \omega t)$ the reflected wave would be $s_R = A \ \sin(kx + \omega t)$, like in Fig. 1.20.

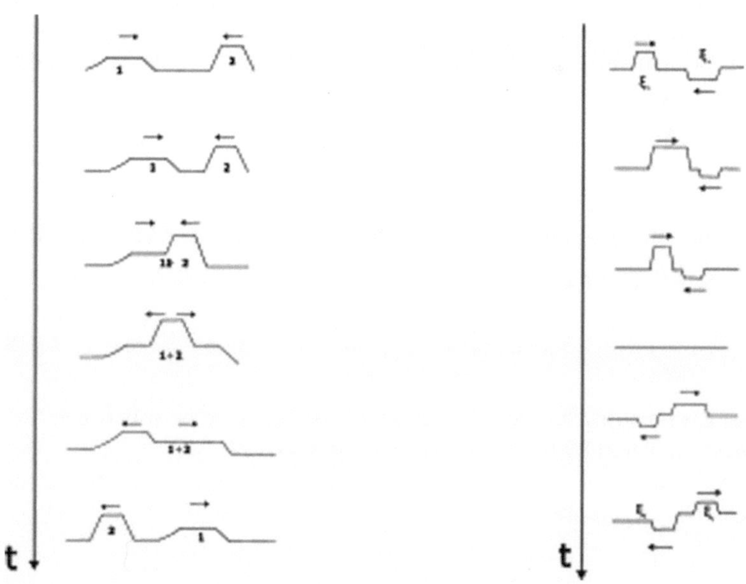

Fig. 1.19 Superimpositions of waves travelling on opposite directions

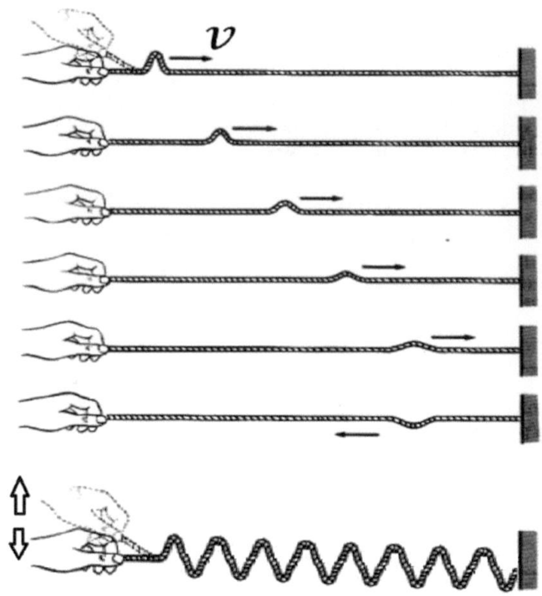

Fig. 1.20 Pulse reflection and standing waves on a rope

Applying the superposition principle the resulting wave is:

$$s_R = 2A\ \sin(kx)\cos(\omega t)$$

and there is no propagation of wave energy between the two constrain of the rope.

Not moving point, call nodes, occur when $\sin(kx) = 0$ that happen when $kx = n\pi$ so for:

$$x = n\frac{\lambda}{2}, \quad n = 1, 2, 3\ldots \tag{1.59}$$

Maximum motion points, called antinodes, occur when $\sin(kx) = 1$ that happen when $kx = 2(m+1)\frac{\pi}{2}$ so for:

$$x = (2m+1)\frac{\lambda}{4}, \quad m = 0, 1, 2\ldots \tag{1.60}$$

If we have a string where a little displacement is given on it between two fixed ends those are naturally fixed and must be nodes so:

$L = n\frac{\lambda}{2} = n\frac{v}{2f}, \quad n = 0, 1, 2\ldots$ therefore, only certain vibration frequencies are possible, their values are:

$$f_n = n\frac{v}{2L} \tag{1.61}$$

The first five are illustrated in Fig. 1.21a. Is the principle of guitar string (Fig. 1.21b) to generate notes of different frequencies changing string tension,

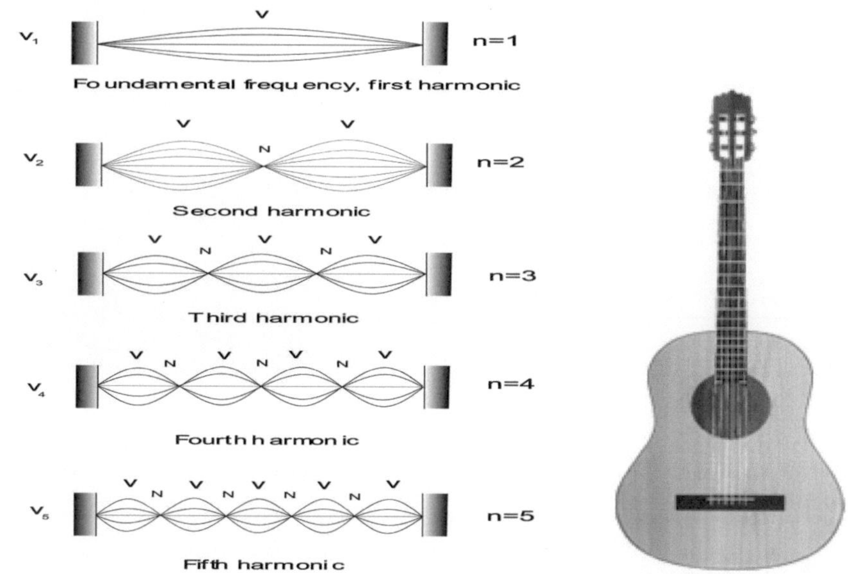

(a) (b)

Fig. 1.21 Stationary waves on string

changing velocity v as expressed by Eq. (1.60), accordingly to Eq. (1.41) relating v with the tension T of the string.

The vibration of a string is transmitted to the wooden structure of the guitar, which in turn transfers the vibration to the surrounding air, generating sound. A similar phenomenon occurs with acoustic waves inside a fluid-filled tube with either open or closed ends. When both ends of the tube are open, antinodes form at each end. In this case, the only allowable frequencies f_n that support standing waves with antinodes at both ends are given by:

$$f_n = n\frac{v}{2L} \tag{1.62}$$

As illustrated in Fig. 1.22a for n = 1, 2 and 3.

If one end of the tube is closed, a node must form at that end, while an antinode forms at the open end. Under these conditions, the only possible frequencies f_m that support standing waves with a node at the closed end and an antinode at the open end are generally given by:

$$f_m = (2m + 1)\frac{v}{4L} \tag{1.63}$$

As illustrated in Fig. 1.22 for m = 0, 1 and 2. Is the principle of pipe organs (see Fig. 1.22c) to generate notes of different frequencies, given by the pipe length L.

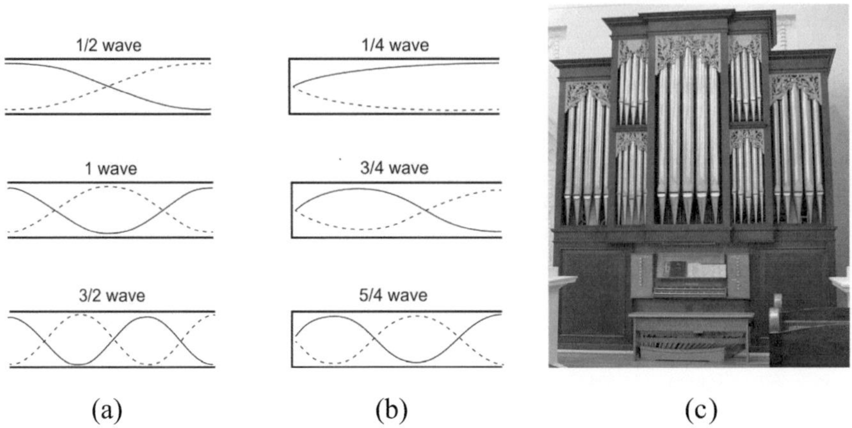

Fig. 1.22 Standing waves inside a tube **a** open at two ends and **b** open at one and closed at the other end

1.7.2 Wave Interference in Time

Let's consider two simple harmonic waves produced by two identical sources S_1 and S_2. In a point P we have, coming from source S_1, the armonic wave $s_1(x, t) = A\sin(kx_1 - \omega t)$ and in same point, from source S_2, the armonic wave $s_2(x, t) = A\sin(kx_2 - \omega t)$, as illustrated in Fig. 1.23.

For superposition principle, in point P, because of the trigonometric identity:

$$\cos\alpha + \cos\beta = 2\cos\left(\frac{\alpha+\beta}{2}\right)\cos\left(\frac{\alpha-\beta}{2}\right) \tag{1.64}$$

we will have:

$$s(x, t) = s_1(x, t) + s_2(x, t) = 2A\cos\left(\frac{k(x_1 - x_2)}{2}\right)\sin\left(k\frac{x_1 + x_2}{2} - \omega t\right).$$

So, the amplitude $\left(2A\cos\left(\frac{k(x_1-x_2)}{2}\right)\right)$ in point P, in time, will depends on phase difference:

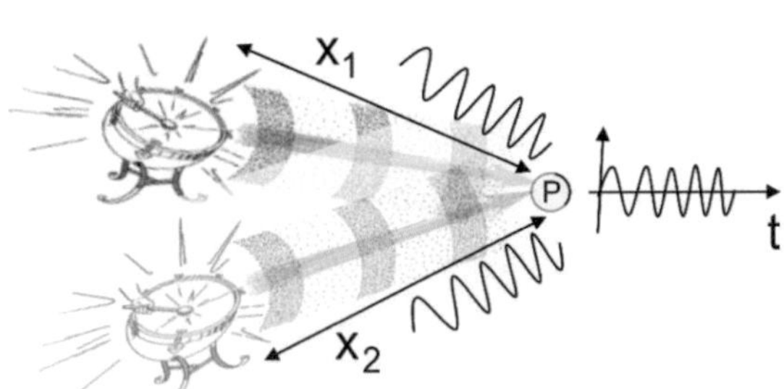

Fig. 1.23 Superimposition of two waves in a point

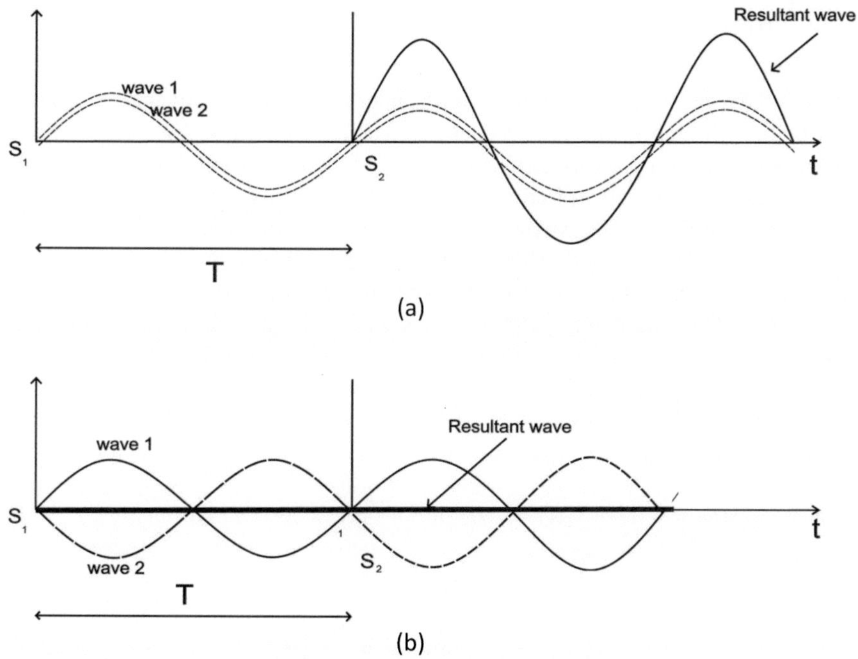

Fig. 1.24 Constructive waves interference (**a**) and destructive interference (**b**)

$\Delta\phi = \phi_1 - \phi_2 = (kx_1 - \omega t) - (kx_2 - \omega t) = k(x_1 - x_2)$. This amplitude s_0 will be:

$$s_0 = 2A\cos\left(\frac{k(x_1 - x_2)}{2}\right) = 2A\cos\frac{\Delta\phi}{2} \tag{1.65}$$

Two extreme cases of interference are those with $s_0 = 2A$ *or* $s_0 = 0$, respectively called constructive and destructive interference, illustrated in Fig. 1.24. Constructive interference happens, in general, when:

$$\Delta\phi = 2\pi n \quad n = 0, 1, 2\ldots \tag{1.66}$$

Or also for

$$\Delta x = n\lambda \tag{1.67}$$

We have destructive interference if:

$$\Delta\phi = (2m + 1)\pi \quad m = 0, 1, 2\ldots \tag{1.68}$$

Or also for:

$$\Delta x = (2m + 1)\frac{\lambda}{2} \tag{1.69}$$

1.7.3 Beats

Beats are wave phenomena resulting from the interference of two waves arriving at the same point with slightly different frequencies, as illustrated in Fig. 1.25. This interference produces a periodic variation in amplitude, perceived as amplitude modulation at a frequency equal to the difference between the two wave frequencies.

As shown in Fig. 1.25a, when two waves of slightly different frequencies reach the same point, they gradually shift out of phase in this point, leading to destructive interference. After a short interval, they realign in phase, producing constructive interference. This alternating pattern creates a periodic fluctuation in amplitude in that point, as depicted in Fig. 1.25b, giving rise to the characteristic "beats" effect.

If $s_1 = A\sin(\omega_1 t)$ and $s_2 = A\sin(\omega_2 t)$ are our two waves, for superposition principle, in point P where they superimpose, once again because of the trigonometric identity equation (1.64) we have:

$$s(t) = s_1(t) + s_2(t) = A(t)\sin(\omega t) = 2A\ \cos(\Omega t)\ \sin(\omega t),$$

were

$$\Omega = \frac{\omega_1 - \omega_2}{2} \text{ and } \omega = \frac{\omega_1 + \omega_2}{2} \tag{1.70}$$

So, the result is a wave of frequency ω and a modulated intensity given by

$$I = I_{max}\cos^2(\Omega t) \tag{1.71}$$

at the modulation frequency Ω. This effect is commonly used to tune a guitar by listening to the "beats" produced between the sound of a string and that of a reference tone. By adjusting the string's tension, and thus its frequency, the beats gradually slow down and eventually disappear when the string and the reference source reach the same frequency. Anyway same effect can be used with any other kind of mechanical

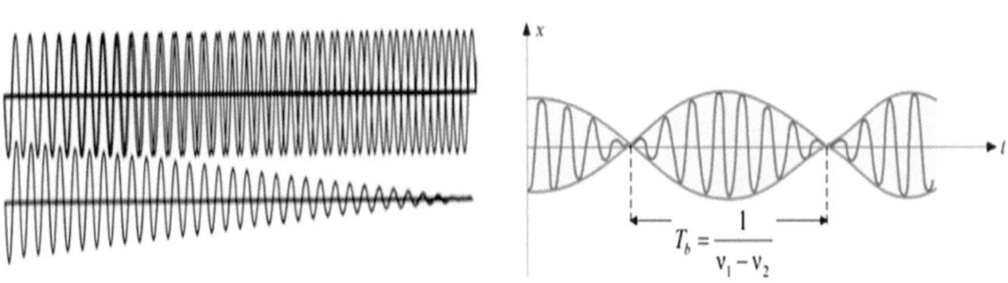

(a) (b)

Fig. 1.25 Beats generated by amplitude modulation from constructive to destructive interference

or electromagnetic waves to compare frequencies, a technique called heterodyne because of the two different frequencies.

1.7.4 Wave Interference in Space

Interferences occur also in space. It is very simple to visualize it on a plane by using circular or liner regular patterns printed on transparent sheets, partially superimposed, like in Fig. 1.26.

Same mathematical formulation of the previously illustrated in time can be done, simply changing the variable x with t and the period T with wavelength λ, obviously in one dimension only.

1.8 Waves Interactions with Aperture and Obstacles

A very useful principle for estimating wave propagation in space is *Huygens' Principle* [1]. It states that every point on a wavefront acts as a source of secondary spherical wavelets. The new wavefront at a new position is found by drawing an envelope around these wavelets. As illustrated in Fig. 1.27, whether for a point source, an arbitrary wavefront shape, or a plane wave, this geometric construction helps visualize the evolution of the wavefront over space. Lines drawn perpendicular to the wavefronts are known as *rays*, and they indicate the direction of wave propagation.

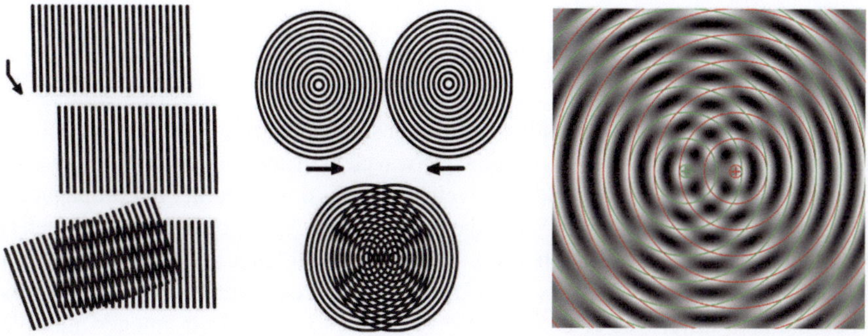

Fig. 1.26 Waves interference in space by superimposing transparent sheet of regular patterns

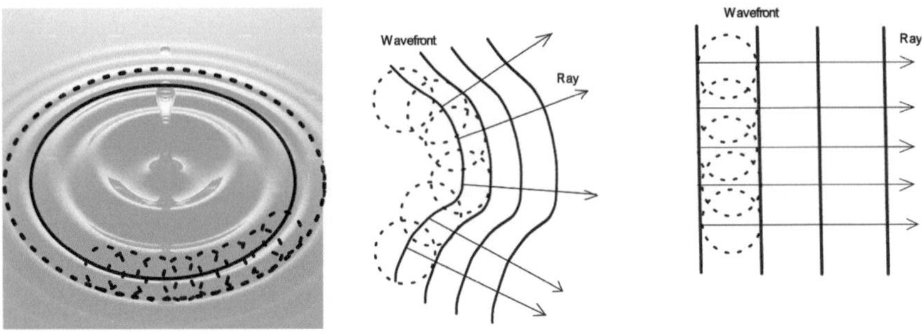

Fig. 1.27 Huygens principle

1.8.1 Diffraction

Huygens' principle helps us understand that when a wave encounters an obstacle made of a material through which it cannot propagate, and this obstacle contains an aperture whose width is approximately equal to or smaller than the wavelength, the aperture acts as a new point source. As a result, the wave spreads out in all directions beyond the aperture. This phenomenon is known as diffraction, and it is illustrated in Fig. 1.28.

If the aperture is significantly larger than the wavelength, Huygens' Principle implies that multiple secondary wavelets are generated across the opening. These wavelets combine to form an envelope that approximates a straight, parallel wavefront, as shown in Fig. 1.28. In this case, diffraction is minimal, and a well-defined shadow region is formed behind the obstacle.

1.8.2 Double Slit Diffraction

If there are two apertures with widths equal to or smaller than the wavelength, each act as a point source emitting waves of the same wavelength and frequency. These waves can interfere constructively or destructively at various points along their propagation paths, as illustrated in Fig. 1.29. When the wave is monochromatic light and a screen is placed at a distance D, as shown in Fig. 1.29b, an interference pattern of alternating bright and dark fringes appears. Constructive interference (bright regions) occurs when the path difference X between the two waves is zero or an integer multiple of the wavelength. Destructive interference (dark regions) occurs when the path difference is equal to half a wavelength or an odd multiple of half a wavelength. Similar interference phenomena can be observed with other types of waves, such as sound or water waves.

From the triangles of Fig. 1.29b it is easy to calculate the difference of travelling length between the two ray X_2 and X_3, for example. If they give positive interference on the screen this difference should be equal to l. And if the screen is far, D is big,

Fig. 1.28 Wave interaction with an hole of different diameter respect to the wavelength

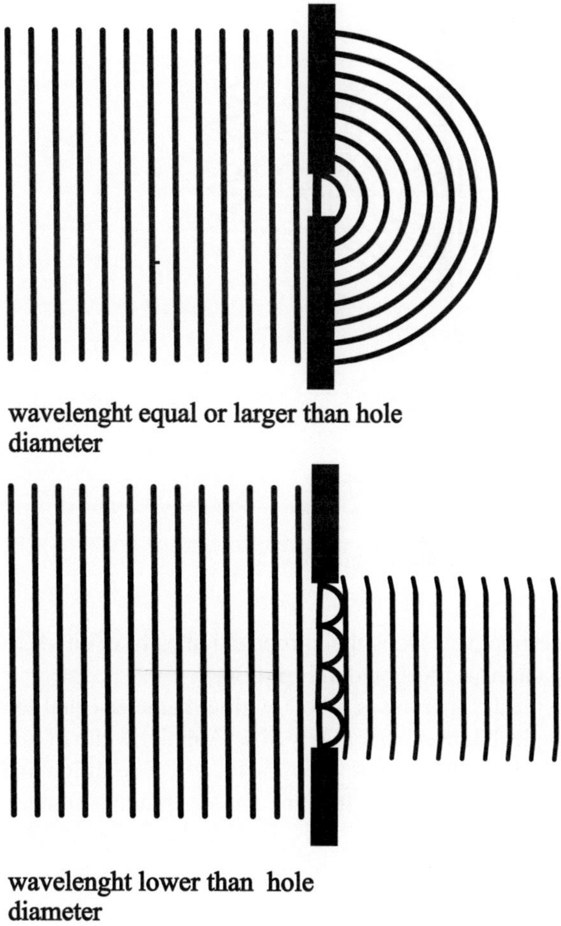

and θ is little so $sin\theta \approx \theta$. From the little triangle in Fig. 1.29b we have Dx = d sin θ = λ = dθ. From Fig. 1.29d if the spacing of the two aperture is very little respect distance D we can approximate the two apertures on a single point, and b distance as same length of the arc of radius D and aperture θ. So, for definition of angles in radiant θ = b/D and therefore spacing b between two light band on the screen is equal to b = λ D/d.

1.8.3 Single Slit Diffraction

A similar, though more complex, phenomenon occurs with a single slit. When the slit is wider than the wavelength, Huygens' Principle tells us that each point across the slit acts as a secondary source. These multiple sources can interfere constructively or

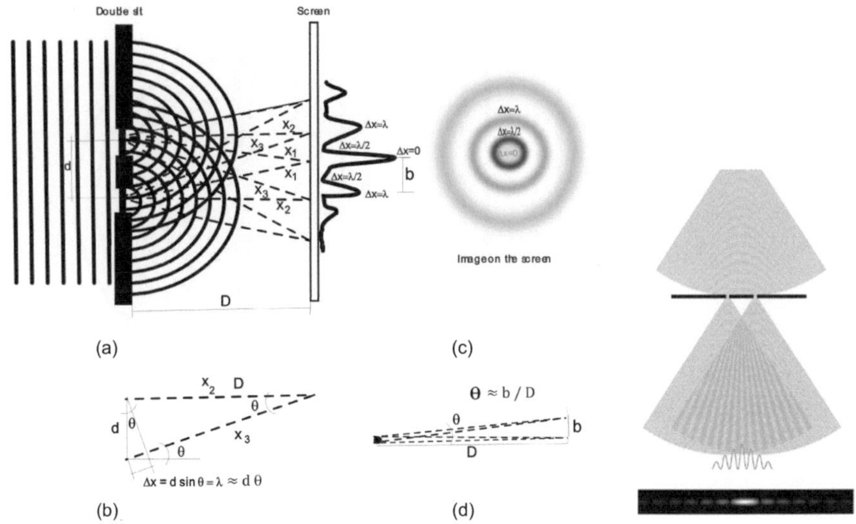

Fig. 1.29 Double slit diffraction

destructively along their propagation paths or on a screen, producing an interference pattern, as illustrated in Fig. 1.30.

If the aperture is a hole with a diameter equal to d this phenomenon produces typical figures named Airy disk. Angular width of airy disks, for little values of θ is given by the $\sin\theta \approx \theta = 2.44(\lambda/d)$. Two different Airy disks, so two points, are indistinguishable if their angular separation is smaller than $\theta_r = \lambda/d$.

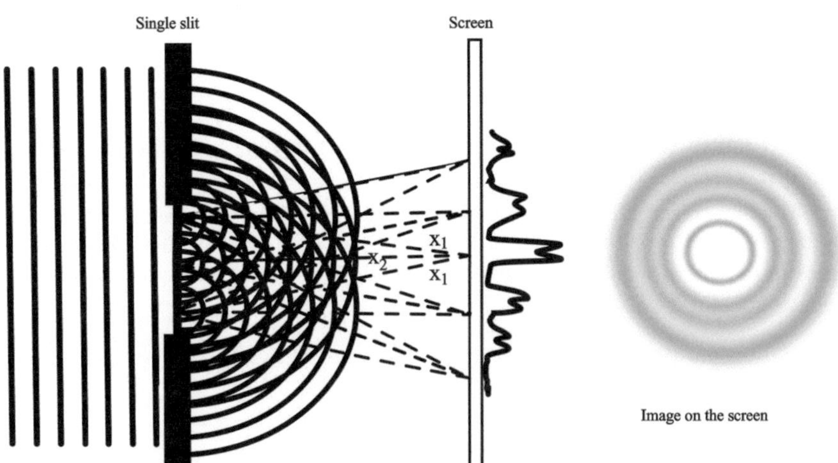

Fig. 1.30 Single slit diffraction

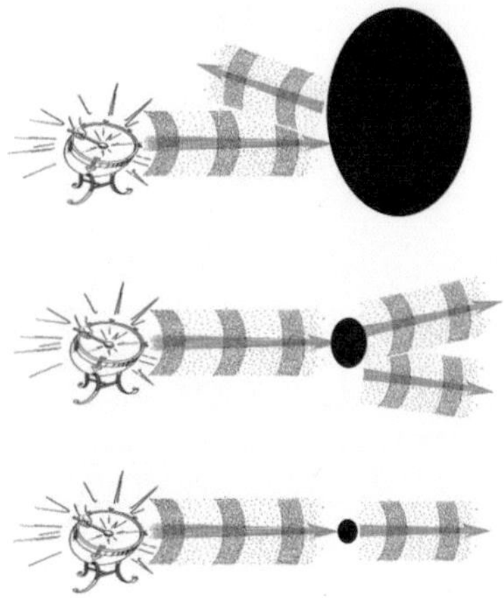

Fig. 1.31 Interaction of obstacles with waves

Using the same approach, based on Huygens' Principle, we can also explain the symmetrical effects observed when a wave encounters an obstacle made of a material through which it cannot propagate. If the obstacle is much larger than the wavelength, the wave is primarily reflected. If the obstacle is comparable in size to the wavelength, the wave undergoes diffraction and possibly refraction, changing its propagation direction. However, if the obstacle is much smaller than the wavelength, it has little to no effect on the wave's propagation, as illustrated in Fig. 1.31.

So, depending on the ratio wavelength/object dimension we could observe that the wave is reflected, refracted or not disturbed at all in its propagation.

1.9 Wave Interaction at Interfaces Between Different Media

When a wave of a single frequency travels through a medium with propagation velocity v_1 and reaches the interface with a second medium where the propagation velocity is v_2, it partially enters the second medium, changing direction, and may also be partially reflected, as illustrated in Fig. 1.32. The change in direction, named refraction, can be calculated using the well-known Snell's Law, which can also be better understood through Huygens' Principle.

In Fig. 1.32, consider ray 1, which reaches point P on the interface first, and ray 2, which reaches point O after a time interval T. Both rays arrive at the flat boundary separating the two media with the same angle of incidence θ_i, measured with respect to the normal. The difference in the distances traveled by the two rays in the first medium is clearly $v_1 T$.

Looking at triangle POQ we can write.

$$\text{PO} \sin \theta_i = \text{v}_1\text{T} \tag{1.72}$$

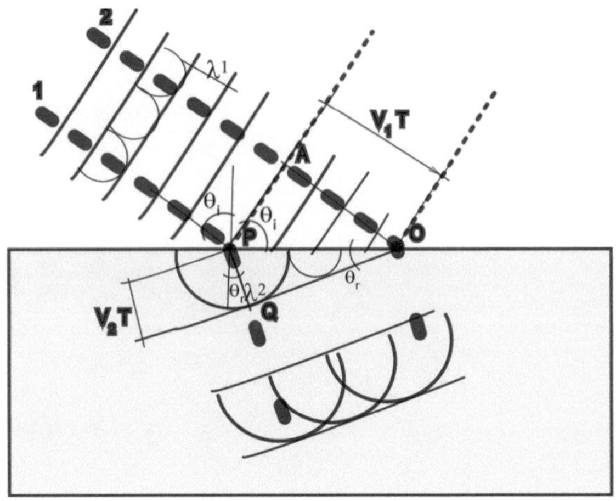

Fig. 1.32 Schnell low by Huygens principle

In same time the ray 2 generate a source in point P where we can draft a circle, accordingly to Huygens principle, of radius λ, if time T is the corresponding wave period. The wavefront generated is tangent to this circle and should pass for the point O where the new circle is not started, because the ray just arrived. So the new warfront inside the second medium is the segment OQ. From the triangle POQ we can write

$$\text{PO}\,\sin\,\theta_r = \text{v}_2\text{T} \tag{1.73}$$

Dividing Eqs. (1.72) by (1.73) we get the Snell's law:

$$\frac{\sin(\theta_i)}{\sin(\theta_r)} = \frac{v_1}{v_2} \tag{1.74}$$

It is easy, again by Huygens principle, to prove that the relation between incident and reflected angles is:

$$\theta_i = \theta_{ref} \tag{1.75}$$

So, reflection occur at same incidence angle to the normal to the surface. This happen in Fig. 1.33a when the separating surface between the two media is flat, with a roughness little respect incident wavelength, but if the surface has a roughness larger than wavelength, we can see it as a composition of different surfaces with different normal to the surface directions, so the different ray are reflected in different directions, all accordingly with (1.75). The situation is that in Fig. 1.33b and the phenomena is called diffusion.

Total reflection phenomena happens when light travels inside a medium with a higher refractive index surrounded by another with a lower refractive index. By Snell's law we can observe that if the incidence angle is large enough the sine of the

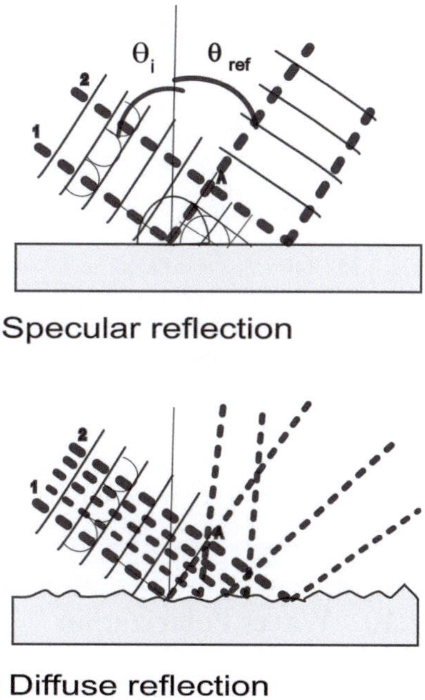

Fig. 1.33 Specular and diffuse reflections

angle of refraction requires to be greater than one, obviously impossible, so we have no refraction. The light in this case is totally reflected by the boundary between the two media. The largest possible angle of incidence which still results in a refracted ray is called the critical angle; in this case the refracted ray travels along the boundary between the two media, its value is (Fig. 1.34):

$$\theta_c = \arcsin\left(\frac{n_2}{n_1}\right). \tag{1.76}$$

When a wave composed of different frequencies an so wavelength arrive to an interface between two media (prism surface of Fig. 1.35) and the second media has different travelling velocity for the different frequency, i.e. refraction index

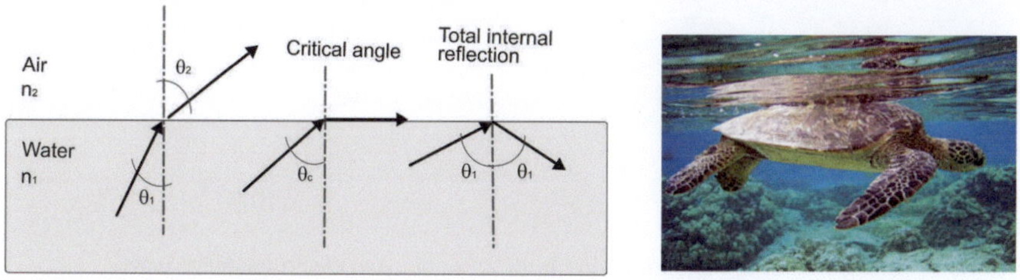

Fig. 1.34 Total internal reflection

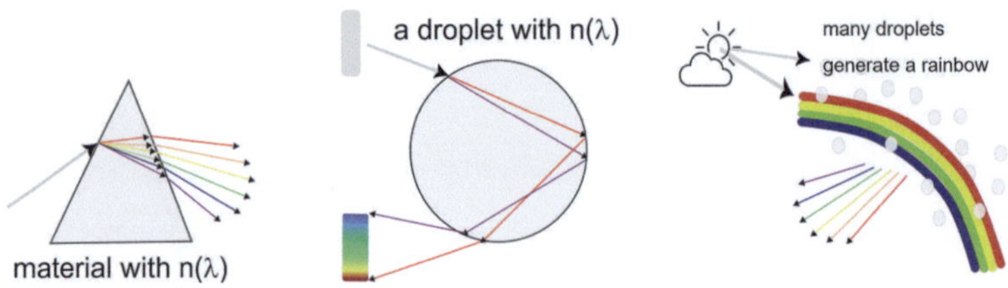

Fig. 1.35 Refraction at different angles of the different colours

change with wavelength as n = n(λ) Snell's law of refraction apply to each different wavelength, giving different refraction angle for each one, so light rays of different wavelength (color for visible light) follow different directions like in Fig. 1.35.

The phenomenon for light wave is called chromatic dispersion and for this reason the droplets of water in the sky, having n changing with λ, generates the rainbow.

1.10 Waves Polarization

Electromagnetic waves (and light in particular) are generated by the oscillation of an electric field **E** and a magnetic field **B**, which are perpendicular to each other and to the direction of wave propagation as previously described. However, the electric field can oscillate in any direction within the plane perpendicular to the propagation axis. During propagation, this direction of oscillation can vary randomly over time and space, resulting in a distribution of electric field orientations, illustrated as a star-like pattern in Fig. 1.36a.

The star of all the possible oscillation direction of the electric field in the plane xy, perpendicular to the z propagation direction, can be expressed by the combination of the two components along x and y as:

$$\mathbf{E}(t) = E_x(t)\mathbf{i} + E_y(t)\mathbf{j} \tag{1.77}$$

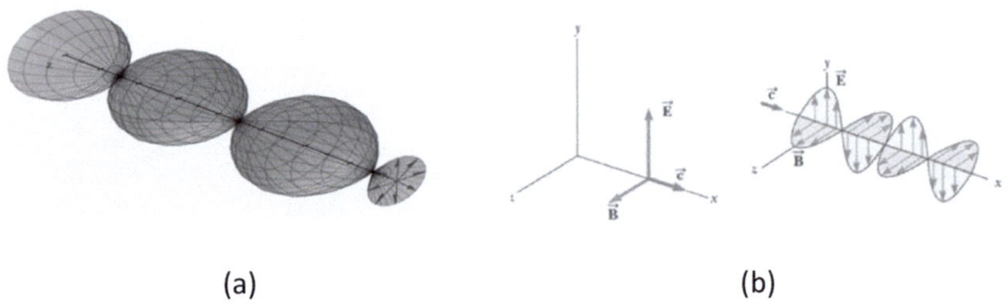

Fig. 1.36 Propagation of **a** unpolarized and **b** linearly polarized electromagnetic waves

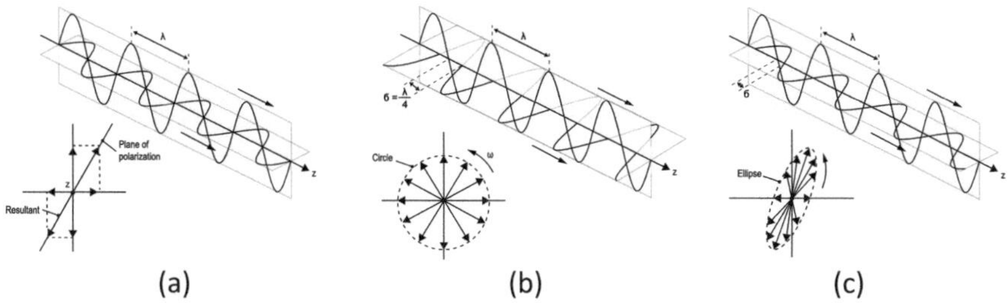

Fig. 1.37 Different kind of polarized light **a** linear, **b** circular, **c** elliptical

For an unpolarized light ray $E_x(t)$ and $E_y(t)$ are arbitrary random functions of the time t. If the unpolarized light ray pass through a polarizing filter the two components can be expressed by:

$$E_x(t) = A_x \cos(\omega t - kz + \varphi_x)$$
$$E_y(t) = A_y \cos(\omega t - kz + \varphi_y)$$

where $\omega = 2\pi/T = 2\pi f$ is the lightwave frequency, $k = 2\pi/\lambda$ is the wave number, φ_x and $\varphi_y$the phases. By summing the two component, if $(\varphi_y - \varphi_x) = 0$ or π we obtain a constant in time resultant vector as in Fig. 1.37a. Delay in time will be $\Delta t = (\varphi_y - \varphi_x)T/2\pi = 0$ or $T/2$, and delay in space will be $\Delta s = (\varphi_y - \varphi_x)\lambda/2\pi = 0$ or $\lambda/2$. Polarization angle will be $\alpha = \text{arctg}(E_y/E_x) = \pm\, \text{arctg}(A_y/A_x)$, defining the plane of oscillation of the electric field.

If at the output of a polarizing filter we have $A_x = A_y$ and $(\varphi_y - \varphi_x) = \pm\pi/2$ the two component of electric field will rotate around a circle while propagating along z direction, so generating a elicoids in its propagation. This condition is named circularly polarized light and is illustrated in Fig. 1.37b. In this case the delay in time will be $\Delta t = (\varphi_y - \varphi_x)T/2\pi = \pm T/4$ and delay in space will be $\Delta s = (\varphi_y - \varphi_x)\lambda/2\pi = \pm\lambda/4$.

If $A_x \neq A_y$ and $(\varphi_y - \varphi_x) = \beta$ the two components of electric field will rotate around an ellipse while propagating along z direction, so generating a helicoid with an elliptical base in its propagation. This condition is named elliptical polarized light and is illustrated in Fig. 1.37c. Ellipse axes are defined by the value of angle s.

There are several methods that can be used to polarize light or other electromagnetic waves:

(1) Polarization by absorption occurs when unpolarized wave passes through a material that only transmits wave oscillating in a specific direction. Such a material acts as a polarizing filter, a common example being a Polaroid filter for the light. These materials contain molecules aligned along a particular axis, which absorb the component of the electric field **E** that oscillates parallel to their alignment. As a result, only the component of **E** perpendicular to the molecular alignment is transmitted. If the polarized light, then passes through a

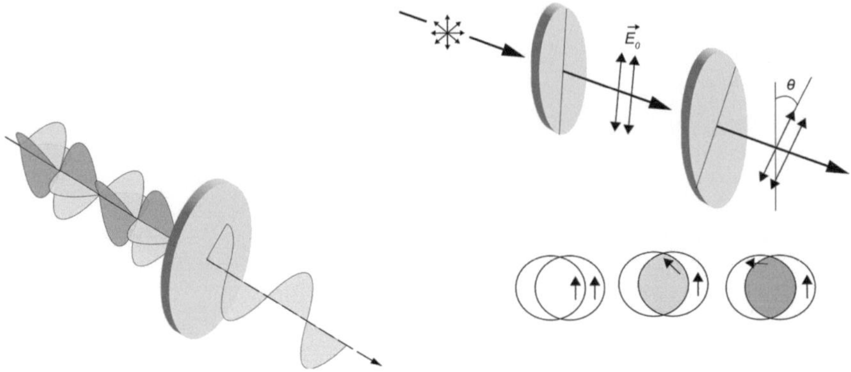

(a) (b)

Fig. 1.38 Polarizing filter and combination polarizer- rotating analyzer

second Polaroid filter, called the analyser, which is rotated by an angle θ relative to the first, the transmitted wave intensity decreases according to Malus' Law (Fig. 1.38):

$$I = I_0 \cdot \cos^2 \theta \tag{1.78}$$

Polarization by reflection occurs when unpolarized waves strikes the surface of a material and becomes partially polarized. For instance, when unpolarized light hits the surface of water, part of the light is absorbed or transmitted into the water, while another portion is reflected. The reflected light becomes partially polarized. At a specific angle, known as Brewster's angle, the reflected light is completely polarized in a direction perpendicular to the plane of incidence.

(2) Polarization by birefringence, as illustrated in Fig. 1.39b, occurs when a single light ray is split into two separate rays upon passing through certain anisotropic materials. Materials such as calcite crystals exhibit this behaviour due to their internal structure, which causes them to have different refractive indices depending on both the direction of light propagation and the orientation of the electric field oscillation.

As a result, the incident light is decomposed into two rays with orthogonal polarizations: the ordinary ray and the extraordinary ray. These two rays experience different refractive angles and travel along distinct optical paths within the crystal. Consequently, they emerge from the material spatially separated, producing two distinct images, as illustrated in Fig. 1.39c.

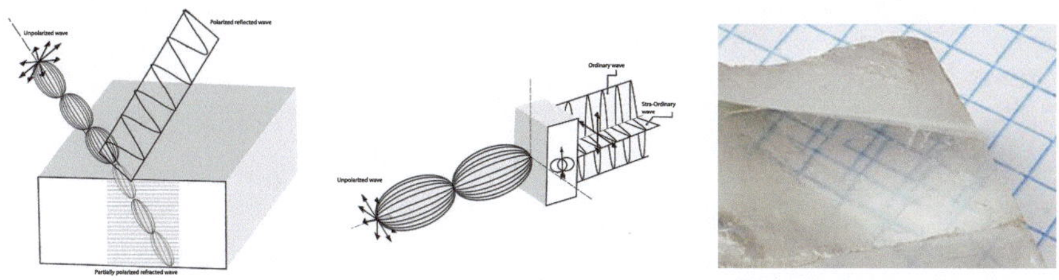

(a) (b) (c)

Fig. 1.39 Polarization by reflection (**a**) and by bi-rifrangent materials (**b**) and (**c**)

1.11 Quantistic Rappresentation of Waves

At the beginning of 20th century Albert Einstein, Plank and other scientists, studying photoelectric effect, i.e. the phenomenon where electrons of a solid material are emitted when is is exposed to light of a certain frequency. The way of this emission occur cannot be explained by Maxwell equations and wave model but only considering an electromagnetic wave as a flow of packets of energy (particles without mass) called photons. Energy of each one of these photons is given by the Plank's law:

$$E = hf \ = hc/\lambda \tag{1.79}$$

where h is a universal constant, whose value is $h = 6.626 \times 10^{-34}$Js, named Plank's constant, $c = \lambda f$ the speed of the equivalent electromagnetic wave, λ its wavelength and f its frequency. It is the base of quantum mechanics, i.e. energy of an electromagnetic wave is not a continuous variable but a flow of energy packets of value hf. This because no electrons are emitted by the material exposed to light of a frequency lower than a certain soil, and because electron emission occur instantaneously, with no need to wait a certain time that a continuous power flow of a wave give to the materials atoms a sufficient amount of energy to the electrons they need to cames out, so the idea that energy travel in packets, named photons.

Assuming this only photons of a high enough frequency, and so energy, can interact with electrons having only some quantized possible levels of energy states. If this happen electrons receiving a photon change their energy of the quantity $E_1 - E_2$ moving from a possible level of energy E_1 to another E_2, or if they change its energy from a level E_2 to another E_1 like in Fig. 1.40, they emit a photon of energy equal to the difference, and so of a well-defined frequency given by Eq. (1.78) as $f = h/(E_2 - E_1)$.

The total energy E_{tot} carried by an electromagnetic "wave", non-existing anymore, is substituted by a photon flow of N_f photons/s as:

$$E_{tot} = N_f E = N_f hf \tag{1.80}$$

Giving a photon flow like in Fig. 1.41.

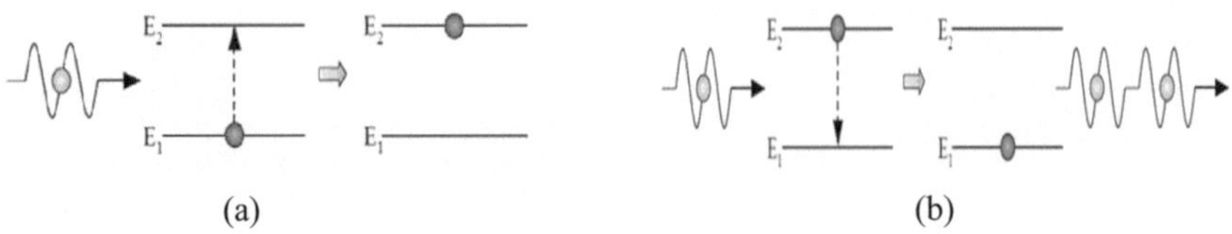

Fig. 1.40 Interaction of a single photon of an electromagnetic wave with energy levels of electrons of a material, **a** single and **b** stimulated emission

Fig. 1.41 Electromagnetic waves as a photon flow accordingly to quantum theory

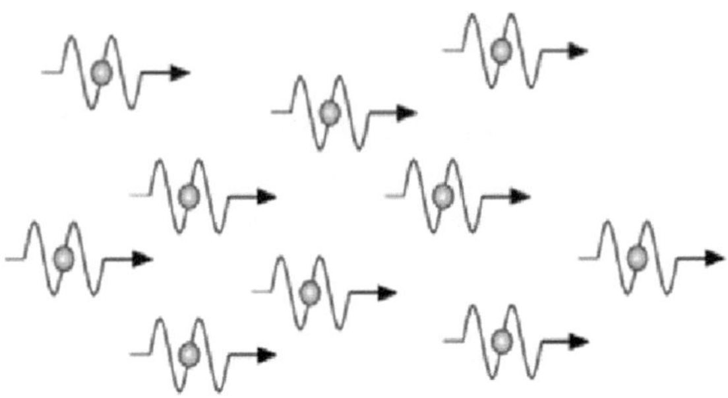

Energy of the single photon of each frequency and wavelength of all the electromagnetic spectrum is given in Table 1.1.

So electromagnetic waves at high frequency, like x or gamma rays, has very high level of energy that can interact with atomic structures, not possible for electromagnetic radio waves at lower frequencies.

1.12 Laser Light Sources

A LASER (Light Amplification by Stimulated Emission of Radiation) is a particular source of electromagnetic waves (or photon flux) having a single, well-defined frequency and so wavelenght. Laser sources operate accordingly to the principles of quantum mechanics, involving the interaction between light and atoms or molecules by photon emission and absorption. Here's a simplified explanation of how a laser works:

1. Energy Absorption (Pumping)

   The process begins with an external energy source (like an electric current, flash-lamp, or another laser) supplying energy (a photon flux) to the laser medium material. The atoms or molecules in the medium absorb these photons, exciting their electrons to higher energy levels.
2. Spontaneous Emission

   - Excited State: When the electrons are in these higher energy states, they are unstable and will eventually return to a lower energy state.

Table 1.1 Electromagnetic quantum energies values from Plank law

Type	Frequecy	Wavelenght	Photon energy
γ—ray	30–300 EHz	1–10 pm	124 KeV–1.24 MeV
hX—ray	3–30 EHz	10–100 pm	12,4–124 KeV
sX—ray	30 PHz–3EHz	10 nm–100 pm	124 eV–12,4 KeV
UV–Ultra violet	3–30 PHz	10–100 nm	1,24 eV–124 meV
Visible	Aroud 300 THz	0.4–0.7 μm	Around 1,5 eV
NIR—Near infrared	30–300 THz	1–10 μm	124 meV–1,24 eV
MIR—Mid infrared	3–30 THz	10–100 μm	12,4–124 meV
FIR—Far infrared	300 GHz–3 THz	100 μm–1 mm	124 meV
EHF—Extremely high frequency (microwave)	30–300 GHz	1 mm–1 cm	1,24 meV–124 μeVeV
SHF—Super high frequency (microwave)	3–30 GHz	1 cm–1 dm	12,4–124 μeV
UHF—Ultra high frequency	300 MHz–3 GHz	1 m–1 dm	1,24–12,4 μeV
VHF—Very high frequency	30–300 MHz	1–10 m	124 neV–1,24 μeV
HF—High frequency	3–30 MHz	10–100 m	12.4–124 neV
MF—Medium frequecy	300 KHz–3 MHz	100 m–1 Km	1,24–12,4 neV
LF—low frequency	30–300 KHz	1–10 km	1,24 neV–124 peV
VLF—Very low frequency	3–30 KHz	10–100 Km	12,4–124 peV
VF—Voice frequency	300 Hz–3 KHz	100 Km–1 Mm	1,24–12.4 peV
ELF—Extremely low frequency	30–300 Hz	1–10 Mm	124 feV–1,24 peV

- Photon Emission: As the electrons drop to a lower energy level, they release energy in the form of photons. This is called spontaneous emission.

3. Stimulated Emission

- If a photon of the right wavelength passes by an excited atom or molecule, it can stimulate the release of another photon. The emitted photon will have the same phase, direction, and wavelength as the stimulating photon, which leads to the amplification of photon flow.
- Chain Reaction: This process triggers a chain reaction where more and more photons are emitted, all coherent (in phase and direction) with each other.

4. Optical Resonance

- Mirrors: The laser medium is placed between two mirrors, forming an optical cavity. One mirror is fully reflective, while the other is partially reflective.

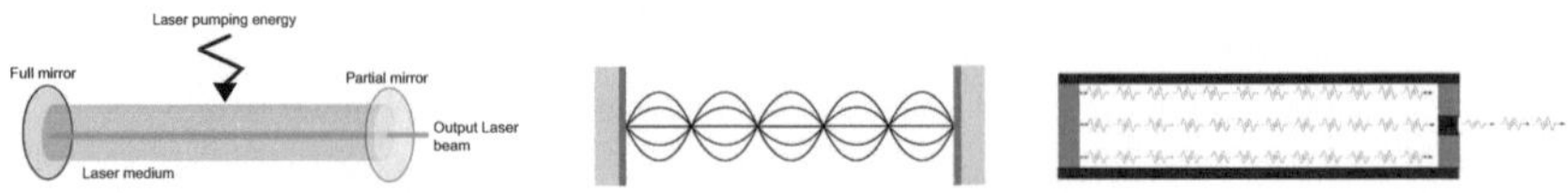

(a) Laser main components (b) Laser photon resonance "like" a guitar string

Fig. 1.42 Basics of laser sources

- Photon reflection: The photons bounce back and forth between these mirrors, stimulating more emissions and amplifying the light within the cavity.

Optical resonance generates inside laser cavity a standing electromagnetic wave, that can be seen as bouncing photons, like the string of a guitar, as illustrated in Fig. 1.42b

5. Laser Output

- Partially Reflective Mirror: Some of the light came out through the partially reflective mirror, producing a coherent and single wavelength photon flow, 1.e. the laser beam.
- Continuous Emission: As long as energy is supplied to the laser medium (through pumping), the laser will continue to emit light.

The main components of a laser source are that illustrated in Fig. 1.42a. It can be seen as a resonance optical cavity, "like" a guitar string with a standing wave, as previously illustrated.

Population inversion is a critical concept in the functioning of a laser. It refers to the condition where a higher number of atoms or molecules in a laser medium are in an excited state (higher energy level) than in the ground state (lower energy level). This condition is essential for the process of stimulated emission to dominate over absorption, which is necessary to generate a coherent laser beam. Main items are:

1. Normal Population Distribution

   Under typical conditions, most atoms or molecules in a medium are in their lowest energy state, known as the ground state. When light passes through this medium, the photons are more likely to be absorbed by atoms in the ground state, raising them to an excited state. This process, known as absorption, prevents the creation of a laser beam.

2. Achieving Population Inversion by pumping mechanism

   To create population inversion, an external energy source (pumping) is used to excite a significant number of atoms or molecules to higher energy levels. The pumping can be achieved through various means, such as electrical discharge, optical pumping with light, or chemical reactions.

   The laser medium typically has three or more energy levels. In a three-level system, atoms are pumped from the ground state to a higher energy state, then quickly drop to a slightly lower, metastable state where they can stay longer. In a four-level system, atoms drop from an excited state to a second excited state (above the ground state), allowing more efficient population inversion.

A key feature of the energy levels used in lasers is the existence of a metastable state, an excited state where atoms can stay for a relatively long time compared to other states. This allows more atoms to accumulate in this state, contributing to the population inversion.

3. Stimulated Emission versus Absorption

   Once population inversion is achieved, there are more atoms in the excited state than in the ground state. When a photon passes through the medium, it is more likely to encounter an excited atom and stimulate the emission of another photon of the same wavelength and phase, rather than being absorbed. This stimulated emission process amplifies light and leads to the coherent laser beam.

4. Importance of Population Inversion

   Without population inversion, the absorption of photons would outweigh the stimulated emission, preventing the laser from amplifying light. Achieving and maintaining population inversion is the key challenge in laser design and operation. Many kind of laser source are nowadays available, thay can be classified mainly in solid active material laser, gas lasers, semiconductor laser illustrated in Fig. 1.43.

The story of laser begun on May 16, 1960, when Theodore H. Maiman operated the first functioning laser using a ruby crystal; and then advancements where made in 1962, when Robert N. Hall demonstrated the first laser diode infrared device (GaAs). The most used gas laser are nowadays that using Helium-Neon, Argon-ion and in industry CO_2 as laser medium. In Table 1.2 the main characteristics of some kind of historical laser sources.

Laser emission can be continuous or pulsed, excitation can be done by electrical discharge pumping, flashlamp pumping, chemical reaction pumping or by another laser pumping.

A laser source generates monochromatic, coherent and collimated (sometime polarized) electromagnetic wave that usually is in the visible range, so is light, but also in the infrared range. Differences between laser and other light sources are illustrated in Fig. 1.44.

Another important characteristic of laser light is its high degree of collimation, meaning that the emitted rays are nearly parallel. The laser beam emerges from

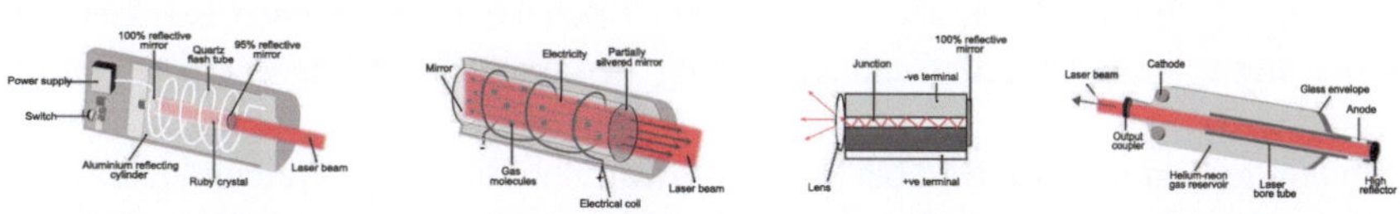

(a) Flash excitation (b) electric excitation (c) Laser diode (d) He-Ne gas laser

Fig. 1.43 Different kind of laser sources

Table 1.2 Some characteristics of the most classical common lasers

Laser type	He–Ne	Ar	Ruby	NdYag	CO_2	AlGaAs InGaAs
Gain material	Gas	Gas	Solid	Solid	Gas	Semiconductor
Wavelength (nm)	632,8	488–514,5	694,3	1064	10600	800–1500
Laser emission	Continous	Continous	Pulsed 350 μs	Continous	Continous	Continous
Optic power	5 mW	20 W	1 J/350 μs	100 W	100–1000 W	100 mW
Beam divergence	1 mrad	1 mrad	5 mrad	5 mrad	5 mrad	20°
Efficency	0,05%	0,05%	0,1 %	0,5 %	10%	60%
Pumping type	Electrical discharge	Electrical discharge	Flash	Flash	Electrical discharge	Charge injection
Cost (Euro)	1500	50000	30000	70000	100000	500

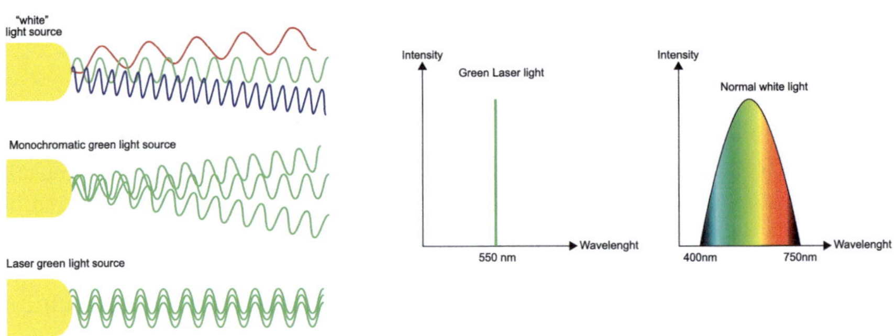

Fig. 1.44 Laser and other light sources

the partially reflective mirror with minimal divergence and maintains this collimation over long distances. This is because the beam interacts with very few atoms or molecules along its path, and only those with electrons capable of transitioning between energy levels that match the photon energy of the laser's single frequency will interact with the beam. This property results in a highly brilliant and low-divergence beam, as illustrated in Fig. 1.45. Laser beam divergence is quantified by the angle $\theta = \lambda/(2\Delta r)$ and increasing of laser spot radius Δr is given by $\Delta r = L \tan \theta$.

In contrast, non-collimated beams are typically produced by semiconductor laser diodes, which emit laser light over a relatively wide angle. These beams exhibit reduced coherence compared to conventional lasers but are characterized by high energy efficiency.

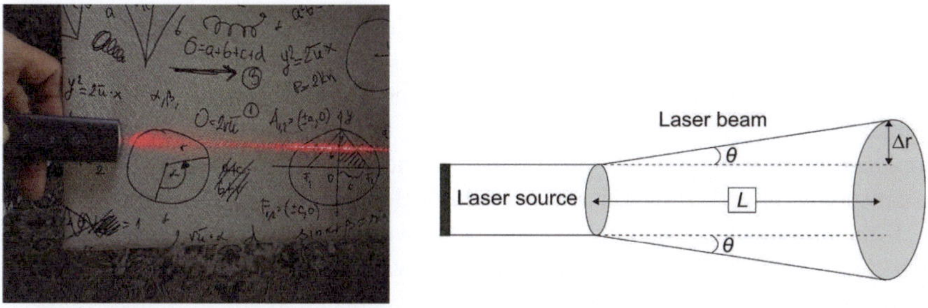

(a) (b)

Fig. 1.45 Laser beam and its divergence

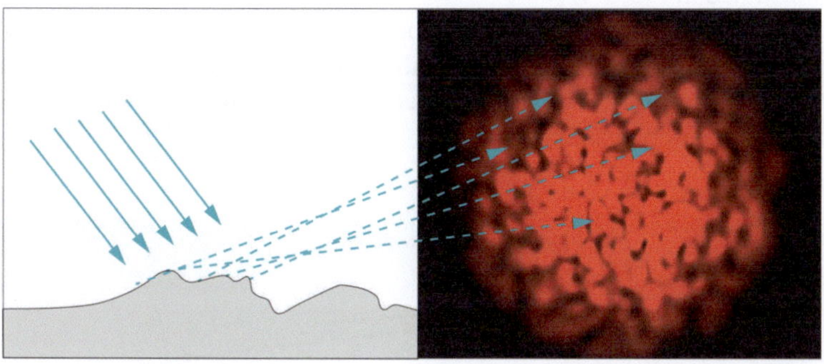

Fig. 1.46 Laser speckle generated by a red laser beam reflected by a rough surface

One phenomenon associated with laser light is called speckle. This effect arises from the random interference of multiple coherent rays (having the same frequency) that differ in phase, often due to reflection from a rough surface. The surface irregularities alter the phase of the reflected waves, producing numerous rays that interfere constructively and destructively randomly at different points on a screen. This results in a granular pattern of fluctuating brightness, as shown in Fig. 1.46, caused by spatial variations in amplitude due to the random interference of the reflected rays.

This speckle effect has many influences, impact and applications in measurement systems based on laser light sources, as we will see in the following chapters.

1.13 Time Delay Measurements Between Waves in Echoes

Measuring the travel time of waves from a source to a reflecting target and back to a receiver is a fundamental method for determining distances. This technique is widely used across various systems employing radio, light, or ultrasonic waves, as will be discussed in the following chapters. In this section, the basic signal processing techniques used to measure the time delay—caused by the wave traveling along a specific path—are introduced.

The Time-of-Flight (ToF) method, which relies on the analysis of reflected wave signals, has become increasingly prevalent in distance and motion measurement applications. It is especially valuable in scenarios requiring a high dynamic range, fast response, good accuracy, and low sensitivity to the surface properties of materials.

If the propagation speed of the wave is known (denoted by v), and the measured time delay is Δt, the distance d traveled by the wave along the straight path from the source to the reflecting target and back can be calculated as:

$$d = v\frac{\Delta t}{2} \tag{1.81}$$

As illustrated in Fig. 1.47.

Given the transmitted wave $s_1(t)$ and the received back-reflected wave $s_2(t)$ after propagation through a path, a classical method for estimating the time delay between these two signals involves the use of the cross-correlation function $R_{s_1 s_2}(\tau)$, which is defined by the following integral [3]:

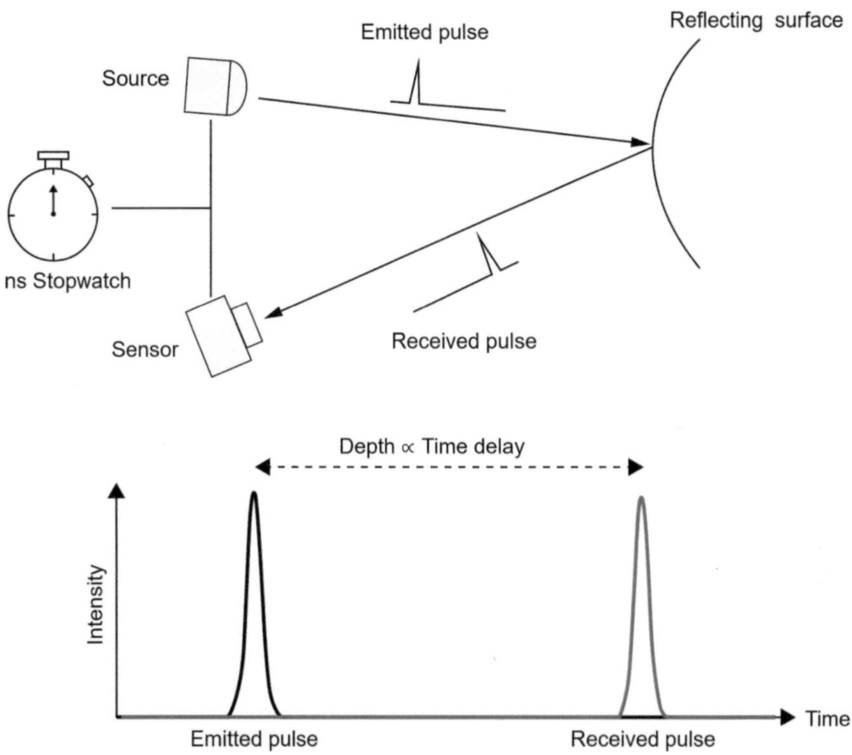

Fig. 1.47 Direct time of flight measurement technique

$$R_{s_1 s_2}(\tau) = \lim_{T\to\infty} \frac{1}{T} \int_0^T s_1(t) s_2(t+\tau) dt \tag{1.82}$$

This definition instructs us to compute the area under the product of one signal, $s_1(t)$, and a time-shifted received signal, $s_2(t + \tau)$, by evaluating their integral over time for all values of τ.

When the time shift τ causes the two signals to overlap significantly, their product becomes large, resulting in a relative maximum in the cross-correlation function. The value of τ at which this maximum occurs indicates the time delay between the two signals, essentially the time shift required to better align one signal with the other in time domain.

A simple example of cross-correlation involves a transmitted square pulse and a received square pulse that arrives after a delay of $\Delta t + \delta t$, as illustrated in Fig. 1.48. Cross-correlation can be visualized geometrically by progressively overlapping the two square pulses. As the overlap increases, the area under their pointwise product also increases linearly, reaching a maximum when the two pulses are perfectly aligned. This results in a triangular-shaped cross-correlation function, with its peak corresponding to a time delay τ equal to $\Delta t + \delta t$, as shown in Fig. 1.48.

In practical scenarios, however, the signals are far more complex. The back-reflected signal often differs significantly from the original transmitted pulse due to distortions, and it is commonly affected by substantial noise. This is especially true when the target is distant, as the back-reflected signal power tends to be very low. Nevertheless, the peak of the cross-correlation function can still indicate the time delay between the transmitted and received signals with high sensitivity. This technique is known as direct Time-of-Flight (dTOF).

Since electromagnetic waves such as light and radio waves travel at extremely high speeds, a bit more of 300,000 km/s in air, accurate measurement of very small time delays (in the nanosecond range or less) is required. For context, in air, light travels about 30 cm per nanosecond, so even minor timing errors can lead to significant inaccuracies in distance measurement.

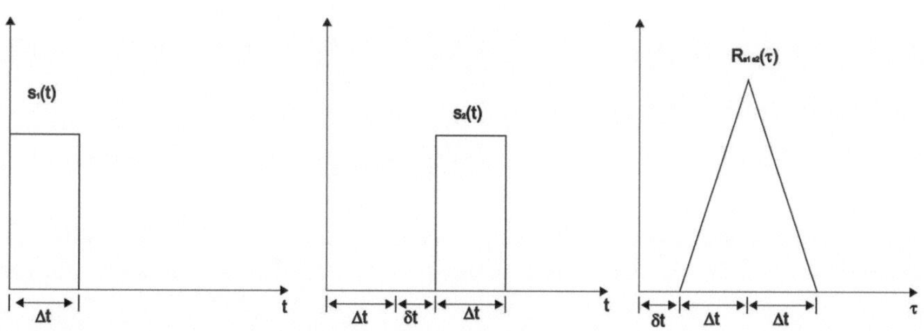

Fig. 1.48 Time delay measurement by crosscorrelation

This technique is not limited to the detection of reflected electromagnetic waves, it is also widely applied to other types of wave reflections, such as sound or ultrasound. For example, animals like bats use echolocation to detect obstacles, and similar principles are used in car ultrasonic parking sensors, sonar systems on ships to measure sea depth, or to detect the position of fish. These applications will be discussed in more detail in the following chapters.

In the case of acoustic waves, such as sound, the speed of propagation in air is a bit more than 300 m/s. Consequently, the corresponding time delays are several orders of magnitude larger than those observed with electromagnetic waves. For instance, a 30 cm travel path corresponds to a delay of approximately 1 μs, which is six orders of magnitude greater than the nanosecond delays observed with light.

One challenge in distance measurement using direct Time-of-Flight (dTOF) arises even when a narrow, focused beam (e.g., a laser) is transmitted in a specific direction. Most targets reflect the incoming wave diffusely, scattering it in multiple directions. As a result, especially at greater distances, the fraction of reflected power received by the sensor can be extremely low and strongly limited by the sensor's angular reception capability.

To address this limitation, alternative time-delay measurement techniques have been developed. One such method involves amplitude modulation of the transmitted wave and the measurement of phase shifts between the transmitted and received signals. This approach, illustrated in Fig. 1.49 is known as indirect Time-of-Flight (iTOF). It is generally more robust to noise and performs better under conditions of low received signal power.

If we use a sine modulating function of the transmitted signal the received ones will be a sine modulated signal, like that emitted, but with a phase shift due to the time it needs to travel along the path, of length 2d.

If the emitted signal is $s_1(t) = \cos(\omega t)$ the received signal will be $s_2(t) = \mathrm{O} + \mathrm{A}\cos(\omega t - \varphi)$, where O is the level of constant signal component, A is a factor giving the back-reflected wave amplitude coming to the point on the object surface and φ the phase delay to find. If we electronically generate a reference signal $S_{ref} = \cos(\omega t - \delta_i)$ with a known variable phase δ_i and multiply this signal with the pixel output $s_2(t)$ by integration of this product we can calculate the average in time measured signal amplitude as:

$$I(\delta_i) = \int s_2(t) S_{ref}(t, \delta_i) dt$$

That gives a result like:

$$I(\delta_i) = P + Q\cos(\delta_i - \varphi)$$

where P, Q and φ are unknown, but using three different reference signals with three phase values δ_1, δ_2, δ_3, the measurements of $I(\delta_1)$, $I(\delta_2)$, $I(\delta_3)$ are sufficient for

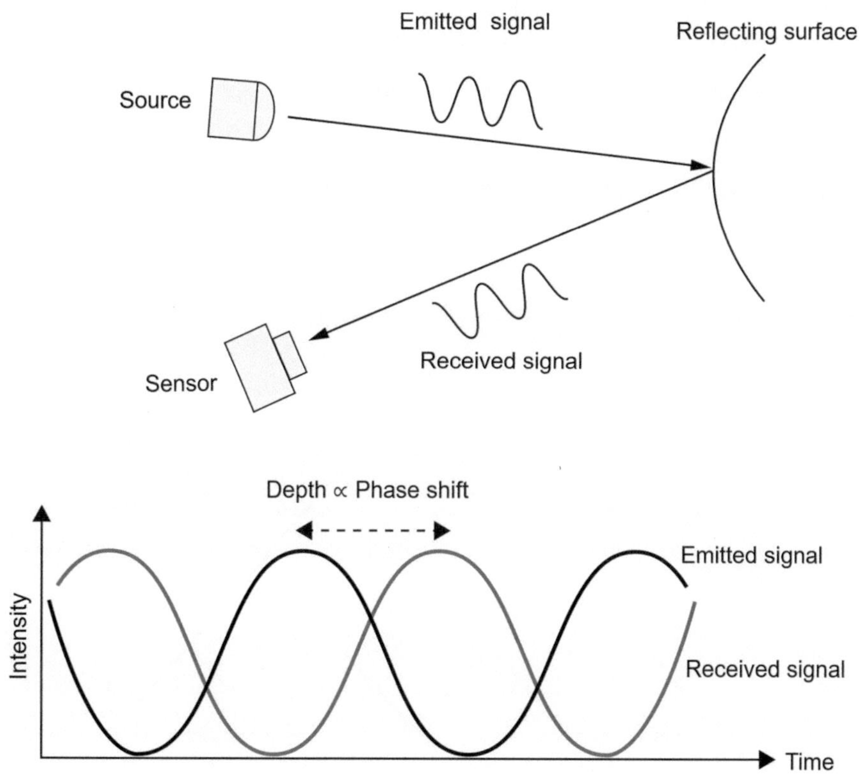

Fig. 1.49 Indirect time of flight (iTOF) measurement techniques

computing the phase φ value and the other two unknowns. Once calculated the phase delay φ, distance of the reflecting target will be:

$$d = c\frac{\varphi}{4\pi f} \tag{1.83}$$

Using for c the speed of electromagnetic waves or light, i.e. the value of about 300.000 Km/s, for a phase shift of half wavelength, so $\varphi = \pi$, using a modulation frequency of $f = 30$ MHz, we obtain a distance of $d = 2.5$ m.

Another way to detect phase shift from transmitted, reflected and then received waves is to use square pulses, i.e. digital signals, instead of sine amplitude modulation [4], as illustrated in Fig. 1.50. Generating a reference square signal $f_R(t)$, by performing multiple correlation with this reference signal, shifted in phase, it is possible to detect the phase delay of the received signal $S_2(t)$ due to travelling along the path.

Using to generate $f_R(t)$ the four phase shifts of $\pi/2$, π, $3\pi/2$, and 2π namely, $f_{R0}(t)$, $f_{R1}(t)$, $f_{R3}(t)$, and $f_{R3}(t)$, we can calculate the four correlations named R_0, R_1, R_2 and R_3 as illustrated in Fig. 1.51.

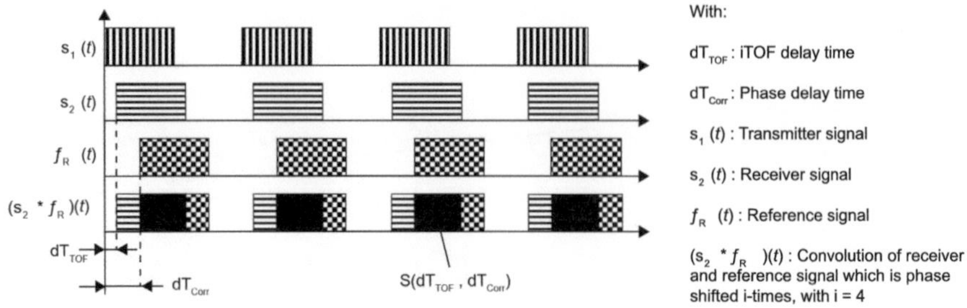

Fig. 1.50 Indirect time of flight estimation using square waves

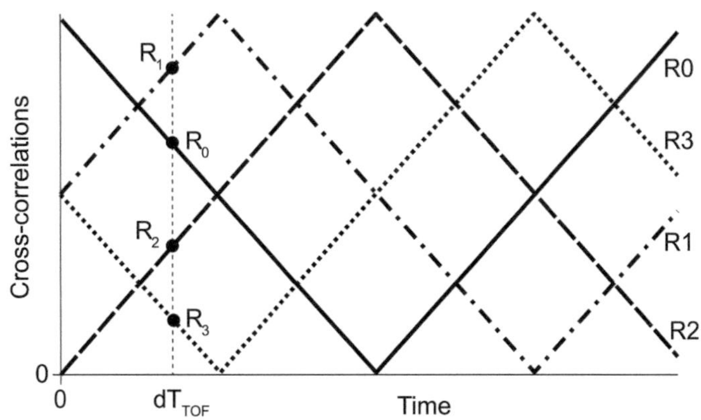

Fig. 1.51 Square signal shifted iTOF technique

Defining the two correlations differences $S = R_2 - R_0$ and $C = R_3 - R_1$, and defining the amplitude parameter $A = \frac{1}{2}(|S| + |C|)$ it is possible to calculate the phase shift φ by the four equations [5]:

$$\varphi = \frac{1}{4}\left(1 + \frac{S + C}{2A}\right)\text{for } S < 0 \text{ and } C \geq 0$$

$$\varphi = \frac{1}{4}\left(3 + \frac{S - C}{2A}\right)\text{for } S \geq 0 \text{ and } C \geq 0$$

$$\varphi = \frac{1}{4}\left(5 + \frac{-S - C}{2A}\right)\text{for } S \geq 0 \text{ and } C < 0$$

$$\varphi = \frac{1}{4}\left(7 + \frac{-S + C}{2A}\right)\text{for } S < 0 \text{ and } C < 0$$

Measurement systems based on the phase shift detection between a transmitted and a received signal, has a physical limitation that is called the ambiguity. Because phase is a periodic information, its value is repeatedly bound to each period. So, distance d is not given by Eq. (1.81) but by the equation:

$$d = \frac{c}{2f}\left(\frac{\varphi}{2\pi} + k\right) \tag{1.84}$$

For all the values k = 0, 1, 2, … In other word we can say that we do not know in witch period of phase delay we are, the first (k = 0), the second (K = 1), or the others.

This can be solved only limiting the measurement range d to only the first period. Measurement range is therefore given by the value of the modulation frequency f, for the maximum shift, so from the above equation with k = 0, $\varphi = 2\pi$ we have:

$$d_{max} = \frac{c}{2f} \tag{1.85}$$

So, measurement range is related to maximum possible phase shift of 2π (5 m for modulation frequency of 30 MHz) an can be improved by reducing the modulation frequency f.

Some other techniques are possible to improve measurement range, for example using two modulations frequencies f_1 and f_2 instead of only one. In this case Eq. (1.85) became:

$$d_{max} = \frac{c}{2(f_1 - f_2)} \tag{1.86}$$

And in this case measurement range d_{max} can be improved a lot by reducing the differences between the two modulation frequencies f_1 and f_2.

1.14 Amplitude Measurement of Waves in Noise by Lock-in Processing

In many measurement applications (RADAR, LIDAR, differential thermography, etc.) a received wave from the reflecting target is, for many reasons, strongly attenuated or sometimes the received signal is very little in amplitude and so strongly corrupted by noise, with a level bigger of one or more order of magnitude, so signal to noise ratio is less than 0,1, and sometimes very lower. Anyway, in these cases, we know the frequency of the signal received but strongly corrupted by noise that is the modulation frequency of the radio, light or thermal wave sent or generated.

Let's assume the signal wave received of interest is:

$$S(t) = A_s \cos(\omega t + \phi_s) + \text{Noise}$$

where A_s is the signal amplitude, ω is the signal angular frequency and ϕ_s the signal phase. We artificially generate a reference signal of same frequency, but different phase:

$$R(t) = A_r \cos(\omega t + \phi_r)$$

If we mixed (multiply) the two signal or adding the two waves arriving in same point, for the moment not considering the noise, we obtain:

$$V_{mix}(t) = (1/2)A_s A_r[\cos(\phi_s - \phi_r) + \cos(2\omega t + \phi_s + \phi_r)]$$

where the first term $\cos(\phi_s - \phi_r)$ is a constant component, the second term $\cos(2\omega t + \phi_s + \phi_r)$ has twice the reference frequency, so filtering above signal with a low pass filter having a cutting frequency below the double of the reference frequency or lower, this remove the noise, and we obtain an output with an amplitude V_{out} equal to:

$$V_{out} = (1/2)A_s A_r \cos(\phi_s - \phi_r) = K\ A_s$$

So a signal with an amplitude proportional to the original signal amplitude we aim to measure without the effects of noise, but the factor K depends on phase shift from the reference signal $(\phi_s - \phi_r)$.

To measure both amplitude and phase, two reference signals with a $\pi/2$ phase shift can be used, expressed by a sine and cosine as:

$$R_1(t) = A_r \cos(\omega t) \quad \text{and} \quad R_2(t) = A_r \sin(\omega t)$$

So multiplying with S(t) and filtering out high-frequency terms and noise we get the two numbers unaffected by noise:

$$I_x = (1/2)A_s\, A_r \cos\, \phi_s \text{ and } I_y = (1/2)A_s A_r \cos\, \phi_s$$

These two numbers give us the real and imaginary part of the complex number representing the s(t) signal, so its amplitude and phase will be:

$$A_s = \frac{2}{A_r}\sqrt{I_x^2 + I_y^2} \quad \phi = \tan^{-1}\frac{I_y}{I_x} \tag{1.87}$$

In general the typical situation is that illustrated in Fig. 1.52, a low amplitude wave plus a much bigger amplitude noise. We know its frequency and we can generate artificially a reference signal and split it into for example two square waves, one in phase with the reference and the other out of phase by $\pi/2$, as illustrated in Fig. 1.52.

By multiplying each one of the two squared reference signals with the received noise+wave but considering only the "clean" wave, as a sine wave, we obtain, if this sine is perfectly in phase with the square reference a reversed sine wave like in Fig. 1.53. Integrating, i.e. low pass filtering give us a level I_x. If the sine is not in phase with the reference square, we obtain a different signal like in Fig. 1.53, and integrating a lower level of I_x.

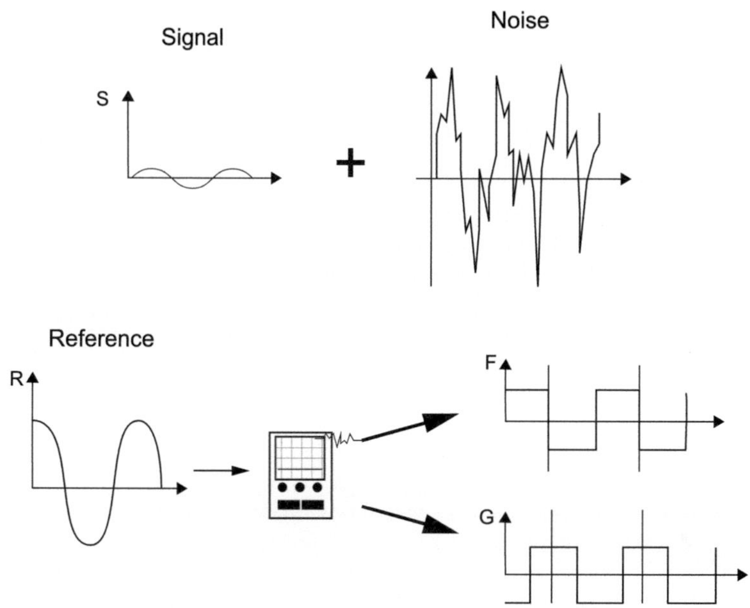

Fig. 1.52 Wave with noise and reference signals generated

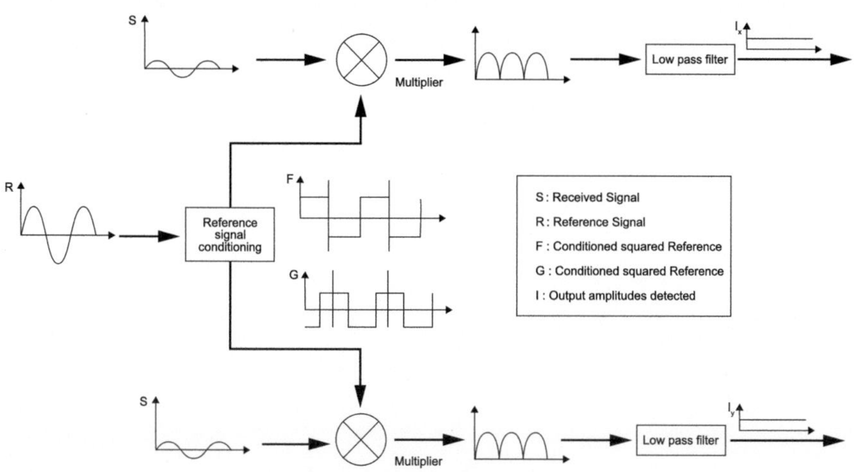

Fig. 1.53 Lock-in wave signal analysis to detect its amplitude

Performing the same with the other square reference signal we obtain a value I_y. In the integration (i.e. low pass filtering) process noise randomly positive and negative theoretically disappear, anyway is strongly reduced giving a not significative contribution in I_x and I_y values. As before, the two values of I_x and I_y are, according to [6], the real and imaginary components of a complex number expressing the original sine signal s(t) in its amplitude and phase, so the complete signal amplitude I and phase ϕ respect to the reference of the signal s(t) will be:

$$I = \sqrt{I_x^2 + I_y^2} \quad \phi = \tan^{-1} \frac{I_y}{I_x} \tag{1.88}$$

Many data processing systems are available to perform the lock in processing, analogic by lock in signal processing amplifiers, digitally on sampled waveforms by many software available in Python, Matlab and many others [7].

References

1. G.B. Thomas, *Calculus*, 14th edn. (Pearson, 2014)
2. S.K. Sahdev, *Electrical Machines* (Cambridge University Press, 2018)
3. H.D. Young, R.A. Friedman, *University Physics* (Addison-Wesley, 1996)
4. G. Rossi, Misure Meccaniche e Termiche, ed. by Carocci (2010)
5. S. Gianlicola, M. Valenti, R. Sala, *A Survay on 3D Cameras: Metrological Comparison of Time-of-Fligt, Structured Light and Active Stereoscopy Technologies* (Springer, Berlin, 2018)
6. K. Creath, V phase-measurement interferometry techniques. Progress Opt. **26**, 349–393 (1988)
7. M. Meade, *Lock-in Amplifiers: Principles and Applications* (Peregrinus, 2018)

Chapter 2
Analogic Image Formation

Abstract The fundamentals of analogic image formation through optical components such as lenses, objectives, mirrors, and prisms are presented. Using the principles of geometric optics and ray tracing, classical mathematical models are derived to describe the typical transformation of coordinates from a three-dimensional real object space to its two-dimensional image on the camera's image plane. The main concepts and techniques employed in analog image processing systems are discussed, with a focus on their role in measurement applications. Essential criteria for selecting camera objectives and ensuring proper analog image generation useful for measurement purposes are also outlined.

2.1 History of Image Formation

To generate an optical image on a plane useful for measurements purposes are necessary: an object, which image it is generated on a plane, called image plane, an analogic light processing system, like a lens, an hole or an objective and a light source. We can consider the object itself like a light source, assuming its surface working as a perfect diffusive surface of the illuminating light, as in Fig. 2.1.

The first technique of image generation on something were seen in China on 1558, named camera obscura. Here a hole was realized on a wall like in Fig. 2.1b and only the light following the direction of the straight line passing for the point in the object and the point where is the hole can pass through the hole and going to the opposite wall, where the image of the point is formed. Image is reversed because the point down on the object sends its light in point up on the image plane and vice versa.

Later lens has been used to form the images on a plane. The invention of the daguerreotype by Louis Daguerre in 1839 marked the beginning of photography. This process involved exposing a silver-coated copper plate to light, which was then developed with mercury vapor to create a permanent image. William Henry Fox Talbot in 1841 invented the calotype process, which used paper coated with silver iodide to create a negative image. This negative could then be used to produce multiple positive prints, laying the groundwork for modern photographic techniques.

G. Rossi, *Mechanical and Thermal Measurements by Images and Waves*, Smart Sensors, Measurement and Instrumentation 51, https://doi.org/10.1007/978-3-032-03464-9_2

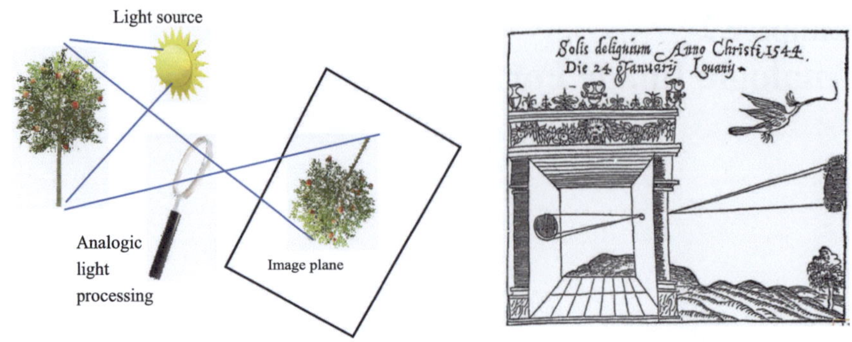

(a) By a lens (b) by a hole

Fig. 2.1 Analogic image formation on a plane

Frederick Scott Archer introduced in 1851 the collodion wet plate process, which involved coating a glass plate with a light-sensitive solution. This process produced a more detailed and stable image than earlier methods and was widely used until the invention of dry plates. Richard Leach Maddox in 1871 invented the gelatin dry plate, which allowed for images to be captured without the immediate need for chemical development, making photography more convenient and portable. Early experiments with color photography included the autochrome process, was developed by the Lumière brothers in 1907, which used dyed grains of starch to create color images. Later, more advanced methods like Kodachrome and Agfacolor revolutionized color photography in the 1930s. The introduction of 35 mm film (1920–1930s) and the development of portable cameras like the Leica revolutionized photography by making it more accessible and affordable. The development of digital imaging technology in the late twentieth century marked a significant improvement in image formation, on same device infinite pictures can be recorded. Digital cameras, using electronic sensors to capture and store images as digital files replace film cameras. Image digitalization will be described in the next chapter. The advent of digital technology led to new possibilities also in image manipulation, storage, and sharing.

Recently the integration of miniaturized digital cameras into smartphones made photography ubiquitous, the new era of communication by images start, in 2007, with the first iPhone.

2.2 Fundamental Models of Image Formation by Geometric Optics

Geometric optic, also known as ray optic, is a part of optic theory that describes light propagation in terms of rays. This model is particularly useful for understanding the main phenomena in systems where the wavelength of light is much smaller than the dimensions of the optical components involved, such as lenses, mirrors, and prisms. Geometric optics simplifies the study of light by assuming that it travels in straight lines and analyzing its interactions with surfaces using simple mathematical rules. The few simple rules of ray optics, demonstrated in the previous chapter, are:

1. Light is represented as rays that travel in straight lines. These rays change direction only when they encounter different media or surfaces characterized by different propagation velocity v and so refraction index, defined as n = c/v, where c is the light speed in void, the maximum possible for every electromagnetic waves and for the light.
2. Reflection: When a light ray encounters a reflective surface (like a mirror), it bounces off the surface. The law of reflection states that the angle of incidence (the angle between the incoming ray and the normal to the surface) equals the angle of reflection (the angle between the reflected ray and the normal).
3. Refraction: When a light ray passes from one medium to another, both transparent (such as from air to water), it bends due to a change in its propagation speed. Snell's Law, already discussed in previous Chap. 1 (1.64), describes this behavior mathematically, generally using not the velocity, as in previous chapter done, but the refractive indices of media n that for light is the ratio between speed of light in vacuum and speed of light in the medium considered, i.e. n = c/v, so Snell's law for light is normally expressed using refractive index of materials where light propagates instead of velocity, so in the form:

$$\frac{\sin(\theta_1)}{\sin(\theta_2)} = \frac{n_2}{n_1} \text{ or } n_1 \sin(\theta_1) = n_2 \sin(\theta_2) \tag{2.1}$$

   where n_1 and n_2 are the refractive index of the two media, and θ_1 and θ_2 are the angles of incidence and refraction, respectively.

Following the basic principles of geometric optics, which form the theoretical foundation for studying image formation, a wide range of ray-tracing software tools have been developed and are now widely available (e.g., https://phydemo.app/ray-optics/simulator/). By using this particular simulator to trace rays emitted from two light beam sources, focused through a lens system and subsequently refracted and reflected by a little grey spherical bubble rising in water, it is possible to obtain the result illustrated in Fig. 2.2.

This principle was proposed in [1] as the foundation for developing a system to measure bubble velocity. The optical setup, which resembles the configuration depicted in Fig. 2.2, functions by analyzing retroreflected light. Specifically, the system captures light pulses back reflected from the bubble's surface and converts them into electrical signals using two receiving photodiodes.

By cross-correlating these two signals as illustraated in Chap. 1, corresponding to the bubble's passage through two focal points separated by a known distance d, the time delay τ between the two events is determined. The bubble's velocity can then be calculated using the simple relation v = d/τ.

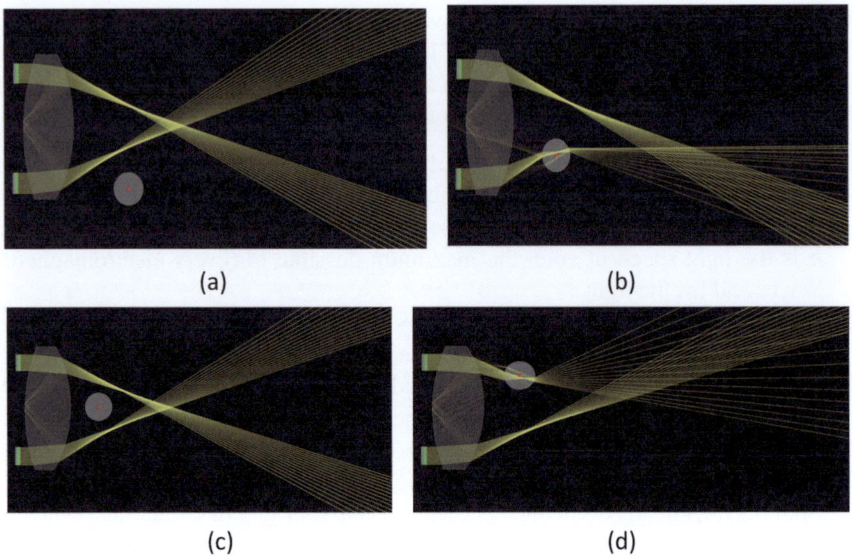

Fig. 2.2 Ray tracing of reflections and refraction from a little grey bubble moving up in front of an optical system with two light sources and a focusing lens system generating two focal points

When every point on an object, such as the surface points of the guitar shown in Fig. 2.3, emits light that scatters in all directions, we can analyze the resulting light paths using the principles of geometric optics. If an opaque plate (a material that does not transmit light) is placed between the object and the image plane, and a small hole or pinhole is introduced at a specific point O on this plate, the system becomes a basic system for image formation. In this setup, the image plane is located at a distance *f* behind the pinhole, while the object (in this case, the guitar) lies in the object plane at a distance *d* in front of the pinhole. According to geometric optics, a light ray emitted from any point R on the object will travel in a straight line through the pinhole at point O and reach a corresponding point I on the image plane. This fundamental principle underlies the operation of the camera obscura: each point on the object is projected through the pinhole along a straight path, resulting in the formation of an inverted image on the image plane.

Considering a generic pair of corresponding points (R, I), and referring to the two similar triangles shown in Fig. 2.3 OO_RQ_R and OO_IQ_I we can derive the following relationship:

$$\frac{y_r}{\mathrm{d}} = \frac{y_I}{\mathrm{f}} \text{ but distance d is the z coordinate of point R so } \frac{y_r}{z_r} = \frac{y_I}{\mathrm{f}} \tag{2.2}$$

From the projection on x–z plane of Fig. 2.3b and the similarity of the two triangles, we also can write:

$$\frac{x_r}{z_r} = \frac{x_I}{\mathrm{f}} \tag{2.3}$$

By combining Eqs. (2.2) and (2.3), we obtain the general fundamental coordinate transformation that maps each point on the real object surface from its 3D coordinates (x_R y_R, z_R) to its corresponding 2D coordinates (x_I y_I) on the image plane:

$$x_I = f\frac{x_R}{z_R} \quad y_I = f\frac{y_R}{z_R} \tag{2.4}$$

As shown in Fig. 2.3, it is evident that a straight line in 3D space, such as the segment RQ_R, is projected as a straight line in the 2D image plane, in this case becoming IQ_I. Although the line remains straight, its length generally changes due to this perspective projection.

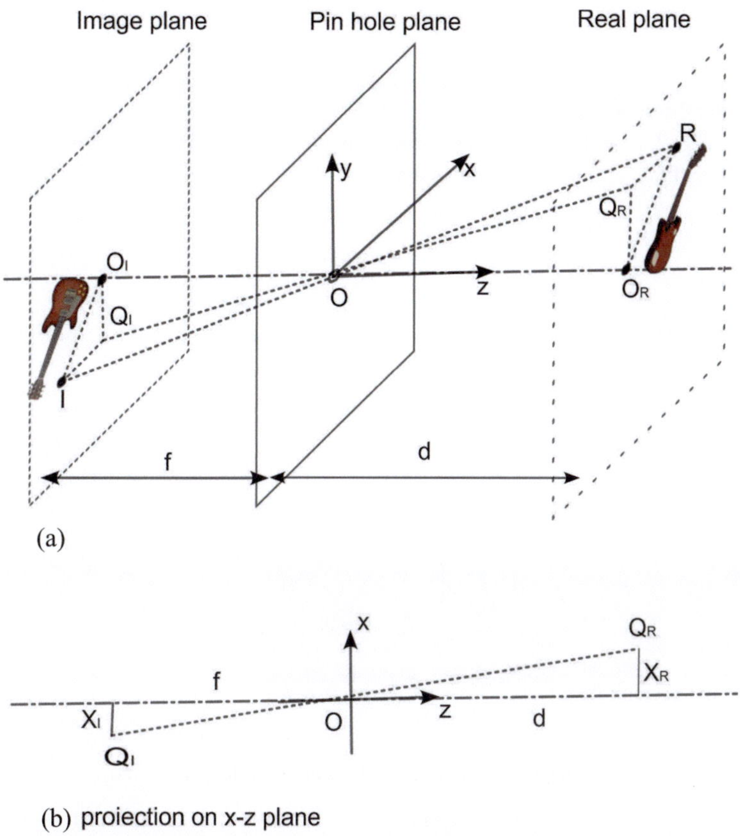

Fig. 2.3 Image formation by a pinhole on a camera obscura image plane

The image magnification m is defined as the ratio between the length of the projected line on the image plane and the corresponding length in 3D real space. For instance, the magnification can be computed using the segment RO_R of the guitar shown in Fig. 2.3, which is projected onto the segment IO_I in the image plane.

Using the two similar triangles ORO_R and OIO_I illustrated in Fig. 2.3, we can derive the following relationship:

$$\frac{IQ_I}{RQ_R} = \frac{f}{d} \text{ but } \frac{IQ_I}{RQ_R} = m \text{ and so } m = -\frac{f}{d} \tag{2.5}$$

In the case of camera obscura projection, the magnification m is typically considered negative because the resulting image is inverted.

To adjust the size of the image on the image plane, one can vary either the object distance d or the length f (i.e., the distance between the pinhole and the image plane). When the object is placed farther from the pinhole (i.e., larger d), the resulting image appears smaller on the image plane (i.e., lower absolute values of m), assuming a fixed distance f. An example of this effect is shown in Fig. 2.4b: the image of train tracks illustrates how transverse lines (such as railway ties) appear progressively smaller as their distance from the pinhole increases. The parallel rails, due to perspective projection, appear to converge toward a single point in the image, this point is known as the vanishing point.

From Fig. 2.4c and pinhole transformation 2.4 it is easy to see that vanishing point coordinates (x_{vp}, y_{vp}) in the plane image are:

$$(x_{vp}, y_{vp}) = \left(f\frac{l_x}{l_z}, f\frac{l_y}{l_z}\right) \tag{2.6}$$

where $(l_x l_y l_z)$ are the coordinates of a generic point on the central line of Fig. 2.4c.

(a) (b) (c)

Fig. 2.4 Vanish points generated on image plane by pinhole projection

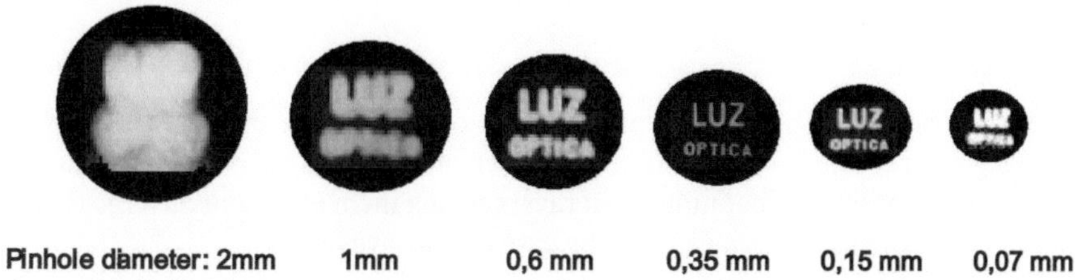

Fig. 2.5 Dimension effects of the hole diameter of a camera obscura pinhole

In the idealized model of the camera obscura, the pinhole is considered a perfect aperture through which only a single light ray from each point on the object passes, ensuring a sharp image. However, if the hole is larger than the wavelength of light, geometric optics still applies, but multiple rays from each object point can pass through the pinhole, as illustrated in Fig. 2.5. This results in image blur due to overlapping projections. To reduce this blurring, the pinhole should be made smaller. However, if the pinhole diameter becomes comparable to, or smaller, than the wavelength of light, the pinhole begins to behave as a point source, causing the light to diffract and spread out in diverging rays. As discussed in the previous chapter, this introduces additional blurring due to diffraction effects. Therefore, an optimal pinhole size must balance the trade-off between geometric blurring (from hole too large) and diffraction blurring (from hole too small).

The optimal hole diameter can be estimated [2] to be $d = 2\sqrt{f\lambda}$, where λ it's the wavelength and f the distance from the hole and the image plane.

In an ideal camera obscura, all points on the image plane appear in perfect focus. However, since only a single ray of light from each point on the object passes through the pinhole, the resulting image has very low light intensity. Consequently, a long exposure time T is required to accumulate enough light energy on the image plane, where a photographic film or digital sensor is positioned, to produce a sufficiently exposed image.

If the optical intensity of light entering through the pinhole is I, and the pinhole has a diameter d, the optical power P passing through the aperture can be expressed as $P = I\pi d^2/4$. The total energy E received by the sensor during the exposure time T is then $E = TP = TI\pi d^2/4$. This equation highlights the trade-off between pinhole size, light intensity, and exposure time in pinhole imaging systems.

To collect more light from each point of a 3D object, enhance image brightness, reduce exposure time needs, lenses have been developed throughout the technological evolution of imaging systems.

2.3 Lens Basics

Lenses are optical components typically made of glass or plastic, characterized by having at least two refracting surfaces, of which at least one is curved. These curved surfaces bend or refract incoming light rays, causing them to either converge (in the case of convex lenses) or diverge (in the case of concave lenses).

To better understand how a lens works, consider the example shown in Fig. 2.6a, which features a lens with two identical spherical surfaces, one on each side. This symmetrical shape helps illustrate the fundamental behavior of light as it passes through a simple biconvex lens.

When parallel rays of light strike a lens, they are refracted twice, once upon entering the lens material and again upon exiting. In both instances, the change in direction is governed by Snell's law. For a lens composed of standard optical glass with a refractive index of $n = 1.5$ and two identical spherical surfaces, ray tracing, performed using the *Phydemo* web application, produces the result illustrated in Fig. 2.6b. The point at which the refracted parallel rays converge after passing through the lens is defined as the focal point F. The position of this point, measured from the optical center O along the optical axis, depends on the curvature radii R_1 and R_2 of the lens surfaces, as well as on the refractive index n of the lens material. Assuming the lens is surrounded by air ($n = 1$), the approximate relationship among these parameters is given by the classical lens maker's formula [2]:

$$\frac{1}{f} = (n-1) \cdot \left(\frac{1}{R_1} + \frac{1}{R_2}\right) \tag{2.7}$$

If we send parallel rays from the other side of the lens we can define a second focal point, symmetrical to the previous, on the other side of the lens, at same distance f from lens optical centre O, as illustrated in Fig. 2.6b.

In Fig. 2.7, ray tracing is applied to two object points, P_1 and P_2, as light from these points passes through the lens. Assuming that point P_1, like all other points on the real object, emits diffuse light in all directions, we can consider three principal rays that help us to understand how the image is formed by a lens:

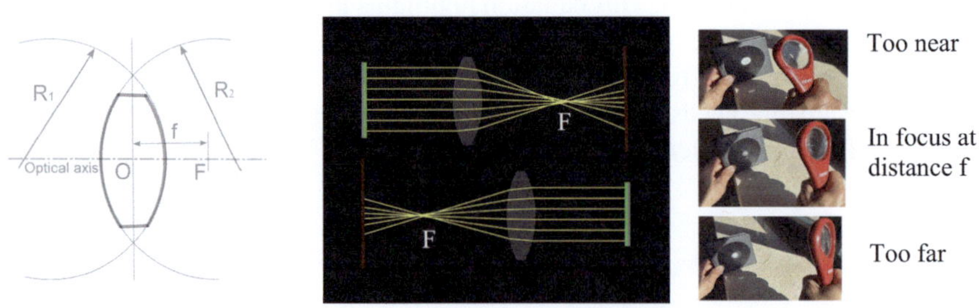

(a) (b) (c)

Fig. 2.6 **a** A simple lens with two spherical surfaces, **b** its two focus on one and on the other lens side determined by ray tracing

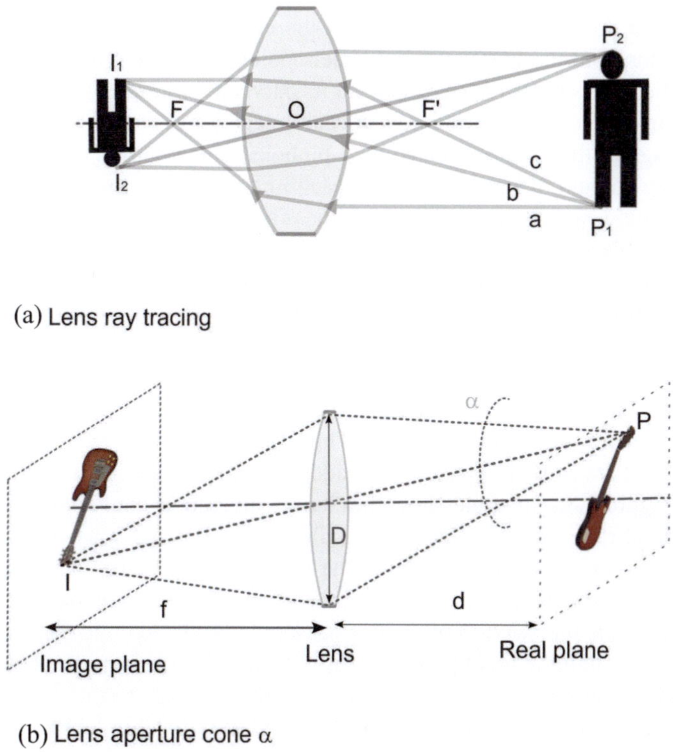

Fig. 2.7 Image formation by a lens

- Ray a, which travels parallel to the optical axis, is refracted by the lens and passes through the focal point F, in accordance with the definition of the focal point.
- Ray b passes through the optical center O of the lens and continues in a straight line without deviation. This behavior is characteristic of all rays that pass through the lens's optical center.
- Ray c is directed toward the second focal point F′ (on the object side). After refraction by the lens, it emerges parallel to the optical axis.

These three rays intersect at point I_1, which marks the image of the object point P_1. All the rays in between a and c from point P_1 arrives on same image point I_1. The same ray tracing apply to point P_2 and to any other point on the object.

As a result, the lens forms a complete inverted image of the object on the image plane, like a pinhole camera, but with significantly greater light intensity, as multiple rays from each object point are focused onto corresponding same point in the image plane.

The primary advantage of a lens over a pinhole, as illustrated in Fig. 2.7b, lies in its ability to collect and focus a significantly greater amount of light. At each image point I on the image plane, all rays within the cone defined by the aperture angle α, originating from the corresponding object point P, are brought to convergence, provided the lens is properly designed and its curvature accurately manufactured.

This convergence of multiple rays at each image point results in an image that is much brighter than that produced by a pinhole, which admits only a single ray of light. The amount of light gathered at each image point is directly related to the aperture angle α, which depends on both the diameter of the lens and the distance between the object and the lens. Consequently, larger lenses are capable of producing brighter images. However, increasing lens diameter introduces greater challenges in manufacturing, as it requires maintaining high optical precision across a larger surface area.

A useful approach to better understand how a lens functions is through the *thin lens model*. Thin lenses are defined as having a thickness that is negligible compared to their curvature radius and diameter. This simplification allows us to focus on how light rays are refracted without considering their detailed path within the lens material. Figure 2.8a illustrates how thin lenses diverge or converge light rays. In this model, all rays that enter the lens parallel to the principal axis converge at the focal point. A ray that passes through the optical center of the lens travels in a straight line without deviation. All other rays from same point intersect at the point determined by the intersection of the line passing through the optical center and the one passing through the focal point. Because the lens is considered extremely thin, internal refraction paths are ignored. This assumption leads to the derivation of a classical and fundamental formula in optics, valid only for thin lenses, that relates the object distance, image distance, and lens focal length.

If we look at Fig. 2.8a the two similar triangle IAF and FOB have same value of ratio of their two cathetus, so $\frac{h}{(i-f)} = \frac{H}{f}$ and from the other two similar triangles IAO and REO we can write for the same corresponding cathetus ratio: $\frac{h}{i} = \frac{H}{r}$. Dividing this relation by the previous one give us $\frac{(i-f)}{i} = \frac{f}{r}$; multiplying both terms of this relation by the product ri give us $ri - rf = if$ and dividing all the terms of this relation by the product (rif) give us $\frac{1}{f} - \frac{1}{i} = \frac{1}{r}$. Rearranging we obtain the classical thin lens formula:

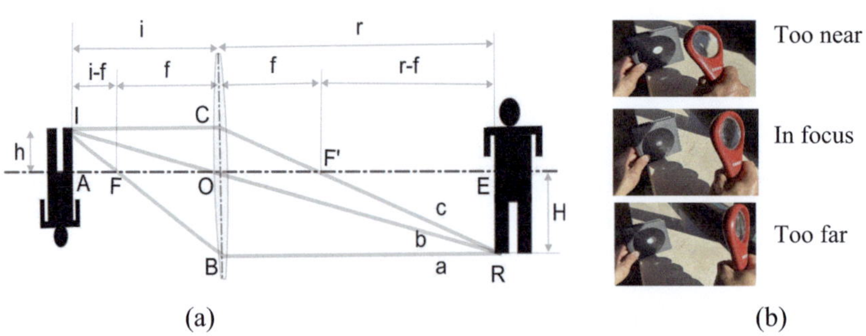

Fig. 2.8 Fundamental model of the thin lenses to generate images and how to find lens focus

$$\frac{1}{f} = \frac{1}{i} + \frac{1}{r} \tag{2.8}$$

This fundamental formula tells us that the reciprocal of focus distance is equal to the sum of the reciprocals of image plane and object distances. If the object is very far to the lens $r = \infty$, and therefore $(1/r) = 0$. From the above thin lens formula we have $\frac{1}{f} = \frac{1}{i}$ so $f = i$. Once again, the focal point of a lens can be determined by observing where it brings parallel rays coming from a far object. Using the sun rays, a typical available far object when the whether is good, they can be used to found the focal point position of a lens, as illustrated in Fig. 2.8b. It is also possible, from Fig. 2.8a, to calculate the image magnification, defined as the ratio of object length in image plane with the real one, so $m = h/H$. From the two similar triangles IOA and REO of Fig. 2.8 we have that $\frac{h}{i} = \frac{H}{r}$, so $\frac{h}{H} = \frac{i}{r}$ and then:

$$m = \frac{h}{H} = \frac{i}{r} \tag{2.9}$$

To describe how a lens transforms coordinates from the 3D real world to the 2D image plane, we can analyze the projection of corresponding points using the principal ray that passes through the optical center O, as illustrated in Fig. 2.9. For any point P on the 3D object, the corresponding image point I on the 2D plane lies along the straight line connecting P to O. This approach simplifies the coordinate mapping process: in an ideal lens (i.e., one with perfect curvature and no aberrations), all rays originating from point P and refracted by the lens converge at a single point I on the image plane. This behavior, previously illustrated in Fig. 2.7b, give us the same geometric projection model of a pinhole imaging system. Therefore, the coordinate transformation performed by a perfect shape lens is equivalent to that of a pinhole. As a result, for each point on the real object with 3D coordinates (x_R y_R, z_R), the corresponding 2D coordinates (x_I y_I) on the image plane are given by the same projection relations:

$$x_I = f\frac{x_R}{z_R} \quad y_I = f\frac{y_R}{z_R} \tag{2.10}$$

2.4 Lens Kind

Lenses can be constructed in various shapes, for different imaging functions. The two main types are illustrated in Fig. 2.10:

1. *Convex lens* (converging lens), thicker at the center than at the edges. It causes incoming parallel light rays to bend inward and converge at a single point known as the *focal point*.

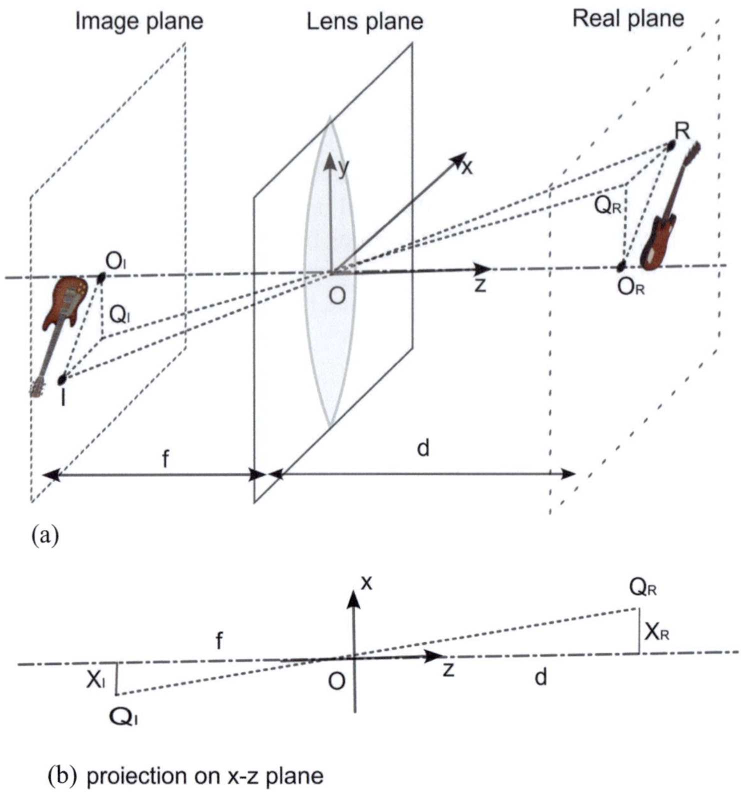

Fig. 2.9 Image formation by a lens

2. *Concave lens* (diverging lens) thinner at the center and thicker at the edges. It causes incoming parallel light rays to spread out, or diverge, as if they were originating from a single point behind the lens, known as the *virtual focal point*.

The shape and curvature of a lens play a crucial role in determining how much light is refracted and where the resulting image is formed. A convex lens can create either a real image, formed when refracted light rays physically converge and can be projected onto a screen, or a virtual image, formed when the rays diverge but appear to originate from a common point behind the lens. In contrast, a concave lens always produces a virtual image. This image is upright, smaller than the actual object, and appears to originate from a point on the same side of the lens as the object. For positive (converging) lenses, when an object is placed at infinity, the incoming light rays are parallel to the optical axis and are brought to a focus at the lens's focal point. Negative (diverging) lenses, on the other hand, cause parallel rays to spread out as if they were emanating from a point located at a distance f from the lens, called the virtual focus, since the rays do not actually pass through that point. Lenses with spherical surfaces are relatively easy to manufacture; however, they do not

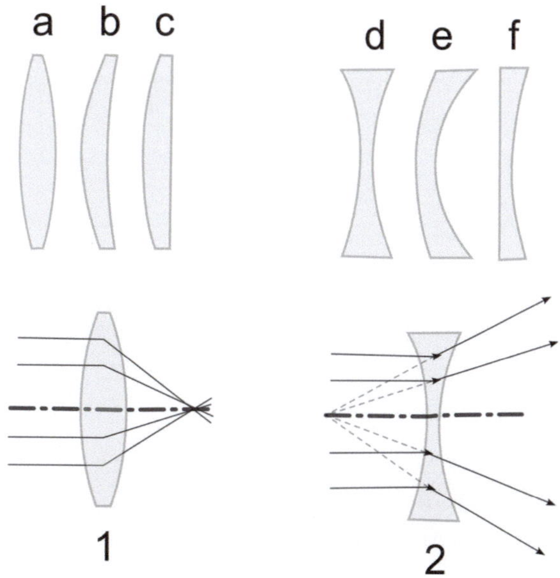

Fig. 2.10 Convex and concave lenses

perfectly focus all incoming parallel rays to a single point. This limitation, known as spherical aberration, can be observed in the ray-tracing simulation performed using the pydemo web application, as shown in Fig. 2.11a. To improve focusing accuracy and reduce aberrations, aspherical lenses, lenses with non-spherical surfaces, can be used. These lenses better control the path of light rays, significantly minimizing spherical aberrations, as illustrated in Fig. 2.11b.

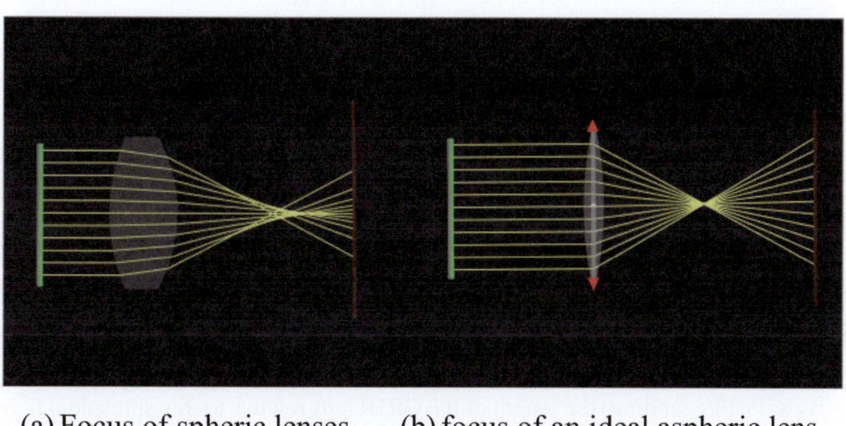

(a) Focus of spheric lenses (b) focus of an ideal aspheric lens

Fig. 2.11 Focusing precision of spheric and aspheric lenses

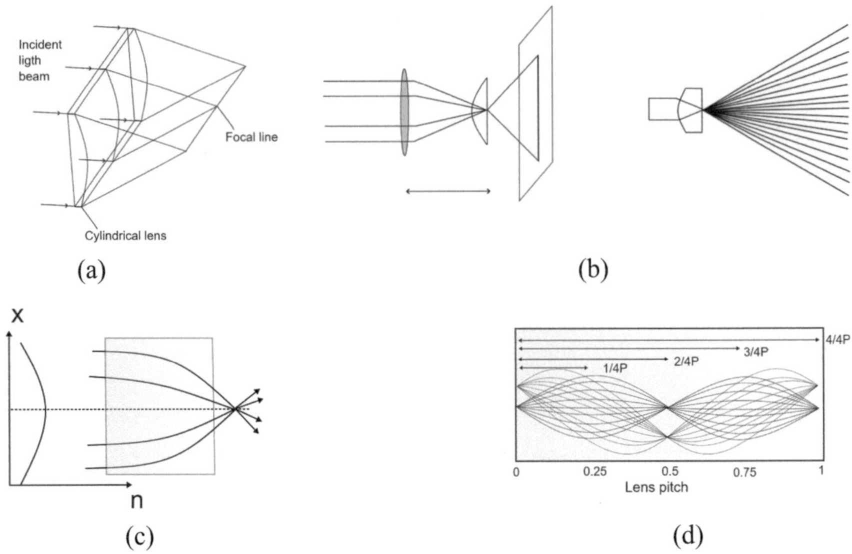

Fig. 2.12 **a** Cylindrical lens; **b** light sheet generation with a cylindrical and a focusing lens; **c** Grin lenses (quarter pitch); **d** Grin lenses of general pitch for focusing, collimation, beam expansion, etc.

Cylindrical lenses consist of one flat surface and one cylindrical surface, which—unlike a spherical surface has curvature in only one direction. As a result, light rays are bent only along that axis, forming a focal line rather than a focal point. Despite the difference in surface geometry, the same optical formulas used for spherical lenses remain applicable to cylindrical lenses. These lenses are commonly used to produce light lines or light sheets, as illustrated in Fig. 2.12b. When combined with spherical lenses, cylindrical lenses can help reduce the thickness of the light sheet while maintaining the same focal diameter. They are also frequently used to expand a collimated laser beam into a thin laser sheet, where the sheet thickness corresponds to the diameter of the original laser beam.

GRIN lenses (short for *GRaded INdex lenses*) are cylindrical glass elements with flat optical input and output surfaces on cylinder basis, but unlike conventional lenses, their refractive index varies radially within the material. This variation follows a symmetric profile $n = n(r)$, where r is the radial distance from the cylinder's axis, and the refractive index reaches its maximum at the center.

When light rays enter the flat surface of a GRIN lens parallel to the cylinder axis, they are continuously bent toward the axis due to the increasing refractive index encountered in successive infinitesimal concentric layers. After crossing the axis, the rays bend away again, resulting in a sinusoidal rays trajectory inside the GRIN lens, as shown in Fig. 2.12d.

By adjusting (cutting) the length of the GRIN lens relative to this sinusoidal light rays path, different optical functions can be achieved. For example:

- A quarter-pitch cutted lenght GRIN lens (length of the lens equal to one-quarter of the sinusoidal period) focuses incoming parallel rays to a point, as a classical lens.

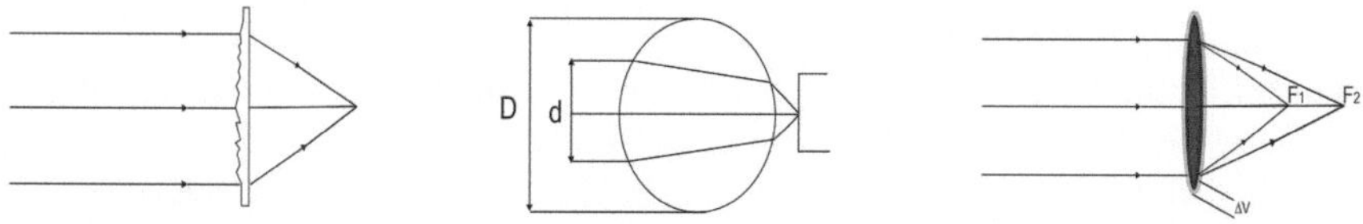

(a) Fresnel lens (b) Spherical ball lens (c) liquid lenses

Fig. 2.13 Fresnel lenses, spherical ball lenses and liquid lenses

- A half-pitch lenght lens can be used to collimates light from a diverging point source because they comes out parallel.
- Full-pitch or other fractional-pitch lenses are used for beam shaping and other purposes.

These lenses are particularly valuable in compact optical systems, and are commonly used for coupling light between flat optical surfaces of components such as LEDs, laser diodes, photodiodes, and optical fibers.

Other specialized types of lenses, typically not used for imaging, but rather for illumination systems or optical coupling between optoelectronic components, include *Fresnel lenses* and *ball lenses*, both illustrated in Fig. 2.13.

Fresnel lenses are particularly effective in applications that require light collection or focusing. Thanks to their flat, segmented structure, they can be manufactured easily and cost-effectively, especially in large plastic sheets. This makes them ideal for use in systems such as solar concentrators, where they focus sunlight onto photovoltaic cells, or in thermal applications that involve heating surfaces. However, their segmented design leads to optical imperfections, making them unsuitable for high-precision imaging tasks.

Ball lenses are compact spherical lenses often used for efficient light coupling, particularly between laser diodes, photodetectors and optical fibers, due to their ability to focus or collimate light in a very compact mechanical assembly.

Another innovative category is *liquid lenses*, shown in Fig. 2.13c. These lenses can dynamically change shape, similar to the human eye's crystalline lens, typically by varying an applied electric voltage. This enables real-time adjustment of the focal length, allowing the system to focus on objects at different distances without mechanical movement. Liquid lenses are increasingly used in compact imaging systems that require adaptability, miniaturization, and fast autofocus, such as mobile devices, medical imaging tools, and industrial inspection systems.

2.5 Lens Depht of Field

Let us consider the optical system of Fig. 2.14a where a non-transparent disk with a hole of diameter D is placed in front of a lens, limiting its optical aperture. This system is defined by two optical parameters, the lens focal length f and the plate diameter D, normally defined by the "stop" number N $= f/\text{D}$, or using the symbol f#=f/D. So, D = f/N or D = f/f#. This means that high values of N generate low diameter of the hole where light passing through, so increasing N reduce image brightness generated by this imaging optical system.

Light coming from all the points of the plane α in Fig. 2.14a generate, across the lens and the stop, image points in plane α', where a not transparent device is placed (a film or a sensor) to form the image, i.e. the image plane. For example, the image of point E is the point E'. Assuming for the lens a model of thin lens we have Eq. (2.8) that relates the distance r of point E from optical center O of the lens with image plane position I from the optical center O with lens focal length f.

If we move the point E to point G in plane β it should generate, for the same thin lens Eq. (2.8), an image point G′ in plane β′, but in the not transparent image plane α' is generated not an image point but a blurred circle of diameter b. Diameter b of this circle can be calculated considering the that for the two similar triangles G′AB and G′ST we can write:

$$\frac{b}{i' - i} = \frac{D}{i'} \text{ so : } b = \frac{(i' - i)D}{i'} \tag{2.11}$$

From thin lens Eq. (2.8) we can calculate i and i′ in terms of distances of the object from the lens r and r′ and lens parameter f, as:

$$i = \frac{rf}{r - f} \text{ and } i' = \frac{r'f}{r' - f} \text{ so } i' - i = \frac{f}{(r' - f)} \frac{f}{(r - f)}(r - r').$$

Substituting above values of i and i′ in Eq. (2.11) we get:

$$b = Df\left|\frac{r - r'}{r'(r - f)}\right|$$

and taking D = f/N finally we have:

$$b = \frac{f^2}{N}\left|\frac{r - r'}{r'(r - f)}\right| \tag{2.12}$$

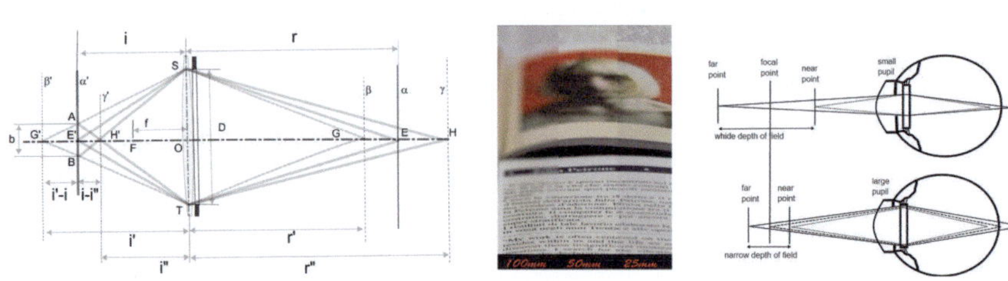

(a) (b) (c)

Fig. 2.14 Lens burring effect and depth of field

So blurring circle diameter b can be reduced by increasing light stop N number, i.e. reducing D diameter (our eyes close the iris to better focusing). Or we can say that little lenses have lower blurring effect, but receive less light, generating darker images. A very high value of stop N number gives us the lowest possible D diameter, arriving to the limit of the hole of a pinhole camera obscura, so practically having no blurring effects, but gives us very dark images generated on the image plane.

The depth of field (DOF) region is defined by the distance range of objects $(r''-r')$ giving a maximum defocus diameter b that can be tolerated on the image plane. It can be the diffraction limit, the grain dimension of a photo film, the sensor resolution on digital image sensors, etc.

From Fig. 2.14, if point E moves to point H not a point but another circle is generated in image plane, same as before. From thin lens formula we have that lens generate an image point in H′ in plane γ', but image plane is on α', so here we have another circle of same diameter b, for a particular H position. From the two similar triangles H′AB and H′ST we can write:

$$\frac{b}{i-i''} = \frac{D}{i''} \text{ so } b = \frac{(i-i'')D}{i''} \tag{2.13}$$

From thin lens Eq. (2.8) we can calculate i and i'' in terms of distances of the object from the lens r and r'' and lens parameter f, as:

$$i = \frac{rf}{r-f} \text{ and } i'' = \frac{r''f}{r''-f} \text{ so } i - i'' = \frac{f}{(r'-f)}\frac{f}{(r-f)}(r-r'').$$

Substituting above values of i and i'' in Eq. (2.11) we get:

$$b = Df\left|\frac{r''-r}{r''(r-f)}\right|$$

and taking D = f/N finally we have:

$$b = \frac{f^2}{N}\left|\frac{r''-r}{r''(r-f)}\right| \tag{2.14}$$

The following relationships are thus obtained:

$$b = \frac{f^2(r-r')}{Nr'(r-f)} \tag{2.15}$$

$$b = \frac{f^2(r''-r)}{Nr''(r-f)} \tag{2.16}$$

So, from the above equations, the depth of field DOF $= r'' - r'$ is

$$\text{DOF} = r'' - r' = \frac{2rf^2bN(r-f)}{f^4 - b^2N^2(r-f)^2} \tag{2.17}$$

In general the terms $b^2N^2(r-f)^2$ can be ignored respect to f^4 and taking the other approximation $(r-f) = r$, because f is generally little respect to r, we have the approximated following classical formula:

$$\text{DOF} = r'' - r' = \frac{2bNr^2}{f^2} \tag{2.18}$$

So DOF increase linearly with N, obviously also linearly with b, decrease a lot, with a square inverse low, with focal distance f and increase a lot, with a square low, with distance r. From the previous Eq. (2.16) we can see that increasing distance r of the real object from the lens reduce the diameter of the circle of confusion, so for a given tolerated circle of confusion value b exist a distance h that set a limit: All objects located at a distance greater than h from the lens generates a lower dimension of b blurring circle and therefore will appear in focus. So taking $r = h$ in Eq. (2.16), considering r'' equal to infinity, the distance h giving a blurring circle equal to b is:

$$h = \frac{f^2}{Nb} + f \tag{2.19}$$

For distance far then h the blurring circle will be little than b so everything far than h will be in focus. The value of h given by Eq. (2.19) is called *hyperfocal distance*, and can be reduced by increasing N, or reducing lens diameter, but this also reduce image brightness, as previously observed. So little lenses can have almost all in focus, similar to the pinhole model, but gives darker images.

To summarize the previously illustrated and quantified concepts, if we take *large openings iris D by using small N numbers or we have large diameter lenses* we will have:

(1) brighter images on image plane.
(2) we can use short exposure times to impress the sensor recording faster the images;
(3) but limited depth of focus (DOF).

If we take *small openings by using large N-numbers or have little diameter lenses* we will have:

(1) darker images on image plane;
(2) we need long exposure times to impress the sensor recording slowly the imaging process;
(3) large depth of focus (DOF).

2.6 Lens Aberrations

Aberrations refer to various forms of image distortion that occur due to unavoidable imperfections in optical systems and materials. These imperfections arise from limitations in the manufacturing process, such as inaccurate glass machining, non-uniform material properties, or imperfect surface shape and finishing.

From a functional perspective, aberrations can be broadly categorized into two main families:

1. Chromatic aberrations
2. Geometric aberrations.

Both types can further be classified as axial (on-axis) or extra-axial (off-axis), depending on how they affect the image formation across the surface of the image plane.

Chromatic aberrations, illustrated in Fig. 2.15a, arise because the focal length f of a lens depends on the wavelength of light due to the dispersion of the material (i.e., the wavelength dependence of the refractive index). Materials like certain types of glass exhibit varying refractive indices for different wavelengths. As a result, different colors come into focus at different points along the optical axis. For example, red light ($\lambda = 0.65$ μm) is focused further away than violet light ($\lambda = 0.45$), leading to color fringing and reduced image sharpness.

The most common geometrical aberrations are caused by spherical lenses, which, as previously noted, do not bring all incoming parallel rays to a single focal point. Instead, they form a characteristic pattern known as a caustic curve, where the rays intersect at slightly different points along a region called the "stick", as illustrated in Fig. 2.15b. This effect results in image blurring and a loss of sharpness. Another type of geometrical aberration is coma, which occurs when light rays arrive parallel from a distant point source but at an angle relative to the optical axis. These rays fail to converge at a single point on the focal plane and instead form an asymmetric, comet-shaped (V-shaped) blur, often with a coloured fringe, shown in Fig. 2.15c. This distortion becomes more pronounced toward the edges of the image.

Another kind of geometrical aberration is the astigmatism, from the Greek *α-* meaning "without" and *στίγμα* (stigma), meaning "mark" or "point". It is an optical aberration found in both single lenses and compound lens systems, including camera

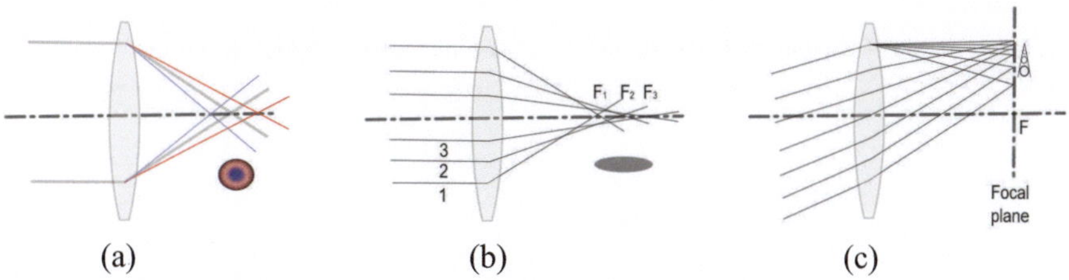

Fig. 2.15 Chromatic and geometrical aberrations

lenses and the human eye. Rays that propagate in two planes intersecting the optical axis at different angles, such as perpendicular planes, come to focus at different points. As a result, the image of a single point appears distorted.

If an optical system affected by astigmatism is used to form the image of a point G, the horizontal and vertical rays on the two planes M and S will come into focus at two different points B_M and B_S, as illustrated in Fig. 2.16. This kind of aberration obviously generate a blurring effect, the image of a point in plane image will be a an ellipse of dimensions depending on the position of the image plane respect the two focal points illustrated in Fig. 2.16.

If we look at images of grid of points instead of a single point a *field curvature* aberration can occurs, more or less evident depending on the quality of the optics.

The image of a flat object with a regular grid on it can be not a regular grid on image plane. Two kind of field distortions can be generated by a lens or by an objective: barrel distortion and a cushion distortion, as illustrated in Fig. 2.17.

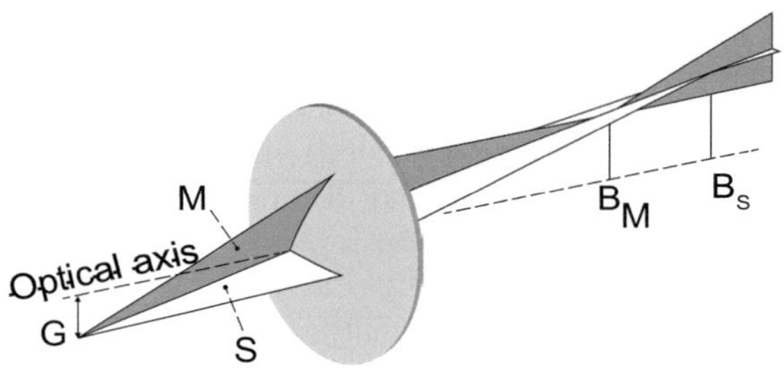

Fig. 2.16 Astigmatism aberration

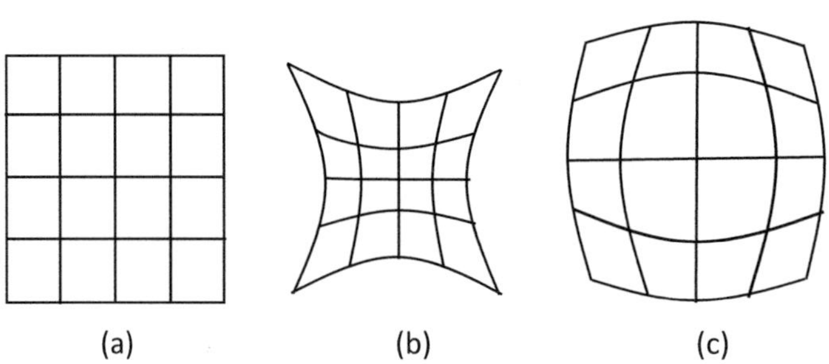

Fig. 2.17 Lens field distortion **a** undistorted, **b** cushion distortion, **c** barrel distortion

2.7 Lens Systems

As discussed earlier, according to the thin lens equation (2.8), focusing an object located at a distance r onto an image plane positioned at distance i from the lens (for a given focal length f) requires maintaining the correct relationship between these distances. One straightforward method to achieve focus is to adjust the image plane distance i. However, in many practical systems, physically moving the image plane can be mechanically challenging or impractical. An alternative is the use of liquid or deformable lenses, which can dynamically adjust their focal length by changing shape. While this approach offers flexibility, liquid lenses generally lack the optical precision of rigid lenses and can be complex to calibrate, especially for high-accuracy measurement applications.

A more robust solution involves employing a system composed of two lenses separated by a variable distance d, as illustrated in Fig. 2.18. By integrating a mechanical adjustment mechanism that alters the spacing d between the lenses, it becomes possible to dynamically control the optical power of the system, enabling precise focusing without needing to move the image plane or rely on deformable optics. In an optical system like the one shown in Fig. 2.18a, the first lens, located at a distance r_1 from the object, forms an intermediate image of point T_1.

One ray travel through the focal point F_1 and is refracted by the lens to become parallel to the optical axis, while another ray passes through the optical center O_1 of the lens and continues undeviated. These two rays intersect at point V, forming a virtual image, since there is no screen or image sensor at that location to capture it.

The rays then proceed toward the second lens, where they are further refracted and brought to convergence at point I_2, which lies on the image plane. Here, where the image sensor is positioned, a real image is formed.

From Eq. (2.9), applied two times to the two lenses, it is easy to derive the total magnification of this optical system as $m = \frac{i_2}{o_2} \cdot \frac{i_1}{o_1}$. This system can be considered equivalent to a single "virtual" lens, designed dashed in Fig. 2.18, with its optical center in O_e.

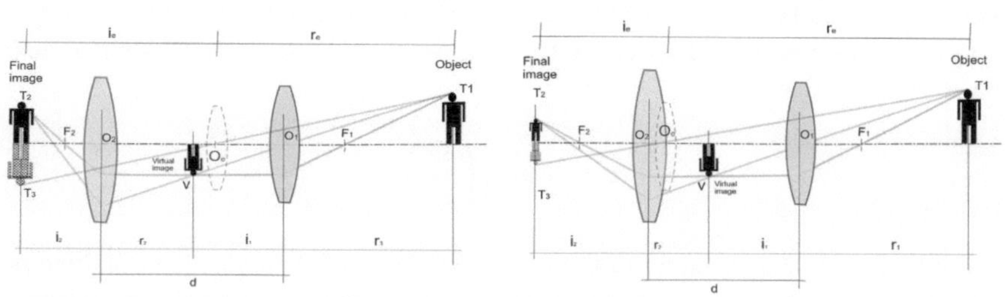

(a) Two converging lens systems

(b) Effect of changing distance between the two lenses

Fig. 2.18 Optical system with two converging lenses

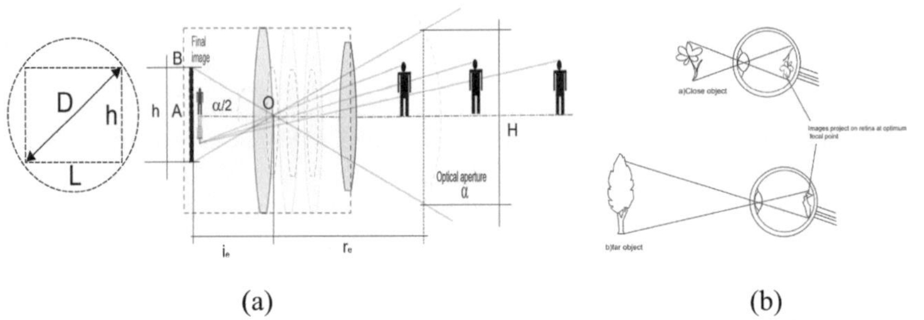

Fig. 2.19 Optical system with two moving lenses and our eyes with only one lens changing shape to obtain same focusing and viewing angle changing effect

This virtual lens generates images of real points (like T_1) tracing the ray passing through its optical center, like a pinhole, onto the image plane at distance i_e, so in point T_3 like the two-lens system, except for the image inversion.

This equivalent lens has therefore ad equivalent focal point given by the lens Eq. (2.8), so by $\frac{1}{f_e} = \frac{1}{i_e} + \frac{1}{r_e}$.

Changing the distance d has the effect of changing the two values i_e and r_e and therefore it is possible accordingly to change the virtual focal distance f_e.

Therefore, using an optical system like that in Fig. 2.19a, by moving relative positions of the two lenses, it is possible to correctly focusing far or near object on the image plane, and changing the viewing angle α. Is the principle used in objectives complex lens systems.

Our eyes obtain the same result changing the "lens" shape, as illustrated in Fig. 2.19b.

Changing viewing angle α means changing the optical aperture of the system. i.e. the maximum real area that can be seen on the image plane where a finite dimension sensor is placed (normally of rectangular shape L × h). If h is the sensor image plane height, H will be the maximum height of the real field of view. Same relation will occur for the other dimension L of the sensor and for its diagonal D that should fit the diameter of the image generated by the (normally circular) lens on the image plane.

From triangle ABO of Fig. 2.19a, because AB = AO tan (α/2), viewing angle α can be calculated as:

$$\alpha = 2arctg\left(\frac{D}{2f}\right) \tag{2.20}$$

Taking h = D and i_e equal to the focal distance f of the equivalent pinhole or equivalent single lens. From viewing angle α the numerical aperture N_A of an optical system is defined as:

$$N_A = n\sin\left(\frac{\alpha}{2}\right) \tag{2.21}$$

Fig. 2.20 Telecentric lens systems and images obtained by a wide field angle or telecentric lens configuration

Where n is the refraction index of the medium, $n = 1$ for void and very close to 1 for air.

Numerical aperture is a typical parameter largely used to define ability of an optical system, an objective or an optical component to collect light.

Lens projection in image plane for a fixed focal distance is like a pinhole projection, so far object are projected on image plane with a lower dimension than same length object nearest to the lens. It is possible to avoid this effect using the telecentric lens configuration of Fig. 2.20.

If we have a two convergent lenses at a particoular distance that gives coincidence of their two focal points, the optical rays arriving parallel from the object, because of focal point definition, will arrive parallel to the focal plane, so object of same high but at different distance from the lenses, generates images of same high on the image plane.

This optical configuration is named "*telecentric*" and is largely used in optical 2D measurement systems by telecentric objectives (e.g. to measure screws diameter and height which pass on a production line not all at same distance from the lenses). Field angle α of a telecentric system is equal to zero so the field of view dimension of telecentic systems is equal to the diameter of the largest lens of the telecentric objective.

Another largely used optical system made of two lenses assembled, is that using an equiconvex lens and a second plano-concave lens as in Fig. 2.21. This system in used to reduce chromatic aberration because the two lenses bending the different colours (wavelength) in opposite directions and therefore cancel out each other. This system, largely used, is named "*achromatic doublet*".

2.8 Objectives

Objectives are essential components in the formation of analog images in devices such as cameras, photographic systems, optical microscopes and telescopes. These elements are sophisticated assemblies of lenses engineered to reduce optical aberrations and distortions, thereby enhancing image quality.

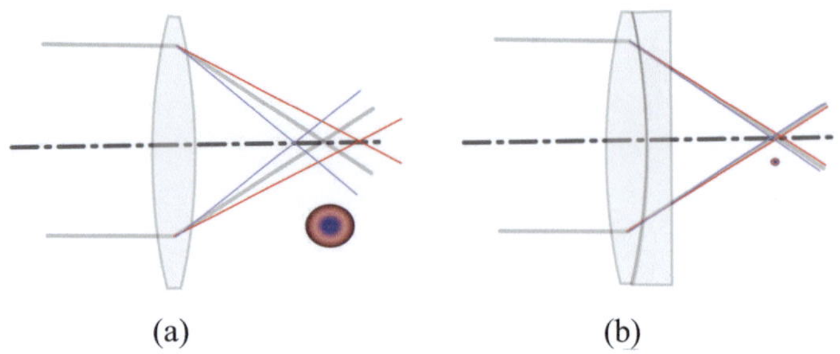

Fig. 2.21 Achromatic doublet

In simpler imaging systems, such as video cameras or smartphones, the objective lens typically features a fixed focus and a fixed aperture without any diaphragm. The diaphragm aperture is characterized by its f-number (or f-stop, N), which is determined by the ratio of the lens's focal length to diaphragm diameter D, above defined as N = f/D. Optical complex systems like photographic cameras, and even the human eye, allow for dynamic adjustment of both diaphragm aperture size and focal length, offering greater control over depth of field and exposure. This is typically achieved using a diaphragms and at least two lens elements or two lens group, as illustrated in the optical configurations of Figs. 2.18 and 2.19.

The sequence of f-numbers of diaphragm aperutre in photographic machines is standardized and follows a geometric progression, as detailed in Table 2.1. Each step in this sequence corresponds to a doubling or halving of the aperture area, depending on whether the aperture is opened or closed. As a result, under uniform lighting conditions, the amount of light reaching the image sensor changes by a factor of two with each incremental change in the f-number (Fig. 2.22).

Consider, for example, an objective lens with a focal length of 100 mm and an aperture of f/2. According to the definition of the f-number N = f/D, where f is the focal length and *D* is the diameter of the iris aperture, we can compute the maximum aperture diameter as D = f/N = 100/2 = 50 mm. Thus, the maximum iris opening for this objective is 50 mm.

When the f-number (also referred to as the stop number) increases along the standard f-stop scale, e.g., from f/2 to f/2.8, f/4, f/5.6, and so on, the aperture area decreases, thereby reducing the amount of light reaching the sensor (as indicated in

Table 2.1 Standard values of stop numbers N and consequent diameter and area for an f = 100 mm objective

Aperture N	1	1,4	2	2,8	4	5,6	8	11	16	22	32	45
Diameter (mm)	100	71.42	50	35.71	25	17.82	12.5	8.93	6.25	4.46	3.13	2.23
Area mm^2	7850	4004	1962	1001	490	250	122	62.5	30.6	15.6	7.69	3,9

Fig. 2.22 Reducing diaphragm diameter of an objective by increasing stop number N, as our eyes

Table 2.1). At the same time, a higher f-number increases the depth of field, resulting in greater sharpness across a wider range of distances in the image.

To regulate the exposure time, the interval during which light reaches the sensor, most cameras employ a mechanical shutter placed between the lens and the sensor (Fig. 2.23). Typical exposure times range from 1/500 s to 1 s, with common values like 1/100 s or 1/50 s depending on lighting conditions and scene dynamics. In digital cameras, this function may also be performed electronically, by the image sensor, eliminating the need for a mechanical shutter.

In video or film applications, the exposure time is intrinsically linked to the frame rate. For instance, a camera recording at 100 frames per second (fps) allows a maximum exposure time of 1/100 s per frame.

Another critical factor is the sensor's sensitivity to light, quantified using the ISO scale (defined by the International Organization for Standardization). Common ISO values include ISO 100, 500, 1000, and beyond. A higher ISO value indicates greater sensor sensitivity, enabling the use of shorter exposure times and/or smaller apertures (i.e., higher f-numbers), which again contributes to a deeper depth of field.

Together, these three parameters, aperture (f-number), exposure time, and ISO sensitivity, form the cornerstone of photographic and imaging system configuration. The optimal selection of these three parameters is essential not only for optimal image capture but also for the selection of lenses, objectives, and sensors across diverse imaging applications.

Objectives consist of complex assemblies of multiple lens groups, as illustrated in Fig. 2.24. The relative movement between these internal lens groups enables two key functions: focusing the image onto a sensor positioned at a fixed distance, and

Fig. 2.23 Mechanical shutter on a camera body between the objective and the image sensor

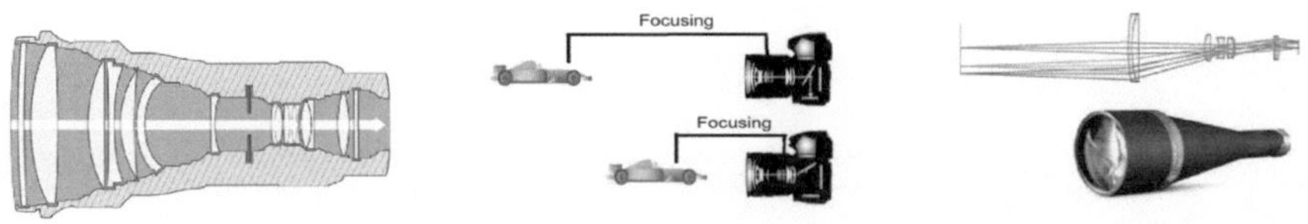

(a) A complex objective (b) Focusing by lens group moving (c) Telecentric objective

Fig. 2.24 Objectives

adjusting the field of view (zooming). This movement can be achieved either manually, by rotating dedicated focus or zoom rings, or automatically, through motorized mechanisms controlled by autofocus systems.

As previously discussed, reducing the aperture diameter (by adjusting the diaphragm) increases the depth of field, resulting in a broader range of distances appearing in focus. However, this also leads to a reduction in image brightness, as a smaller aperture allows less light to reach the image sensor.

A typical photographic objective, such as the one shown in Fig. 2.24a, demonstrates this complexity. Light entering the objective passes through the front element, cross various internal lens groups, and exits through the rear element before reaching the sensor. These internal groups are designed with specific functions:`

- Some correct optical aberrations such as chromatic dispersion and diffraction,
- Others can tilt or shift for image stabilization,
- And yet others can move longitudinally to enable focusing.

The diaphragm, positioned within the lens system, regulates how much light is transmitted through the objective and controls the exposure by modulating light intensity arriving on the image plane.

In Fig. 2.24b, the displacement of a lens group within a photographic objective is illustrated. This internal movement allows the optical system to adjust focus and capture sharp images of objects at different distances.

Regardless of their complexity, optical objectives can be functionally approximated by two key parameters: focal length and objective lens diameter. A larger diameter enables the lens to collect more light, which enhances image brightness, improves contrast, and supports better performance in low-light conditions. When an objective is paired with a sensor of defined physical dimensions, width × height, and diagonal D, it must produce an image circle that is at least large enough to fully cover the sensor's diagonal, as previously shown in Fig. 2.19. This condition determines the limit of the field of view, which is further illustrated in Fig. 2.25. Fixed focal length objectives have a single, constant focal length, commonly 35 mm, 50 mm, or 135 mm. In contrast, zoom objectives offer a range of focal lengths, such as 15–35 mm or 24–70 mm, providing greater versatility for imaging. Standard objectives, typically with a focal length around 50 mm, offer a field of view similar to the human eye and are ideal for general-purpose imaging. As depicted in Fig. 2.25, lenses with shorter focal lengths deliver a wider angle of view, making them suitable for landscapes and interior imaging. Conversely, longer focal lengths, typically 85 mm or more, yield a narrower field of view, allowing photographers to magnify distant subjects, such objectives are known as telephoto, as illustrated in Figs. 2.26 and 2.27.

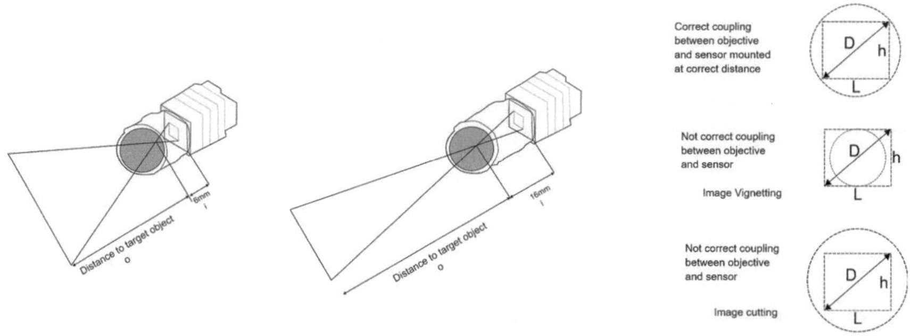

Fig. 2.25 Objective field of view and correct/incorrect coupling between objective and image sensor on camera body

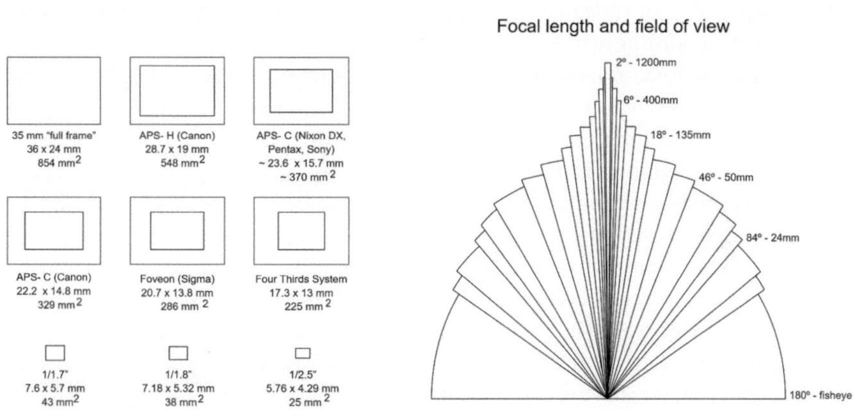

Fig. 2.26 Some sensors dimensions and field of view for a 35 mm full frame sensor

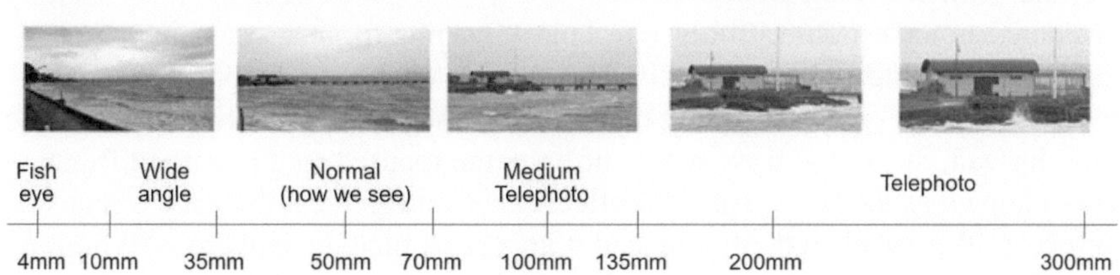

Fig. 2.27 Picture taken by objectives of different focal length on same location but different field of view

To avoid incorrect mounting of objectives on camera body where the sensor is placed many industrial standards exists for both named F-Mount, C-Mount, etc. Same kind of objective-camera standard can be mounted together coupling correctly each other, avoiding vignetting or image cutting.

The fundamental relationship for selecting an appropriate objective is the same as that used for a simple lens. If f is the focal length of the objective, equivalent to a single lens, i is the distance from the optical center of the objective to the image sensor (located inside the camera body), and r is the distance from the optical center to the object, then the following relation take place:

$$\frac{1}{f} = \frac{1}{i} + \frac{1}{r} \tag{2.22}$$

And the magnification m on image plane will be:

$$m = \frac{d_i}{d_r} = \frac{i}{r} \tag{2.23}$$

where d_i is a dimension on image plane and d_r the corresponding dimension on the real plane where object is. Therefore, if we need to take images of a field of view α we must choose an objective with a focal f and a camera sensor with diagonal dimension D, accordingly to Eq. (2.20).

One common issue related to non-uniform light intensity across the image field in optical systems is known as vignetting. This phenomenon occurs becasuse the amount of light reaching the sensor from oblique (off-axis) angles can be reduced compared to that coming from axial direction, as illustrated in Fig. 2.28. In a typical lens system mounted within a cylindrical housing (objective tube), light entering the system at steeper angles may be partially blocked by internal mechanical components such as lens mounts, diaphragm edges, or barrel walls. As a result, the peripheral regions of the image receive less light than the central region. A practical example of vignetting can be observed by photographing a uniformly illuminated white surface, such as a white wall. In the resulting image, the corners and edges may appear darker than the center, despite the even lighting conditions. This gradient in brightness is not due to sensor sensitivity but to optical geometry and light path limitations within the objective. Vignetting can vary depending on lens design, aperture setting and sensor size, and while it is sometimes used deliberately for artistic effect, but it is generally considered an undesirable artifact in technical imaging for measurement purposes.

Even when using a perfect objective lens, completely free from optical aberrations, the focused image of a distant point source on the image plane is not a true geometric point. Instead, due to the wave nature of light, the focused spot forms a diffraction pattern known as the Airy disk. This effect becomes significant when the spot size approaches the wavelength of light, and it imposes a fundamental physical limit on resolution.

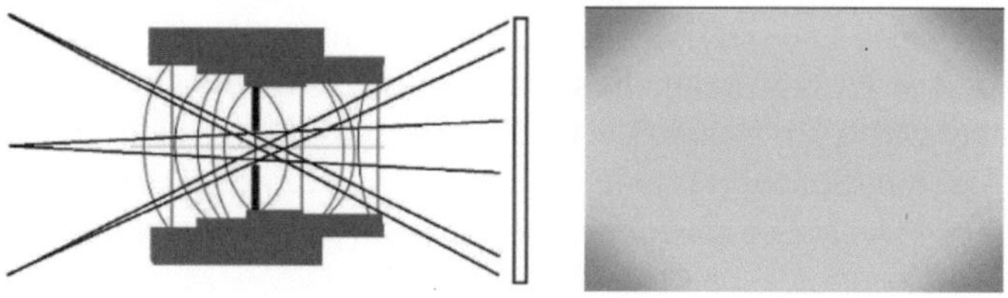

Fig. 2.28 Objective vignetting caused by light shadowing

As discussed earlier through Huygens' principle (Chap. 1, Sect. 1.8), diffraction occurs because light behaves as a wave. Therefore, no lens, regardless of its optical perfection, can focus light to an infinitely small point. Instead, the smallest possible image spot is limited by the diffraction phenomena, setting a boundary on how finely details can be resolved.

According to the Rayleigh criterion [3] and as illustrated in Fig. 2.29, for a circular aperture (such as an objective lens), the minimum achievable diameter δ of the Airy disk on the image plane, when imaging an object at infinity, is given by [3], so $\delta = 1.22\,\lambda f/\mathrm{D}$, where λ is the wavelength of the light, f is the focal length of the objective, D is the diameter of the objective diaphragm aperture (or the diameter of the lens if no diaphragm is used) [2].

This diffraction-limited spot size δ can be interpreted as the minimum diameter of the so-called circle of confusion, and it may be substituted directly into Eqs. (2.17) or (2.18) to estimate the depth of field of an objective. In such cases, taking $b = \delta$ provides a physically grounded constraint on the system's ability to render objects acceptably sharp, based on the inherent diffraction limit of the optical setup.

When two point sources are imaged through an optical system, each produces an Airy disk on the image plane. According to the Rayleigh criterion [3], if the centers of these Airy disks are separated by less than δ, the diameter of the central maximum, they overlap excessively, and the two points become indistinguishable, appearing as a single blurred spot.

This value δ thus defines the minimum resolvable distance between two points, and represents the diffraction-limited resolution of an optical system with aperture diameter D and focal lenght f. Given the relationship between aperture diameter and

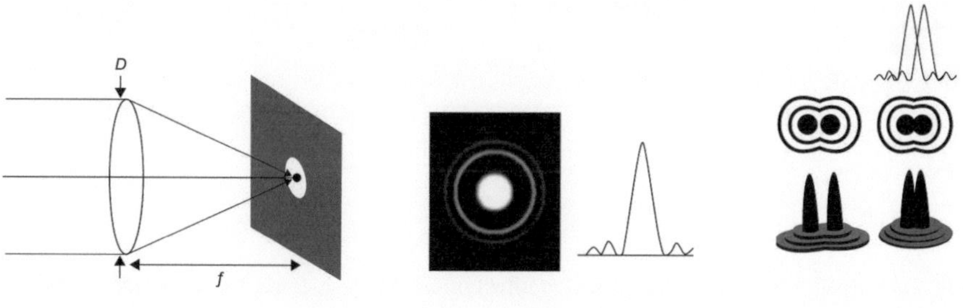

Fig. 2.29 Airy disks as image of points generated by a lens or by an objective because of diffraction

f-number N (where $D = f/N$), the resolution limit of an objective lens can also be expressed as $\delta = 1.22\lambda N$, assuming the system is free of aberrations. For example, using red light with a wavelength $\lambda = 0.63$ µm and an objective with an f-number $N = 2.8$, the diffraction-limited resolution becomes $\delta = 1.22 \times 0.63$ µm$\times 2.8 \approx 2.15$ µm. This means that the system cannot resolve two points separated by less than 2.15 µm on image plane under ideal perfect conditions.

Notably, resolution improves as the f-number N decreases, which corresponds to larger aperture diameters and larger diameter lenses. In this sense, an optical objective acts as a spatial low-pass filter: it attenuates high spatial frequencies, i.e. little spacing between points, little image details. That is, if the input to the lens is a sharp light transition or a small high-contrast spot (a step function), the output on the image plane will be a blurred and broadened version of the input, as illustrated in Fig. 2.30a.

When multiple light spots are positioned closer together than the diffraction limit δ, their respective Airy patterns begin to overlap, as illustrated in Fig. 2.30b. This overlap results in image blur and a loss of detail, making it difficult to distinguish individual features. A similar blurring effect occurs at sharp transitions between dark and light regions, particularly when the transition occurs over a spatial scale comparable to the wavelength of light. In such cases, if the boundary between black and white in the image is narrower than the wavelength, diffraction causes the edge to appear softened or smeared, diminishing the clarity of the transition. To quantitatively assess the performance of an objective lens, standardized test patterns are commonly employed. These patterns typically consist of alternating black and white bars with 100% contrast. As the spacing between the lines decreases, corresponding to increasing spatial frequency, the image formed by the lens exhibits reduced contrast, due to

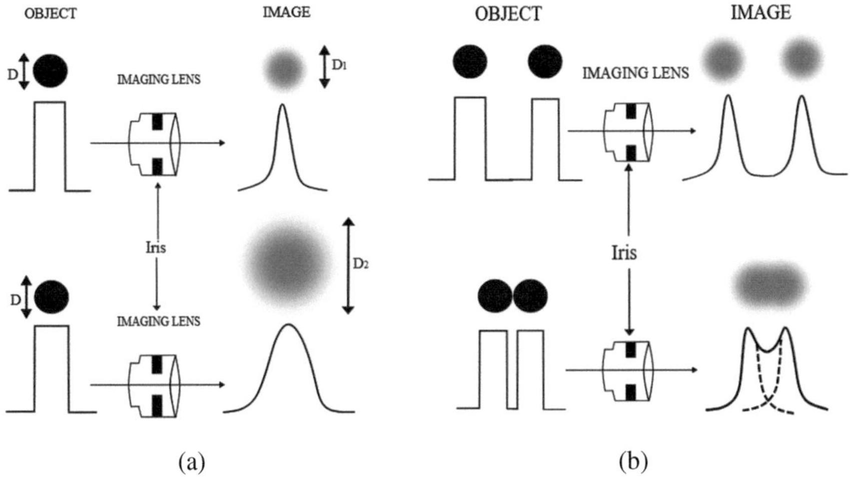

Fig. 2.30 Spatial frequency low pass filter effect of an objective, lower for large lens diaphragm aperture

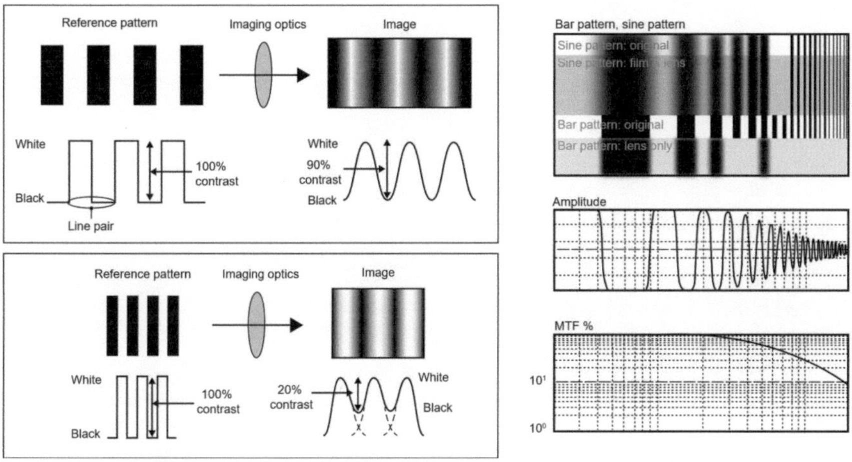

Fig. 2.31 Modulation transfer function of an objective

diffraction-induced blur. This degradation in contrast is described by the Modulation Transfer Function (MTF), a fundamental metric in optical engineering. The MTF expresses the system's ability to reproduce contrast as a function of the image spatial frequency content (for example in terms of line/mm, dots/mm) and is defined as MTF = (output image contrast)/(input image contrast), where the contrast is measured as the difference between the maximum and minimum gray levels in the input and output images. As the spatial frequency increases (i.e. the lines or points become more closely spaced), the MTF value decreases, reflecting the system's diminishing ability to maintain image detail. This behavior underscores the influence of diffraction as a limiting factor in the resolution and contrast reproduction capabilities of any optical system (Fig. 2.31).

The percentage of contrast in the output image can be defined as:

$$C\% = 100\frac{(I_{\max} - I_{\min})}{(I_{\max} + I_{\min})} \tag{2.24}$$

where I is the gray level intensity on the image. Maximum gray level is for black and minimum for white.

Modulation transfer function is typically a given specification of an objective as in Fig. 2.32 as a function of contrast respect to the number of lines per mm on the testing reference image, if they are a lot, they are thin and overlap in the image plane due to blurring. Modulation transfer function is therefore a fundamental objective property to generate good images on the image plane. For telescopes gives the possibility to resolve two far and tiny star points, for microscopes to resolve small details of objects. For imaging systems used for mesurements give us the ultimate limit in spatial measurement resolution.

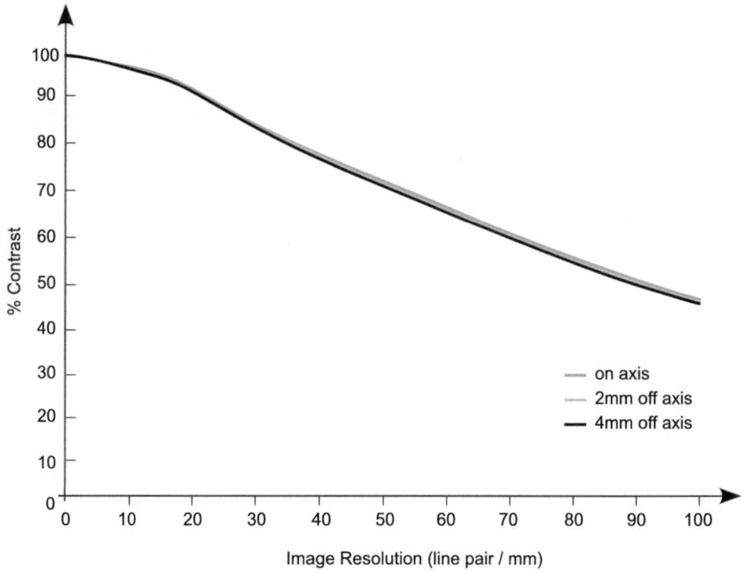

Fig. 2.32 A typical modulation transfer function of an objective

A fundamental parameter that combines the key characteristics of an optical system is the Exposure value (E_V), which quantifies the amount of light that reaches the sensor on the image plane. The exposure value depends on three main factors:

- the exposure time Δt in seconds (given by shutter speed),
- the aperture (expressed as the f-number N),
- the ISO sensitivity S of the image sensor.

It can be expressed by the following relation:

$$E_v = \log_2\left(\left(100\,N^2\right)/\left(\Delta t\ S\right)\right)$$

This combined parameter is essential for achieving proper image brightness under various lighting conditions. This parameter combination effects are illustrated in Fig. 2.33.

As previously discussed, the aperture parameter N not only controls the amount of light entering the lens but also significantly affects the depth of field. On the other hand, high shutter speeds are required to effectively freeze the motion of fast-moving subjects, to avoid shades. If a subject is moving at velocity v, any exposure time of Δt seconds will result in a motion shade at least equal to the displacement $v\Delta t$. To avoid this, a maximum permissible Δt must be defined based on the expected motion velocity v in the scene.

When a specific combination of aperture N and shutter exposure time Δt is chosen, whether to achieve a desired depth of field or to freeze motion, the third critical factor used to ensure proper exposure is the ISO sensitivity S of the image sensor.

Fig. 2.33 Exposure parameters and images obtained under low, correct and over exposition (pictures by Giuliano Carotti)

However, while increasing the ISO sensitivity value S can help maintain sufficient brightness under constrained aperture and shutter exposure time settings, it typically comes at a cost. High ISO sensitivity often lead to image degradation, such as increased noise, reduced sharpness, and loss of detail. The severity of these effects depends on the type of imaging medium, whether film, CCD, or CMOS sensors, each having distinct sensitivity characteristics and noise behavior, but all giving this problems for hight sensitivity values.

Therefore, choosing the right ISO value requires a trade-off between image brightness and quality, and must be carefully balanced based on the nature of the scene, freezing motion requirements, and optical parameters.

2.9 Objective Projection Equations

By considering an idealized objective, equivalent to a simple thin lens or even a pinhole camera model, free from optical aberrations and diffraction effects, it becomes possible to define transformation equations that map 3D real-world coordinates to 2D image plane coordinates [4].

As illustrated in Fig. 2.34, three coordinate systems are involved:

- (x_i, y_i): the 2D coordinates on the image plane,
- (x_c, y_c, z_c): the 3D coordinates of object points relative to the camera coordinate system,
- (x_w, y_w, z_w): the 3D coordinates of object points relative to the world coordinate system, which may be translated and rotated with respect to the camera's reference.

The objective is to define the mathematical transformation that relates each real-world point (x_w, y_w, z_w) to its projected image coordinates (x_i, y_i) on the camera's image plane, as seen through the lens.

To establish this relationship, we begin by modeling the optical system as an ideal pinhole camera with a focal length f. In this simplified model, the camera objective having a focal lengh f is replaced by a pinhole located at a distance f from the image

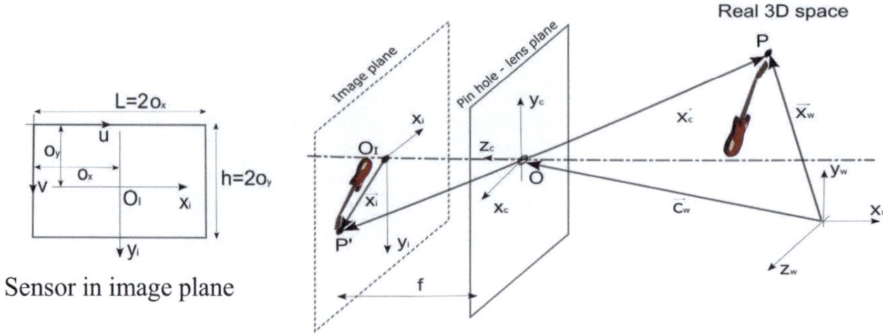

Fig. 2.34 Coordinate systems for 3D real word to 2D image plane transformation definition

plane. The image of a 3D point P in space is then formed at the intersection of the line connecting point P to the optical center O (where the pinhole is) with the image plane, as shown in Fig. 2.34.

This model provides the basic geometry for perspective projection equations definition, which are also the basis for camera calibration, photogrammetry, and computer vision techniques.

As previously established for the pinhole model and thin lens approximation, so using Eq. 2.4, we can express the transformation from the 3D coordinates of a real-world point P, defined in the camera coordinate system as (x_c, y_c, z_c), with the origin O located on the optical axis, to the corresponding 2D coordinates (x_i, y_i) on the image plane, defined with its origin O_I also lying along the optical axis. The perspective projection is given by:

$$x_i = f\frac{x_c}{z_c};\ y_i = f\frac{y_c}{z_c} \tag{2.25}$$

where:

- f is the focal length of the lens (distance from the optical center to the image plane),
- z_c is the depth of the point P along the optical axis,
- (x_i, y_i) are the projected image coordinates in the 2D plane.

This transformation describes the fundamental central projection geometry of ideal imaging systems and forms the basis for many computer vision and photogrammetry applications.

Images are normally formed in a rectangular image sensor of dimensions h × L and coordinates origin on it are generally taken on the top-left corner of the sensor as (u, v), so translated of the half height o_y and the half length o_x of the sensor, as in Fig. 2.33. We have therefore:

$$u = f\frac{x_c}{z_c} + o_x;\ v = f\frac{y_c}{z_c} + o_y \tag{2.26}$$

To perform the transformation, it is necessary to know the values of f, o_x, and o_y, which are referred to as the *intrinsic parameters* of the camera. These parameters are determined by the characteristic of the objective f and by the two image sensor dimensions. So only three numbers, sometimes four using two different focal lenght for x and y.

The transformation equations are inherently nonlinear due to the division by the depth coordinate z_c. However, it is possible to linearize these equations by introducing an additional dimension through a scaling factor, this is achieved using the technique of *homogeneous coordinates.*

In homogeneous representation, a 2D point (u, v) is expressed as a 3D point (u', v', w'), where the third coordinate w' is an arbitrary non-zero value. The condition $w' \neq 0$ must be satisfied. The actual image coordinates are then recovered by normalizing:

$$u = \frac{u'}{w'};\ v = \frac{v'}{w'}$$

For $w' = 1$ we have obviously $u' = u$, and $v' = v$. This added third coordinate can be seen as a scale factor. This method allows the use of linear transformations in projective geometry and simplifies the mathematical camera models.

For 3D points it is also possible an homogeneous representation using 4D vectors, once again adding a fictitious coordinate w'. In vectorial notation we can express this homogeneous representation of a 3D point (x, y, z) as:

$$X = \begin{bmatrix} x \\ y \\ z \\ 1 \end{bmatrix} = \begin{bmatrix} w'x \\ w'y \\ w'z \\ w' \end{bmatrix} = \begin{bmatrix} x' \\ y' \\ z' \\ 1 \end{bmatrix} = X'$$

Taking $w' = z_c$ the above perspective transformation operated by the objective on camera image sensor can be therefore written in matrix linear form as:

$$\begin{bmatrix} u \\ v \\ 1 \end{bmatrix} = \begin{bmatrix} u' \\ v' \\ w' \end{bmatrix} = \begin{bmatrix} z_c \cdot u \\ z_c \cdot v \\ z_c \end{bmatrix} = \begin{bmatrix} fx_c + z_c o_x \\ fy_c + z_c o_y \\ z_c \end{bmatrix} = \begin{bmatrix} f & 0 & o_x & 0 \\ 0 & f & o_y & 0 \\ 0 & 0 & 1 & 0 \end{bmatrix} \cdot \begin{bmatrix} x_c \\ y_c \\ z_c \\ 1 \end{bmatrix} \tag{2.27}$$

Equivalent in matrix form to Eq. (2.26).

Matrix $\boldsymbol{M}_{int} = \begin{bmatrix} f & 0 & o_x & 0 \\ 0 & f & o_y & 0 \\ 0 & 0 & 1 & 0 \end{bmatrix}$ it's called intrinsic matrix, given by the calibration matrix with added a column of zero. Equation (2.27) can be written in a compact equivalent matrix form as:

$$\boldsymbol{u}' = \boldsymbol{M}_{int} \cdot \boldsymbol{X}_c' \tag{2.28}$$

Next step is to define the transformation of coordinates from camera reference system (x_c, y_c, z_c) to real world (x_w, y_w, z_w) reference system. This coordinate transformation is a roto-translation defined by the displacement vector $\mathbf{c}_w$ in Fig. 2.34 and by a relative rotation matrix **R** between the two reference systems. As known [5] **R** is an orthonormal matrix, so its transpose matrix is equal to its inverted matrix, and the product of a matrix with its transpose is equal to one.

Matrix **R** and vector $\mathbf{C}_w$ are the so-called extrinsic parameters of the camera. Vector $\mathbf{X}_c$ of the coordinates in camera reference system can be expressed in terms of coordinates in world reference system with the rotation **R** of the displaced world coordinate ($\boldsymbol{X}_w - \boldsymbol{C}_w$), by the relation:

$$\boldsymbol{X}_c = \boldsymbol{R}(\boldsymbol{X}_w - \boldsymbol{C}_w) = \boldsymbol{R}\boldsymbol{X}_w - \boldsymbol{R}\boldsymbol{C}_w = \boldsymbol{R}\boldsymbol{X}_w + \boldsymbol{t} \tag{2.29}$$

where $\mathbf{t} = -\boldsymbol{R}\boldsymbol{C}_w$. Above equation can be expressed also as:

$$\boldsymbol{X}_c = \begin{bmatrix} x_c \\ y_c \\ z_c \end{bmatrix} = \begin{bmatrix} r_{11} & r_{12} & r_{13} \\ r_{21} & r_{22} & r_{23} \\ r_{31} & r_{32} & r_{33} \end{bmatrix} \begin{bmatrix} x_w \\ y_w \\ z_w \end{bmatrix} + \begin{bmatrix} t_x \\ t_y \\ t_z \end{bmatrix}$$

And using homogenous coordinates as:

$$\boldsymbol{X}'_c = \begin{bmatrix} x_c \\ y_c \\ z_c \\ 1 \end{bmatrix} = \begin{bmatrix} r_{11} & r_{12} & r_{13} & t_x \\ r_{21} & r_{22} & r_{23} & t_y \\ r_{31} & r_{32} & r_{33} & t_z \\ 0 & 0 & 0 & 1 \end{bmatrix} \begin{bmatrix} x_w \\ y_w \\ z_w \\ 1 \end{bmatrix}$$

So, roto-translation in homogeneous coordinates is defined by the matrix

$$\boldsymbol{M}_{est} = \begin{bmatrix} r_{11} & r_{12} & r_{13} & t_x \\ r_{21} & r_{22} & r_{23} & t_y \\ r_{31} & r_{32} & r_{33} & t_z \\ 0 & 0 & 0 & 1 \end{bmatrix},$$

named extrinsic parameter matrix.

In compact form we can write the transformation from world coordinates to camera coordinates as:

$$\boldsymbol{X}'_c = \boldsymbol{M}_{est} \cdot \boldsymbol{X}'_w \tag{2.30}$$

Substituting this last Eq. (2.30) in Eq. (2.28) we obtain the complete projection transformation:

$$\boldsymbol{u}' = \boldsymbol{M}_{int} \cdot \boldsymbol{M}_{est} \cdot \boldsymbol{X}'_w = \boldsymbol{P} \cdot \boldsymbol{X}'_w \tag{2.31}$$

Defining the product of intrinsic matrix and extrinsic matrix as the complete projection matrix **P** the previous equation can be also written as:

$$\begin{bmatrix} u' \\ v' \\ z' \end{bmatrix} = \begin{bmatrix} p_{11} & p_{12} & p_{13} & p_{14} \\ p_{21} & p_{22} & p_{23} & p_{24} \\ p_{31} & p_{32} & p_{33} & p_{34} \end{bmatrix} \begin{bmatrix} x_w \\ y_w \\ z_w \\ 1 \end{bmatrix}$$

So the homogeneous coordinates on image plane are equal to the product of the intrinsic and extrinsic matrix and the homogeneous coordinates in real world, or the product of the projection matrix **P** with the homogeneous coordinates in real world. To know the transformation we need to know all the 12 coefficients p_{ij} of the projection matrix. This is possible by an experimental technique named camera calibration.

Projection matrix acts on homogeneous coordinates, so:

$$\begin{bmatrix} u' \\ v' \\ z' \end{bmatrix} = k \begin{bmatrix} u' \\ v' \\ z' \end{bmatrix}$$

And therefore

$$\begin{bmatrix} p_{11} & p_{12} & p_{13} & p_{14} \\ p_{21} & p_{22} & p_{23} & p_{24} \\ p_{31} & p_{32} & p_{33} & p_{34} \end{bmatrix} \begin{bmatrix} x_w \\ y_w \\ z_w \\ 1 \end{bmatrix} = k \begin{bmatrix} p_{11} & p_{12} & p_{13} & p_{14} \\ p_{21} & p_{22} & p_{23} & p_{24} \\ p_{31} & p_{32} & p_{33} & p_{34} \end{bmatrix} \begin{bmatrix} x_w \\ y_w \\ z_w \\ 1 \end{bmatrix}$$

This means that an image on image plane is the same of a little object near to the objective or of an object of same shape, bigger but far to the objective, as illustrated in Fig. 2.35.

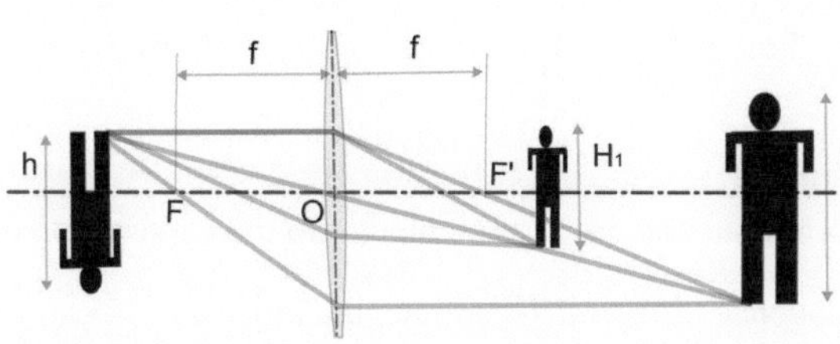

Fig. 2.35 Scale factor indetermination: two men of different height at different distance from the optical center of a lens or an objective gives an image of same height on the image plane

2.10 Camera Calibration

Camera calibration refers to the experimental process of determining the parameters of a camera's projection matrix, which maps 3D real-world points to 2D image coordinates. Specifically, this involves estimating the 12 coefficients of the camera projection matrix.

The process typically involves constructing a system of equations based on a set of known 3D points (x_w, y_w, z_w) in the world coordinate system and their corresponding 2D image coordinates (u, v). Since each point correspondence contributes two equations (one for each image coordinate), a minimum of six points is needed to yield twelve equations, sufficient to solve for the 12 unknowns p_{ij}. However, a deeper analysis reveals that the camera projection matrix can be factorized into two distinct components:

- The intrinsic matrix $\mathbf{M}_{int}$, typically an upper triangular matrix that encapsulates internal camera parameters. Under common assumptions (e.g., square pixels, no skew), this matrix is defined by four parameters: the focal length and the coordinates of the optical center (principal point).
- The extrinsic matrix $[\mathbf{R}|\mathbf{t}]$, which encodes the rotation $\mathbf{R}$ and translation $\mathbf{t}$ that transform coordinates from the world reference frame to the camera reference frame.
  - $\mathbf{R}$ is a 3×3 orthonormal matrix representing rotation, fully described by three parameters (e.g., Euler angles or axis-angle representation).
  - $\mathbf{t}$ is a 3×1 vector representing the position of the world origin as seen from the camera, requiring other three parameters.

Thus, while the projection matrix nominally has 12 elements, the actual number of independent parameters is only 9, four from the intrinsic matrix and five from the extrinsic transformation (since rotation and translation together contribute six, but one overall scale factor is arbitrary in projective geometry).

In practice, camera calibration is performed by capturing images of a known geometric calibration pattern, such as 2D checkerboard, 3D cubes or dot grid, as shown in Fig. 2.36. For each image, corresponding points are identified:

- Their real-world coordinates (x_w, y_w, z_w), defined in a coordinate system fixed to the calibration object;
- Their image coordinates (u, v), as projected onto the camera's image plane.

These correspondences are used to estimate the intrinsic and extrinsic parameters by solving a system of equations using numerical optimization techniques such as least squares fitting or bundle adjustment. These methods minimize the reprojection error, which is the difference between the observed image points and the projected points calculated from the estimated model.

We can therefore write, from Eq. (2.31), for each identified i^{th} point on image plane, the coordinate transformation in matrix form as:

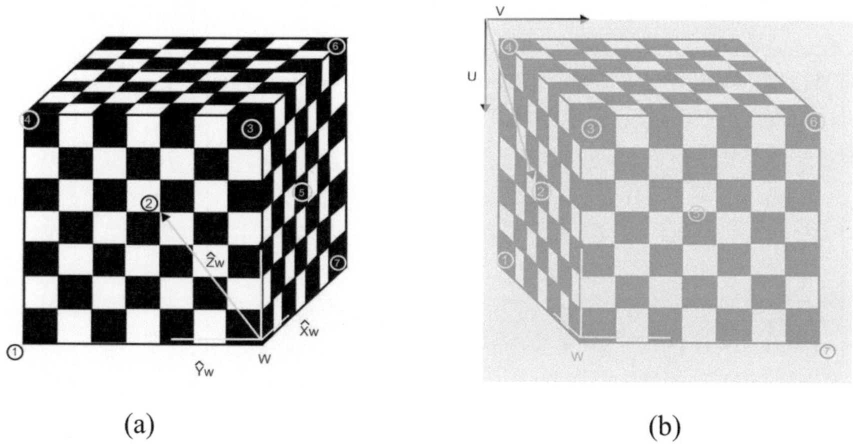

Fig. 2.36 Camera calibration by a known geometry object: **a** a known 3D geometry object, **b** its the captured 2D image of it

$$\begin{bmatrix} u^i \\ v^i \\ 1 \end{bmatrix} = \begin{bmatrix} p_{11} & p_{12} & p_{13} & p_{14} \\ p_{21} & p_{22} & p_{23} & p_{24} \\ p_{31} & p_{32} & p_{33} & p_{34} \end{bmatrix} \begin{bmatrix} x_w^i \\ y_w^i \\ z_w^i \\ 1 \end{bmatrix}$$

where all the vector coefficients (u^i, v^i, 1) and (x^iw, y^iw, z^iw, 1) are known and the pij matrix coefficients are the unknown to find. From the above matrix equation it is possible to write, for each identified corresponding point in calibration object and in its image, the two eqation linear system:

$$u^i = \frac{x_w^i p_{11} + y_w^i p_{12} + z_w^i p_{13} + p_{14}}{x_w^i p_{31} + y_w^i p_{32} + z_w^i p_{33} + p_{34}}$$

$$v^i = \frac{x_w^i p_{21} + y_w^i p_{22} + z_w^i p_{23} + p_{24}}{x_w^i p_{31} + y_w^i p_{32} + z_w^i p_{33} + p_{34}}$$

In the case of a single point correspondence between a 3D world point and its 2D image projection, we obtain two equations involving the 12 unknowns p_{ij} of the camera projection matrix. However, by repeating this process for additional points, we can generate a sufficient number of equations to solve for these unknowns.

For example, identifying six distinct 3D-2D point correspondences yields a total of 12 independent equations, which is just enough to determine the 12 elements of the projection matrix (up to an overall scale factor). This results in a linear system of equations with the form:

$$\begin{bmatrix} x_w^1 & y_w^1 & z_w^1 & 1 & 0 & 0 & 0 & 0 & -x_w^1 u_1 & -y_w^1 u_1 & -z_w^1 u_1 & -u_1 \\ 0 & 0 & 0 & 0 & x_w^1 & y_w^1 & z_w^1 & 1 & -x_w^1 v_1 & -y_w^1 v_1 & -z_w^1 v_1 & -v_1 \\ \vdots & \vdots & \vdots & \vdots & \vdots & \vdots & \vdots & \vdots & \vdots & \vdots & \vdots & \vdots \\ x_w^i & y_w^i & z_w^i & 1 & 0 & 0 & 0 & 0 & -x_w^i u_i & -y_w^i u_i & -z_w^i u_i & -u_i \\ 0 & 0 & 0 & 0 & x_w^i & y_w^i & z_w^i & 1 & -x_w^i v_i & -y_w^i v_i & -z_w^i v_i & -v_i \\ \vdots & \vdots & \vdots & \vdots & \vdots & \vdots & \vdots & \vdots & \vdots & \vdots & \vdots & \vdots \\ x_w^{12} & y_w^{12} & z_w^{12} & 1 & 0 & 0 & 0 & 0 & -x_w^{12} u_{12} & -y_w^{12} u_{12} & -z_w^{12} u_{12} & -u_{12} \\ 0 & 0 & 0 & 0 & x_w^{12} & y_w^{12} & z_w^{12} & 1 & -x_w^{12} v_{12} & -y_w^{12} v_{12} & -z_w^{12} v_{12} & -v_{12} \end{bmatrix} \begin{bmatrix} p_{11} \\ p_{12} \\ p_{13} \\ p_{14} \\ p_{21} \\ p_{22} \\ p_{23} \\ p_{24} \\ p_{31} \\ p_{32} \\ p_{33} \\ p_{34} \end{bmatrix} = \begin{bmatrix} 0 \\ 0 \\ 0 \\ 0 \\ 0 \\ 0 \\ 0 \\ 0 \\ 0 \\ 0 \\ 0 \\ 0 \end{bmatrix}$$

In practice, more than six points are typically used to overdetermine the system, allowing for a least-squares solution that minimizes the effect of measurement noise and improves robustness.

Numerous software tools are available to perform camera calibration based on the approach described above. Many of these tools are freely accessible through popular platforms such as Python libraries (e.g., OpenCV), MATLAB toolboxes, and other specialized calibration software packages.

Once the camera projection matrix has been estimated, it can be decomposed into its intrinsic and extrinsic components by leveraging the mathematical properties of matrix operations. Specifically, the projection matrix **P** can be expressed as the product of:

- the intrinsic matrix $\mathbf{M}_{\text{int}}$, a typically upper triangular matrix containing the internal parameters of the camera (such as focal length and principal point coordinates),
- the extrinsic matrix $[\mathbf{R}|\mathbf{t}]$, which includes a rotation matrix **R** (orthonormal) and a translation vector **t** that describe the camera's pose relative to the world coordinate system.

According to linear algebra, such a matrix product admits a unique decomposition under certain conditions. From the calibrated projection matrix coefficients $\mathrm{p_{ij}}$, it is therefore possible to extract:

- Intrinsic parameters, including the focal length f, and the coordinates of the principal point $(\mathrm{o_x}, \mathrm{o_y})$ giving the two dimensions of the image plane.
- Extrinsic parameters, which define the position and orientation of the object or scene with respect to the camera.

However, it is important to note that this formulation is based on the ideal pinhole camera model, which assumes an aberration-free optical system. In practice, real lenses introduce distortions that deviate from this ideal behavior. To account for these, an additional distortion calibration step is typically performed by capturing images of known geometric patterns (e.g., checkerboards or dot grids). This allows the estimation of distortion correction coefficients.

One widely adopted method for modeling lens distortion is the Brown–Conrady distortion model (Brown 1966), which incorporates both:

- Radial distortion, caused by the curvature of the lens, leading to effects such as barrel or pincushion distortion.
- Tangential distortion, resulting from lens misalignments, such as off-center lens elements or tilt during manufacturing or assembly.

To correct for these distortions, the transformation between the original distorted image coordinates (u, v) and the rectified (undistorted) coordinates (u_r, v_r) is described by a set of equations that apply the distortion model. These equations enable the correction of the captured image, restoring accurate geometric representations of the scene. A general relationship between the distorted and rectified coordinates can be written as follows:

$$u_r = \left(1 + k_1 r^2 + k_2 r^4 + k_3 r^6\right)u + 2p_1 uv + p_2\left(r^2 + 2u^2\right)$$

$$v_r = \left(1 + k_1 r^2 + k_2 r^4 + k_3 r^6\right)v + 2p_2 uv + p_1\left(r^2 + 2v^2\right)$$

where the distortion terms depend on the radial distance from the image center and the specific distortion coefficients and where:

$$\mathrm{r} = (\mathrm{u} - \mathrm{o_x})^2 + \left(\mathrm{v} - \mathrm{o_y}\right)^2$$

The effects of these distortions and possible corrections are illustrated in Fig. 2.37.

So other five distortion coefficient (k_1, k_2, k_3) and (p_1, p_2) should be estimated by calibration to correct field distortion aberration effects. The problem of sperical aberration, cromatic aberration, astigmatism etc., still remain. They cause that a point (u, v) is not"a point" but a blurred circle. Their effects can be minimized only by using high quality optics, and choosing correctly the objective for our measurement system.

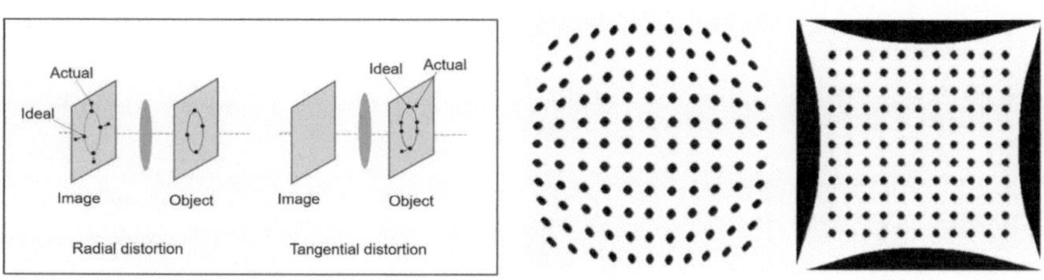

Fig. 2.37 Radial and tangential distortions and their possible corrections by calibration

2.11 Optical Components

Reflecting optical components, commonly referred to as *mirrors*, form images solely through the process of reflection. Mirrors reflect incident light rays according to the law of reflection, which states that the angle of incidence equals the angle of reflection, both measured relative to the normal of the reflecting surface.

Like lenses, mirrors can be used to converge or diverge light rays and to form both real and virtual images. The simplest type of mirror is the plane (flat) mirror, as illustrated in Fig. 2.38. It reflects the rays emitted or scattered by each point on an object symmetrically across the mirror surface, producing a virtual image that appears behind the mirror. This image is upright, of the same size as the object, and appears at the same distance from the mirror as the object itself. The magnification of a plane mirror is $m = AB/A'B' = h/h'$, so equal to 1, indicating that the image height equals the object height.

More complex are spherical mirrors, such as convex mirrors, typically portions of a spherical surface with a vertex V and center of curvature C as shown in Fig. 2.38. Using ray tracing, it becomes evident that convex mirrors form a virtual, upright (not inverted), and diminished image, that is, the image is smaller than the object, with magnification less than 1.

To understand this, consider two principal rays:

- Ray r_1, originating from point B, travels toward the mirror along a path that coincides with a radius of the sphere, passing through the center C. Since this ray strikes the surface perpendicularly, it reflects back along the same path, now labeled i_1. The extension of this reflected ray is shown as a dashed line that passes through point C.
- Ray r_2 travels parallel to the optical axis toward the mirror. According to the mirror's focal properties, it is reflected in direction i_2, such that its extension passes through the focal point F, just as with lenses.

The intersection point of the extensions of the reflected rays i_1 and i_2 determines the position of the virtual image point B′. This image cannot be projected onto a screen, as it exists only as a perceived point behind the mirror surface.

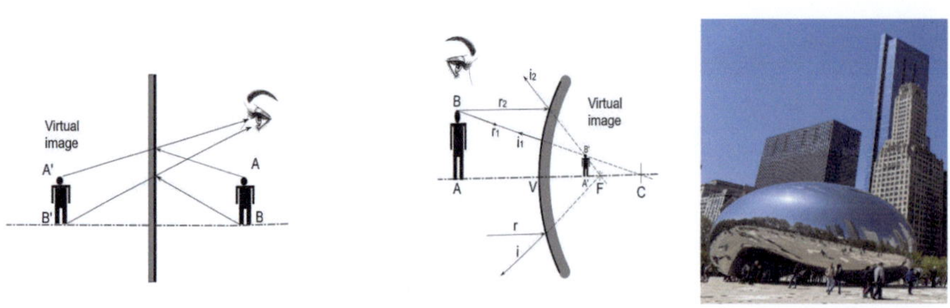

Fig. 2.38 Flat and convex mirrors

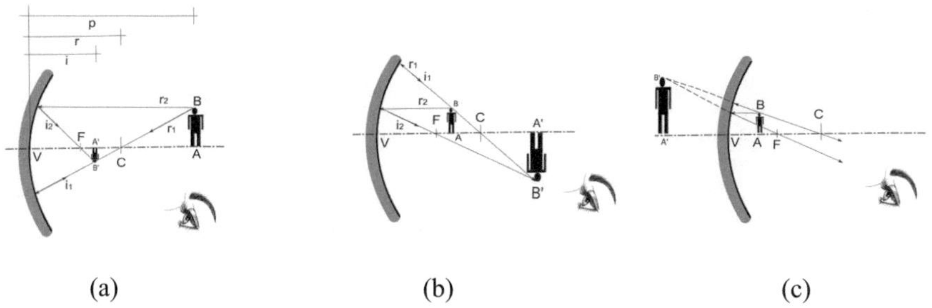

Fig. 2.39 Concave mirrors

Using the same ray tracing approach, so following the ray going parallel to the optical axis therefore converging to the focus and the ray not deflected passing through the curvature center, we can see that for concave mirrors, when the object reflected by the mirror is farther from the mirror more than its curvature center C, as in Fig. 2.39a image generated is inverted and reduced (m < 1); when the object reflected by the mirror is between focus F and curvature center C, as in Fig. 2.39b image generated is inverted and bigger (m > 1); when the object reflected by the mirror is between focus F and vertex V, as in Fig. 2.39c image generated is upright, virtual and bigger (m > 1).

It is also possible to verify, in a similar way as previously done for lenses, that for a concave mirror with object far from the mirror more than point C, as in Fig. 2.39a, if we call p the distance VA, r the distance VC i.e. the radius of the mirror, i the distance A'V we have:

$$\frac{1}{p} + \frac{1}{i} = \frac{2}{r} = \frac{1}{f}$$

So mirrors can be used instead of lenses for image processing in the so called reflective optic systems instead of refractive optic systems.

Another widely used category of optical components is the *optical prism.* Prisms are typically made of optical-grade glass and consist of flat, polished surfaces joined at precise angles. Unlike mirrors, which reflect light from a surface, prisms manipulate light through both reflection and refraction as it passes through or bounces within the prism material. The simplest type of prism is the right-angle prism, often shaped like a half-cube with one face inclined at 45° relative to the adjacent surfaces, as illustrated in Fig. 2.40. In this configuration, the inclined face acts like a mirror under total internal reflection conditions. Such a prism can be used to redirect a light beam, typically bending it by 90°, as shown in Fig. 2.40, or to reflect the beam back in the direction from which it came (a retroreflector configuration). Because of their high precision, durability, and optical stability, prisms are commonly employed in imaging systems, laser setups, and optical instruments for beam steering, image rotation, and alignment correction.

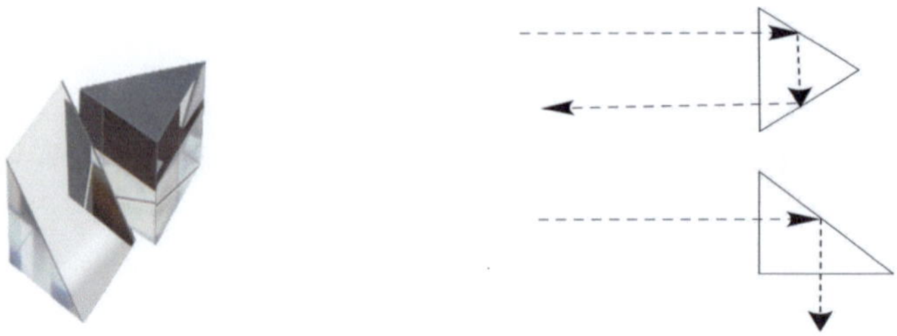

Fig. 2.40 Optical prisms

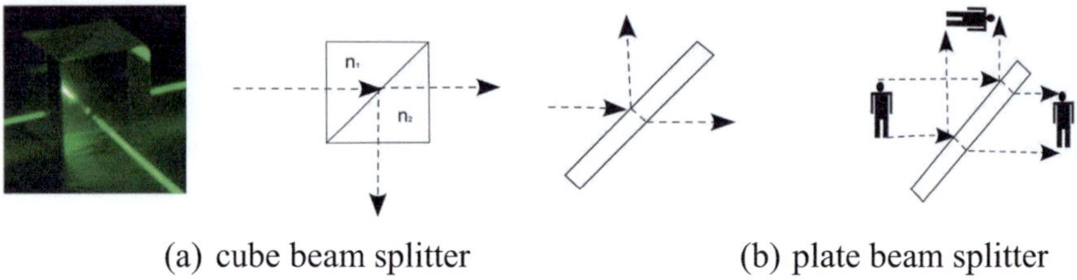

Fig. 2.41 Beam splitters

Another widely used category of optical components is the *beam splitter*. A beam splitter divides an incoming light beam into two separate beams, typically by a combination of reflection and refraction.

One common design consists of two prisms made from materials with different refractive indices (n_1 and n_2), optically bonded along their diagonal surfaces, as illustrated in Fig. 2.41a. By carefully selecting the refractive indices, the system can be engineered so that a beam incident at a specific angle (e.g., 45° relative to the surface normal) is split such that half of the light continues along the original direction, while the other half is deflected by 90°. This configuration ensures a 50:50 power split between the transmitted and reflected components.

An alternative and simpler implementation uses a thin plate of optical glass, as shown in Fig. 2.41b. In this case, a partially reflective coating on the surface allows for a portion of the light to be reflected while the remainder is transmitted.

Beam splitters are essential in various optical systems. They are commonly used to:

- Split a laser beam into two paths for interferometry or metrology,
- Create duplicate optical paths in imaging systems,
- Generate two images from a single object, for example in stereo vision or heads-up displays.

Because of their versatility and precision, beam splitters are integral components in laser optics, microscopy, fiber optics, interferometers and optical instrumentation.

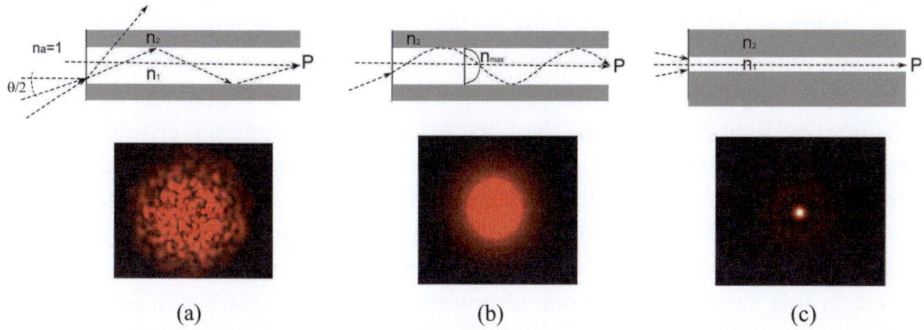

Fig. 2.42 Optical fibers and their output light spots

Another widely used category of optical components is the *optical fiber*. Optical fibers function as waveguides for light and are composed of two concentric cylindrical layers of glass:

- the core, a narrow central cylindrical region through which light propagates,
- the cladding, a surrounding cylindrical layer with a lower refractive index than the core, as illustrated in Fig. 2.42.

When a light ray enters the input flat polished face of the fiber core, it is refracted according to Snell's Law and directed into the core. As the ray travels through the fiber and reaches the core-cladding interface, it undergoes total internal reflection, provided that the angle of incidence at this interface is greater than the critical angle. In this case, no refraction into the cladding occurs, and the light continues to propagate within the core through successive internal reflections.

This guiding mechanism will occur only if the angle of incidence at the fiber entrance is within a specific limit, defined by the numerical aperture Na of the fiber. The maximum angle at which light can enter the fiber and still be guided is called the acceptance angle, denoted θ. Light entering at an angle less than $\theta/2$ relative to the fiber axis will be successfully confined within the core. The acceptance angle θ is given by the following relationship [6]:

$$Na = sin(\vartheta) = \sqrt{n_1^2 - n_2^2}$$

where:

- n_1 is the refractive index of the core,
- n_2 is the refractive index of the cladding.

Optical fibers are fundamental in telecommunications, medical imaging (endoscopy), sensing applications, and laser delivery systems, owing to their ability to transmit light over long distances with minimal loss and high precision.

Above described fibers are a type of optical fiber in which the refractive index changes abruptly, like a step function, from one value in the core to a lower value in the cladding, named step-index fibers. This sharp boundary enables total internal reflection, guiding light within the core.

However, step-index fibers are prone to a phenomenon called modal dispersion. Consider two light rays entering the fiber: one traveling along the central axis and another, a marginal ray, reflecting repeatedly at steep angles, following the longest permissible path that still satisfies total internal reflection. Because these paths have different lengths, the rays often arrive at the same output point P out of phase, potentially causing destructive interference and a reduction in light intensity.

Due to this variation, some output points experience constructive interference, while others experience destructive interference, resulting in a random speckle-like light intensity distribution, as shown in Fig. 2.42a. This effect is similar to the speckle patterns observed when coherent laser light is reflected from rough surfaces (as discussed in Chap. 1).

To overcome limitations given by speckle in some fiber optics applications, graded-index fibers were developed. In these fibers, the refractive index of the core gradually decreases from the center toward the cladding, like the previously described GRIN lenses. This gradient causes light rays to continuously bend rather than reflect sharply, following a sinusoidal path. Although marginal rays still travel a longer physical path, they pass through regions of lower refractive index, where the speed of light is higher. In contrast, rays traveling closer to the fiber's center follow shorter paths but pass through regions with a higher refractive index, and thus travel more slowly. As a result, rays with different paths arrive at the output point P in phase, producing constructive interference and yielding a smooth light distribution with maximum intensity at the center, gradually decreasing toward the edges, as shown in Fig. 2.42b.

A third type is the single-mode (monomode) fiber, which features an extremely narrow core, typically around 1–5 μm in diameter, comparable to the wavelength of light (0.4–0.7 μm). This narrow dimension permits the propagation of only a single light mode, effectively eliminating modal dispersion. The result is a clean output beam with a single, sharply defined central spot, as illustrated in Fig. 2.42c.

Optical fibers are extensively used in telecommunications for transmitting optical signals over long distances with minimal loss. In addition, bundles of optical fibers are employed in medical imaging and industrial inspection tools, such as endoscopes and boroscopes, allowing images to be transmitted along curved paths, as shown in Fig. 2.43.

Last but not least, an essential category of optical components used in imaging systems are the *optical filters* illustrated in Fig. 2.44. These filters are typically thin plates made of specialized materials designed to selectively transmit or block specific wavelengths of electromagnetic radiation, whether in the visible, ultraviolet (UV), or infrared (IR) range, while allowing others to pass through.

Optical filters play a critical role in applications such as photography, microscopy, machine vision, astronomy, and laser technology, where precise control over the spectral content of light is required. By modifying the light that reaches a sensor, camera objective , or observer, filters help enhance contrast, isolate spectral bands,

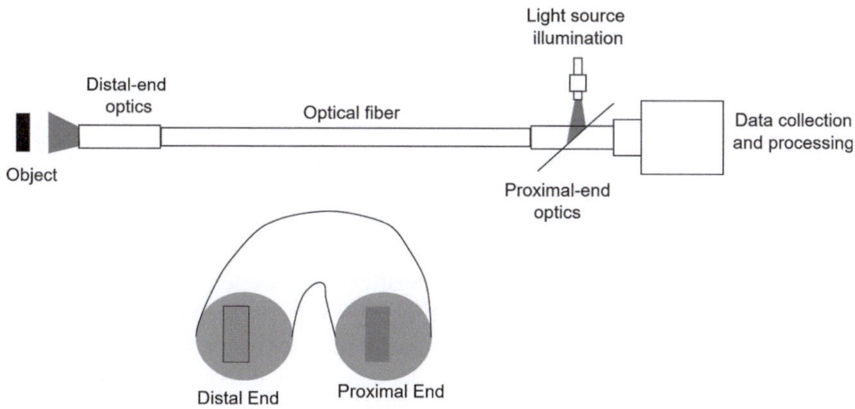

Fig. 2.43 Applications of optical fibers in imaging by fiber boundles

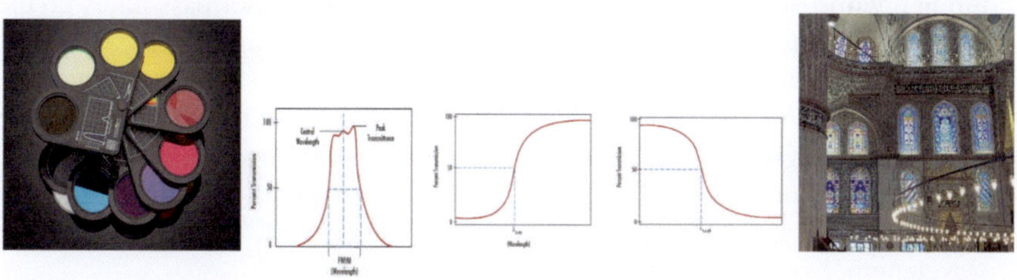

(a) Commercial filters (b) band-pass (c) high-pass (d) low-pass (e) an application

Fig. 2.44 Optical filters (**a**), their different possible optical transfer function (**b–d**) and an application in Santa Sofia windows, in Istambul (**e**)

or prevent sensor saturation. Can be mounted in front of an objective for various purposes, including:

- Color selection: to allow only specific wavelengths (colors) to pass through, enabling monochromatic imaging or highlighting certain features in a scene.
- Light intensity reduction: using neutral density (ND) filters, which uniformly reduce the intensity of all wavelengths without altering color balance, useful for long-exposure photography or preventing overexposure.
- Polarization control: using polarizing filters, which transmit only light waves oscillating in a specific direction. These are useful for reducing glare, enhancing contrast, or revealing features not visible under unpolarized light.

Through these capabilities, optical filters are indispensable tools in both scientific and artistic imaging, enabling selective visualization and measurement of light across the electromagnetic spectrum.

There are various types of optical filters, classified based on how they work or their intended function:

- Absorptive that absorb certain wavelengths of light while transmitting others. The materials used, like colored glass or dyes, are designed to block specific portions of the spectrum. Absorptive filters are generally less expensive but may heat up over time as they absorb light.
- Reflective (dichroic) use thin-film coatings that reflect unwanted wavelengths of light rather than absorbing them. These filters can provide very sharp cutoffs between transmitted and blocked wavelengths and are highly efficient for specific applications, like laser optics or fluorescence microscopy.
- Neutral density (ND) that reduce the intensity of all wavelengths of light equally, effectively dimming the light without affecting its spectral composition. They are commonly used in photography and optical testing to reduce light exposure.
- Bandpass transmit light only within a specific range of wavelengths, blocking wavelengths outside of that range. They are used in scientific instruments like spectrometers to isolate specific portions of the light spectrum.
- Longpass and shortpass allow wavelengths longer than a certain cutoff to pass through while blocking shorter wavelengths. Conversely, shortpass filters do the opposite, transmitting shorter wavelengths and blocking longer ones.
- Polarizing allow light waves of a specific polarization to pass through while blocking waves of other polarizations. They are often used in imaging to reduce reflections and glare, as well as in 3D glasses and optical instruments.
- Notch filters block a specific, narrow band of wavelengths and allow all others to pass. These are used in applications where it is necessary to eliminate a single wavelength, like laser systems or fluorescence imaging.

References

1. G. Rossi, A new intensity modulation based fiber optic probe for bubble shape detection, velocity and diameter measurements. Rev. Sci. Instrum. J. **67**(7), 2541–2544 (1996). Ed American Institute of Physics, Argonne, IL, USA
2. G. Giusfredi, *Phisical Optics* (Springer, 2019)
3. M. Born, E. Wolf, *Principles of Optics*, 6th edn. (Pergamon Press, Oxford, 1980)
4. P. Zanuttigh, G. Marin, C. Dal Mutto, G.M. Cortelazzo, *Time-of-Flight and Structured Light Depth Cameras–Technology and Applications* (Springer, 2016)
5. M.J. Kidger, *Fundamental Optical Design* (SPIE Press, 2002)
6. C.S. Williams, O.A. Backlund, *Introduction to the Optical Transfer Function* (SPIE Press, 2002)

Chapter 3
Image Digitalization

Abstract Analog images represent the intensity distribution of electromagnetic waves within the visible spectrum, specifically wavelengths ranging from 0.4 to 0.7 μm, commonly referred to as light. These waves can also be interpreted as a flux of energy packets named photons. In the previous chapter, we explored how to select appropriate optical components to accurately project analog images onto a plane, for measurements purposes. To convert the incoming photon flux at this plane into electrical signals, and subsequently into a two-dimensional digital output (a numerical matrix), the most commonly used devices are photodiodes. These are typically arranged in arrays, such as Charge-Coupled Devices (CCD) or Complementary Metal–Oxide–Semiconductor (CMOS) sensors array. This chapter focuses on the processes of image sampling and quantization as performed by these sensor arrays. Furthermore, spatial Fourier transform and convolution operations applied to two-dimensional signals (i.e., the image matrices) are discussed. These mathematical tools are essential for determining the optimal sensor array parameters needed to achieve high-quality digitization of analog images for measurement purposes.

3.1 Light Intensity Conversion to Electrical Signals

As discussed in previous chapters, light consists of electromagnetic waves with wavelengths ranging approximately from 0.4 to 0.7 μm, corresponding to the visible spectrum, from red to violet. Light can also be described as a photon flux, denoted as N_f photons per second, where each photon is a pack of energy in the range of approximately 3–5 eV, as determined by Planck's law from the above wavelength range by using Eq. (1.79) in Chap. 1. Most sensors used to convert light into electrical signals operate based on the photoelectric effect, which occurs when photons interact with materials, primarily silicon for light sensors, producing a measurable output electric current.

When a photon of a light ray strikes a silicon atom, as illustrated in Fig. 3.1, it can transfer its energy to an electron. If the photon's energy exceeds the material's band gap, this interaction excites the electron, freeing it from its atomic bond. The

G. Rossi, *Mechanical and Thermal Measurements by Images and Waves*, Smart Sensors, Measurement and Instrumentation 51, https://doi.org/10.1007/978-3-032-03464-9_3

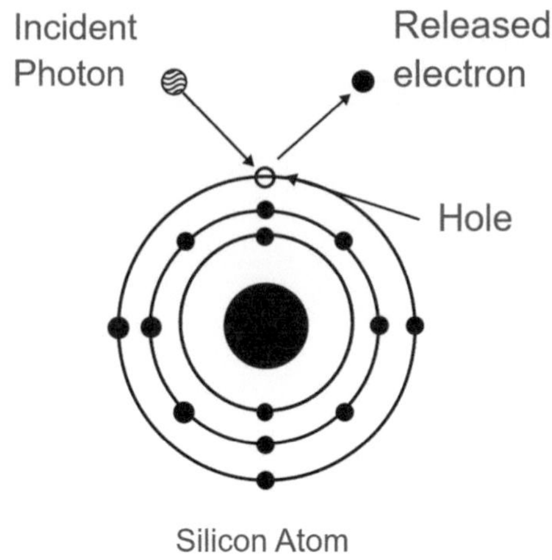

Fig. 3.1 Interaction of a photon with a silicon atom

result is the generation of a negatively charged free electron and a corresponding positively charged "hole" left behind in the silicon lattice. A continuous stream of incoming photons produces many such electron–hole pairs. If these charges are generated within an electric field, such as that created by a p–n junction in a semiconductor device, they can be separated and directed, resulting in a measurable electric current. In this way, incident light energy is converted into electrical energy. The core challenge in designing light sensors lies in efficiently collecting the photogenerated charge carriers and accurately measuring either the resulting current or the total charge accumulated over a defined period of light exposure. To achieve this, advanced and often complex structures made from silicon and other semiconductor materials have been developed to optimize charge collection, transport, and readout.

The most basic and widely used sensor for converting light into an electrical signal is the silicon photodiode, shown in Fig. 3.2. This device is built upon a standard p–n junction made from silicon.

The p-type region is created by doping silicon with elements such as boron, which introduces an excess of holes, positively charged carriers. Conversely, the n-type region is doped with elements like phosphorus, which contribute additional electrons, negatively charged carriers.

At the interface between these two regions, a depletion region forms. This area is devoid of free charge carriers and is characterized by a built-in electric field arising from the imbalance of charges. It is this internal electric field that enables the photodiode to separate and collect the electron–hole pairs generated when photons are absorbed, making it a fundamental component in light detection and optical sensing technologies.

When light (photons) strikes the surface of a silicon photodiode, its energy is absorbed, provided that the photon's energy exceeds the bandgap of silicon (approximately 1.1 eV). This absorption excites electrons from the valence band to the conduction band, resulting in the generation of electron–hole pairs.

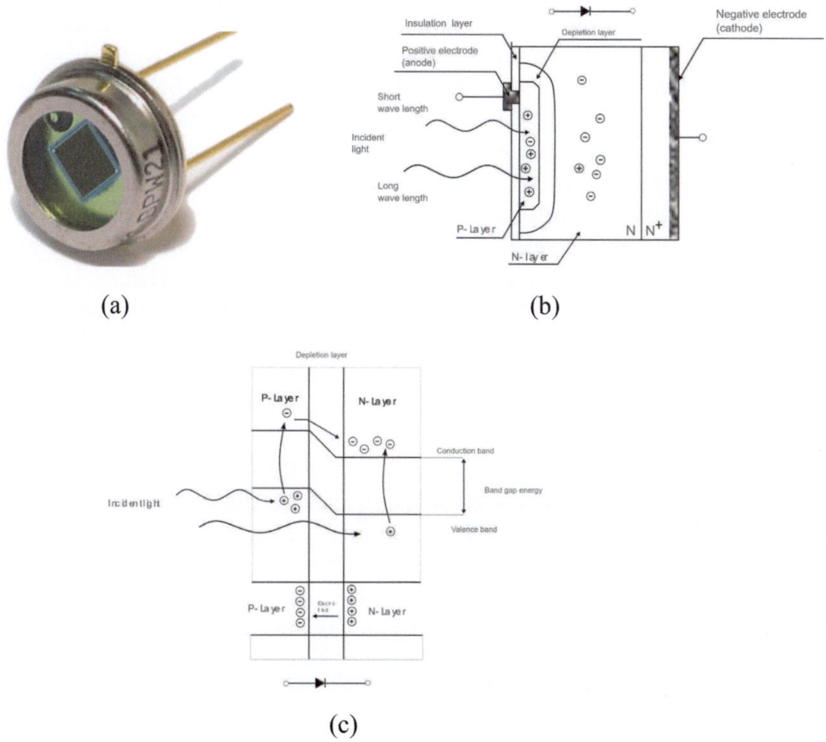

Fig. 3.2 Photodiodes picture (**a**), its structure (**b**) and its working principle (**c**)

These charge carriers are primarily generated within the depletion region, where a built-in electric field is present due to the p–n junction, as shown in Fig. 3.2. This electric field plays a crucial role: it rapidly separates the electrons and holes before they can recombine. Electrons are driven toward the n-type region, while holes move toward the p-type region, establishing a net movement of charge that gives rise to an electric current. The magnitude of this current is proportional to the intensity of the incident light: higher light intensity generates more electron–hole pairs, thus producing a stronger current. This current flows through the external circuit via the photodiode's electrodes and, when amplified, enables the photodiode to function effectively as a light detector.

Photodiodes are typically operated under reverse bias conditions, which widen the depletion region. This enhances sensitivity to light, reduces junction capacitance, and improves the device's response speed. In reverse bias, the diode does not conduct unless light is present. The electric field across the depletion region remains strong, further promoting charge separation and minimizing recombination, thereby maximizing photocurrent generation.

A more advanced variant of the photodiode is the avalanche photodiode (APD). These devices operate under a higher reverse voltage that causes internal gain through the process of impact ionization. The multiplication factor, and hence the gain, is proportional to the applied voltage. APDs offer higher sensitivity, faster response times, lower dark current, and a better signal-to-noise ratio (SNR) than standard photodiodes. They typically exhibit a spectral response in the range of 0.2–1.15 μm.

For the detection of extremely low light levels, Single-Photon Avalanche Diodes (SPADs) are employed. These are specialized APDs that operate in Geiger mode, wherein a single photon can trigger a self-sustaining avalanche breakdown, producing an output of up to a million electrons from a single photon event. SPADs are highly sensitive and ideal for applications requiring single-photon detection.

To extend the range of spectral sensitivity, alternative semiconductor materials are used. For example, InGaAs (Indium Gallium Arsenide) provides enhanced sensitivity in the near-infrared region and is well-suited for high-speed, low-light detection. InSb (Indium Antimonide) is employed for sensing in the mid-infrared range, particularly between 3 and 5 μm.

From an electrical standpoint, a photodiode can be modeled as a current source I_L (representing the photocurrent generated by incident photons) in parallel with an ideal diode. This model also includes a junction capacitance C_j, a shunt resistance R_{sh}, and a series resistance R_s, as illustrated in Fig. 3.3. The output voltage V_o and current I_o are typically measured across a load resistor R_L or an amplifier connected to the photodiode's output, enabling signal processing and readout.

Main characteristics of photodiodes are:

1. *Sensitivity*: It is the ratio between output sensor current I_o and collected total light power P on its sensitive surface area,

$$S = I_o/P \tag{3.1}$$

so it depends also on the wavelength of the incident light. Silicon photodiodes are most sensitive to light in the visible and near-infrared (NIR) spectrum, with a sensitivity curve $S(\lambda)$ typically ranging from 400 to 1100 nm.

2. *Photosensitive area*: Is the area A of light collection of the photodiode. Output current of a photodiode increase with area for a given light intensity I = P/A

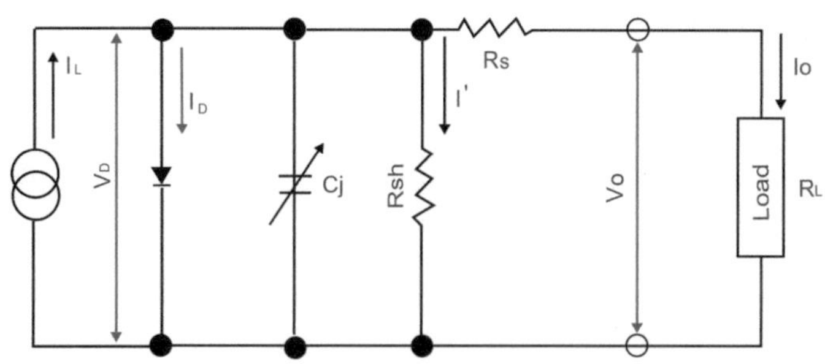

Fig. 3.3 Electrical parameters of the equivalent circuit of a photodiode

because $I_o = S\,P = S\,I\,A$. So larger photodiodes give a better (higher) output current signal for a given light intensity.

3. *Spectral Range*: This refers to the range of light wavelengths to which the photodiode is sensitive. It is typically represented as a sensitivity versus wavelength curve, illustrating how the device responds across different parts of the spectrum. In photodiode datasheets, the spectral range is often specified using three key wavelength values: the minimum and maximum wavelengths at which the device remains responsive, and the peak wavelength where the photodiode exhibits maximum sensitivity.
4. *Bandwidth* is typically specified in datasheets as the –3 dB bandwidth (BW), which corresponds to the frequency at which the photodiode's response falls to 70.7% (or –3 dB) of its peak value. For silicon photodiodes, this bandwidth often exceeds 300 MHz, which is sufficient for most high-speed imaging and optical communication applications. The bandwidth of a photodiode defines the range of optical signal modulation frequencies over which the device can accurately respond. It represents the photodiode's ability to follow rapid changes in illumination and is primarily influenced by its junction capacitance C_j and the load resistance R_L, as shown in the equivalent circuit model in Fig. 3.3.

   While the photoelectric conversion process, the generation of electron-hole pairs by absorbed photons, occurs almost instantaneously, the diffusion and transit times of charge carriers through the semiconductor structure introduce inherent delays. These delays limit how quickly the photodiode can respond to variations in the optical signal level, thereby reducing the effective bandwidth.

   In general, larger-area photodiodes exhibit higher junction capacitance, which increases the RC time constant of the equivalent circuit reducing the bandwidth. This creates a fundamental trade-off: larger photodiodes having a high sensitivity produce stronger electrical signals due to increased light collection but at the cost of slower response times and lower bandwidth.

   The bandwidth also governs the temporal response of the photodiode to sudden changes in light intensity, for example, transitioning from darkness to illumination. The rise time t_d, defined as the time it takes the output signal to increase from 10% to 90% of its final value, can be estimated using simplified semi-empirical relations derived from the RC time constant and carrier transit dynamics as:

$$t_d = 0,35/B_W \tag{3.2}$$

5. *Linearity*: Is the deviation from straight line of the relation between light intensity and output current, typically given in percentage of maximum output. Silicon photodiodes have a very good linearity.
6. *Quantum Efficiency*: Is the ratio of the number of electron–hole pairs generated with the number of incident photons. It can be given in percentage by the relation

$$Q_E = 100 \times (1240\,S\,/\lambda) \tag{3.3}$$

where S is the sensitivity in A/W and λ the wavelength in nm.

7. Dark current refers to a small leakage current that flows through the photodiode even in complete darkness, in the absence of incident light. It is primarily caused by thermally generated charge carriers within the depletion region, as well as other intrinsic noise sources in the semiconductor material.

   Minimizing dark current is critical for enhancing the signal-to-noise ratio (SNR) of the photodiode, especially in low-light detection applications. A high dark current can obscure weak optical signals, reducing the accuracy and sensitivity of the device. In some cases, dark current is represented in terms of Noise Equivalent Power (NEP), a measure of the minimum detectable optical power that yields a signal equal to the noise level. If I_r is the dark current and S is the sensitivity of the photodiode (typically expressed in A/W), the NEP can be approximated as $NEP = I_r/S$. Therefore, photodiode signal to noise ratio will be:

$$SNR = I_o / I_r = I_o / (NEP \times S). \tag{3.4}$$

   Above equation can be futher detailed [2] considering that $I_o = P\,S\,M$, where M is the photodiode intrinsic gain and $I_r = (2eB_W((P+P_B)S+i_r)FM^2 + (4kTB_W)/R_L)^{1/2}$ where F is the detector excess noise factor, P_B the background light optical power, R_L the load output resistance, e the electron electric charge, k the Boltzmann constant and T the temperature. So the equation:

$$SNR = \frac{PSM}{\sqrt{2eB_w((P + P_B)S + I_r)FM^2 + (4kTB_w/R_L)}} \tag{3.5}$$

   Illustrate the main parameters affecting the signal to noise ratio of a photodiode.

3.2 CCD and CMOS Image Sensors

Charge-Coupled Device (CCD) image sensors were invented in 1970 by Willard Boyle and George E. Smith at AT&T Bell Laboratories. Their groundbreaking work in the development of this technology earned them the Nobel Prize in Physics in 2009.

At their core, CCDs are arrays of photodiodes arranged in a grid, with each photodiode coupled via charge transfer, which gives rise to the name *Charge-Coupled Device*. Each photodiode serves as the light-sensitive element of a matrix cell, commonly referred to as a pixel. When exposed to light, each pixel converts incident photons into stored electrical charge, electrons, as illustrated in Fig. 3.4.

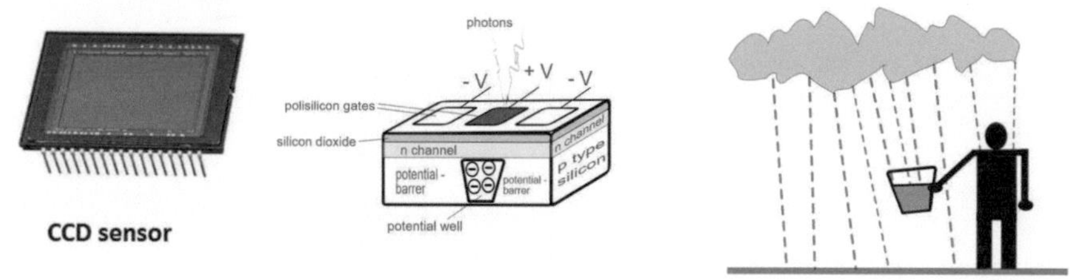

Fig. 3.4 Working principle of digital image sensor photodiode arrays: The rain analogy

Unlike simple photodiodes that produce a continuous photocurrent, CCD photodiodes act more like buckets collecting raindrops. An applied positive potential on a light transparent conductive gate above the silicon surface forms a potential well that traps and holds the photo-generated electrons within the silicon substrate, the pixel "bucket". These electrons remain stored in place until they are read out sequentially via the CCD's charge-transfer mechanism.

Assuming an exposure time Δt and a photon flux N_f (in photons per second), the incident optical power is given by $P = N_f \cdot h \cdot f$, where h is Planck's constant and f is the frequency of the incident light. This optical energy is converted into electrical charge, resulting in a total accumulated charge Q within each pixel, which corresponds to the light intensity received during the exposure time, quantified as:

$$Q = e\, N_f Dt\, Q_E, \tag{3.6}$$

where Q_E is the previously defined *quantum efficiency* (the fraction of electrons generated for each arriving photon) and e is the charge of the electron. So, a N_f photon flux generates an electron flux equal to $Q_e\, N_f$, and in an exposure time Δt a total number of electrons given by:

$$N_e = Q_e N_f Dt \tag{3.7}$$

where the negative part of the charge is $Q = e\, N_e$. In terms of light intensity, $I = P/A$, where P is the light power arriving on the photodetector area A, we can also say that if the single photodiode of the matrix is a little square of dimension d, so of area d^2, the number of electrons generated on it in Δt seconds of exposure will be:

$$N_e = Q_e I\, d^2 Dt. \tag{3.8}$$

The positive voltage, initially applied to retain electrons within each pixel "basket," can also be used, by sequentially varying its value across adjacent pixels, to transfer the stored electrons from one pixel to the next, as illustrated in Fig. 3.5a. This controlled movement of charge is the fundamental principle behind charge transfer in CCD image sensors.

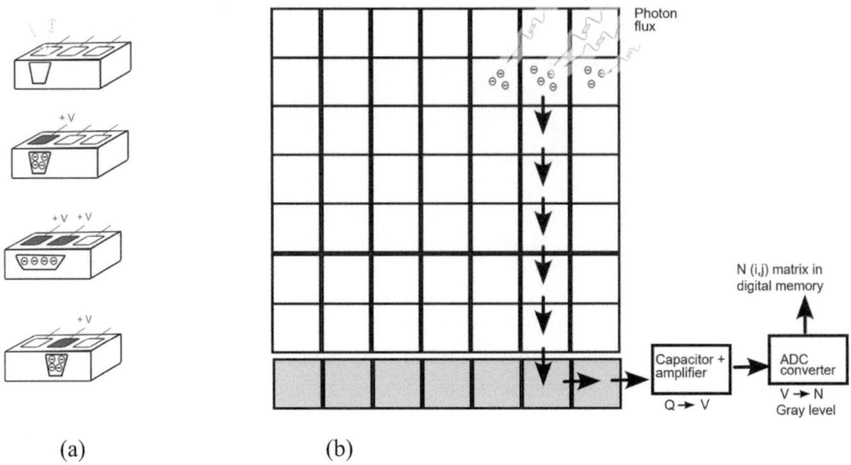

Fig. 3.5 Transferring electrons in a CCD array by sequentially modulating potentials across adjacent regions

To read out the light-induced charge accumulated in each pixel of a CCD image sensor, various charge transfer strategies can be employed. In a full-frame CCD, the entire pixel matrix is exposed to light, and after exposure, the rows of pixels are sequentially shifted downward into a dedicated readout register, a horizontal row shielded from further light exposure. Once a row enters this register, the charge in each pixel is read out sequentially: it is first converted into a voltage, then digitized by an analog-to-digital converter (ADC), as shown in Fig. 3.5b. The resulting digital values are stored in memory as a matrix, thereby reconstructing the captured image by this matrix discretization.

This readout process takes time, and during this interval, it is crucial that the image sensor is not exposed to additional light, as this could result in smearing or image artifacts. Light exposure can be prevented either mechanically, closing a shutter, or electronically, by disabling the photosensitive function. However, the time needed to shift and digitize the entire pixel matrix poses a limitation for high-speed imaging, where rapid frame capture is required.

To address this constraint, alternative CCD architectures have been developed, including the frame-transfer CCD and the interline-transfer CCD:

- In a frame-transfer CCD (Fig. 3.6b), the entire image from the photoactive array is rapidly shifted into an adjacent storage array of the same size that is completely shielded from light. Once the transfer is complete, the photoactive array can immediately begin capturing a new image, while the stored data is sequentially read out and digitized. This parallelization significantly reduces the delay between

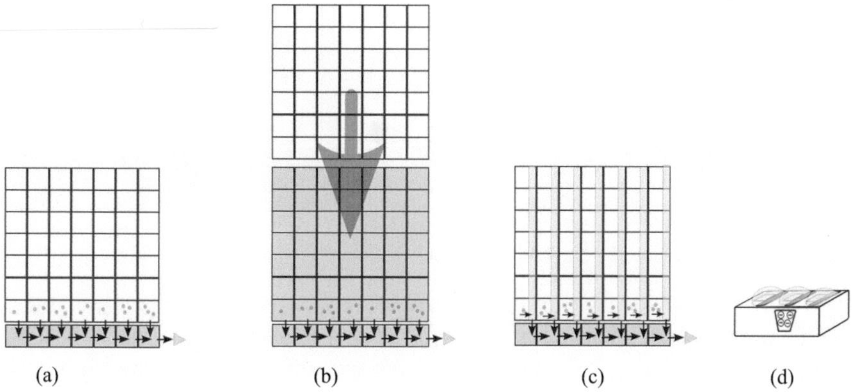

Fig. 3.6 Different techniques of pixel readout on CCD

exposures, improving overall frame rate. However, this design increases sensor complexity: either the sensor must be made physically larger to accommodate the storage array, or the individual pixels must be miniaturized, which may degrade image quality.

- The interline-transfer CCD (Fig. 3.6c) achieves similar performance improvements. In this architecture, each pixel includes a light-shielded vertical shift register adjacent to the photosensitive area. After the exposure ends, the accumulated charge is rapidly transferred to this masked region, which is then shifted vertically into the readout register. Like the frame-transfer CCD, this enables the photosensitive area to begin a new exposure immediately, even while the previous image is being digitized. The main trade-off, however, is reduced pixel sensitivity, since part of each pixel is dedicated to the masked transfer region, thereby decreasing the light-sensitive surface area.

To mitigate this reduction in light-sensitive surface area, microlens arrays can be integrated on top of the sensor (Fig. 3.6d). These tiny lenses focus incoming light onto the active portion of each pixel, effectively increasing the light-collection efficiency and enhancing the sensor's overall sensitivity and performance.

Each pixel in a CCD image sensor has a finite charge storage capacity, known as the full well capacity. This value is determined by the physical dimensions of the pixel and its design parameters, and it defines the maximum number of electrons the pixel can accumulate in response to incident photons.

When the incoming light intensity is too high, the resulting photon flux may generate more electrons than the pixel can hold. Once the full well capacity is exceeded, the pixel becomes saturated. At this point, any excess charge cannot be retained and may begin to spill over into adjacent pixels.

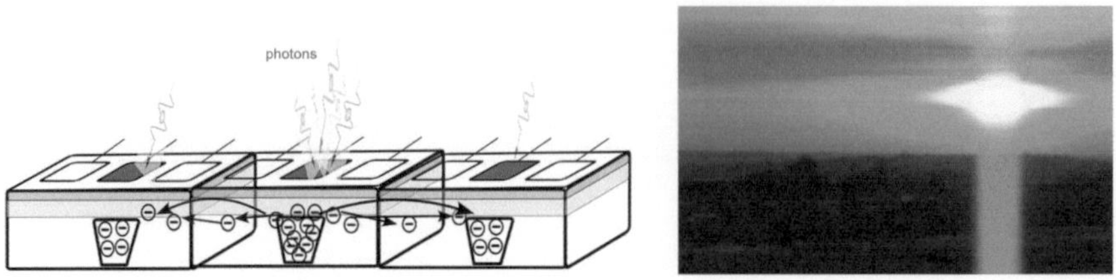

Fig. 3.7 Blooming caused by pixel saturation and electron diffusion over potential barriers

This overflow phenomenon is known as blooming, and it can cause noticeable image artifacts. Because CCDs transfer charge sequentially row by row during the readout process, excess charge from a saturated pixel typically spreads vertically, along the same column, leading to streaks or bright trails that extend from overexposed regions. As illustrated in Fig. 3.7, blooming results in vertical smearing or striping in the final image, significantly reducing image quality and obscuring image areas.

To mitigate the effects of blooming in CCD sensors, many designs incorporate anti-blooming mechanisms. One common approach involves the use of interline transfer drains integrated into the sensor matrix. These structures act as overflow channels, diverting excess electrons away from saturated pixel wells before they can spill into neighboring pixels. By draining surplus charge, these anti-blooming systems help preserve image quality and prevent vertical streaking caused by pixel saturation.

In recent years, Complementary Metal-Oxide Semiconductor (CMOS) technology has emerged as a widely adopted alternative for digital image sensors. Although CMOS technology dates back to the 1950s, well before the invention of CCDs, it was not until around 2009 that CMOS image sensors achieved the sensitivity and accuracy required for high-quality imaging applications.

Like CCDs, CMOS sensors use photodiodes to detect light and convert photons into electrons. However, their key innovation lies in the pixel-level architecture. In CMOS sensors, each pixel includes integrated electronic components, typically comprising at least three transistors, a configuration known as 3 T-CMOS, as illustrated in Fig. 3.8b. These transistors enable in-pixel charge-to-voltage conversion and local amplification, resulting in several notable advantages over traditional CCDs:

- Faster readout speeds, since there is no need to transfer charge across the entire array. Instead, each pixel's voltage can be accessed directly through electronic switching.
- Lower power consumption and reduced noise, owing to localized signal processing and less charge movement.
- Parallel readout and digitization, with one analog-to-digital converter (ADC) per column, dramatically increasing the speed of data acquisition compared to the serial readout architecture of CCDs.

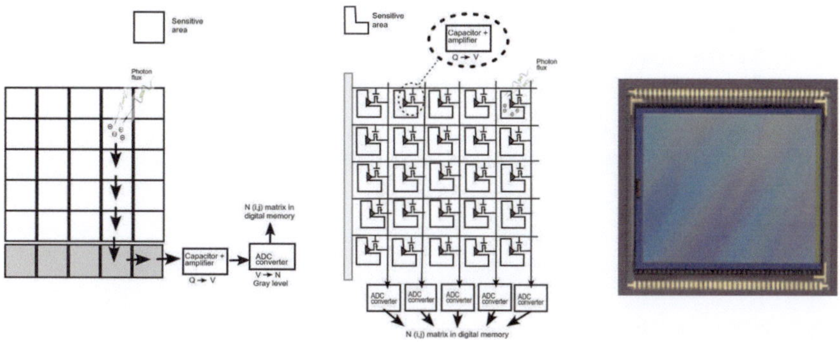

(a) CCD (b) CMOS (c) CMOS sensor picture

Fig. 3.8 CMOS sensors

Despite these benefits, CMOS sensors are inherently more complex due to the additional circuitry embedded within each pixel. These components occupy valuable surface area that would otherwise be photosensitive, leading to a lower fill factor. As a result, longer exposure times may be required in low-light environments, potentially offsetting some of CMOS's speed advantages.

Additionally, CMOS sensors typically operate with a rolling shutter mechanism, in which rows of pixels are exposed sequentially, from top to bottom, rather than all at once. While this enables high-speed imaging, it can also introduce motion artifacts (such as skew or wobble) in scenes with fast-moving objects.

By incorporating an additional capacitor and two extra transistors into each pixel, CMOS sensors can be designed to support a global shutter mechanism. In this architecture, the voltage representing the light-induced charge is temporarily stored locally within each pixel on the added capacitor. This allows the entire pixel array to capture light simultaneously, while the previously stored values are read out and digitized in parallel.

Unlike rolling shutter systems, which expose the sensor row by row, a global shutter enables all pixels to begin and end exposure at the same moment. This simultaneous exposure is critical for accurately capturing fast-moving subjects or high-speed scenes, where any temporal delay across rows could lead to distortions.

The difference between rolling and global shutter performance becomes especially apparent in dynamic environments. Rolling shutters often produce visual artifacts such as skew, wobble, or "jello" effects, as the scene may change between the exposure of the top and bottom rows. In contrast, a global shutter preserves the true geometry of moving objects by freezing the entire scene in a single instant. This distinction is illustrated in Fig. 3.9a, where the image captured with a rolling shutter exhibits noticeable distortion, while the global shutter image maintains spatial accuracy, as illustrated in Fig. 3.9b.

While some CMOS sensor technologies support global shutter operation, the majority of CMOS image sensors in widespread use today still rely on a rolling shutter

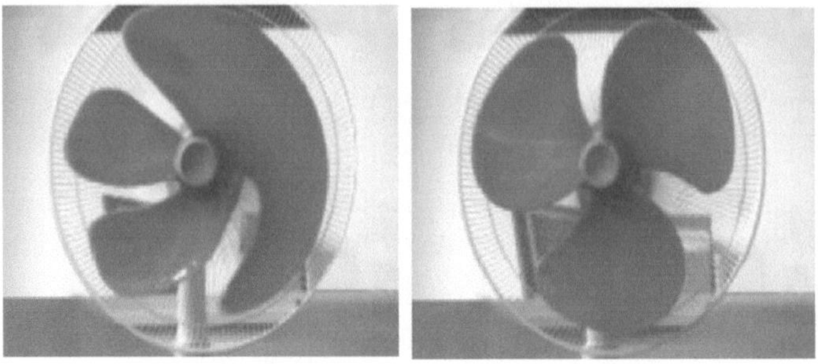

(a) (b)

Fig. 3.9 Effects of rolling (**a**) and global (**b**) shutter

mechanism. Rolling shutter offer several important advantages for applications that do not require high-speed imaging:

- Higher sensitivity, due to simpler and more light-efficient pixel structures.
- Faster frame rates, especially in high-resolution modes.
- Lower manufacturing costs, making them ideal for consumer-grade devices.
- Simpler pixel architecture, which contributes to compact design and energy efficiency.

In a rolling shutter system, the sensor captures the image row by row, sequentially from top to bottom. Although each row is read very quickly, typically within 10 microseconds, a full-frame image consisting of 1,000 rows would result in an overall temporal gap of around 10 ms between the first and last row. While this delay can cause temporal artifacts in high-speed scenarios, it is generally acceptable for standard imaging tasks, including photography, video capture, and general-purpose imaging.

Due to their combination of performance, cost-efficiency, and design flexibility, CMOS sensors have become the dominant technology in a wide range of applications, including smartphones, digital cameras, webcams, automotive imaging systems, and industrial vision equipment.

One of the additional strengths of CMOS architecture is its structural similarity to memory arrays: each pixel stores a voltage value that can be accessed directly. This capability enables the selective readout of specific parts of the sensor, known as Regions of Interest (ROI). By reading only a defined subregion of the image array, for example only a region that change between frames (a car moving) as in Fig. 3.10b, the system can significantly increase the readout speed and frame rate, while reducing the volume of data processed. This makes ROI readout particularly useful for real-time monitoring, object tracking, or targeted analysis in high-performance imaging applications.

(a) (b)

Fig. 3.10 Region of interest and its applications

A practical example of Region of Interest (ROI) readout can be found in traffic control cameras used for automatic license plate recognition. These systems focus exclusively on the portion of the image where the vehicle's license plate appears. By electronically addressing and reading only this relevant subregion, the system can achieve high-speed image capture and processing, enabling real-time identification without the need to read or store the entire image frame. This significantly reduces data load and improves system efficiency in time-critical applications.

One of the fundamental advantages of CMOS sensor architecture is its ability to individually address each pixel. The voltage corresponding to the light-induced charge is stored on a capacitor within the pixel and can be directly read out, without charge transfer through adjacent pixels, as required in CCD sensors. This pixel-level readout not only supports features like ROI selection but also inherently eliminates the risk of blooming, as excess charge does not propagate beyond its originating pixel (Fig. 3.11).

Compared to CCD sensors, CMOS sensors generally offer a larger full well capacity, meaning each pixel can store a greater number of photo-generated electrons. This higher capacity makes CMOS sensors less prone to saturation and blooming, enabling them to more effectively capture scenes with high dynamic range. As a result, a single frame can contain both bright and dark regions, with each pixel accurately recording varying light intensities without compromising image quality.

A major breakthrough in CMOS imaging technology occurred in 2016, when Photometrics introduced the first back-illuminated CMOS (BI-CMOS) sensor. This

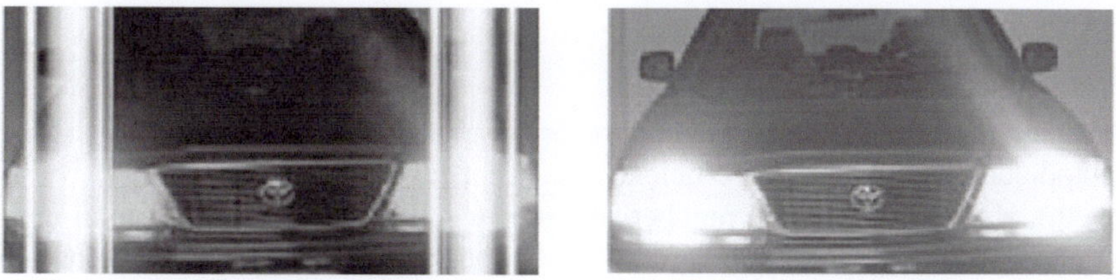

(a) (b)

Fig. 3.11 Saturation and blurring effects on **a** CCD and **b** CMOS sensors

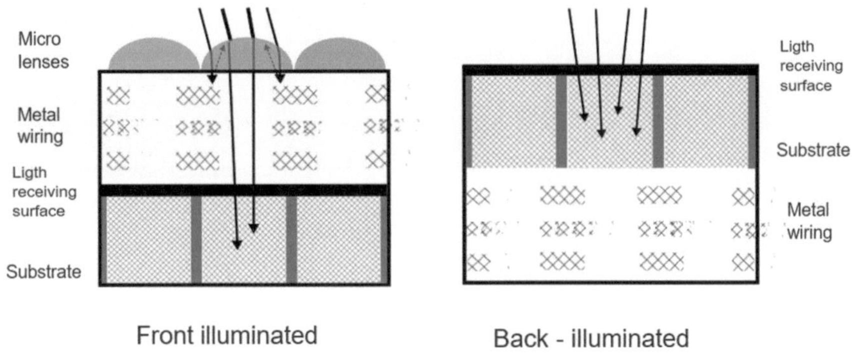

Fig. 3.12 Back illuminated CMOS sensors

innovation significantly improved light sensitivity while maintaining the key advantages of CMOS architecture, including high-speed readout, low power consumption, and random pixel access.

Traditional front-illuminated CMOS sensors have their electronic circuitry, including transistors and metal interconnects, positioned on the same side as the incident light. This design can partially block incoming photons, reducing the fill factor, or the proportion of each pixel that is sensitive to light.

In contrast, the back-illuminated CMOS design, illustrated in Fig. 3.12, reverses this architecture. Light enters from the rear side of the silicon wafer, reaching the photodiodes directly, without interference from front-side circuitry. This structural reversal dramatically increases the fill factor, up to 95% of the pixel area, allowing more photons to be captured.

The result is a substantial improvement in sensitivity, especially in low-light conditions or high-speed imaging scenarios where shorter exposure times are essential. BI-CMOS sensors effectively combine the superior image quality of CCDs with the speed and flexibility of CMOS, making them ideal for demanding applications such as astronomy, biomedical imaging, and scientific instrumentation.

3.3 Color, Linear and Polarized Sensors

A flux of photons with varying energies (or wavelengths) generates electrons in the photodiodes with a quantum efficiency that depends on the photon energy (wavelength). However, while the number of generated electrons reflects the total incident energy, the spectral information, i.e., the specific wavelengths of the incoming photons, is lost in the process.

In other words, each pixel's photodiode integrates the entire wavelength distribution of the incoming light over the exposure time, converting this cumulative energy into an accumulated global charge. The final sensor output is a grey level (or intensity value) that is proportional to the total charge, without distinguishing between colors or wavelengths.

To obtain spectral information, i.e. the intensity of different wavelengths, the most common approach is to place a bandpass filter in front of the sensor. This filter allows only a narrow range of wavelengths to reach the photodiodes, ensuring that the resulting charge corresponds specifically to that portion of the spectrum.

To produce a color image, it is necessary to acquire at least three grayscale images, each corresponding to a different wavelength band. This is typically achieved using three bandpass filters: one centered on the red region of the spectrum, one on green, and one on blue. Each filtered image captures the intensity (gray level) of light within its respective spectral band.

The resulting color image is reconstructed by performing a linear combination of the three primary color intensity levels, R (Red), G (Green), and B (Blue), using appropriate weighting factors k_1,k_2,k_3 The composite color intensity C can be expressed as:

$$C = k_1R + k_2G + k_3B$$

These weights can be tuned based on the desired color space or the sensor's response characteristics, enabling accurate color representation in the final image.

There are two main methods for implementing this triple-band filtering. One approach involves the use of beam-splitting optics illustrated in Cap 2, such as a dichroic mirror system, to divide the incoming image into three separate optical paths. Each path is then passed through a different bandpass filter (red, green, or blue), directing the filtered light onto three corresponding sensors. This configuration allows for the simultaneous capture of the red, green, and blue grayscale images, as illustrated in Fig. 3.13.

This method is often used in high-precision applications such as scientific imaging, microscopy, and broadcast-grade video cameras, where color accuracy and resolution are critical.

An alternative and widely used approach to capturing color images involves placing miniaturized color filters directly over each pixel in the sensor array, as shown in Fig. 3.14. In this configuration, individual pixels are made selectively sensitive to

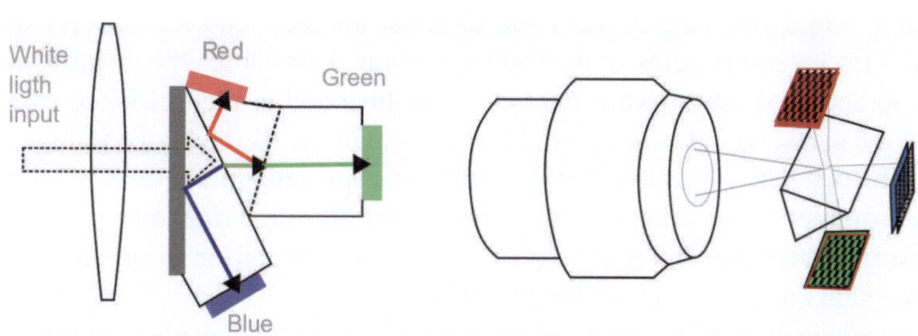

Fig. 3.13 Colour image digital generation by combining tree digital images acquired trough colour filters

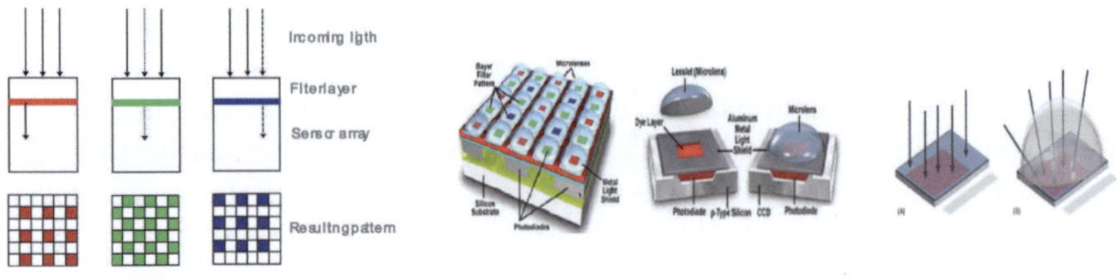

(a) (b) (c)

Fig. 3.14 Color image digital sensors with microlenses to improve fill factor

red, green, or blue light by overlaying them with corresponding microscopic band-pass filters. As a result, each pixel records only one color component of the incoming light. For example, some pixels detect only red light, others blue, and others green. This means that the sensor does not capture full color information at every pixel location, effectively reducing the native spatial resolution for each individual color channel.

To reconstruct a full-color image, image processing algorithms are used to perform demosaicing, which involves interpolating the missing color values at each pixel based on the known values of neighboring pixels. This process generates three complete color planes, red, green, and blue, by estimating the missing components for each pixel. The result is a full-resolution color image, albeit with some trade-offs in color fidelity and sharpness, depending on the quality of the interpolation algorithm.

Despite this limitation, the use of color filter arrays (CFAs), such as the Bayer filter pattern, remains the most common solution in consumer and professional digital cameras due to its simplicity, compactness, and cost-effectiveness and it is similar to what our eyes do, where the number of green sensors is about twice the number of red and blue sensors.

As previously noted, the spectral sensitivity of typical silicon-based sensing elements used in CCD and CMOS image sensors extends from approximately 400–1000 nm. In contrast, the human eye is sensitive to wavelengths in the range of 400–700 nm, with peak sensitivity in the yellow-green region, around 580 nm.

Because silicon sensors are also responsive to near-infrared wavelengths beyond the visible range, particularly from 700 to 1000 nm, they tend to overrepresent the red and infrared components of the light spectrum. This causes the resulting digital images to appear shifted toward red tones, as illustrated in Fig. 3.15a, compared to the way the scene would be perceived by the human eye as illustrated in Fig. 3.15b.

This mismatch can lead to color inaccuracies, especially in scenes where infrared light is prominent. To correct for this, infrared-cut filters are often placed in front of the sensor to block unwanted wavelengths and bring the sensor's response closer to the perceptual colour balance of human vision.

For specialized imaging applications, multispectral cameras are employed using a similar principle as conventional colour cameras, but instead of three channels (red,

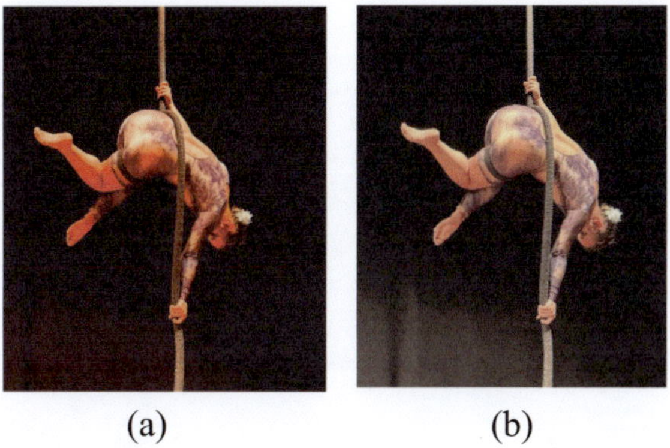

Fig. 3.15 Different color sensitivity of silicon digitizing arrays (**a**) and human eyes (**b**)

green, and blue), they utilize more than three image splits or multiple bandpass filters to capture a broader range of spectral information. These advanced imaging systems are designed to record light across a wide portion of the electromagnetic spectrum, providing not just a single intensity value per pixel, but a spectral distribution that describes how the light is composed at different wavelengths.

Unlike standard digital cameras, which are limited to the visible spectrum (approximately 400–700 nm), multispectral cameras can detect non-visible bands such as near-infrared (NIR), short-wave infrared (SWIR), and ultraviolet (UV). They acquire image data across multiple discrete spectral bands, typically ranging from 3 to 12 bands, effectively producing 3–12 grayscale values per pixel. More advanced systems can capture dozens of bands, enabling much richer spectral detail.

At the next level of spectral imaging, hyperspectral cameras go further by capturing a full high-resolution spectrum for each pixel. A typical hyperspectral system might cover the 400–1000 nm spectral range, using up to 200 narrow bands, which corresponds to a spectral resolution of about 3 nm. The result is a data cube (also called a hyperspectral cube), where each frame represents one wavelength band, and each pixel contains a detailed spectral signature.

Leading companies in this field include SPECIM (www.specim.com), JAI (www.jai.com), and Spectral Devices (www.spectraldevices.com).

Applications of multispectral and hyperspectral cameras are in agriculture for assessing crop health, soil properties, and plant stress; environmental monitoring for detecting pollutants, mapping plant species, and monitoring land cover; remote sensing and aerial imaging on drones and satellites for land surveying, disaster assessment, and climate studies; for biomedical and tissue research performing non-invasive diagnostics and tissue analysis; for art conservation and cultural heritage to reveal hidden layers in paintings, manuscripts, and artifacts without physical contact; for security and surveillance by performing material detection, facial recognition, and enhanced scene analysis.

Despite their powerful capabilities, these systems come with certain limitations such as high cost, due to complex optics and sensors; large data volumes, requiring substantial storage capacity and processing power; complex calibration,

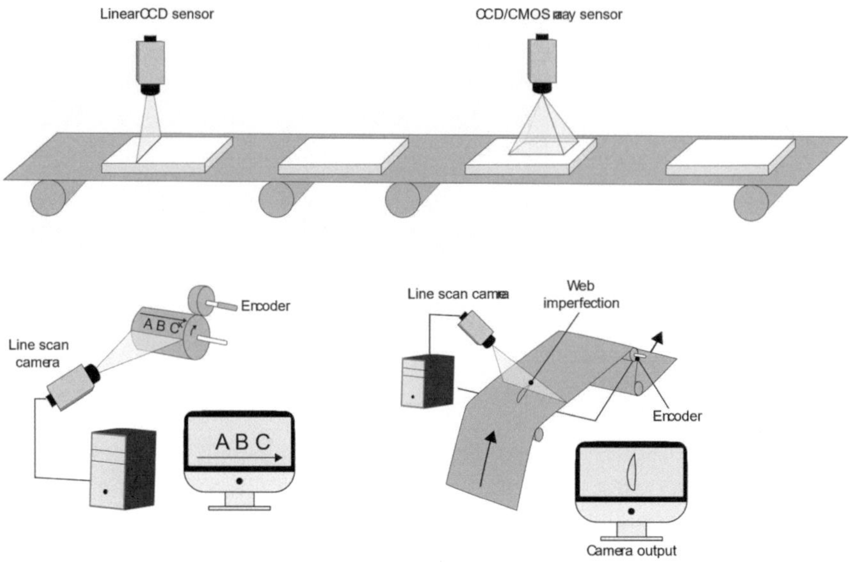

Fig. 3.16 CCD and CMOS linear arrays and their applications

as accurate spectral data often demands precise environmental adjustments and application-specific tuning.

Camera image sensors can be implemented not only as two-dimensional matrices, but also as linear arrays, consisting of a single line or a few lines of pixels, typically two or three. These are known as linear array image sensors, as illustrated in Fig. 3.16.

Linear arrays are particularly useful for applications involving moving objects, where the image is built up line by line as the object passes in front of the sensor or as the sensor itself moves across the scene. This approach allows for high-resolution imaging with lower hardware complexity and reduced costs compared to full-frame matrix sensors. By capturing data one line at a time, linear sensors can also achieve faster readout speeds and are well-suited for industrial inspection, document scanning, and high-speed imaging systems.

Instead of placing colour bandpass filters over each pixel in a digitizing sensor array, it is also possible to use polarizing filters, optical elements that transmit only the component of light oscillating in a specific direction. These filters allow the sensor to measure the polarization state of incoming light.

Typically, a group of four adjacent pixels is used, each covered with a polarizing filter oriented at a different angle: 0, 45, 90, and 135°, as shown in Fig. 3.17. This arrangement enables the sensor to capture light intensity values corresponding to each polarization direction: I_0, I_{45}, I_{90}, and I_{135}.

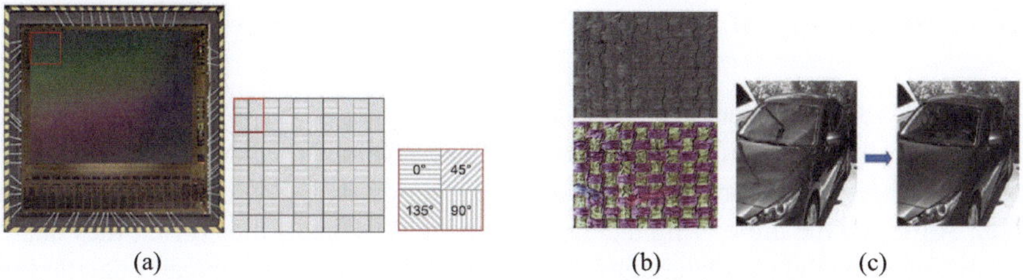

Fig. 3.17 Polarizing detection arrays and their possible applications

By combining these four intensity measurements, it is possible to analyze the polarization state of the incident light and calculate parameters such as the polarization angle and degree of polarization. This type of cameras (Fig. 3.17), known as a polarization-sensitive cameras, are used in applications such as material inspection, surface characterization, biomedical imaging, and enhanced object detection under challenging lighting conditions. Production companies of polarization cameras are Sony, IDS, Teledyne-Flir, Photron, Baumer, etc.

The polarization angle ϕ can be calculated by the relation:

$$\phi = \frac{1}{2} \cdot \arctan\left(\frac{I_{45} - I_{135}}{I_0 - I_{90}}\right) \tag{3.9}$$

And the linearly polarized fraction LP% of the incident light is given by:

$$LP\% = \frac{\sqrt{(I_0 - I_{90})^2 + (I_{45} - I_{135})^2}}{(I_0 + I_{90})} \tag{3.10}$$

Polarized light cameras have a wide range of practical applications, particularly in scenarios where traditional imaging systems struggle due to glare or reflective surfaces. One common use is in seeing through glass or other reflective materials. These surfaces tend to reflect light that is polarized in a specific direction. By capturing only the component of light polarized perpendicularly to the reflected glare, polarized cameras can effectively suppress reflections, revealing details hidden behind the reflective surface, as illustrated in Fig. 3.17c.

Another valuable application is the inspection of fiber-reinforced materials as illustrated in Fig. 3.17b. These materials often reflect polarized light aligned with the orientation of the embedded fibers. By analyzing images that represent the polarization angle of reflected light, it becomes possible to visualize fiber patterns, identify irregularities, and detect misalignments, which may signal structural defects or weaknesses in the material. This technique enhances quality control in manufacturing processes, particularly in aerospace, automotive, and composite materials industries. Photoelasticity applications of polarized light cameras for deformation measurements will be described in details in Chap. 7.

3.4 Parameters of Digital Imaging Sensors Array

The image sensors discussed above generate outputs in the form of vectors or, more generally, matrices composed of numerical values (i.e., gray levels). These values are directly proportional to the electric charge accumulated or voltage at each pixel. Although the underlying technologies of these sensors may differ, they share several fundamental characteristics. These common attributes, typically detailed in the datasheets of digital camera image sensors, include the following:

(a) *Pixel dimension.* The size of individual pixels plays a crucial role in determining the resolution and performance characteristics of an image sensor. Smaller pixels enable higher spatial resolution within a fixed sensor length L or height H, allowing for the capture of finer image detail. However, this increase in resolution often comes at the expense of other performance factors, such as sensitivity and full well capacity. Full well capacity refers to the maximum number of electrons (or the total charge) a pixel can store before reaching saturation. It is commonly illustrated by analogy to the volume of a bucket collecting a "rain" of photons. This capacity depends directly on the pixel's physical dimensions and the camera's operating voltage. Larger pixels typically have greater full well capacities, making them less susceptible to saturation and more effective at capturing a wider dynamic range, especially under high illumination conditions. Conversely, they result in lower resolution due to a reduced number of intensity measurements per unit area. State-of-the-art sensors now feature pixel sizes as small as 1 μm, while more conventional design commonly use pixel sizes ranging from 4 to 20 μm. Fig. 3.18 illustrates the typical relationship between pixel size and full well capacity for square pixels. Selecting the appropriate pixel dimension thus requires balancing the trade-off between spatial resolution and other sensor characteristics such as sensitivity, dynamic range given by full well capacity, and signal-to-noise ratio.

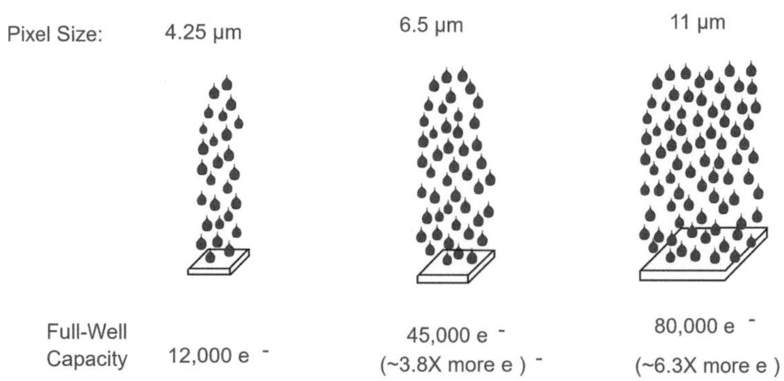

Fig. 3.18 Typical pixel size and full well capacities

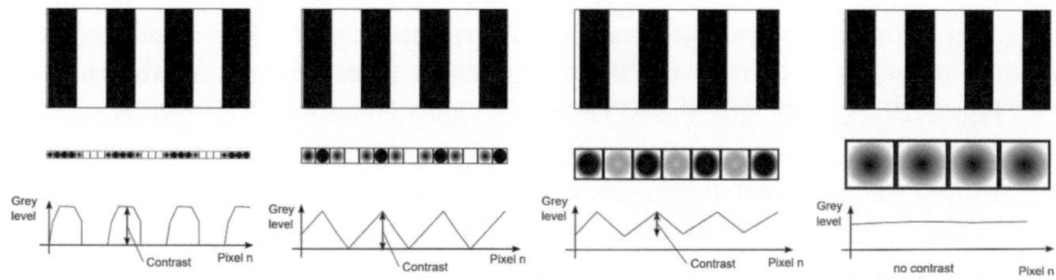

Fig. 3.19 Pixel dimension and loss of contrast when reference line width on image plane is equal to pixel dimension

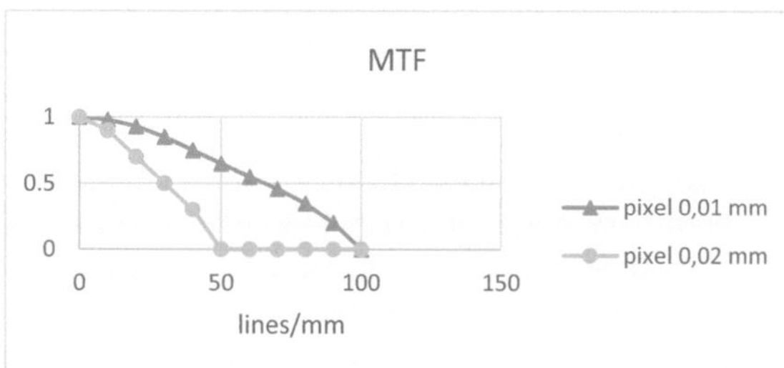

Fig. 3.20 Pixel dimension and digital modulation transfer function

Pixel size plays a critical role in determining the digital Modulation Transfer Function (dMTF) of an image sensor, a concept that parallels the discussion of MTF in analogue optical systems presented in Chap. 2, where the resolution limiting factor was the diffraction-defined spot size. The dMTF quantifies the ability of the image sensor to preserve contrast in the digitized images and is typically evaluated using high-contrast patterns such as alternating black and white stripes (as illustrated in Fig. 3.19). The resulting contrast is measured from the range of grey levels captured in the digital output image.

The sensor's ability to reproduce contrast accurately is governed by the relationship between pixel size and the spatial frequency of the input pattern on the image plane, in the real image spatial frequency is scaled by the magnification m parameter defined in previous Chap. 2. Spatial frequency is commonly expressed as the number of line pairs per millimeter or, equivalently, as line width. Each pixel integrates the light falling within its area, often encompassing both black and white regions, resulting in an averaged grey value. This averaging effect introduces blur, analogous to the blurring introduced by diffraction in analogic optical systems.

Maximum contrast is achieved only when individual pixels are fully covered by a single tone, either black or white, in the analogue image. As shown in Fig. 3.19, when the line spacing of the black and white stripes matches the pixel size on the image plane, each pixel receives an equal portion of black and white light. This results in uniform grey pixels across the image and leads to a complete loss of contrast in the digitized output. In such cases, the original line pattern becomes indistinguishable, effectively erased by the spatial averaging effect imposed by the pixel structure.

It is therefore straightforward to understand that the MTF associated with the pixel dimensions of the digitizing sensor array behaves as illustrated in Fig. 3.20. Specifically, the MTF falls to zero when the pixel size equals, or exceeds, the spacing between adjacent lines in the analogue image. In other words, when the spatial frequency of the image in the image plane matches the sampling frequency of the pixel array, the sensor is no longer able to distinguish the alternating pattern, resulting in complete contrast loss.

For example, consider a pixel size of d = 0.01 mm. If the image projected onto the sensor image plane contains a pattern with 100 line pairs per millimeter, each line has a width of 1/100 = 0.01 mm, identical to the pixel size. In this configuration, each pixel integrates equal portions of black and white areas, leading to a uniform grey output with zero contrast. As a result, the line pattern is entirely lost in the digitized image, resulting in a constant grey level.

(b) *Dynamic range* of a sensor is defined as the ratio between the highest and lowest light intensities it can reliably detect. It quantifies the sensor's ability to capture both very bright and very low light intensity signals within the same image. Specifically, the maximum measurable intensity level, L_{max}, corresponds to the threshold beyond which the sensor saturates, while the minimum detectable intensity, L_{min}, lies just above the dark noise level, representing the lowest intensity level signal distinguishable from background noise.

A higher dynamic range indicates the sensor's capability to handle a wider range of brightness levels, enabling it to capture detail in both highlight and dark regions. This characteristic is crucial in applications requiring high image fidelity under challenging lighting conditions. The dynamic range D_R is typically expressed on a logarithmic scale as:

$$D_R = 20\log\left(\frac{L_{max}}{L_{min}}\right)dB \tag{3.11}$$

Pixels are typically engineered so that their dynamic range closely matches the full input range of the analog-to-digital (AD) converter. This range is defined by the ratio between the maximum and minimum representable grey levels, which is determined by the converter's bit depth. Specifically, for an AD converter with a resolution of n bits, the total number of grey levels is 2^n, and thus the dynamic range can be expressed as (Lmax/Lmin) = 2^n. This alignment ensures that the sensor makes full use of the available digital representation, optimizing contrast resolution and minimizing quantization errors.

The following table presents typical dynamic range values for common medium level imaging sensors, including the human eye, which, notably, remains unmatched in its ability to perceive an exceptionally wide range of brightness levels (Table 3.1). Recently some high level performances scientific cameras approach the dynamic range of human eye.

(c) *Resolution* is defined by the total number of pixels arranged in its matrix. For a sensor with N rows and M columns, the total number of pixels is N × M. This parameter directly influences the level of detail that can be captured in an

Table 3.1 Dynamic range of medium level image sensors

Sensor	$B_{max} : B_{min}$	dB
Human eye	1,000,000:1	120
HDR display	200,000:1	106
Digital camera	4096:1	72.2
Film camera	2948:1	66.2
Digital video	45:1	33.1

image. Figure 3.21 illustrates the dramatic increase in sensor resolution over the past 25 years. In the early stages of digital imaging technology, around 1996, image sensors typically have 256 rows times 512 columns, so 0,13 megapixels. By 2013, sensor resolutions had advanced to 16 megapixels. Today, that trend continues, with cutting-edge sensors surpassing 50 megapixels and continuing to evolve, driven by demands for higher image quality and detail in fields ranging from consumer electronics to scientific imaging.

(d) *Fill factor* is defined as the ratio between the total light-sensitive area of all the pixels and the overall surface area of the sensor. In practical terms, it represents how much of the sensor area is actively used for light detection. Due

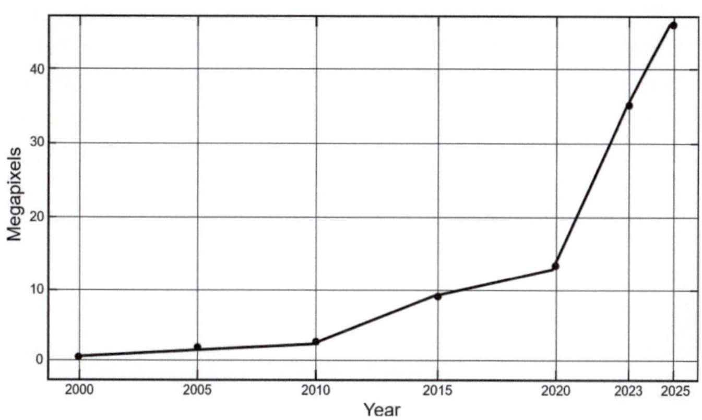

Fig. 3.21 Improvement of sensor array resolution in last 25 years

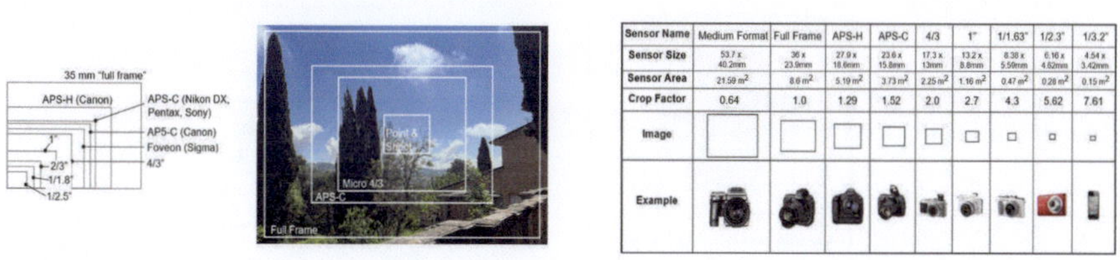

(a) (b) (c)

Fig. 3.22 Image sensor typical dimensions

to the presence of integrated electronics, signal transmission lines, and other circuit elements within each pixel, not all of the pixel area can be dedicated to capturing light. As a result, the fill factor is often less than 100%. To improve the fill factor, microlenses are commonly employed. These tiny lenses are placed over each pixel to focus incoming light onto the active photosensitive region, effectively compensating for the reduced sensitive area and enhancing overall light collection efficiency.

(e) *Sensor matrix dimensions* are its length L and height H, resulting in a total sensor area of L × H. Larger sensors generally offer superior image quality, as they are capable of capturing more light and are often used with larger, high-aperture lenses that enhance optical performance. However, increased sensor size also leads to higher manufacturing costs and more expensive optics. Common sensor formats are shown in Fig. 3.22, beginning with the widely recognized full-frame format of 36 × 23.9 mm, which is derived from the traditional 35 mm photographic film standard.

Ideally, the sensor's diagonal dimension $D = \sqrt{(L^2 + H^2)}$ should match the diameter of the lens's image circle. If the sensor diagonal is smaller than the image circle, a portion of the optical field is unused, effectively cropping the image. Conversely, if the sensor is larger than the image circle, vignetting may occur, resulting in a loss of the corners of the image, as previously discussed.

An additional key parameter in camera image sensor design is the aspect ratio, defined as the ratio of the height H to the width L of the sensor's active area where the image is formed. The most common aspect ratios include 4:3, frequently used in photography and industrial imaging, and 16:9, the standard for most video formats.

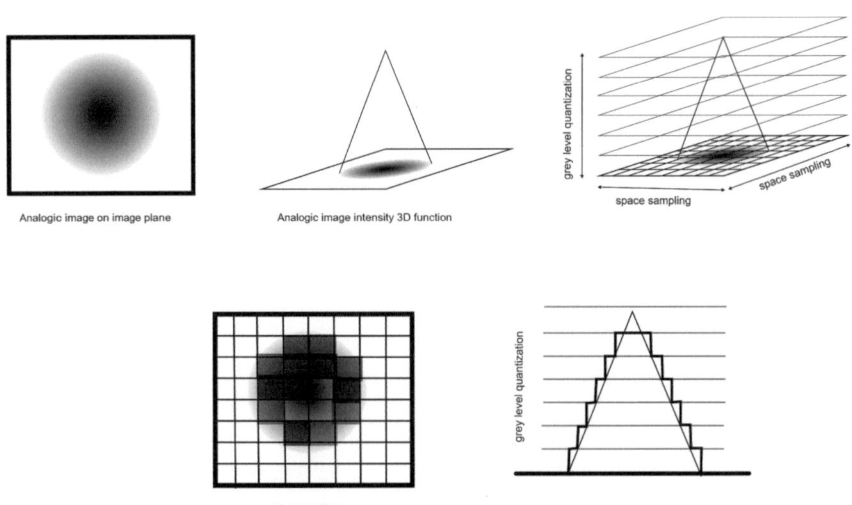

Fig. 3.23 Sampling and quantization operated by a digital image sensor

The combined effect of resolution, sensor size, and aspect ratio plays a crucial role in determining image quality. While increasing pixel count can theoretically enhance resolution, true improvements are constrained by the optical diffraction limit, a fundamental analogue limitation discussed in the previous chapter. Once pixel dimensions become smaller than this limit, further miniaturization does not yield higher effective resolution, as the system cannot resolve details finer than what diffraction allows in the image plane.

To achieve meaningful resolution gains, it is necessary to project a larger analogue image onto the image plane and capture it using a larger sensor. This enables the use of more pixels that remain above the diffraction threshold, ensuring that each pixel contributes to actual image detail rather than oversampling blurred information.

The crop factor is another important concept, defined as the ratio between the diagonal of a full-frame sensor (36 × 23.9 mm) and the diagonal of the sensor in use. Sensors smaller than full-frame exhibit a crop factor greater than 1, meaning they capture a narrower field of view with the same lens, resulting in an image that appears magnified, as illustrated in Fig. 3.22b. The crop factor significantly influences lens selection, as it affects framing, effective focal length, and depth of field, making it a critical consideration in both photography and imaging system design.

(f) *Quantization resolution.* Image digitalization consists of two fundamental processes: spatial sampling and intensity quantization. Spatial sampling involves dividing the continuous image plane into discrete elements, pixels, as determined by the finite dimensions of the sensor's pixel matrix, as previously discussed. This process defines the spatial resolution of the image and determines how finely image details can be captured.

Intensity quantization, on the other hand, refers to the conversion of the continuous range of light intensities into a finite set of discrete digital values.

These values are typically represented as binary integers, and their precision is determined by the bit depth of the sensor's analog-to-digital converter (AD). A common AD configuration is 8-bit, which enables $2^8 = 256$ distinct grey levels, often referred to as the "256 gray scale". However, modern imaging systems increasingly utilize higher bit depths, such as 12-bit (4,096 levels) or 14-bit (16,384 levels), which allow for greater dynamic range, finer gradations in brightness, and improved image fidelity.

This enhanced quantization is particularly important in demanding fields such as medical imaging, astronomy, professional photography and measurement purposes by images where subtle variations in light intensity carry critical information.

Figure 3.23 visually illustrates these two key concepts:

1. Bidimensional spatial sampling, representing the division of the image into a spatial grid of pixels.
2. Intensity level quantization, depicting how each pixel's captured light intensity is mapped to a discrete grey level number in the digital domain.

(g) *Frame rate* is the number of complete image frames (i.e., full matrices of pixel data) that a camera can capture per second. It is typically expressed in frames per second (fps) and is a key performance metric in both video and high-speed imaging applications.

The following factors influence the achievable frame rate.

- Pixel sensitivity: Pixels with lower sensitivity require longer exposure times to accumulate sufficient charge, which can limit the maximum frame rate, especially under low-light conditions. So for high speed imaging intense lighting is essential, if possible.
- Charge or voltage transfer speed: The speed at which the collected charge or voltage from each pixel is read out and transferred to the analog-to-digital (AD) converters directly affects the sensor's throughput.
- AD converter performance: Both the conversion speed and the number of parallel AD converters play a crucial role. Faster converters and parallelized architectures allow for quicker digitization of pixel data, thereby supporting higher frame rates.

In essence, the maximum frame rate is determined by the interaction between pixel exposure requirements, readout speed, and digitization efficiency. Optimizing these elements is essential for balancing frame rate and image quality in demanding imaging systems.

Figure 3.24 illustrates the current state-of-the-art in high-speed, high-resolution imaging systems, showcasing the best combinations of frame rate and resolution achieved in modern cameras.

(h) *Linearity* is given by the amount of deviation of input–output relation of pixels from a straight line. The same previously defined for photodiodes.

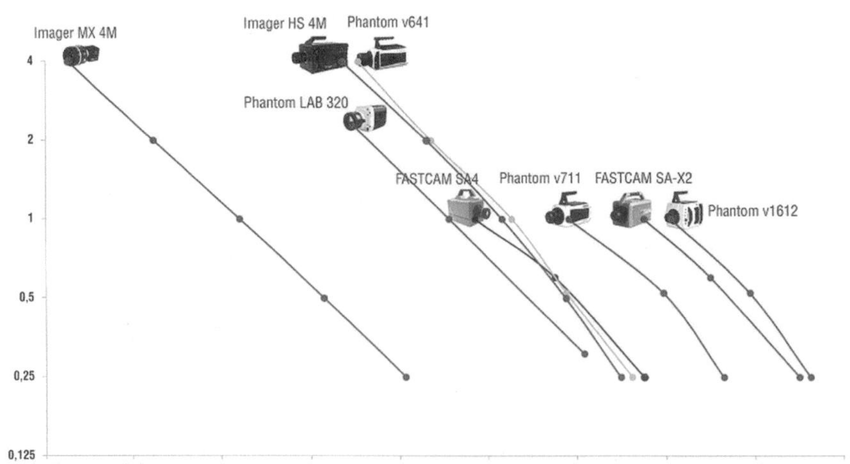

Fig. 3.24 Resolution and frame rate of high speed cameras Mpixel Vs Frame rate in KHz, From a LaVision Inc company depliant

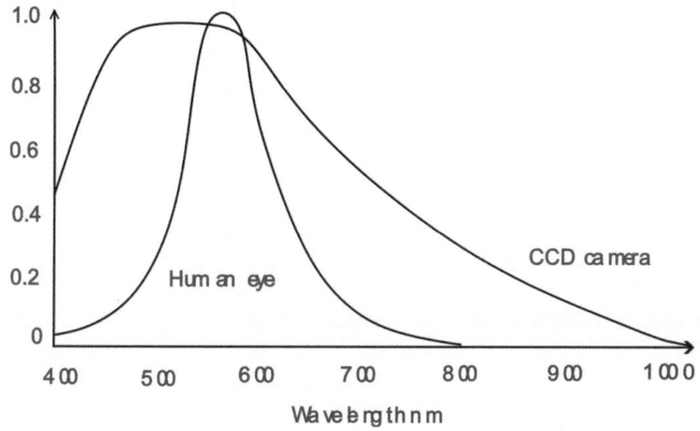

Fig. 3.25 Relative sensitivity of a typical camera with silicon sensors versus the human eye

(i) *Quantum Efficiency* is the ratio of the number of electron–hole pairs generated with the number of incident photons. The same previously defined for photodiodes.
(j) *Sensitivity* is the same characteristic previously defined for a single photodiode. Is the ratio between output charge or voltage generated by the pixel, or output number given by AD converter, and input intensity of light.
(k) *Spectral Range* defines the range of light wavelengths to which the pixels are sensitive. It characterizes the sensor's ability to detect and respond to different parts of the electromagnetic spectrum. This range is typically represented as a sensitivity curve, showing the relative pixel response as a function of wavelength, as illustrated in Fig. 3.25.

The sensitivity of pixels arises from quantum efficiency $\mathrm{Q_E}(\lambda)$, that is the ability to convert a stream of photons, a stream of light or a light intensity into a stream of electrons output. A photon flux $\mathrm{p}(\lambda)$ will generate as output an electron flux $I(\lambda)$ on each individual pixel given by:

$$\mathrm{I}(\lambda) = \mathrm{Q_E}(\lambda)\mathrm{p}(\lambda) \tag{3.12}$$

And the total electron flow I will be the sum of the contribution at each wavelength given by the integral:

$$I = \int_0^\infty Q_E(\lambda)p(\lambda)d\lambda \tag{3.13}$$

From the total electron flow I, for a pixel dimension equal to d (supposed square) and using an exposure time Δt, it is possible to define the brightness of the image B as

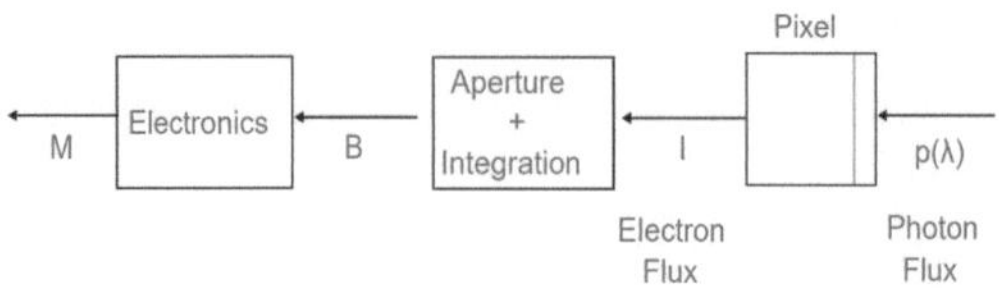

(a)

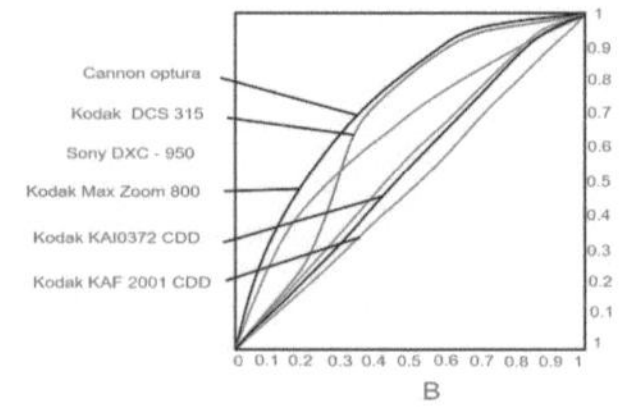

(b)

Fig. 3.26 Photon flux conversion into measured digital brightness levels

$$B = I\, d^2 \Delta t \tag{3.14}$$

The electronics of the camera convert B values to the output grey level digital number M, i.e. the measured digital brightness. The whole process is illustrated in Fig 3.26.

Typical curves relating B and M for some common kinds of cameras are illustrated in Fig 3.26b, and most of all are nonlinear. Assuming that the response of the camera is linear, to increase the brightness of the scene it is possible to combine multiple frames taken of the same image (M_1, M_2, ... M_n) with different exposures e_1, e_2 ... e_n and thus obtain an overall combined image M_f as:

$$M_f = \sum_1^n M_i \tag{3.15}$$

This technique is widely employed in smartphone photography to enhance image quality. The camera captures a series of images ($M_1, M_2, \ldots M_n$) at different exposure levels (e.g., e_1, e_2, ...), and then sum them all into a single image. As illustrated in Figure 3.27, this process effectively balances lighting across the scene, preserving detail in both highlights and shadows. The result is a final image with enhanced contrast and detail, particularly beneficial in higher dynamic range (HDR) lighting conditions where there is a wide variation in brightness.

This technique allows for an effective improvement in dynamic range by a factor of n, where n is the number of images captured at different exposure levels. However, the resulting image is averaged and does not represent directly the actual grey level values present in the original scene. As such, it is generally not suitable for quantitative measurements where accurate pixel intensity grey level values are required.

Moreover, if the subject or camera moves between exposures, the alignment of frames can be compromised, potentially leading to geometric distortions or artifacts. While motion compensation algorithms exists and can help mitigate these issues, they do not typically produce results with sufficient accuracy for metrological applications. Anyway in some conditions it can be useful.

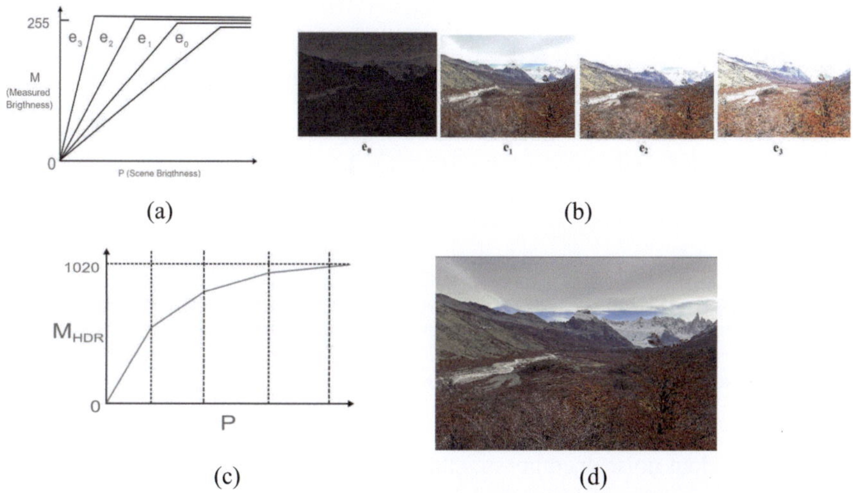

Fig. 3.27 Combining multiple exposures images (**b**) into one (**d**) by summing

(l) *Noise.* Noise in image sensors arises from a variety of complex physical and electronic phenomena. In general, noise is defined as any unwanted variation in the signal that occurs during the processes of image acquisition, conversion, transmission, or processing, ultimately degrading image quality. Noise sources in image sensors can be broadly classified into the following categories:

- Photon shot noise arises from the inherently random and discrete nature of photon arrival, an effect rooted in the quantum properties of light. Even under steady illumination, photons do not arrive at perfectly regular intervals but rather follow a statistical distribution. As illustrated in Fig. 3.28, consider a light source that produces, on average, 3 photons per second. In practice, this might mean that 6 photons arrive during the first second, 3 in the second, and none in the third. Although the average photon rate is $(6 + 3 + 0)/3 = 3$ photons/second, the number fluctuates from one moment to the next. These fluctuations are governed by a Poisson distribution, which describes the probability of a given number of photon arrivals in a fixed time interval.

  For average photon arrival rates exceeding approximately 10 photons per second, the Poisson distribution can be closely approximated by a Gaussian (normal) distribution, simplifying analysis.

  Notably, photon shot noise is the only type of noise that depends directly on the scene itself, as it is linked to the intensity of incident light. Higher light levels generate more photons, and thus greater absolute variation, though the relative noise may decrease. This phenomenon introduces a fundamental limit to image precision, especially under low-light conditions, where photon counts (and thus signal-to-noise ratios) are low.

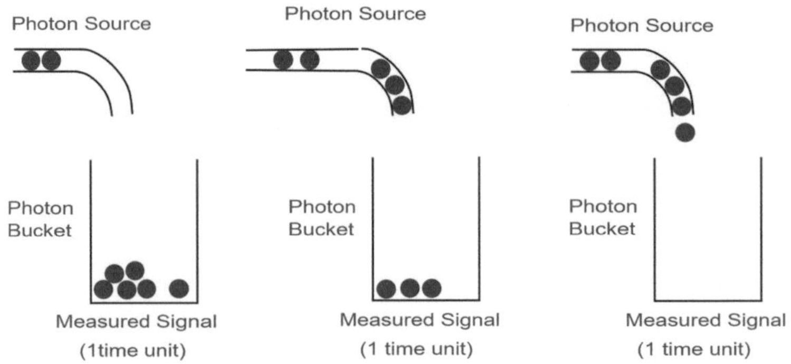

Fig. 3.28 Noise generated by discrete photon flow

Readout noise encompasses the errors introduced during the conversion of the optical signal into digital values, and it is independent of the image intensity, unlike photon shot noise. It can be broadly divided into two main components:

Electronic noise (Fig. 3.29a): This originates from the sensor's internal analog circuitry, such as amplifiers, transmission lines, and other components involved in signal readout prior to analog-to-digital (AD) conversion. It includes thermal noise, flicker noise, and other electronic interferences inherent to semiconductor devices.

Quantization noise (Fig. 3.29b): This results from the discretization process during analog-to-digital conversion. When a continuous analog signal is mapped onto discrete digital levels, a resolution error is introduced. The magnitude of this error is determined by the quantization step, denoted as $\Delta = 1/2^n$, where n is the number of bits of the AD converter. A higher bit depth reduces Δ, thereby minimizing quantization noise.

Readout noise is typically modeled as Gaussian, due to the central limit behaviour of the various underlying random processes.

- Even when the sensor shutter is closed and no photons are reaching the sensor, a measurable noise component known as dark current noise still exists. As previously discussed in relation to photodiodes, this noise is caused by thermally generated electrons within the depletion region of the sensor's semiconductor material. These thermally induced charge carriers accumulate over time, mimicking a signal even in complete darkness. Dark current noise (Fig. 3.29c) is fundamentally linked to the electronic properties of the sensor and increases with temperature and exposure duration. It represents a baseline noise level that must be accounted for in low-light or long-exposure applications, where its influence can become significant.
- Another form of noise arises from defective pixels (Fig. 3.29d), which are present to some extent in virtually all sensor arrays. These pixels may consistently report incorrect values, such as being stuck at a fixed intensity (hot or

(a) Electrical noise (b) quantization noise (c) random thermal (d) defective pixels

Fig. 3.29 Different kind of noise

dead pixels) or showing abnormal response behavior. However, this type of noise is considered a fixed-pattern noise, meaning it remains spatially consistent across images. As such, it can be characterized and corrected through calibration procedures. Once identified, the defective pixel pattern can be subtracted or replaced using interpolation techniques to minimize its impact on the final image.

3.5 Image Fourier Analysis

Jean Baptiste Joseph Fourier is the mathematician who probably gave the greatest impetus to technical culture between the eighteenth and nineteenth centuries. Fourier analysis and the concepts related to it are still applied in all branches of science and technology.

The basic concept he proposes is very simple: Any periodic function can be considered as a weighted sum of infinite sinusoids at different frequencies. A periodic function is a function, defined from $-\infty$ to $+\infty$, which repeats the same after a certain length λ, called spatial period. This length λ is the minimum repetition length, since clearly the function repeats even after 2λ, 3λ, …If every periodic function can be reproduced as a linear combination of sines and cosines, the sine function acquires considerable importance. It is the simplest periodic function and for Fourier it is the basic building block that constitutes all periodic functions.

For a light intensity I(x) changing periodically along x coordinate, we can write for this basic building block the expression:

$$\mathrm{I(x)} = \mathrm{I_A} \sin\,(\omega\,\mathrm{x}\,) \tag{3.16}$$

where $\mathrm{I_A}$ is the maximum positive or negative light intensity, $\omega = 2\pi/\lambda$ is the angular frequency and $\mathrm{u} = 1/\lambda$ is the spatial frequency, as illustrated in Fig. 3.30.

As previously mentioned, any periodic wave, such as a square wave, can be reconstructed by summing a series of sine and cosine waves: the first having a frequency equal to the inverse of the square wave's spatial period (the fundamental spatial frequency), the second at twice the frequency with a reduced amplitude, the

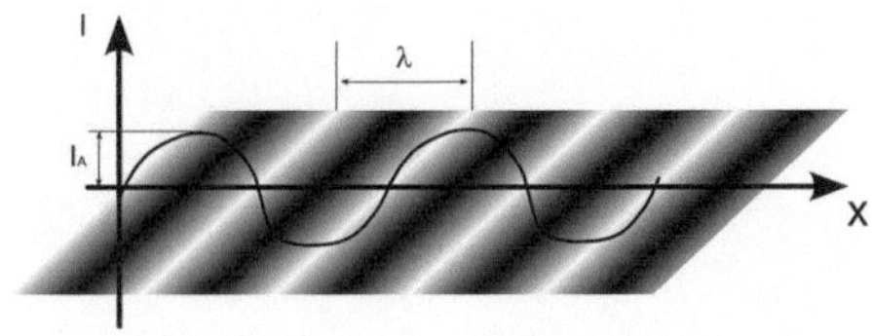

Fig. 3.30 Basic building block for images reconstruction by Fourier analysis

third at three times the frequency with even lower amplitude, and so on, progressively adding sine and cosine waves of higher frequencies and diminishing amplitudes. This process is illustrated in Fig. 3.31.

For one-dimensional images, represented as grey-level intensity distributions I(x) along the x-axis, i.e., a grey-level profile, similar to the waveforms discussed in Chap. 1, Fourier theory states that we can reconstruct the complete grey-level distribution I(x) by summing a series of sinusoidal components. These components are sine and cosine functions weighted appropriately, allowing us to express I(x) as a sum of sinusoidal functions using the general form:

$$I(x) = a_0 + \sum_{n=1}^{\infty} (a_n \cos(\omega_n x) + b_n \sin(\omega_n x)) \tag{3.17}$$

Or using the Eulero relation $e^{i\theta} = \cos\theta + i\,\mathrm{sen}\theta$, in the equivalent exponential complex form:

$$I(x) = \sum_{-\infty}^{\infty} X_n e^{i\omega_n x} \tag{3.18}$$

where $\omega_n = 2\pi n/\lambda; f_n = \omega_n/2\pi$ and coefficients a_0, a_n, b_ne X_n may be calculated by the integrals:

$$\begin{aligned} a_0 &= \frac{1}{\lambda}\int_0^{\lambda} I(x)\cdot dx \; a_n = \frac{2}{\lambda}\int_0^{\lambda} I(x)\cos\omega_n x\cdot dx \\ X_n &= \frac{1}{\lambda}\int_0^{\lambda} I(x)\cdot e^{-i\omega_n x}\cdot dx \; b_n = \frac{2}{\lambda}\int_0^{\lambda} I(x)\sin\omega_n x\cdot dx \end{aligned} \tag{3.19}$$

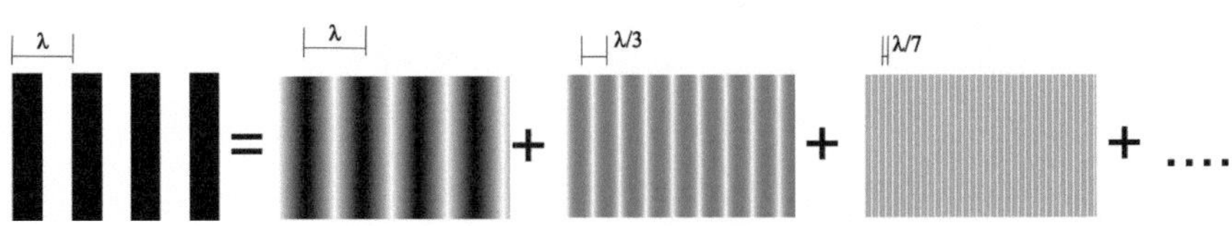

Fig. 3.31 Reconstruction of an image of strips by summing sine image functions of decreasing spatial period and intensity

Fourier series can be seen as a vector of complex numbers $[a_n, b_n]$, each one for a certain spatial frequency of the nth sine or cosine function. Each of this complex number can also be expressed by its modulus $M_n = \left(a_n^2 + b_n^2\right)^{1/2}$ and phase $\varphi_n = \text{arctg}\,(b_n/a_n)$, so by the complex vector $[M_n, \varphi_n]$.

From the vector of complex coefficients $[a_n, b_n]$, or equivalently $[M_n, \varphi_n]$, one complex number for each spatial frequency $f_n = n/\lambda$ for $n = 1, 2, 3$ …., it is possible to reconstruct the image I(x) using the Fourier series expression provided above. In essence, having the function I(x) or its corresponding vector of complex coefficients conveys the same information. This vector is known as the *image spectrum:* a set of complex numbers that represents the image in the frequency domain. It indicates the amplitude and phase of each sinusoidal component (i.e., each spatial frequency) required to reconstruct the grey-level distribution of the mono-dimensional image along the spatial axis.

This set of sinusoidal components, e.g., how much of 2 cycles/mm, how much of 7 cycles/mm, and so on is referred to as the *frequency content* of the image. For periodic images, this spectrum consists of a discrete set of complex numbers corresponding to harmonics at multiple spatial frequencies $f_n = n/\lambda$, with $n = 1, 2, 3,\ldots$

What above said above is valid for periodic images, but it is possible to generalize to any kind of image using integrals (continuous sums of continuous frequencies) instead of simple sums of discrete frequencies, defining the so-called Fourier transformation pair, given by the two integrals:

$$F(u) = \int_{-\infty}^{+\infty} I(x)e^{-i2\pi ux}dx \tag{3.20}$$

$$I(x) = \int_{-\infty}^{+\infty} F(u)e^{i2\pi ux}du \tag{3.21}$$

The direct Fourier transform integral, which maps the spatial domain x to the spatial frequency domain u, yields a complex-valued continuous function F(u). This function has both a real part $R_e(u)$ and an imaginary part $I_m(u)$, or equivalently, a magnitude M(u) and a phase $\phi(u)$. As previously stated, these components are related by the identities $M^2 = {R_e}^2 + {I_m}^2$ and $\tan\phi = I_m/R_e$. Therefore, the Fourier transform can be expressed in the following form:

$$F(u) = R_e(u) + i\,I_m(u) \tag{3.22}$$

And make plots of it by the two plots of $R_e(u)$ and $I_m(u)$ or plotting magnitude M(u) and phase $\phi(u)$. Fourier transform integrals needs to sum positive and negative frequencies and image grey levels, in general.

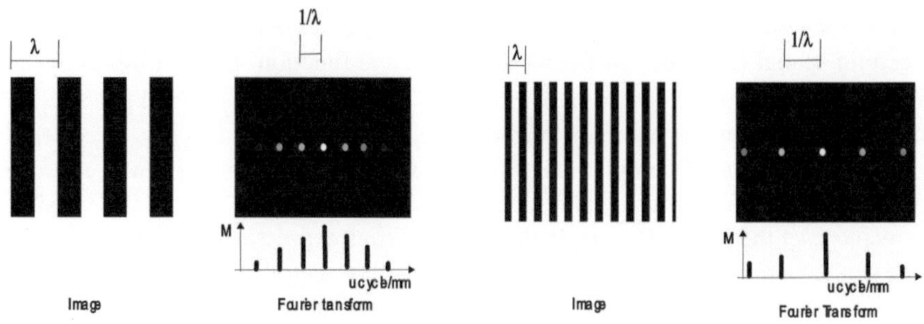

Fig. 3.32 Fourier transformations of two strip images

Many software, also many freeware, are available on internet to calculate direct and inverse Fourier transformation. If we use one of it (https://turbomaze.github.io/JS-Fourier-Image-Analysis/) for two strip images, we can get the results illustrated in Fig. 3.32 in terms of M(u) Fourier transform amplitude.

It is possible to see that Fourier transformations of these images gives only discrete points spaced in frequency domain u of n/λ, this because images are periodic functions and that a reciprocal relation exists between space domain x and spatial frequency domain u, i.e. strip spaced by λ mm, gives spot in frequency domain spaced by $1/\lambda$ cycle/mm. So very narrow spaced strip of 0.01 mm gives a high spatial frequency of 100 cycle/mm of the sine series of n × 100 spatial frequencies we need to sum, from n = 1 to N, to reconstruct it, starting from its spectrum.

Taking as grey level distribution I(x) a sinusoidal function as

$$\mathrm{I(x)} = \mathrm{I_A} \sin\left(2\pi(1/\lambda)\,\mathrm{x}\right) \tag{3.23}$$

and calculating the Fourier transform by the integral (3.20) the result obtained can be expressed in terms of the Dirac delta function, a function defined to be 1 only for u = 0 and 0 at any other point. so $\delta(\mathrm{u} + 1/\lambda)$ is 0 only when u = $-1/\lambda$ and $\delta(\mathrm{u} - 1/\lambda)$ is 0 only when u = $1/\lambda$. Using this delta function we can express the result of the integration as:

$$\mathrm{F(u)} = \tfrac{1}{2}\,\mathrm{I_A i}\left[\delta\left(\mathrm{u} + (1/\lambda)\right) + \delta\left(\mathrm{u} - (1/\lambda)\right)\right] \tag{3.24}$$

Therefore, the result consists of two vertical lines with amplitudes equal to that of the sinusoid, symmetrically positioned with respect to the origin. This is illustrated in both amplitude and phase in Fig. 3.33.

Negative frequency has no physical meaning, as negative frequencies do not exist in a physical sense. However, it holds mathematical significance because functions in both the spatial and frequency domains are defined over the entire range from $-\infty$ to $+\infty$. Consequently, negative frequencies are essential for the inverse Fourier transform to reconstruct the spatial domain signal from its frequency domain representation.

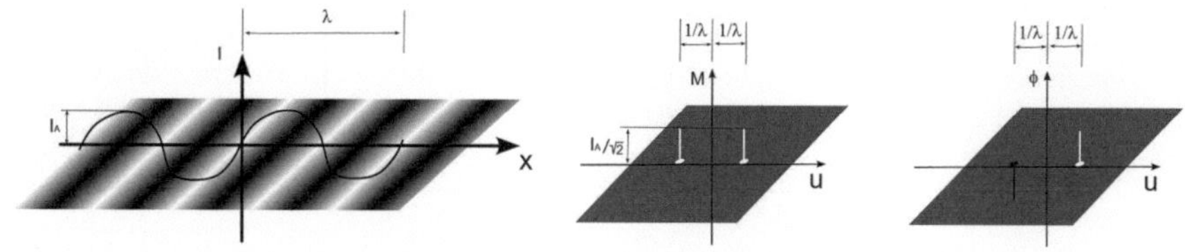

Fig. 3.33 Fourier transform of a sine grey level distribution

If we compute the Fourier transforms of single rectangular lines of varying widths, as shown in Fig. 3.34, each with width L and amplitude 1, the result of the Fourier integral will be the continuous sinc function, given by:

$$F(u) = L\left[\frac{\sin(\pi uL)}{(\pi uL)}\right] = \text{Lsinc}(\pi uL) \tag{3.25}$$

Grey level distribution of this function is illustrated in Fig. 3.34, that goes to 0 in two symmetric points at spatial frequencies coordinates $-\pi / L + \pi / L$ so having a width $2\pi/\text{L}$, proportional to the reciprocal of rectangular grey level width L.

We have therefore once again a reciprocal relation between spatial length dimensions and spatial frequency domain dimensions. i.e. thin strips have a wide Fourier transformation, so they need more frequencies to be represented, and thick strip have thin Fourier transformation, so they need less frequencies to be represented. The frequency content is in this case continuous because the image is non periodic.

If we take a constant grey level distribution such as I(x) = 1, making the Fourier integral (3.20) to obtain its transform in frequency domain we obtain no frequencies because the result is a delta function F(u) = δ(u) giving only a pulse at zero frequency, as in Fig. 3.35a. If we do the contrary making the Fourier transformation of a little, infinitesimal dot, expressed analytically by a delta function I(x) = δ(x) we get the result F(u) = 1, so very little dots contain all the frequencies with equal amplitudes, as illustrated in Fig. 3.35b.

If we take a smooth strip having a gaussian grey level distribution given by:

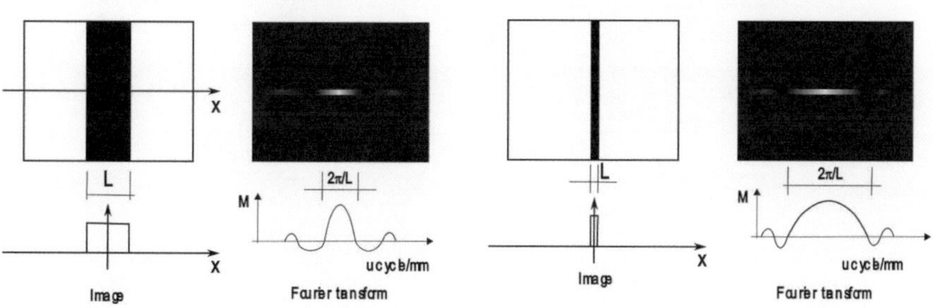

Fig. 3.34 Fourier transformations of two single line images

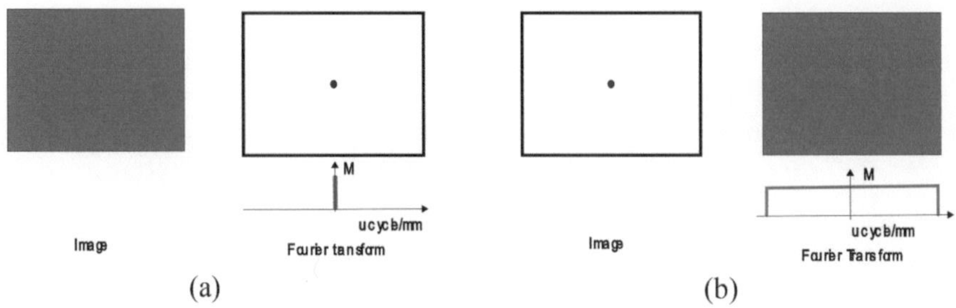

Fig. 3.35 Fourier transform of a constant grey level image is a little dot (**a**) and of a little dot is a constant gey level (**b**)

$$I(x) = e^{-ax^2}$$

And we calculate the Foureir transform integral using Eq. (3.20) we get:

$$F(u) = \sqrt{\frac{\pi}{a}} e^{-\frac{\pi^2 u^2}{a}}$$

The result is illustrated in Fig. 3.36

Main properties of Fourier transformation are illustrated in Table 3.2

The Fourier transform can be used to reduce noise in an image, a process based on filtering. Consider an image of a stripe affected by noise. The image and its corresponding Fourier transform are shown in Fig. 3.37a. By applying a band-pass filter with a Gaussian spatial grey-level distribution, which, as previously noted, also results in a Gaussian shape in the frequency domain, we can effectively suppress noise. If the filter is sufficiently wide to cover the main frequency components of the stripe pattern while attenuating the higher frequencies associated with noise, then

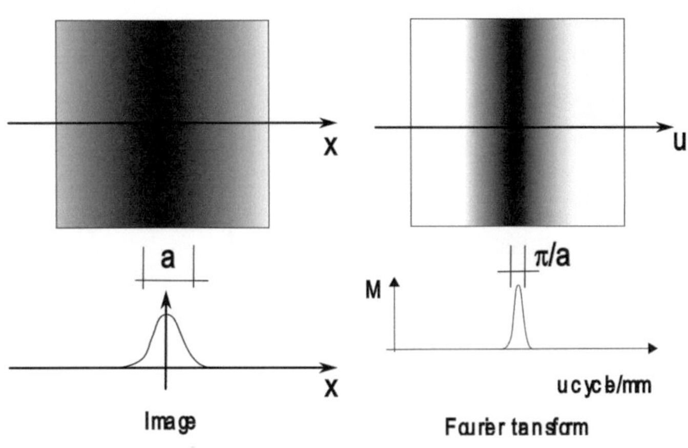

Fig. 3.36 Fourier transform of a gaussian smoothed grey level strip

Table 3.2 Some properties of Fourier transformation

Property	Spatial domain	Frequency domain
Linearity	$\alpha f_1(x) + \beta f_2(x)$	$\alpha F_1(u) + \beta F_2(u)$
Scaling	$f(ax)$	$\frac{1}{\lvert a \rvert} F\left(\frac{u}{a}\right)$
Shifting	$f(x - a)$	$e^{-i2\pi ua} F(u)$
Differentiation	$\frac{d^n}{dx^n}(f(x))$	$(i2\pi u)^n F(u)$

multiplying the spectrum in Fig. 3.37a by the Gaussian filter spectrum shown in Fig. 3.37b yields the filtered spectrum in Fig. 3.37c. Performing the inverse Fourier transform on this filtered spectrum produces the denoised image.

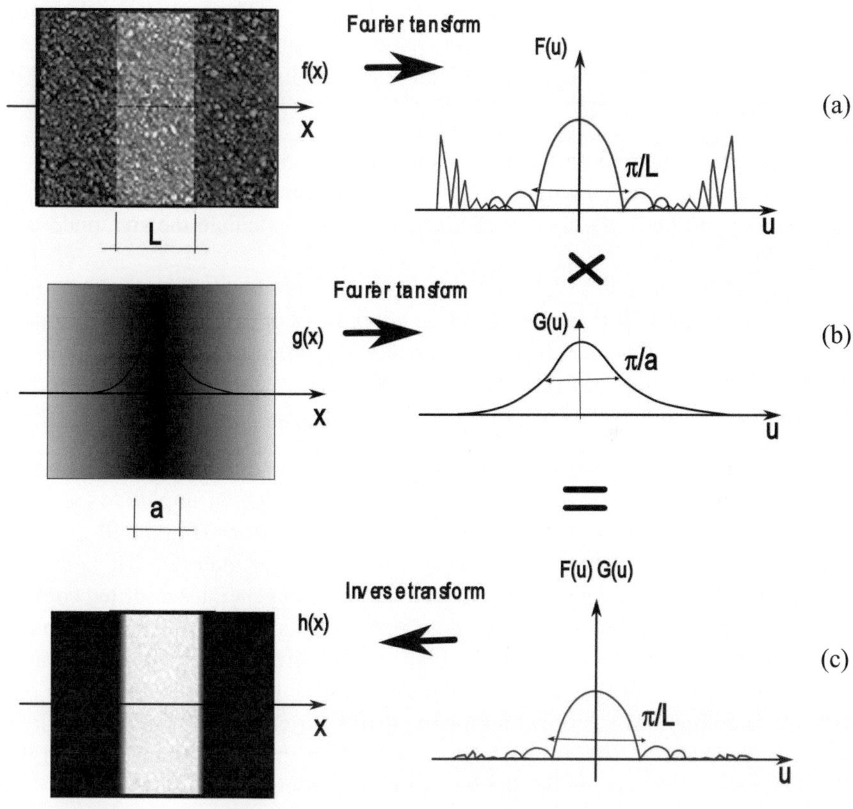

Fig. 3.37 Filtering in frequency domain for noise reduction

3.6 Spatial Signal Convolution

Another fundamental concept to understand digitization of images and image processing basics is convolution. Considering two functions f(x) and g(x), the convolution integral function h(x) of f and g is defined as follows:

$$h(x) = f(x) * g(x) = \int_{-\infty}^{+\infty} f(\tau)g(x-\tau)d\tau \tag{3.26}$$

And indicated by an $*$ operator between the two functin f and g.

Convolution integral is therefore the total area under the product of the two functions with changed abscissa and shifted one to the other of x. The convolution function is this area calculated for each x shifting value. So, from the above analitical definition the operations to do in order to obtain the convolution function, illustrated in Fig. 3.38, are the following:

(a) Change variable to both functions f and g from x to τ and flip g to $-\tau$
(b) Translate the function g by x respect to the function f.
(c) Overlap and multiply the two obtained functions, calculate the area under the product function.
(d) Do the same for each x value, and the result of the product will be non-zero only for x less than L/2, if the two function are pulses not overlapped their product will be zero and the area below the product (the integral) also will be zero.
(e) Area is the convolution value for the x shift value of g respect to f.
(f) Plot the area values vs x, i.e. the convolution function.

By performing the operations described above, it becomes evident that the convolution of a band-limited function, whose Fourier transform has an amplitude spectrum as shown in Fig. 3.39a, with a series of Dirac delta functions (Fig. 3.39b), results in the replication of the original frequency band at regular intervals. This effect is illustrated in Fig. 3.39c, where each Diract delta funcion generates a shifted copy of the original spectrum, leading to a periodic repetition across the frequency domain.

Convolution is closely related to the Fourier transform through the fundamental Convolution Theorem [3]. The convolution of two gray-level distributions, f(x) and g(x), can be computed by simply multiplying their Fourier transforms F(u) and G(u), respectively. That is, H(u) $=$ F(u) $\cdot$ G(u). To recover the convolution result in the spatial domain, h(x), one applies the inverse Fourier transform to H(u).

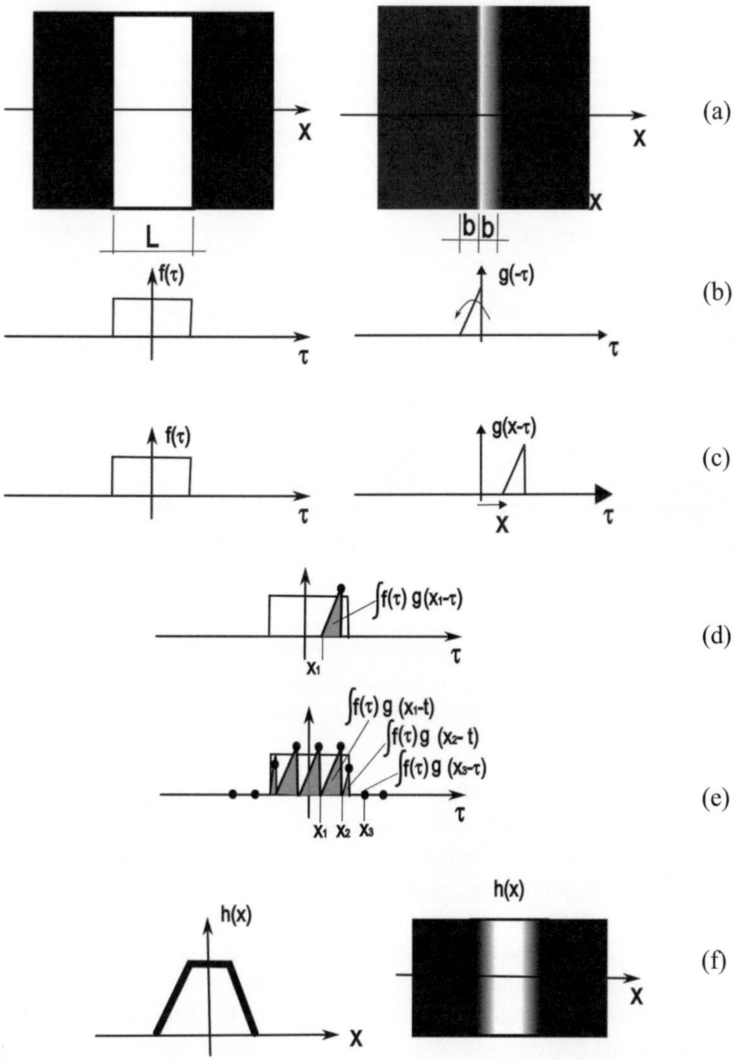

Fig. 3.38 Convolution

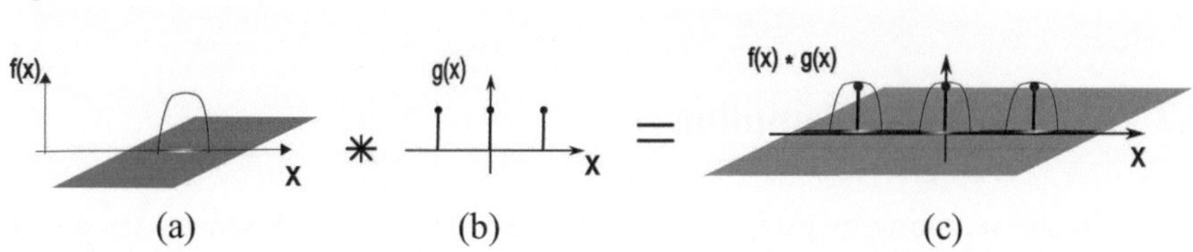

Fig. 3.39 Convolution of a band-limited grey level profile with a pulse series

Table 3.3 Convolution theorem

Spatial domain	Frequency domain
Convolution > $h(x) = f(x) * g(x)$	< Multiplication $H(u) = F(u)\ G(u)$
Multiplication > $h(x) = f(x)\ g(x)$	< Convolution $H(u) = F(u) * G(u)$

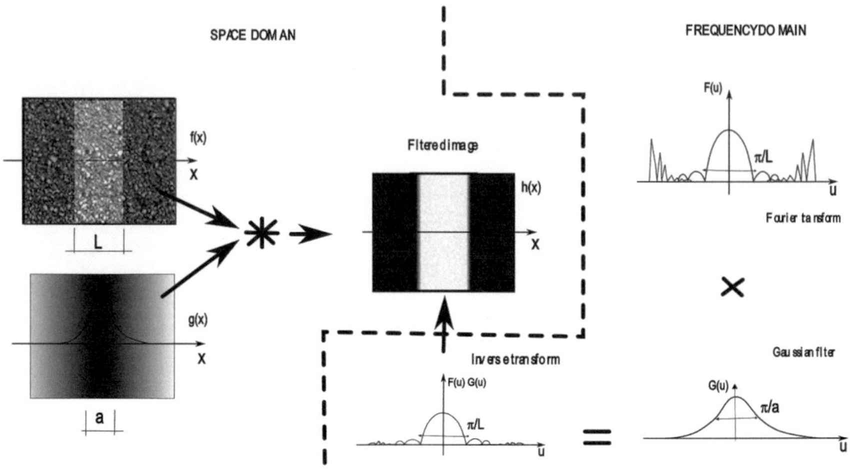

Fig. 3.40 Image filtering for noise reduction by convolution

Conversely, the Fourier transform of the product of two gray-level distributions, $h(x) = f(x) \cdot g(x)$, is equal to the convolution of their respective Fourier transforms $H(u) = F(u) * G(u)$, where $*$ denotes convolution in the frequency domain. Convolution theorem enunciates are summarized in Table 3.3.

The convolution theorem tells us that a product in the frequency domain corresponds to a convolution in the time domain and vice versa, so, convolution can be used instead of filtering in frequency domain to remove noise from images, as previously done by Fourier transform multiplication of image and filter functions. To obtain the same results we can perform the convolution of the spatial grey level profile with a gaussian grey level profile, as illustrated in Fig. 3.40.

3.7 Spatial Signal Sampling

Spatial image sampling by using a linear array of CCD or CMOS sensors, as previously illustrated in Fig. 3.23, can be seen mathematically as the product of the image grey level profile $f(x)$ with a series of equispaced of x_0 delta functions of amplitude equal to 1, i.e. the sampling function $s(x)$:

$$s(x) = \sum_{(n=-\infty)}^{(n=\infty)} \delta(x - nx_0) \tag{3.27}$$

As illustrated in Fig. 3.41.

According to the convolution theorem, the Fourier transform of the sampled gray-level distribution is given by the convolution of the Fourier transforms of the original continuous (analog) gray-level distribution and the sampling function. The Fourier transform of the sampling function is:

$$S(u) = \frac{1}{x_0} \sum_{n=-\infty}^{n=\infty} \delta(u - \frac{n}{x_0}) \tag{3.28}$$

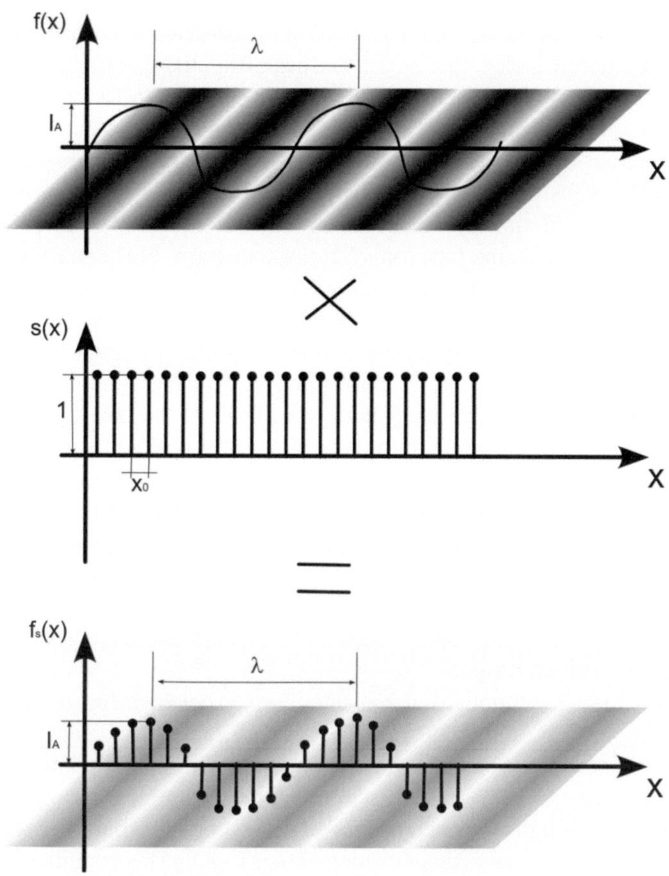

Fig. 3.41 Mathematical representation of digital image sampling

So sampled function in space domain will be:

$$f_s(x) = f(x)s(x) = f(x) \sum_{n=-\infty}^{n=\infty} \delta(x - nx_0) \tag{3.29}$$

And according to the convolution theorem, in frequency domain, we have:

$$F_s(u) = F(u) * S(u) = F(u) * \frac{1}{x_0} \sum_{n=-\infty}^{n=\infty} \delta\left(u - \frac{n}{x_0}\right) \tag{3.30}$$

This is illustrated in Fig. 3.41.

The convolution of the double-peak Fourier transform of a simple sine function with the periodic pulse train (i.e., the Fourier transform of the sampling function) results in a series of paired peaks, one pair for each pulse in the sampling function's spectrum. This occurs because the convolution is non-zero only where the supports of the two functions overlap, that is, where a peak from the sine function aligns with a pulse from the sampling spectrum. This interaction leads to the replication of the sine spectrum at regular intervals, as shown in Fig. 3.42.

To reconstruct the original sine wave gray-level distribution, one performs an inverse Fourier transform restricted to a limited frequency band, specifically, the region around zero frequency that contains only the central pair of peaks, corresponding to the original sine function's Fourier transform. This band-limiting operation effectively eliminates the replicated spectra and allows accurate reconstruction of the original analog signal. The result of this process is illustrated in Fig. 3.42.

However, if the spatial sampling interval exceeds the wavelength of the original sine wave, the outcome changes significantly, as illustrated in Fig. 3.43. In this case, the convolution between the Fourier transform of the sine function and the spatial sampling function results in replicated spectral peaks that are closer to the origin than those of the original signal.

As a consequence, when the inverse Fourier transform is applied using only the two peaks closest to zero frequency, the reconstructed signal appears as a sine wave with a longer wavelength, that is, a lower spatial frequency, than the original. This phenomenon is a manifestation of aliasing, where insufficient spatial sampling resolution causes high-frequency components to be misrepresented as lower-frequency ones in the reconstructed signal.

In general, the convolution in the frequency domain between the Fourier transform of a sine gray-level profile, with wavelength λ and spatial frequency $f = 1/\lambda$, and the Fourier transform of a spatial sampling function with sampling interval x_0 (thus, sampling frequency $f_s = 1/x_0$) produces a double-peak pattern, as shown in Fig. 3.44. It is evident from the figure that a frequency window centered around zero can contain only the signal frequencies without overlap only if $f_s > 2f$. This condition defines the fundamental *Nyquist criterion*, which ensures that the spatial sampling frequency is chosen correctly to avoid aliasing, only if it is at least the double of the maximum grey-level distribution spatial frequency.

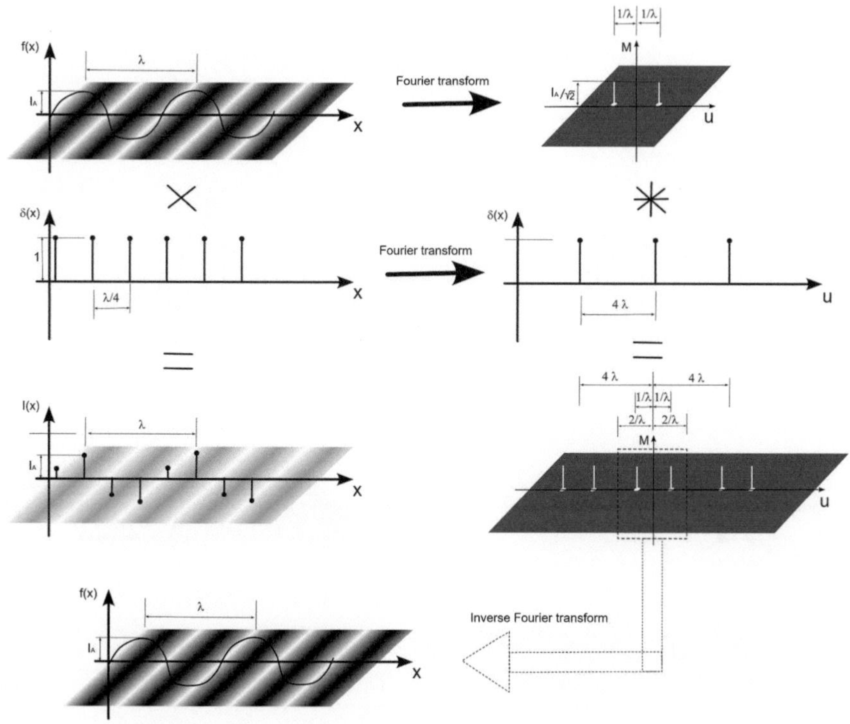

Fig. 3.42 Digital image sampling in time and frequency domain

For a band limited non periodic signal as in Fig. 3.45 same "reflection" of high to lower frequency due to convolution with the sampling function spectrum, named aliasing, occur if $f_s < 2f$.

So, if we have a not periodic grey level distribution, with spectrum limited to an f value, in order to avoid aliasing effects it will be necessary a sample frequency $f_s > 2f$.

For a band-limited, non-periodic signal, as illustrated in Fig. 3.45, a similar phenomenon occurs: high-frequency components are reflected into lower frequencies due to the convolution with the spectrum of the sampling function. This effect (aliasing) arises when the sampling frequency f_s is less than twice the maximum signal frequency f, i.e., when $f_s < 2f$.

To prevent aliasing in the sampling of a non-periodic gray-level distribution whose spectrum is limited to a maximum frequency f, it is essential to satisfy the Nyquist condition $f_s > 2f$.

This ensures that the replicated spectra do not overlap in the frequency domain, thereby allowing for accurate reconstruction of the original signal from its sampled version.

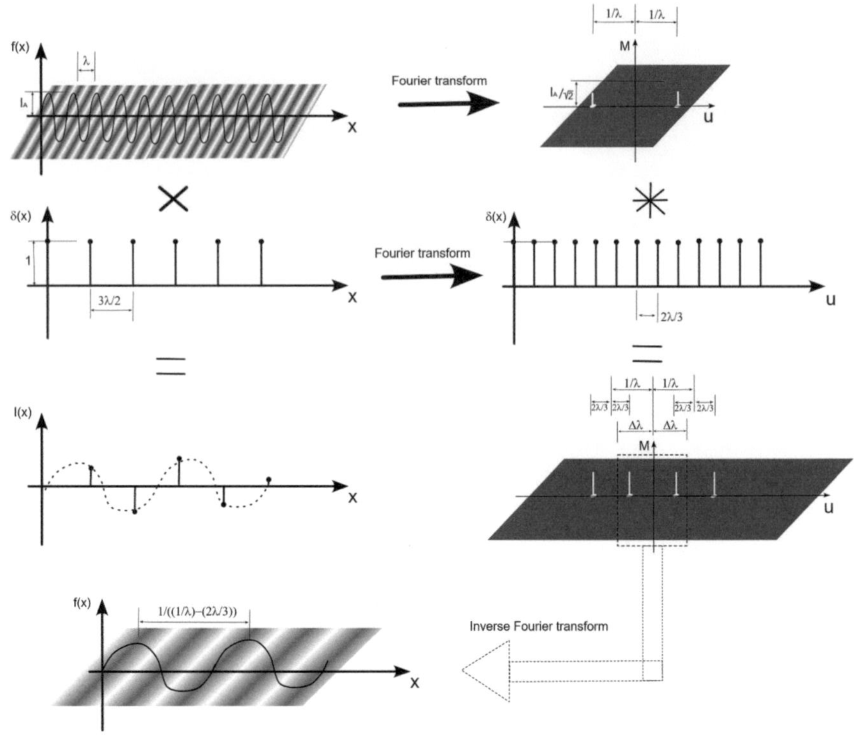

Fig. 3.43 Aliased digital image sampling in time and frequency domain

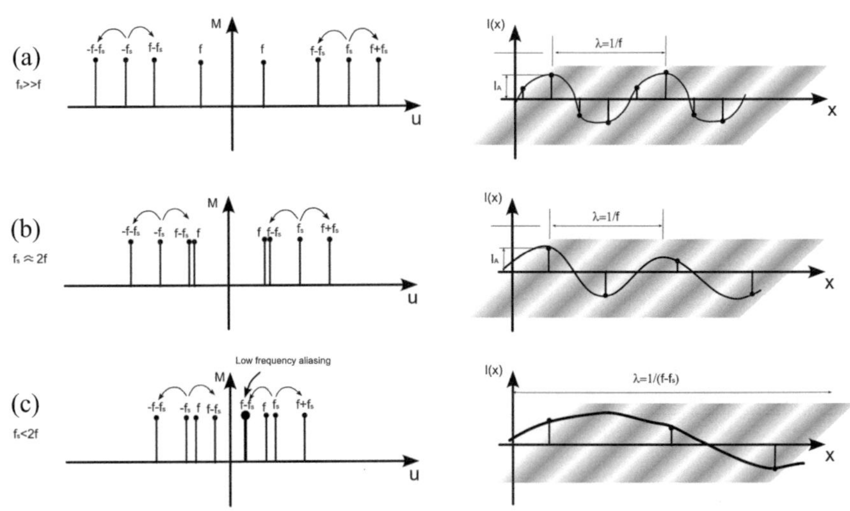

Fig. 3.44 Nyquist criterion for choosing spatial sample frequency f_s

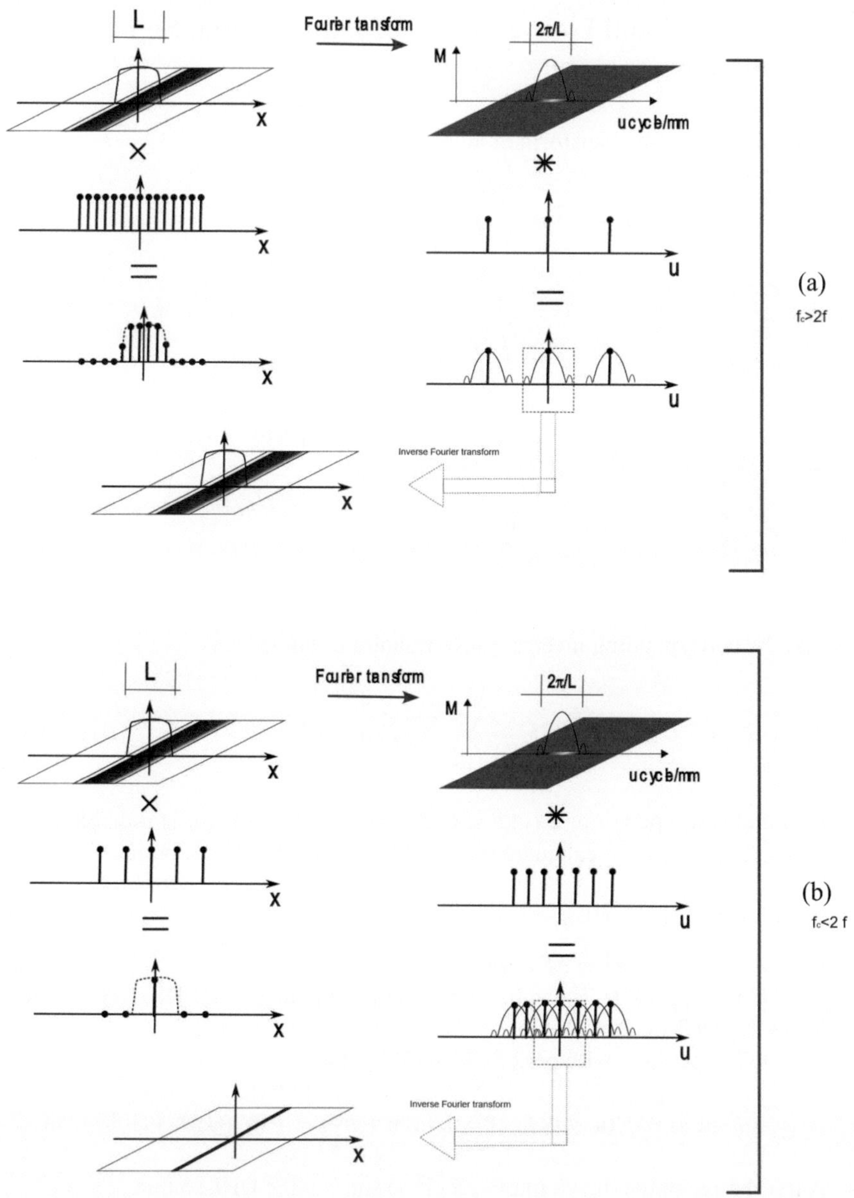

Fig. 3.45 Sampling a band limited grey level distribution using **a** $f_s > 2f$ or **b** $f_s < 2f$

3.8 Bidimensional Fourier Transform and Spatial Signal Sampling

Bidimensional Fourier transform of an image I(x, y) is defined as [3]:

$$F(u, v) = \int\int I(x, y)e^{-2\pi i(ux+vy)}dxdy \tag{3.31}$$

And inverse Fourier transform as:

$$I(x, y) = \int\int F(u, v)e^{2\pi i(ux+vy)}dudv \tag{3.32}$$

But if we have digital sampled images I(m, n) of MxN pixels bi-dimensional discrete Fourier transform (DFT) is defined as [3]:

$$F(u, v) = \sum_{m=0}^{M-1}\sum_{n=0}^{N-1} I(m, n)e^{-i2\pi\left(\frac{um}{M}+\frac{vn}{N}\right)} \tag{3.33}$$

And the corresponding inverse transformation is defined as:

$$I(m, n) = \frac{1}{MN}\sum_{u=0}^{M-1}\sum_{v=0}^{N-1} F(u, v)e^{i2\pi\left(\frac{um}{M}+\frac{vn}{N}\right)} \tag{3.34}$$

If M and N are power of 2 (like 256, 512, 1024, 2048 etc.) it is possible to use an efficient algorithm to calculate this double sum named Fast Fourier Transform (FFT).

Main properties of DFT are:

- Linearity $F\left[af + bg\right] = aF + bG$
- The DFT is separable, it can be obtained by computing a 1D DFT on the rows, then a 1D DFT on the columns
- The DFT is periodic with periods M and N so that

$$F(u, v) = F(u + kM, v) = F(u, v + lN) = F(u + kM, v + lN)$$

- A translation on the image implies a phase shift on the DFT so that

$$F\left[f(m - m_0, n - n_0)\right] = F(u, v)\exp\left(-j2\pi\left(\frac{um_0}{M} + \frac{vn_0}{N}\right)\right)$$

- A translation on the image implies a phase shift on the DFT.

Bidimensional convolution H(i, j) of two digitized images F(i, j) and G(k, l) is defined by:

$$H(i,j) = \sum_{k=-m}^{m} \sum_{l=-n}^{n} F(i-k, j-l)G(k,l) \tag{3.35}$$

For bidimensional Fourier transform the convolution theorem is also valid, so:

$$F\{f * g\} = F \cdot G \tag{3.36}$$

$$F\{f * g\} = F \cdot G \tag{3.37}$$

Convolution of G with F is equal to convolution of F with G, so convolution is commutative:

$$F * G = G * F \tag{3.38}$$

Convolution of two continuous 2D functions F(x, y) and G(x, y), where x and y are continuous values (real numbers), is given by:

$$H(F * G)(x,y) = \int_{-\infty}^{\infty} \int_{-\infty}^{\infty} F(x-\tau, y-\nu)G(\tau,\nu)d\tau d\nu \tag{3.39}$$

An example of convolution of two images is illustrated in Fig. 3.46. It can be obtained with one of the many available software or freeware (for example using python libs as illustrated in https://vincmazet.github.io/bip/). It is possible to see that convolution of an image with little dots is the simple replication of the image around each dot, as previously explained for 1D gray-level distributions convolved with delta functions.

According to Fourier, images can be represented as the sum of weighted sine or exponential functions, which, as previously discussed, are complex sinusoidal components derived from Euler's formula. These complex exponentials serve as the building blocks of the image. For example, the image shown in Fig. 3.47 is simply the result of summing three sine-wave-based grey level functions.

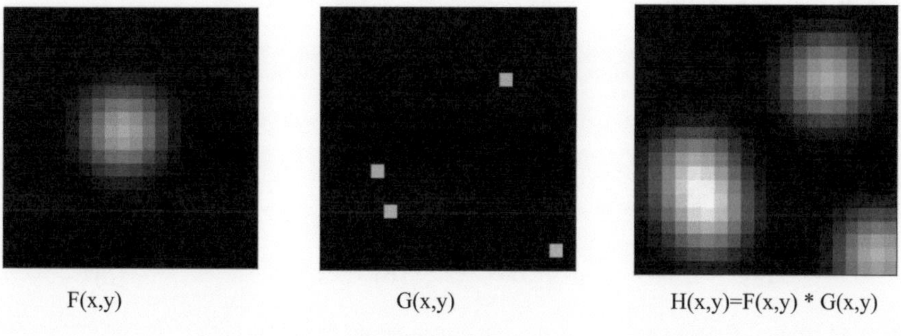

Fig. 3.46 An example of convolution of two images

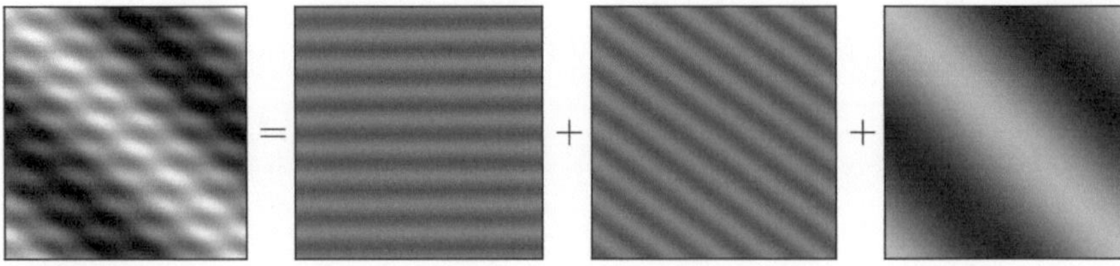

Fig. 3.47 Tree sine components of a simple image

Complex exponentials in two dimensions can be represented as sinusoidal functions characterized by four key parameters: period, amplitude, phase, and orientation, as illustrated in Fig. 3.48. While these 2D sinusoids are mathematically more intricate than the 1D sine functions used in Fourier series, the underlying concepts remain fundamentally similar. The sinusoidal components of a two-dimensional Fourier transform can be described by the following expression:

$$f(m,n) = \frac{A}{2}\left\{cos\left[\frac{2\pi}{\lambda}\left(\frac{n}{N}cos\theta - \frac{m}{M}sin\theta\right) + \phi\right] + 1\right\} \tag{3.40}$$

And its amplitude A, wavelength λ, phase delay ϕ and orientation θ parameters are illustrated in Fig. 3.48a. An example of Fourier transform of a picture is given in Fig. 3.49.

Just as in the one-dimensional Fourier transform, a reciprocal relationship exists between the spatial and frequency domains. This means that small features in the spatial domain correspond to broad frequency spectra, while large, smooth regions correspond to narrow frequency bands, an effect clearly illustrated in Fig. 3.50 for little and larger dots.

In general, the phase component of an image transform carries most of the significative information. For instance, if we swap the phase components of the transformation of the two images of Fig. 3.51a and make the antitrasformation we get the

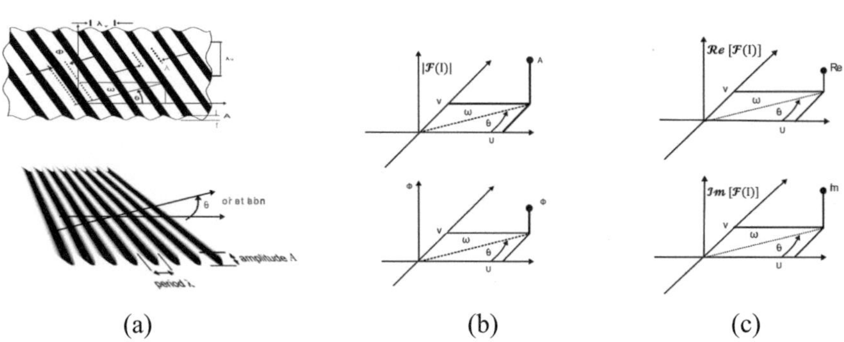

Fig. 3.48 Bidimensional Fourier transform sine components and parameters

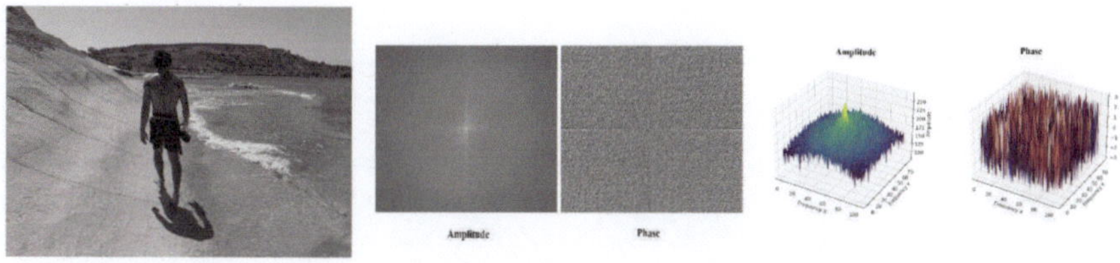

Fig. 3.49 Example of an image digital Fourier transform in amplitude and phase

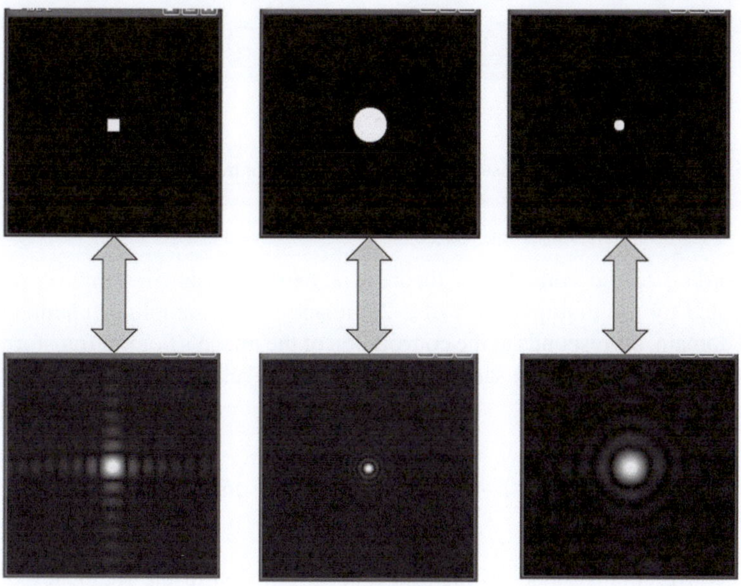

Fig. 3.50 Reciprocal relation between space and frequency domain

result of Fig. 3.51c. This highlights the dominant role of phase in preserving image structure and identity.

By selectively removing (filtering) portions of an image's frequency spectrum, it is possible to isolate either high or low frequency components. This process allows for effects such as smoothing the image or enhancing its edges, as illustrated in Fig. 3.52.

Sampling 2D images, as for 1D images, can be mathematically expressed as multiplication of the image F(r, c) with a 2D Dirac delta sampling function as illustrated in Fig. 3.53 and mathematically expressed by:

$$samp_N\{F\}(m,n) = F(r,c) \cdot \sum_{j=-\infty}^{\infty} \sum_{k=-\infty}^{\infty} \delta(m - jN)\delta(n - kN)$$

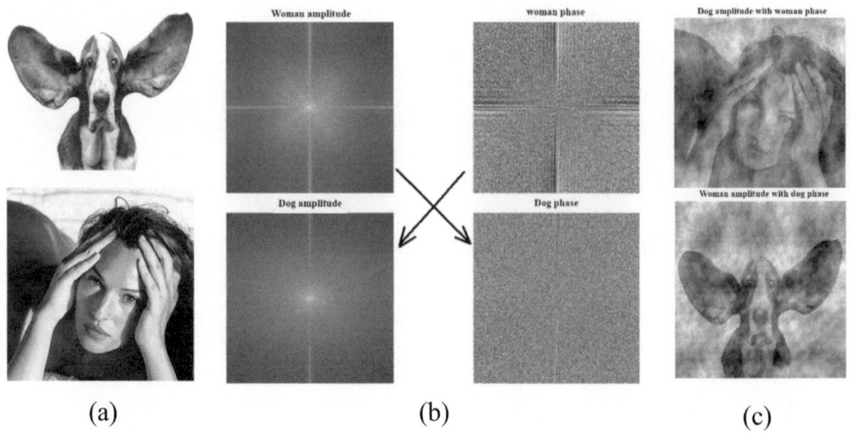

Fig. 3.51 Example of Fourier transform exchanging of phase or magnitudes between two different images

According to the convolution theorem, a two-dimensional sampling process, represented by the multiplication of a 2D image with a sampling function in the spatial domain, corresponds to the convolution of the image's Fourier transform with the Fourier transform of the sampling function. This relationship can be expressed as:

$$samp_{\frac{1}{N}}(u,v) = \sum_{j=-\infty}^{\infty}\sum_{k=-\infty}^{\infty} \delta\left(u - \frac{j}{N}\right)\delta\left(v - \frac{k}{N}\right)$$

The convolution of a spectrum with a pulse function in the frequency domain results in a shifted copy of the original spectrum centered at the position of the pulse. When this occurs for each sampling pulse of the sampling function, it generates a periodic repetition of the spectrum at intervals of 1/N in the frequency domain, as described by the following expression:

$$\mathcal{F}\{I\}(u,v) * samp_{\frac{1}{N}}(u,v) = \sum_{j=-\infty}^{\infty}\sum_{k=-\infty}^{\infty} \mathcal{F}\{I\}(u,v) * \delta\left(u - \frac{j}{N}\right) * \delta\left(v - \frac{k}{N}\right)$$

Therefore, to prevent overlapping of the replicated spectra caused by convolution, thus avoiding aliasing, the spatial bandwidth of the image's spectrum must be limited. Specifically, the spectral content should be confined within a radius r < 1/(2N), as illustrated in Fig. 3.54.

This leads to the fundamental criterion for selecting the pixel size of the image digitization sensor in relation to the image's spectral spatial bandwidth radius r. To accurately capture the image without aliasing, the two-dimensional sampling frequency 1/N must be at least 2r.

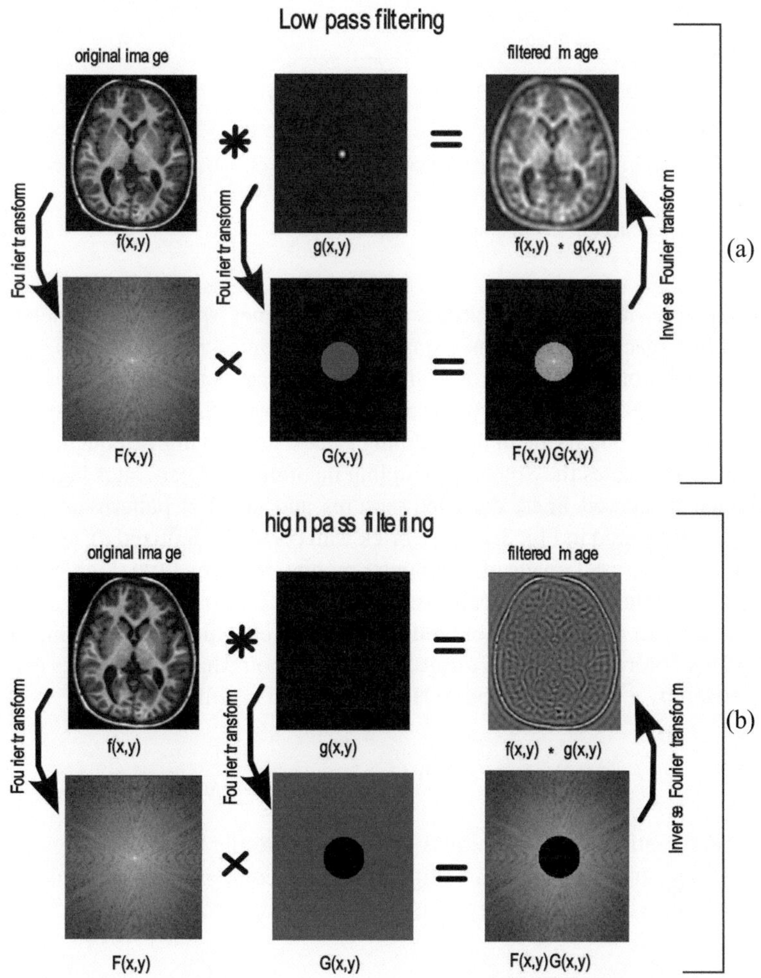

Fig. 3.52 2D filtering for smoothing or highlighting borders of an image

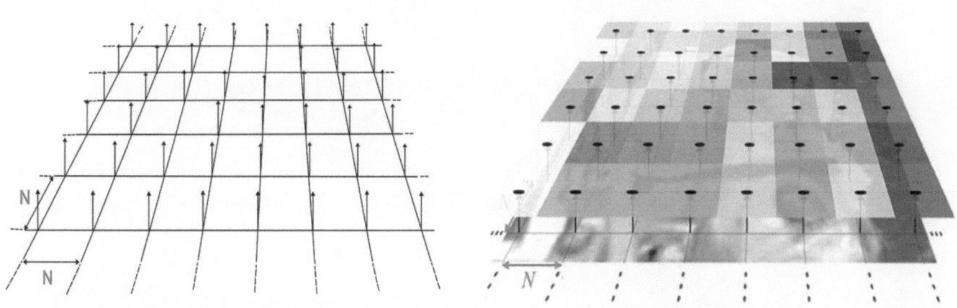

(a) Sampling function (b) sampled image

Fig. 3.53 Sampling of 2D images as multiplication by a 2D Dirac delta function

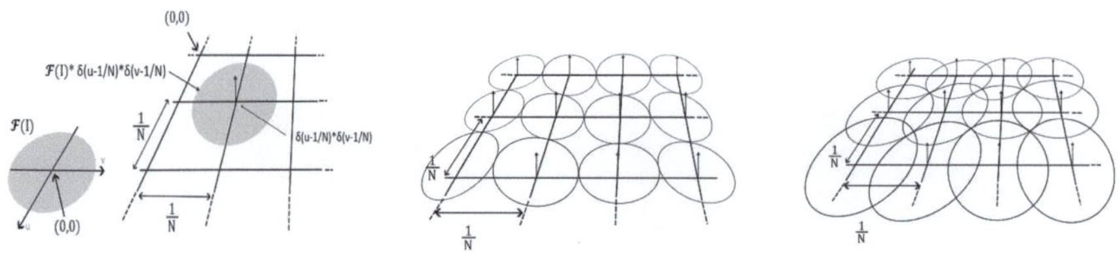

Fig. 3.54 Aliasing effect in 2D sampling

For example, if the original image consists of lines spaced l millimeters, the projected line spacing on the sensor image plane becomes *m*v millimeters, where *m* is the magnification factor of the optical system. Consequently, to meet the sampling requirement, the pixel resolution (and corresponding pixel size) must be at most *m*l/2.

Aliasing arises due to the undersampling (1/N < 2r) of a high-frequency content images, which violates the Shannon sampling theorem. This generates jagged edges on diagonal or curved lines, distorted textures and unatural patterns (e.g., moiré patterns) as illustrated in Fig. 3.55 for a brick wall correctly digitized a), aliased in b) and for a typical aliased image of a computer monitor made by little (high frequency) digital pixels of the digital computer screen.

This effect can be mitigated, but reducing image resolution, by applying an anti-aliasing filter before the sampling array. Such a filter should have a frequency response characterized by a low-pass function H(u, v), defined as follows:

$$H(u, v) = 0 \text{ for } u > \frac{1}{2N} o \; v > \frac{1}{2N}$$

In general, anti-aliasing techniques rely on pre-filtering the image, slightly blurring it before sampling, to remove high-frequency components that could lead to aliasing. As previously discussed, this can be achieved through methods such as

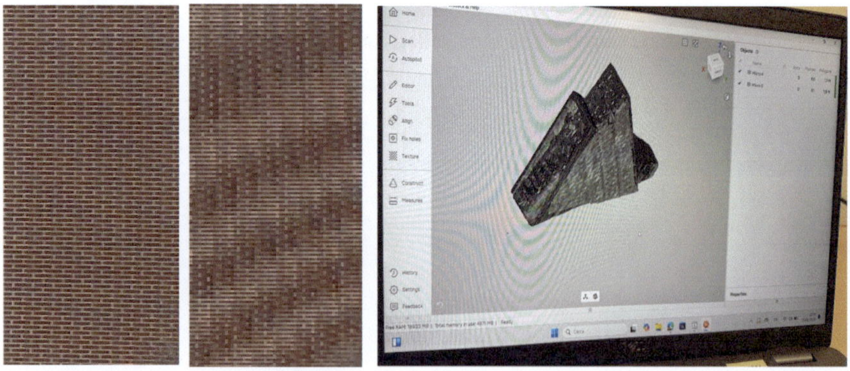

(a) (b) (c)

Fig. 3.55 Examples of aliased images

Gaussian blur (via convolution), box filters that average neighboring pixel values, or other more advanced filtering techniques.

Additionally, post-sampling methods can be employed to reduce aliasing artifacts. These include resampling the image at a higher resolution followed by down-sampling, subpixel interpolation, and various numerical correction techniques designed to smooth and refine the image after digitization.

Anyway, for measurement purposes by image processing it is better to choose correctly optics and image sensor resolution with the above given criteria to avoid aliasing effects.

References

1. D. Litwiller, CCD versus CMOS. Photon. Spectra **35**(1), 154–158 (2001)
2. T. Dalsa, The evolution of cmos imaging technology. https://www.teledynedalsa.com/public/mv/appnotes/EvolutionofCMOSTechnologywp.pdf (2011)
3. R.C. Gonzalez, R.E. Woods, *Digital Image Processing* (Addison-Wesley, Reading, Massachusetts,1992)
4. K.R. Castleman, *Digital Image Processing*, 2nd edn. (Prentice-Hall, Englewood Cliffs, New Jersey, 1996)
5. Image s spa. [Online]. Available: https://www.imagesspa.it/telecamere/
6. G. Rossi, Misure meccaniche e termiche. Basi teoriche e principali sensori e strumenti (2010)
7. J.C. Russ, *The Image Processing Handbook*, 2nd edn. (CRC Press, Boca Raton, Florida, 1995)
8. I.T. Young, J.J. Gerbrands, L.J. van Vliet, *Fundamentals of Image Processing* (1995)

Chapter 4
2D Measurements by Images

Abstract From high-quality analogue images that are properly digitized, many dimensional and shape-related measurements can be extracted using image processing techniques. These measurements play a crucial role across various domains, including the verification of dimensional tolerances in mechanical components, quality control in industrial manufacturing, support for civil engineering applications, and processes such as product design and reverse engineering. Additional applications extend to medical diagnostics, physical and materials analysis, precision agriculture, the development of digital twins, and the documentation and study of artworks. This chapter presents the foundational principles of performing 2D measurements through image-based methods. It addresses critical elements such as appropriate lighting strategies, core image processing algorithms for edge detection, and procedures for computing distances between points along edges to determine object dimensions. Furthermore, it explores techniques for detecting and analyzing key points, often referred to as blobs, examining their features, characteristic "signatures," shapes, and classical geometric descriptors.

4.1 Introduction

Everything we perceive around us consists of images formed by reflected light except for light sources themselves, such as the sun, lamps, flames, or extremely hot surfaces emitting electromagnetic waves in the visible range, as will be better explained in Chap. 8 by Plank's law. Therefore, to get images of objects good for measurements we need a light source.

Since prehistoric times, humans have developed light sources to illuminate their surroundings, especially at night or in dark environments like their homes in caves. Fire was the earliest form of artificial light. One of the first portable lighting systems was the torch, created by wrapping resinous materials around sticks to produce a sustained flame. Around 4500 BCE, oil lamps emerged, using animal fat or plant oils with wicks to maintain a controlled flame. By approximately 500 BCE, candles made from tallow or beeswax became common.

G. Rossi, *Mechanical and Thermal Measurements by Images and Waves*, Smart Sensors, Measurement and Instrumentation 51, https://doi.org/10.1007/978-3-032-03464-9_4

The invention of gas lighting in the late eighteenth century revolutionized urban life. Coal gas-fueled lamps began to illuminate public streets in the early nineteenth century. Paris, for example, became known as "La Ville Lumière" (The City of Light) due to its extensive gas lighting.

During the same century, Humphry Davy demonstrated the first electric arc lamp. Later, Thomas Edison (1879) and Joseph Swan (1878) independently developed practical incandescent light bulbs, which emitted light by heating a filament with an electric current until it glowed.

In the 1930s, fluorescent lamps were introduced. These used phosphors to convert ultraviolet light generated by an internal electric arc into visible light. The 1950s saw the development of halogen lamps, an improved version of incandescent bulbs that produced brighter light and had longer lifespans.

A breakthrough came in the 1960s with the invention of light-emitting diodes (LEDs), which marked a departure from traditional heat-based light sources. LEDs work by converting electrical energy directly into photons through electroluminescence. By the 1990 and 2000s, advances in semiconductor technology made LEDs highly energy-efficient and suitable for widespread use. Today, they have largely replaced conventional lighting due to their efficiency, often ten times greater than traditional bulbs, and long operational life.

In battery-powered portable lighting systems, the use of LEDs extends operational duration by a factor of ten compared to incandescent bulbs or other traditional light sources. This dramatic improvement in efficiency has led to the widespread replacement of older technologies, such as acetylene gas lamps, which were commonly used for work in mines or caves, with LED torches. Remarkably, this transition occurred only about 20 years ago, despite over 200 years of technological progress following three industrial revolutions and the development of numerous energy sources and lighting systems.

The key reason behind this shift is that LEDs provide a significantly more efficient method of converting electrical energy into light. Unlike previous technologies, they require far less energy to produce sustained illumination, making them ideal for applications where energy availability is limited but long-duration lighting is essential. This serves as an important technical lesson: high level performances of a system or a machine can be obtained not only increasing its energy content but also using energy better, more efficiently.

Light intensity, or luminous intensity, is one of the seven fundamental quantities in the International System of Units (SI). It was most recently redefined in 2019 based on fixed values of fundamental physical constants [1]. Specifically, the unit of luminous intensity, the candela (cd), is defined by assigning a fixed numerical value to the luminous efficacy of monochromatic radiation of frequency 540×10^{12} Hz (green light), denoted as K_{cd}. This value is set at 683 $l_m \cdot W^{-1}$, which is equivalent to $cd \cdot sr \cdot W^{-1}$ or, in base SI units, $cd \cdot sr \cdot kg^{-1} \cdot m^{-2} \cdot s^3$.

The kilogram, metre, and second are each defined in terms of fundamental physical constants: the Planck constant (h) for the kilogram, the speed of light in a vacuum (c) for the metre, and the hyperfine transition frequency of the cesium-133 atom ($\Delta\nu Cs$) for the second [1].

One candela corresponds approximately to the luminous intensity emitted by a common candle. It quantifies the amount of luminous flux (measured in lumens) emitted per unit solid angle in a specified direction from a light source.

The solid angle, measured in steradians (sr), is a dimensionless quantity. It is defined as the ratio between the surface area (*A*) of a spherical cap and the square of the radius (*R*) of the sphere as $\Omega = A/R^2$. This formulation describes how much of the sphere's surface is "subtended" by a given conical beam of light originating from the sphere's center.

For example if we have a typical compact fluorescent light bulb of 25 W electrical power that gives a total of around 2000 lumens; if that light is radiated equally in all directions on an half sphere (i.e. over 2π steradians), it will have an intensity of $I_v = 2000/2\pi = 318$ cd. If focused into a 20° beam (0.095 steradians), the same light bulb would have an intensity of around 18,000 cd.

To have good images of an object surface where we aim to perform measurements we need a proper light source, if possible optimized and correctly chosen for the specific object shape and surface optical characteristics [2]. We can say that the quality of the image we get depends on:

1. Objectives and optics used for analogic image formation
2. Digitalization and sampling camera system and sensor chosen
3. Illumination system
4. Object surface optical characteristics.

In previous chapters we have seen the first two, here we will go more in depth on the second two.

The effects of lighting on an object's surface can be analysed using geometric optics techniques, such as ray tracing. Various ray tracing software tools are available for this purpose, including advanced rendering softwares. One of the most widely used free, powerful and open-source software tools for rendering is Blender.

At each point on a surface, light can be reflected, refracted (according to Snell's Law), absorbed, or transmitted depending on the material's optical properties. Reflection can occur in two primary forms: specular (mirror-like) and diffuse (scattered). For most practical measurement applications, the surfaces involved are not perfectly smooth; they typically exhibit roughness greater than the wavelength of visible light, resulting in predominantly diffuse reflection.

From a measurement standpoint, the ideal imaging condition is a perfectly diffusive surface. Such a surface behaves like a uniform light emitter at every point, making it easier to model and measure accurately. This assumption underlies the image formation models discussed in Chap. 2, which involve lenses, mirrors, and optical objectives.

A perfectly diffusive surface condition can be achieved in two ways:

1. Using short-wavelength light sources, which are less sensitive to small-scale surface roughness.
2. Modifying the surface roughness, for example, by applying matte paint or other surface treatments, possible when such modifications are practical and do not interfere with the measurement objectives.

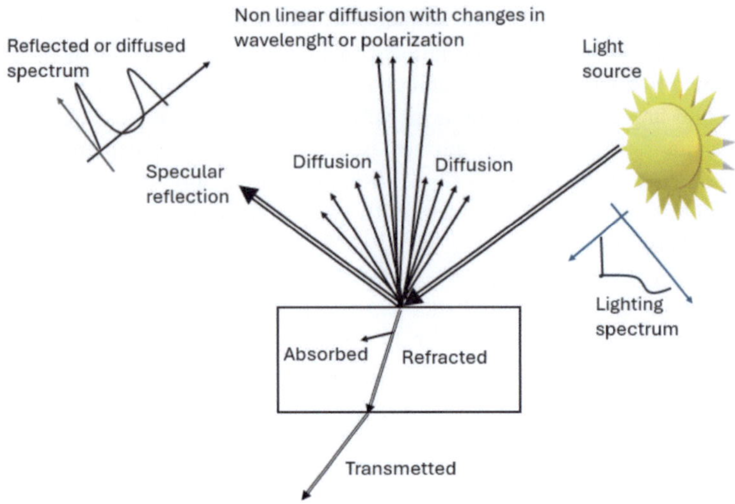

Fig. 4.1 Interactions of the surface of an object with a light source

This control of surface optical behaviour is essential for obtaining precise and reliable measurements in imaging systems.

Image of an object is given by the diffused light from its surface that changes in intensity, spatial distribution and spectral composition, as illustrated in Fig. 4.1, depending on:

1. Characteristics of lighting source:
   - Intensity and spectral bandwidth of lighting.
   - Direction and intensity distribution of illuminating rays.
2. Characteristics of the surface:
   - Color, i.e. the wavelength that the surface mainly reflects accordingly to its "optical spectral wavelength transfer function" i.e. the ratio between the amplitude of each reflected wavelength and that of the incident wavelength on the surface. Surface roughness and therefore reflection or diffusion of the illuminating rays.

In case of incorrect illumination problems such as reflections, hot spots, shadows, low contrast, blooming (high exposure), or dark images (low exposure) or images with low signal to noise ratio can occur making images not useful for measuring purposes.

4.2 Light Sources

The most used lighting sources to produce images good for measurement purposes [2] are:

(1) Filament halogen lamps.
(2) Fluorescent lamps.
(3) Light emitting diodes (LED).

And sometimes also the traditional incandescent lamps are used. Light sources are characterized mainly by:

(a) Light spectrum generated (partially specified sometimes by lighting color temperature)
(b) Directionality, i.e. ray beam angular intensity distribution
(c) Efficiency in terms of the ratio between output light intensity generated to the absorbed electrical power
(d) Heat generation and dissipation efficiency
(e) Life of the light source
(f) Stability, i.e. lighting intensity over time I(t)
(g) Purchase and operational cost.

4.2.1 Filament Halogen Lamps

Halogen lamps operate on the same fundamental principle as traditional incandescent lamps: a tungsten filament is heated to high temperatures by an electric current flowing on it, emitting light similarly to the glowing gases of a flame because of its high temperature. Electromagnetic emission of infrared radiation, light and ultraviolet generated by hot surfaces will be further explained in the next Chap. 8, through Planck's Law. What distinguishes halogen lamps is the presence of a halogen gas, typically iodine or bromine, inside the bulb. This gas participates in a chemical reaction known as the halogen cycle, in which vaporized tungsten from the heated filament reacts with the halogen gas and is subsequently redeposited onto the filament. This regenerative process significantly reduces filament degradation, extends the lamp's lifespan, and helps maintain the clarity of the bulb by preventing darkening due to tungsten deposits.

Because the halogen cycle requires higher operating temperatures (around 3000 K), the outer envelope of a halogen lamp is made of quartz rather than ordinary glass, as quartz can withstand these elevated temperatures without deforming or failing.

While halogen lamps are more efficient and longer-lasting than standard incandescent bulbs, they are still less efficient than modern LED or fluorescent lighting systems. A significant portion of the energy consumed by halogen lamps is lost as heat, and they also emit considerable amounts of ultraviolet (UV) and infrared (IR) radiation, which can pose safety and thermal management concerns in certain applications. Main advantage of halogen lamp is the very high intensity of light generated by little bulbs.

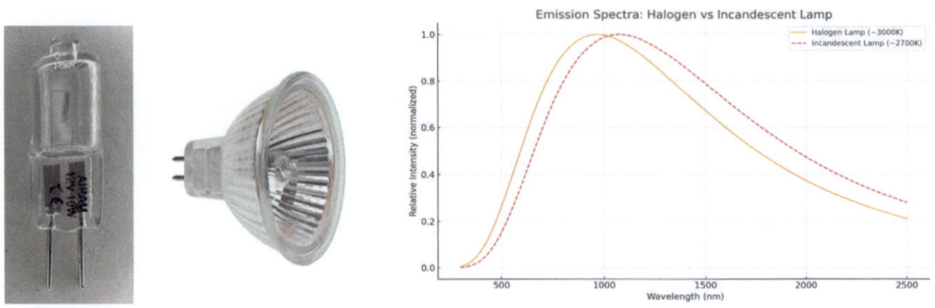

(a) (b)

Fig. 4.2 Halogen lamp ant their typical emission spectrum compared with that of standard incandescent bulbs

In Fig. 4.2b, the emission spectra of a halogen lamp and a standard incandescent lamp are simulated based on Planck's law (Eq. 8.3), using the grey body approximation for the emitting tungsten filament of the lamp, operating at approximately 3000 K. The halogen lamp exhibits a smoothly increasing spectral distribution of radiated intensity across wavelengths from 300 to 2500 nm, with a peak in the near-infrared region around 1000 nm, just outside the visible range.

For comparison, the spectrum of a conventional incandescent lamp, illustrated by the dashed curve, is also shown. It corresponds to a tungsten filament at a lower operating temperature (around 2700 K), which shifts the spectral distribution further into the red and infrared regions. As a result, the light produced appears warmer and softer, with reduced intensity in the blue part of the spectrum.

This comparison highlights the advantages of halogen lamps in terms of luminous efficacy and correlated colour temperature. Operating at higher temperatures, halogen lamps emit more energy in the visible and near-infrared regions, resulting in a whiter, brighter light output and improved energy efficiency compared to standard incandescent lamps.

4.2.2 *Fluorescent Lamps*

Fluorescent lamps works by exciting mercury vapor inside a glass tube to produce ultraviolet (UV) invisible light, which then stimulates a phosphor coating inside the tube to emit visible light. They are known for their energy efficiency and long lifespan compared to incandescent bulbs because there is no filament stressed by temperature cycles. The glass tube contains low-pressure mercury vapor and an inert gas (argon or krypton). Two electrodes located at each end of the tube generate an electric arc. A phosphor coating covers the inside of the tube and converts UV light into visible light. A ballast regulates the current and provides the initial voltage needed to start the lamp, to ionize the gas inside the tube, creating an electric arc between the electrodes.

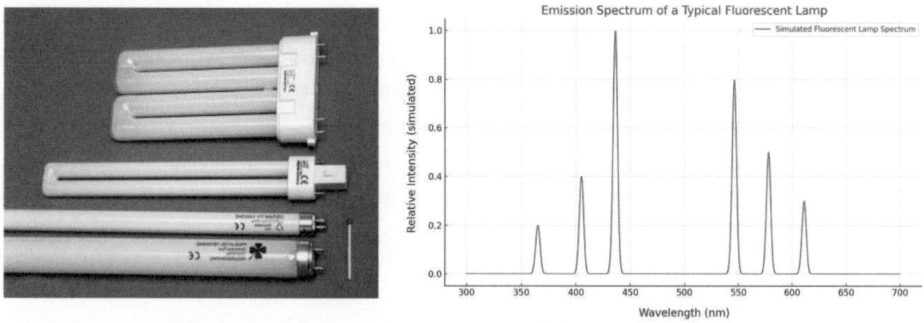

(a) (b)

Fig. 4.3 Fluorescent lamp and their typical spectrum of emitted light

Fluorescent lights use significantly less electricity power than incandescent bulbs, producing more lumens per watt. They typically last 7,000 to 15,000 h, far outlasting traditional incandescent lamps. Cost of the lamp is higher but better efficiency and longer life can give economic advantages in applications. They produce less heat compared to incandescent bulbs or halogen lamps. Can emit small amounts of UV light.

Unlike halogen lamps, fluorescent lamps have a spectrum with pronounced peaks of emission, as illustrated in Fig. 4.3b. The sharp peaks at specific wavelengths are due to emissions from mercury vapor and phosphor coatings. These peaks are primarily in the visible range, providing efficient lighting with distinct colour characteristics.

4.2.3 Light Emitting Diodes Light Sources

The discovering and technical evolution of the Light Emitting Diode (LED) is a tale of scientific curiosity, engineering brilliance, and global transformation in the way we light our world with a strong impact in eco-sustainability and energy resource saving. The LED's roots go back to 1907, when British experimenter H. J. Round of Marconi Labs discovered electroluminescence, the light emission from a solid material, while working with silicon carbide (SiC) and a cat's-whisker detector. He observed a faint yellow glow, but the phenomenon remained a curiosity with no practical application. Only in 1962, Nick Holonyak Jr., working at General Electric, developed the first visible-spectrum LED. It emitted red light, and Holonyak is widely recognized as the "father of the LED." His invention was a breakthrough because earlier LEDs only emitted infrared light and were mainly used in remote controls. Shortly after, other researchers developed green and yellow LEDs, but the elusive blue LED remained out of reach for decades.

The missing piece for having white light LED sources was blue light. In the 1990s, Shuji Nakamura of Nichia Corporation, alongside Isamu Akasaki and Hiroshi Amano, successfully created a high-efficiency blue LED using gallium nitride (GaN). This achievement led to the ability to combine red, green, and blue LEDs, or to use a blue LED with a yellow phosphor coating, to create white light useful for lighting. Their contribution was so critical that the three were awarded the 2014 Nobel Prize in Physics.

With the full colour spectrum now possible, white LEDs became practical and highly efficient. Their advantages include:

- High energy efficiency (up to 80% less energy use than incandescent bulbs)
- Long lifespan (25,000–50,000 h)
- Cooler operation
- Compact shape
- No toxic materials.

LEDs now replace incandescent and fluorescent lights in homes, offices, street lighting, vehicle headlights, is a key technology for TVs and smartphones, and are the most used lighting devices for industrial imaging for measurement purposes.

A Light Emitting Diode (LED) is a semiconductor device that emits light when an electric current passes through it. The principle behind its operation is electro-luminescence, where light is emitted because of electron–hole recombination in the semiconductor material. The structure of a LED is about the same already seen in Cap 3 for photodiodes: a p–n junction made from a direct-bandgap semiconductor material (typical for infrared LEDs is gallium arsenide (GaAs), for red or green led is gallium phosphide (GaP), or for blu LED gallium nitride (GaN)), where the p-type region contains an abundance of holes (positive charge carriers) and a n-type region that contains an abundance of electrons (negative charge carriers). The LED Active Layer is the region near the p–n junction where light is emitted during recombination. LEDs are typically encased in a transparent epoxy resin to protect the device and guide the emitted light. A voltage is applied across the LED in the forward bias direction, electrons from the n-type region and holes from the p-type region are pushed toward the junction. This reduces the width of the depletion region, allowing current to flow through the junction. At the p–n junction, electrons from the n-region combine with holes from the p-region. During this recombination, the electrons release energy in the form of photons. The energy of the emitted photon corresponds to the bandgap energy of the semiconductor material E_g so frequency f and wavelength λ are given by the Planck's low $f = E_g/h$ and $\lambda = c/f = c\ h/E_g$, where c is the speed of light. So the emitted light's wavelength λ (and hence its color) depends on the bandgap energy reciprocal of the semiconductor material $1/E_g$, therefore larger bandgap energy materials generates shorter wavelength, so blue or ultraviolet light, and smaller bandgap energy materials generates longer wavelength, i.e. red or infrared light.

In Fig. 4.4b the spectrum of a typical white LED used for lighting is illustrated. The spectrum shows two key features:

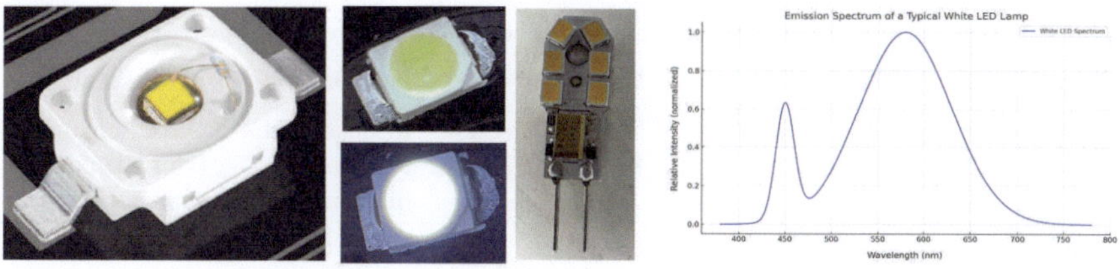

(a) (b)

Fig. 4.4 A White LED and its typical light emission spectrum

- A sharp peak around 450 nm, representing the blue light from the LED chip.
- A broad emission band centered around 580 nm, produced by the phosphor yellow coating that converts part of the blue light into yellow–red wavelengths.

Together these emissions combine to create the LED white light. This characteristic spectrum differs from the continuous blackbody-like emission of an halogen or incandescent lamps, illustrating how LEDs achieve energy-efficient lighting with tailored spectral output.

Unlike incandescent bulbs, LEDs directly convert electrical energy (electrons) into light (photons) with minimal heat generation, making them highly efficient.

Light emission of LED's is very fast, it does not require time for heating something, like a filament, at high temperature, it is a direct conversion of an electron flow to a photon flow so very fast light switching or intensity modulation is possible, ideal for high-speed imaging applications.

The color and spectrum of the light emitted can be precisely controlled by choosing the semiconductor material and by using proper phosphor coatings for obtaining the required spectral emission. LEDs emit light in a specific direction, reducing the need for reflectors or diffusers, and can be easily coupled to optical fibers, mirrors, lenses. LEDs are solid-state devices, making them robust and with a very long working life.

In Table 4.1 the main characteristics of the different light sources are summarized.

4.3 Lighting Geometry

Lighting geometry, i.e., the angular intensity distribution of illuminating light rays, plays a crucial role in enhancing the visibility of shapes, details, and key surface features of the object being measured [2]. When appropriately configured, lighting can enable simple yet effective analogue image processing techniques that quickly yield the desired results. This often eliminates the need for complex digital processing, which may not always provide the same low level of uncertainty.

Table 4.1 Comparison of halogen, LED, incandescent, and fluorescent lamps

Characteristic	Halogen	LED	Incandescent	Fluorescent
Light Source Type	Incandescent (with halogen gas)	Solid-state (Light Emitting Diode)	Incandescent (tungsten filament)	Gas-discharge (with mercury vapor)
Luminous Efficacy	12–24 lumens/watt	80–120 + lumens/watt	10–15 lumens/watt	35–100 lumens/watt
Color Temperature	2,800–3,200 K (warm to neutral)	2,700–6,500 K (warm to cool white)	~2,700 K (warm white)	2,700–6,500 K (varied)
Lifespan	2,000–4,000 h	15,000–50,000 + hours	750–2,000 h	7,000–15,000 h
Warm-up Time	Low	Instant	Low	Few seconds
Heat Emission	High	Low	Very high	Moderate
Energy Efficiency	Moderate	Very high	Very low	High
Initial Cost	Low to moderate	Moderate to high	Very low	Low to moderate
Running Cost	Moderate	Very low	High	Low

4.3.1 Direct Illumination

We have a direct illumination setup when light rays travel straight from the source to the surface of the object and are then reflected within the camera's field of view. The distribution of the reflected intensity is directly influenced by the source's own intensity distribution, typically provided in its technical specifications, such as the polar diagrams shown in Fig. 4.5, which gives the intensity of rays across different angles. Light sources can vary in form, including point, area, or line sources. The reflected light distribution can be analysed using ray tracing software, or by Blender, as previously illustrated in Chap. 2, to aid in selecting and designing the most effective lighting configuration.

The primary advantages of direct illumination are its high luminance and uniform light distribution when the light source is positioned at an adequate distance from the target area. Direct illumination can be employed to create three different lighting conditions, such as bright field, dark field, or back-lighting dark field, as illustrated in Fig. 4.6.

Bright Field lighting

In this configuration, the light reflected from the object falls directly within the camera's field of view. This approach yields good results when imaging diffusive surfaces. However, reflective or complex-shaped surfaces may produce misleading back reflections, resulting in shadows or high-intensity reflection regions (hot spots) on the camera's image plane. Figure 4.7 illustrates examples of both successful and problematic outcomes using bright field lighting.

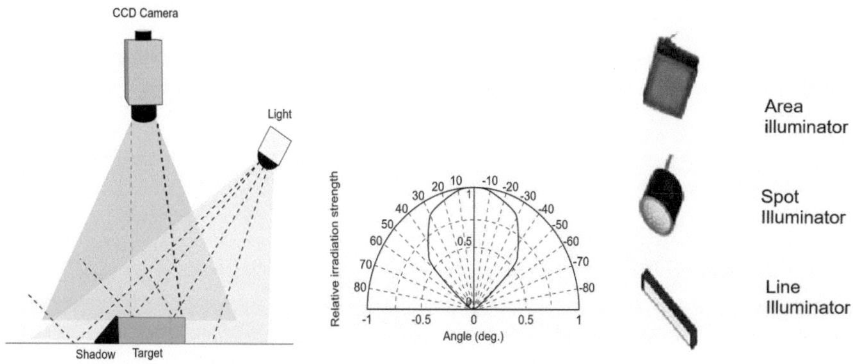

Fig. 4.5 Direct illumination geometry and possible light sources

Fig. 4.6 Direct illumination geometry possible to obtain bright field, dark field or back lighting

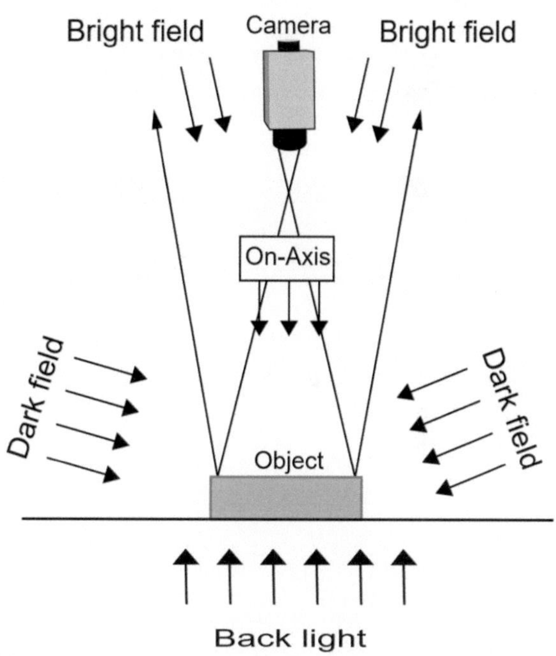

Dark Field lighting

In this configuration, the lighting rays arrive to the object surface at an angle close to 90° relative to the camera's optical axis, causing the main light reflections to fall outside the camera's field of view. As a result, the image is primarily composed of shadows, while only the light reflected by small surface defects such as those shown in Fig. 4.8 are reflected inside the camera field of view. This technique effectively highlights such imperfections.

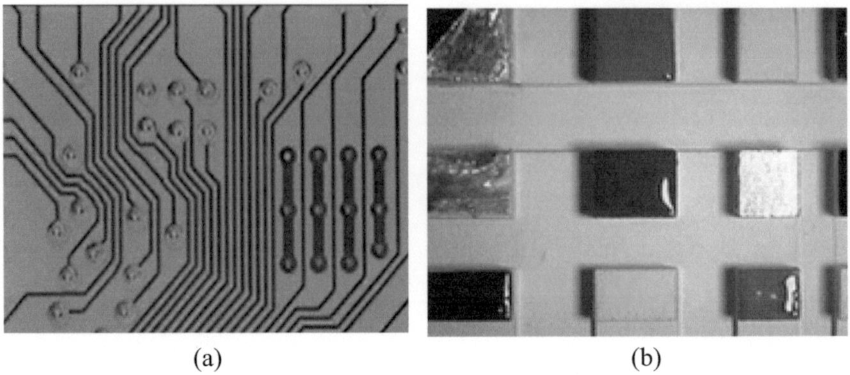

Fig. 4.7 Bright field lighting with **a** good and **b** not good image because of reflections, hot spots and shadows

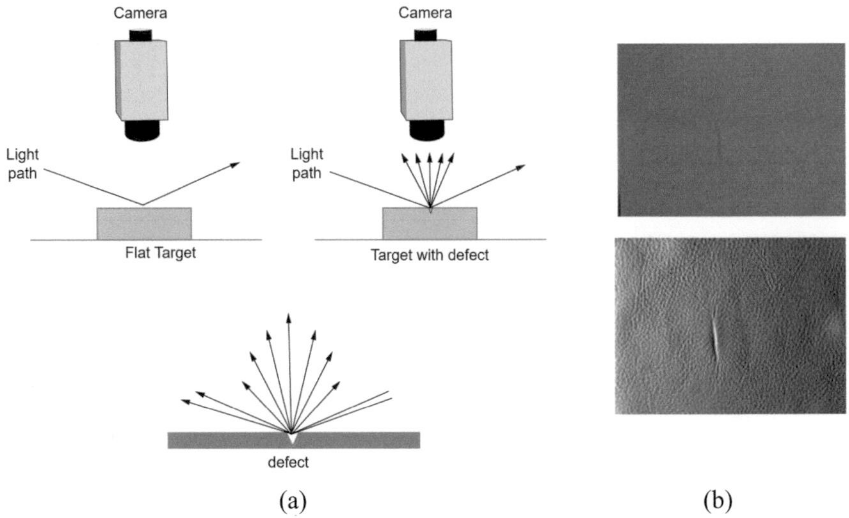

Fig. 4.8 Dark field illumination scheme and example

This lighting technique is especially useful for analyzing flat surfaces to highlight their micro-morphology. It is also commonly used to detect and measure surface defects such as abrasions, scratches, and cracks and is particularly well-suited for reflective materials.

Dark back lighting

In a backlighting dark field setup, the object is positioned between the light source and the camera, which captures the object's silhouette or shadow as illustrated in Fig. 4.9a. This technique can utilize either a distant point light source or an area

(a) (b) (c)

Fig. 4.9 Back lighting scheme (**a**) and examples of images obtained using a baklighting panel (**c**)

illuminator pad, as in Fig. 4.9c. For moving objects, such as those on a production line, a linear light source paired with a line-scan camera is commonly employed.

In backlighting, the object appears as a dark silhouette against a bright background, which takes on the color of the light source. This setup maximizes edge contrast, remove shadows, making it particularly effective for measuring the size, shape, or profile of an object, for example the diameter of a hole or the fluid level inside transparent bottles or tubes. Figure 4.9b shows two images of a bottle captured using bright field and backlighting techniques. It is evident that the liquid level inside the bottle is more clearly visible and measurable with backlighting. Similarly, Fig. 4.9c presents images of a pen, a mechanical component and a coin under both lighting conditions, over an illuminating pad on and off. The component profiles are more distinct and without shadows with backlighting, allowing for easier and more accurate dimensional measurements. Much of the image processing in these cases is effectively performed by the lighting setup itself, without the need for complex digital processing. To minimize light diffusion effects that may blur the object's edges, as shown in Fig. 4.10a, a collimated backlight source can be used. This can be achieved either with a distant point light source that emits parallel rays or by using telecentric optics for both illumination and image capture, as detailed in Chap. 2 and illustrated in Fig. 4.10b.

As an example of application is illustrated in Fig. 4.11 a system, developed in 2016 by the author and collaborators, as part of the project 'Open-framework quality control vision tool for fast operating automation systems'. The system was designed in cooperation with the Italian company Mar-Tech to perform high-speed quality control of mechanical components, enabling approximately 10 dimensional measurements per part at a rate of 100 parts per minute. It operates directly on the production line using multiple imaging stations taking images along different directions while components move. During development, various optical configurations were tested. As illustrated, Fig. 4.12b demonstrates the superior quality achieved using coaxial backlighting combined with telecentric imaging, and reduction in processing time. In contrast, Fig. 4.12a lacks telecentric illumination, resulting in refractive distortions on the screw profile that hinder defect detection reduce measurement accuracy, needs a more complex and time consuming image processing.

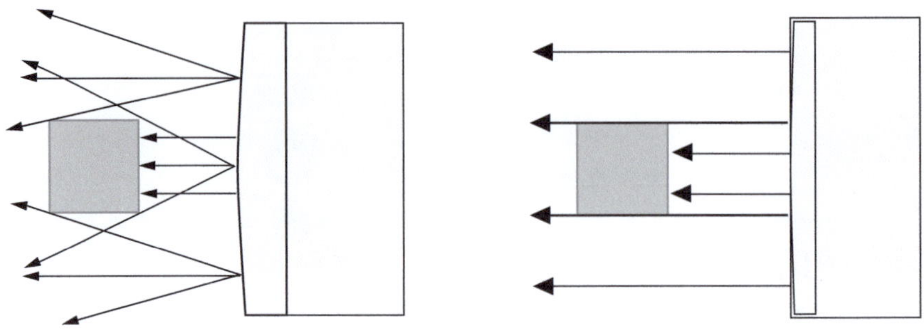

(a) (b)

Fig. 4.10 Normal (**a**) and coaxial (**b**) back lighting

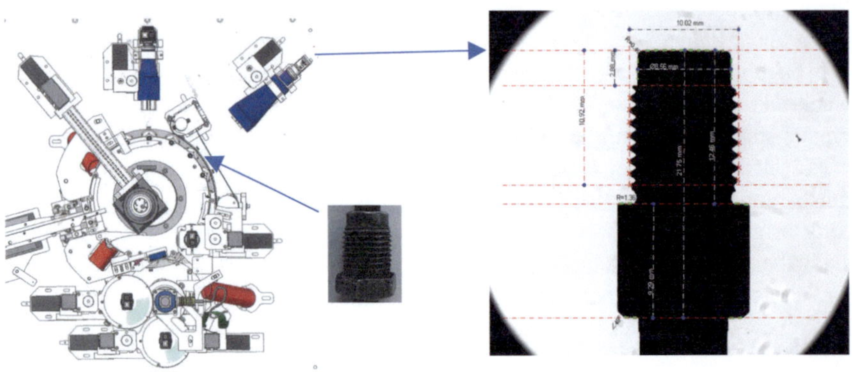

Fig. 4.11 A fast automatic quality control system of mechanical components by multiple imaging stations

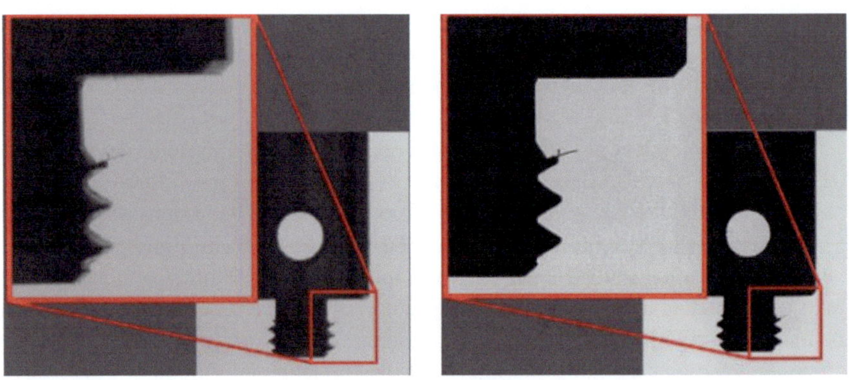

(a) (b)

Fig. 4.12 Results obtained using a quality control system of mechanical components by direct (**a**) or back lighting with telecentric optics (**b**)

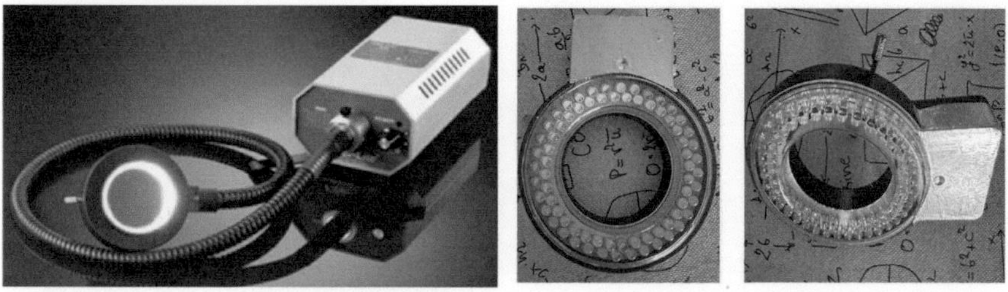

(a) (b)

Fig. 4.13 Halogen lamps with fiber optic boundle and LED ring illuminators

4.3.2 Ring Illumination

Ring illuminators, positioned on the camera objective or integrated into the object's support, are commonly used in optical measurement systems for quality control, and machine vision applications to implement either bright field or dark field lighting. For instance, halogen lamps can be coupled with fiber optic bundles to produce a specific high-intensity light distribution, such as a ring of light sources, as shown in Fig. 4.13. This type of illumination is especially valuable in situations where exposure time is limited by component motion, such as in fast moving components along a production line or in high-speed measurement applications, providing the necessary brightness to have a sufficient exposure level with low aperture times.

Ring illuminators can also be implemented using LED sources arranged in circular arrays to achieve the same lighting geometry, as shown in Fig. 4.13b. Depending on the configuration and angle of illumination, both bright field and dark field lighting schemes can be realized. Figures 4.14a, b illustrate two images of the same coin captured under these two different lighting conditions, highlighting the contrasting visual effects each method produces.

4.3.3 Indirect and Diffused Illumination

As previously illustrated, direct lighting of an object's surface can produce unwanted high-intensity reflection areas, commonly known as hotspots, due to the non-uniform intensity distribution of the light source itself and/or to the surface reflection coefficient, as shown in Fig. 4.15.

By placing a partially transparent optical component between the light source and the object, it is possible to achieve a more uniform distribution of light intensity on the object's surface through absorption and refraction. Additionally, light from one or more sources can be diffused by means of diffusive reflecting surfaces, such as spherical or cylindrical domes, which scatter the light evenly, as illustrated in Fig. 4.16. This setup can reduce hot spot due to not uniform lighting sources.

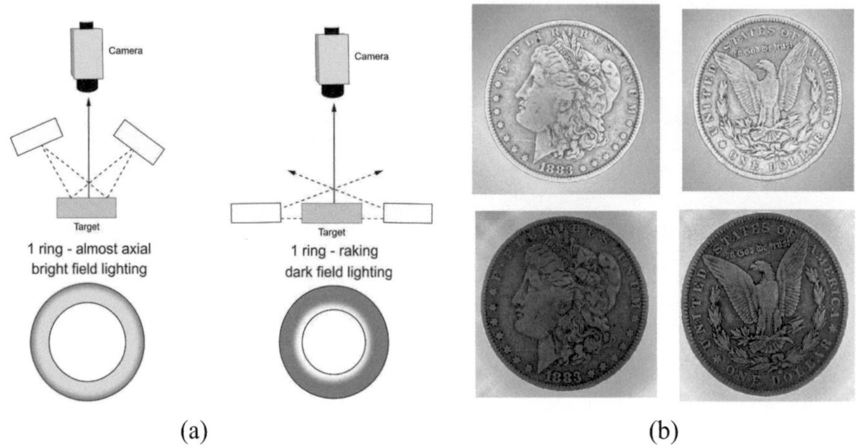

Fig. 4.14 Ring illuminators for bright or dark field lighting scheme and examples

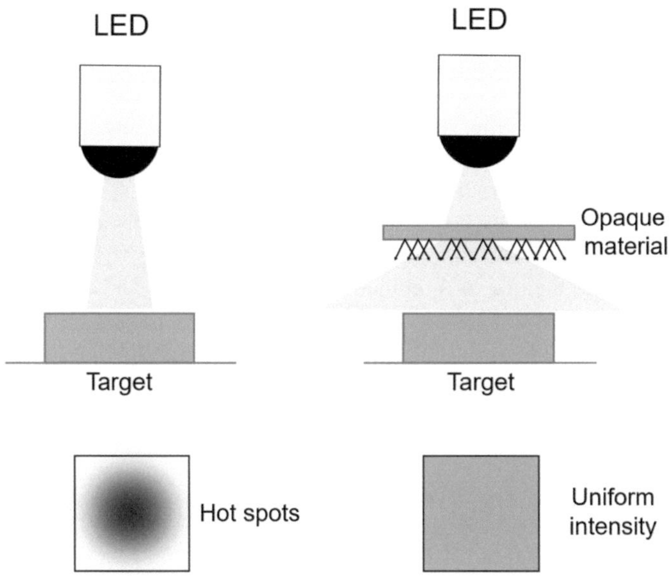

Fig. 4.15 Using a partially transparent material to diffuse light distribution of a not uniform lighting source

Dome-based diffuse reflections can be combined with light diffusion through transparent materials to create what are known as cloudy day illuminators, as shown in Fig. 4.17. These systems are particularly effective for imaging highly specular or reflective surfaces like that of polished metals. The resulting illumination mimics natural light conditions on a cloudy day, where sunlight is scattered uniformly in all

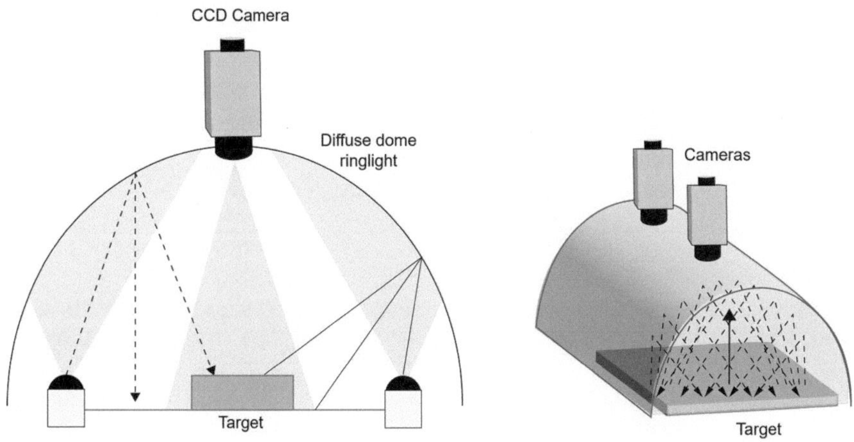

Fig. 4.16 Light diffusion by diffusive reflecting domes

directions by the clouds. This eliminates shadows and hotspots, ensuring uniform lighting across the entire surface of the object.

For example, in Fig. 4.17c the image of a steel ball is taken by direct lighting and in Fig. 4.17b by a cloud illuminator reducing hot spot due to specular reflections.

Another effective method for managing shadows on object surfaces with sharp or deep cavities is the use of indirect coaxial lighting, as illustrated in Fig. 4.18. This technique employs a beam splitter, an optical component previously described in Chap. 2, to redirect light from the source coaxially to the receiving optics onto the object's surface, while simultaneously allowing the back-reflected light to travel along the same optical axis toward the same beam splitter. Anyway, as previously

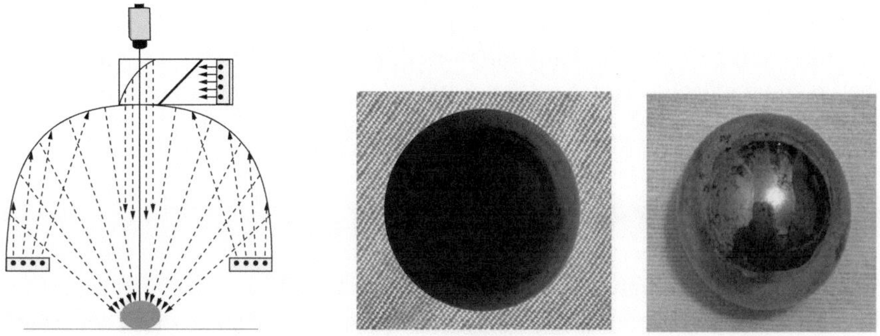

(a) (b) (c)

Fig. 4.17 Cloud day illuminators geometry and example

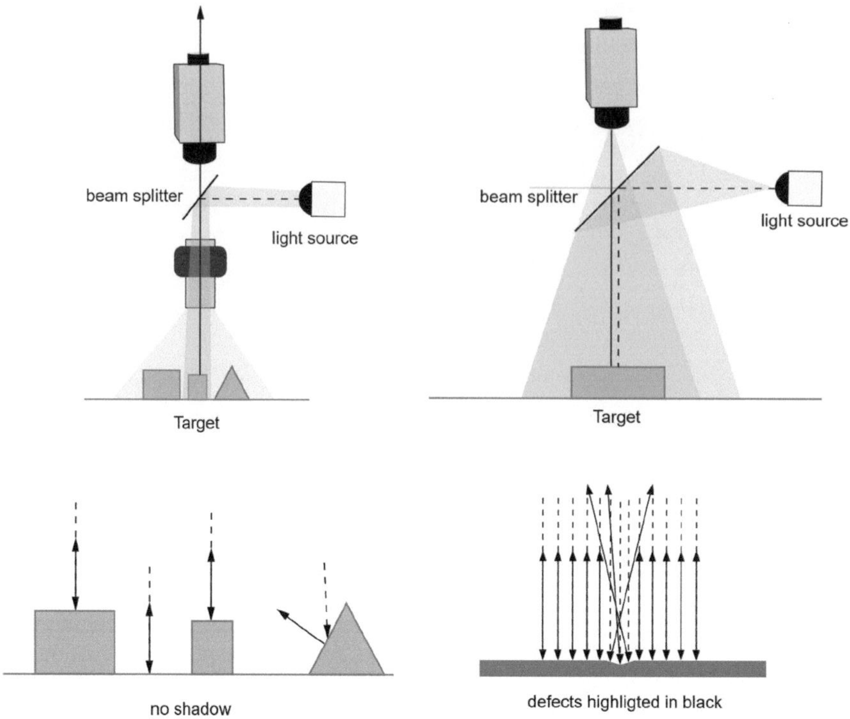

Fig. 4.18 Coaxial lighting by a beam splitter

illustrated the beam splitter divide an image in two, so reduce both typically at 50% the illuminating lighting power and the received one.

Coaxial illumination can be combined with a diffuser. It provides uniform illumination over the entire area of the object, does not generate shadows, and reduces local reflections. It is a very useful setup for inspecting electronic boards or for inspecting labels on high reflective surfaces (Fig. 4.19b).

4.3.4 Illumination by Polarized Light

Lighting by polarized light, as previously discussed in Chap. 1, can be obtained by placing a polarizing filter over the light source. An additional polarizing filter, called an analyzer, can be mounted on the objective of the receiving imaging system, as shown in Fig. 4.20a. The analyzer can be oriented to allow only light polarized in its direction to pass through. As a result, rays reflected, polarized in other directions, are effectively blocked, enhancing image contrast and minimizing unwanted glare.

For example, Fig. 4.20b presents pairs of images captured with and without the use of polarizing filters. These comparisons clearly demonstrate how reflections from

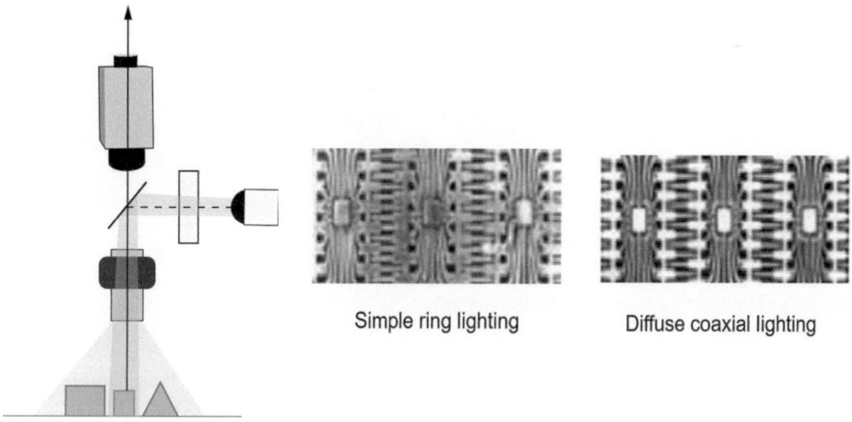

(a) (b)

Fig. 4.19 Diffuse coaxial lighting

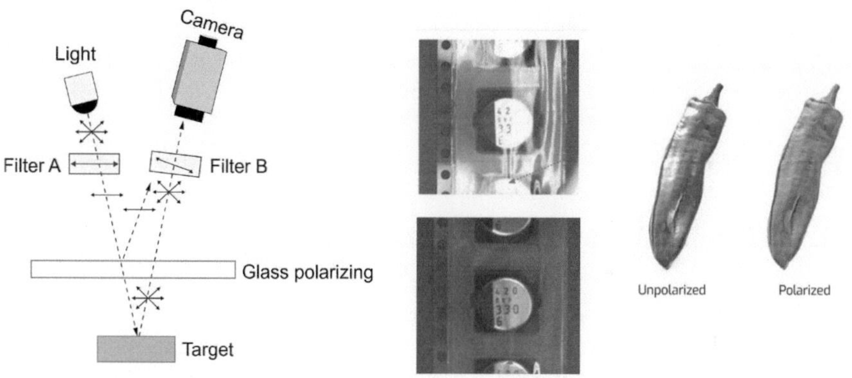

(a) (b)

Fig. 4.20 Polarized lighting: receiving optics scheme (**a**) and examples (**b**)

materials that reflected polarized light can be significantly reduced. This technique makes it possible to see through transparent surfaces such as glass or plastic sheets, whose polarized reflections would otherwise obscure underlying details and object shapes when using standard lighting.

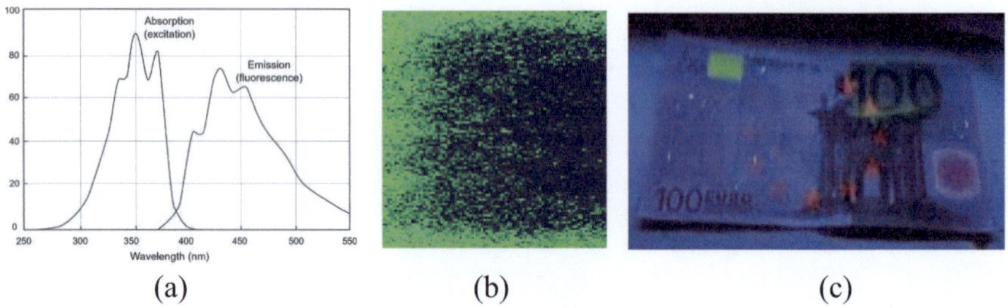

Fig. 4.21 UV lighting principle and applications

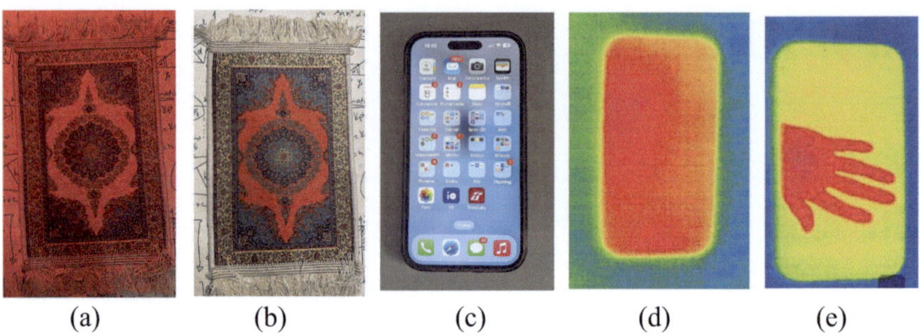

Fig. 4.22 Infrared lighting

4.3.5 Illumination by Ultraviolet Light (Black Light)

Certain materials, surfaces, paints, or particles, when illuminated with light of a specific wavelength, can reflect back light at a different wavelength. For example, under ultraviolet (UV) illumination at around 350 nm, known as black light because it is invisible to the human eye, some particles exhibit the fluorescence phenomenon, emitting visible light at wavelengths around 450 nm, as illustrated in Fig. 4.21. This phenomenon allows only the fluorescent particles to be visualized, enabling precise measurement of their shape and motion for example within a fluid, with image processing techniques illustrated in the following chapters.

4.3.6 Illumination by Infrared Light

When imaging of an object's surface is compromised by issues such as reflections, hotspots, or interference from ambient light sources, particularly within the visible spectrum, a possible solution is to use infrared (IR) illumination.

By employing IR light sources along with either a thermal camera or a standard CCD/CMOS camera equipped with high-pass filters, it is possible to capture images exclusively in the infrared range, effectively eliminating visible light disturbances. Examples are presented in Fig. 4.22: (a) shows an infrared-lighted carpet, (b) its corresponding image in visible range, in Fig. 4.22 (c) a smartphone illuminated with visible light, in (d) the same smartphone under infrared illumination where all the colour details on the screen disappered, and in (e) the infrared reflection of a hand, placed in front of the same smartphone, generated by his high infrared reflecting surface.

An infrared image is typically a grayscale image, as the sensor integrates the incoming photon flux across the entire infrared spectrum, removing specific wavelength information. Since infrared wavelengths are longer than those of visible light, the diffraction-limited resolution discussed in Chap. 2 is lower. Additionally, thermal cameras often have lower spatial resolution compared to standard visible-range cameras, as explained in Chap. 8. However, as previously noted, CCD and CMOS sensors exhibit good sensitivity in the near-infrared (NIR) range. This makes them suitable alternatives to thermal cameras for capturing images in the NIR spectrum, even when the scene is not visible to the human eye.

4.3.7 Illumination by Structured Light

One of the simplest methods to highlight and visualize the position of a point on a flat, uniform, or even irregular surface is by using a laser pointer. This technique creates the simplest possible structured light pattern using a narrow beam that produces a bright reflection spot (a single point) on the captured image by cat's eyes, as shown in Fig. 4.23a. Another widely used structured lighting technique involves generating a light sheet, typically by passing a laser beam through a cylindrical lens. This lens spreads the beam into a thin plane of light, oriented perpendicular to the lens axis, as described in Chap. 2. The resulting light sheet, illustrated in Fig. 4.23b, reveals

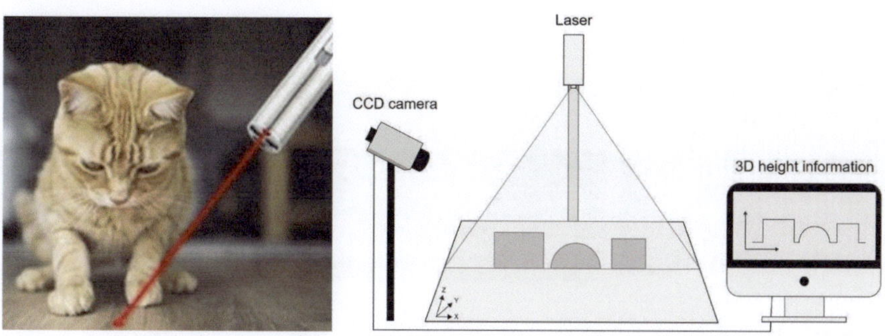

(a) (b)

Fig. 4.23 Laser line and plane lighting to see points and profiles of surfaces

the intersection between the sheet and the object's surface, effectively outlining its profile or cross-section.

In this case, the structured lighting technique produces a line on the image that corresponds to the object's surface profile.

More generally, structured lighting refers to the projection of specific geometric light patterns onto an object's surface with the goal of identifying spatial information such as point positions, profiles and complete surface shapes.

Typical structured light patterns, as shown in Fig. 4.24, include lines, point arrays, circular patterns, and more. These patterns can be easily generated using slide projectors, video projectors, or dedicated structured lighting systems.

An example of the results obtained through structured lighting is shown in Fig. 4.25, where multiple points or surface profiles are highlighted on a uniformly coloured objects. This technique enhances the visibility of the surface's complete shape, allowing for a more detailed description and, as will be illustrated in the following chapters, helps to perform fast and accurate shape measurement by this preliminary image analogue processing by lighting.

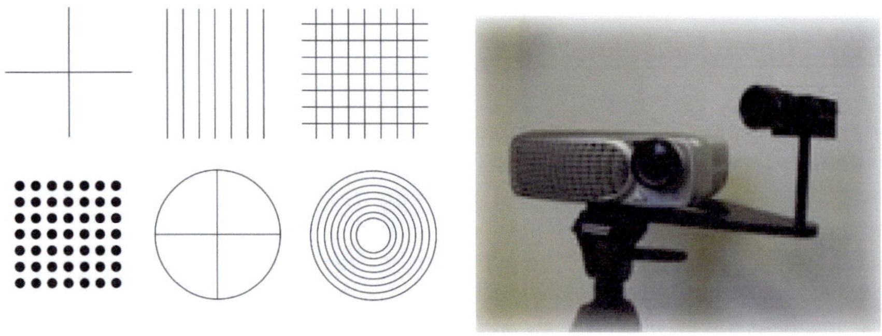

Fig. 4.24 Structured lighting patterns

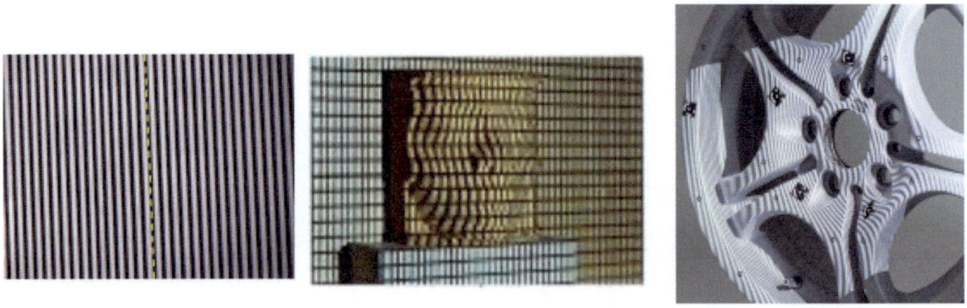

Fig. 4.25 Structured lighting effects

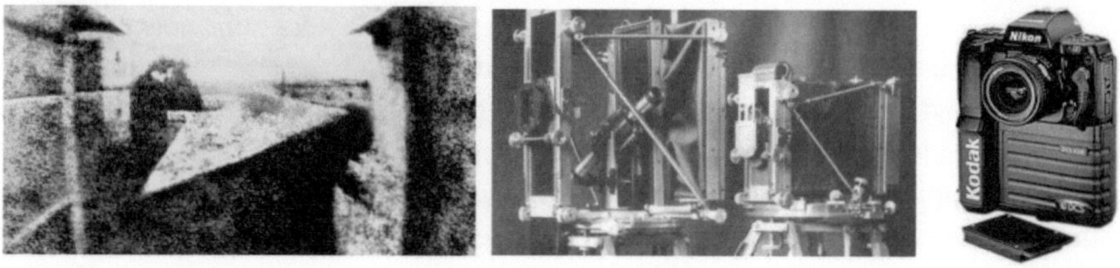

(a) (b) (c)

Fig. 4.26 First picture on paper and first photogrammetric system

4.4 Photogrammetry

The first name used for indicating the techniques and methods aimed at obtaining measurements by pictures is "photogrammetry". This practice originated in the first half of the nineteenth century and saw its first use in architecture to measure the shape of buildings and territories with images. The first photograph on paper dates to 1826 by Joseph Nicèphore Nièpce (Fig. 4.26a), and is a view from his study window in Le Gras, but the discovery is attributed to the year 1839 (Niépce and Daguerre).

The first use in the field of measurement came in 1850 (Laussedat and Meydenbauer), in the field of architecture, with the development of a special measuring camera (Fig. 4.26b) and special analysis methods for reconstructing buildings shape.

It was not until the new century that the first photogrammetric analogic camera was released by Kodak (DCS420) in 1975, designed for measuring purposes. Later, with the advancement of technology and digitisation, in the 1990s, the first digital systems in this field came onto the market.

4.5 Image Calibration for 2D Measurements

In the previous chapters, we described how to correctly capture an analog image on the image plane and how to digitize it properly. The result is a numerical matrix representing the image, which we can analyze to perform measurements. However, our ability to identify the position of features in the image is limited to discrete pixel coordinates, meaning we can only express distances in terms of pixel counts [3].

To convert these pixel-based measurements into real-world units, such as millimeters or meters, we need at least a scale factor k (mm/pixel). More generally, we must know the complex relationship between the real-world coordinates of points on an object's surface and their corresponding digital coordinates (i, j) in the matrix.

As discussed in Chap. 2, there are mathematical models that link real-world coordinates to image plane coordinates. These models can be derived by capturing images of objects with known geometries. Additionally, we have examined how to incorporate the main sources of field aberration introduced by the camera's lenses and objectives into these models to improve measurement accuracy.

The process of determining the relationship between pixel coordinates and real-world coordinates by imaging known geometries is called camera calibration. Main mathematical models and available software to perform camera calibration has been described previously in paragraph 2.10 of Chap. 2.

In this chapter, we focus on 2D measurement techniques that can be performed using a single camera [4]. As previously discussed, if the camera is not equipped with a telecentric objective, the size of objects projected onto the image plane will vary depending on both the focal length of the lens and the distance between the object and the camera. Distant objects appear smaller in the image, and the dimensions of an object are also affected by its orientation relative to the camera's optical axis, perspective distortion, and lens aberrations.

As described in paragraph 2.10 of Chap. 2, the calibration relationship depends on the relative position between the object and the camera. For accurate measurements, this relative configuration must remain the same during both the calibration and measurement phases, or at least vary within a limited and controlled range. In general, the calibration setup must replicate the actual measurement conditions.

Calibration is typically performed using reference targets with precisely known geometries, such as dot grids, crossline patterns, or chessboards. It is essential to ensure the use of a high-quality calibration target, a vibration-free mounting system for the camera, and stable conditions for both the calibration plate and the object during measurement.

By analysing images of the reference target, it is possible to estimate the camera calibration matrix, which encapsulates both the projection geometry and the effects of lens distortions. This matrix enables the transformation of pixel coordinates into real-world dimensions, thus making precise measurements possible directly from digital images (Fig. 4.27).

One of the most relevant aspect to evaluate by the calibration is the measurement uncertainty. This can be estimated by performing length measurements on an object using the image processing system and then comparing the results to those obtained using a reference measurement method with significantly lower uncertainty. Generally, the reference method should have an uncertainty at least ten times lower than

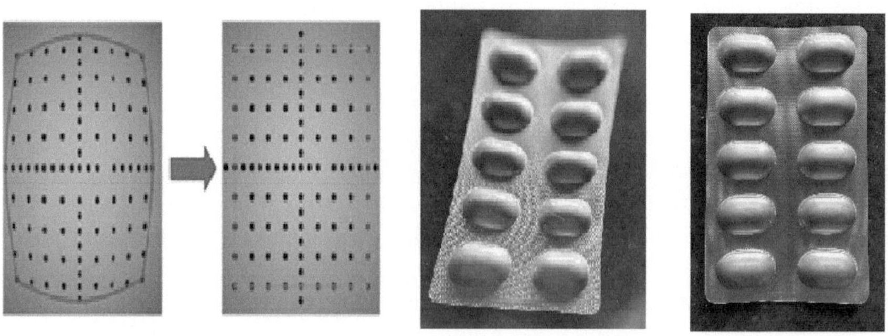

Fig. 4.27 Calibration correction for 2D dimensional measurements

the target uncertainty of the system to calibrate. Alternatively, a reference object with a known dimension and certified uncertainty, again at least ten times more precise, can be used.

For example, if we aim to demonstrate that our image-based measurement system has an uncertainty of 10 μm, we must validate it using a reference with an uncertainty of less than 1 μm, as a general rule.

The calibration uncertainty of the image-based system depends on all components of the measurement chain. This includes the optical quality of the lenses, lighting conditions, the resolution of the digitization process in space, quantization resolution in amplitude, image processing software, and the calibration model parameters used. Each of these factors contributes to the overall system accuracy and must be carefully optimized and controlled to ensure reliable and traceable measurements.

If the system benefits from high-quality lighting, excellent optics, a grey level digitalization camera with more than 12 bits, and the image plane is perpendicular to the optical axis (or telecentric lenses are used) then a simplified estimation of measurement uncertainty can be made. In such optimal conditions, the primary source of uncertainty can be attributed to the spatial resolution of the digital image.

Let's assume the following parameters:

(1) The camera sensor has a physical width (or diagonal) of L.
(2) The sensor digitizes this width into N pixels, resulting in a spatial resolution of L/N.
(3) The lens provides a magnification of m at the object distance, resulting in a field of view of width (L m) on the object.

The image processing software includes sub-pixel interpolation, enabling point localization with a resolution of 1/n pixel, i.e., dividing each pixel into n interpolated positions.

Under these assumptions, a rough estimation of the discretization uncertainty, denoted as U_c, can be expressed as:

$$U_c = (L\ m)/(n\ N) \tag{4.1}$$

This formula provides an approximate estimate of the system's spatial uncertainty, based on the combined effects of sensor resolution, magnification, and interpolation accuracy, and can be useful to choose at least the camera main parameters of number of pixels and sensor dimension for an application requiring at least the uncertainty U_c.

So, for example, on a field of view (L m) of 1 m, subpixel interpolation of a factor of $n = 10$, having a digitizing sensor array with 2048 pixel in width we can estimate an uncertainty due only to spatial digitalization resolution equal to $U_c = 1/20480$ m, so of about 50 μm.

Camera calibration should be performed periodically, as the optical and imaging system including both the illumination setup and the camera can be subject to changes over time that may compromise calibration accuracy. These changes may arise from mechanical shocks, optical misalignments, contamination (such as dust or smudges on lenses), or material degradation of optical components. Even minor variations in these elements can significantly affect the reliability of the calibration and, consequently, the measurement uncertainty.

In cases where the object has a complex geometry, or where high resolution and measurement accuracy cannot be achieved even with high-quality optics and proper calibration, it becomes necessary to adopt 3D measurement techniques. These methods provide additional spatial information and are better suited for capturing intricate surfaces or depth-related features that 2D imaging systems cannot accurately resolve.

4.6 Edge Detection

Taking measurements on images means determining the distance between two points of two lines within the image, or rays of fitting circles, or other geometrical primitives defined by image processing [3]. Lines can be used to define edges of the object, generally borders between the pixels representing object surface and the pixels representing the background, as in Fig. 4.28.

In general, an edge can be defined as a region in an image where there is a step change in grey level values.

While edges commonly arise from differences in grey level between an object's surface and its background, they can also result from various other factors, including:

- Discontinuities in surface orientation
- Variations in surface reflectivity, which may cause changes in colour or grey level
- Lighting inconsistencies, such as shadows or bright reflection spots (hot spots)
- Deliberately introduced features through structured lighting methods (e.g., projected points or lines).

Edges can take the form of lines, circles, or more complex curves. When analyzing an edge by moving along a direction perpendicular to it, the grey level typically changes in a pattern, such as the one illustrated in Fig. 4.29.

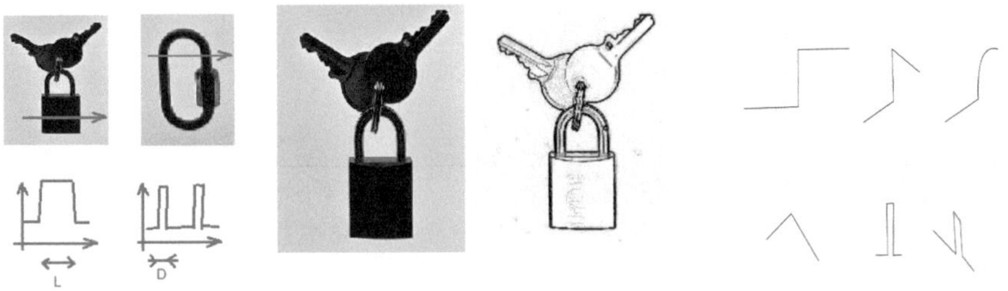

(a) (b) (c)

Fig. 4.28 Edges in images

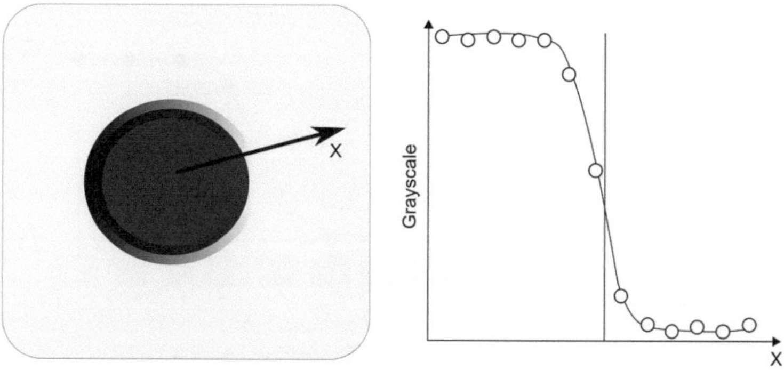

Fig. 4.29 Grey level distribution along a radius crossing an edge of a circular object

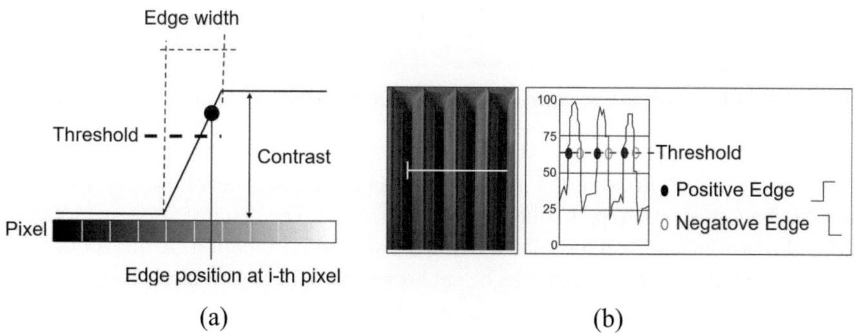

Fig. 4.30 Edge parameters

Edges are characterized by some parameters [3]: lower gray level value, higher value and their difference, named *edge contrast*, and its *width*, the number of the pixel from lower to higher value, as illustrated in Fig. 4.30a. Edge can be also *positive* or *negative*, as illustrated in Fig. 4.30b.

If there is no contrast, then no edge can be detected. An edge can only be defined if the contrast exceeds a minimum threshold, which must be at least above the image's noise level to ensure reliable detection.

Edge positions can typically be detected using one of the following four main approaches:

1. *Thresholding an interpolated grey-level profile along a line by setting a specific intensity level (threshold).*
2. *Setting a global threshold to the image histogram to separate edges based on overall intensity distribution.*
3. *Identifying local maxima and minima in the first derivative of the image, which correspond to points of rapid intensity change.*

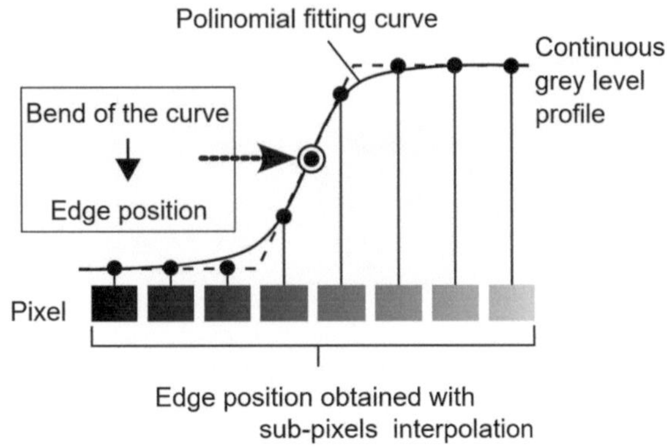

Fig. 4.31 Edge position by interpolation of grey level step along a line

4. *Locating zero-crossings in the second derivative of the image, which also indicate the presence of edges.*

(1) *Thresholding an interpolated grey-level profile along a line by setting a specific intensity level.*

If we define a line crossing the edge in an image, the grey level profile along this line can be a positive step, as illustrated in Fig. 4.31. The discrete grey level values along the profile can be interpolated using a polynomial function. A threshold level (often called the *soil level*) can then be set between the two grey levels of the step, at the midpoint or at concavity change point. This threshold intersects the interpolated curve at a specific point, and the abscissa (x-coordinate) of this intersection define the edge position.

This is the most basic and fundamental technique to find the edge position also useful to define a model to estimate uncertainty in position measurement of the edge. So, if we have different grey level profile on some edges like in Fig 4.32b uncertainty in edge position detection can be estimated as follow.

As illustrated in Fig. 4.34a, let us consider the following parameters:

- ΔI: the contrast of the edge,
- I_W: the width of the edge,
- L_n: the noise level, which may arise from various sources. For instance, discretization noise can be considered as $L_n = 1/2^n$, where *n* is the number of bits of the sensor's A/D converter. If interpolation is used, the noise may be lower, e.g., $L_n = 1/(k\ 2^n)$, where *k* is a sub-level resolution factor (like sub-pixel resolution). However, in practical scenarios, noise may be higher due to other noise sources.

Based on similarity of the two triangles abc and ABC in Fig 4.34a, we have $U_s/L_n=I_W/\Delta I$, so the uncertainty in edge position, denoted as U_s, can be estimated in unit length as:

$$U_s = 1\, L_n I_W / \Delta I \tag{4.2}$$

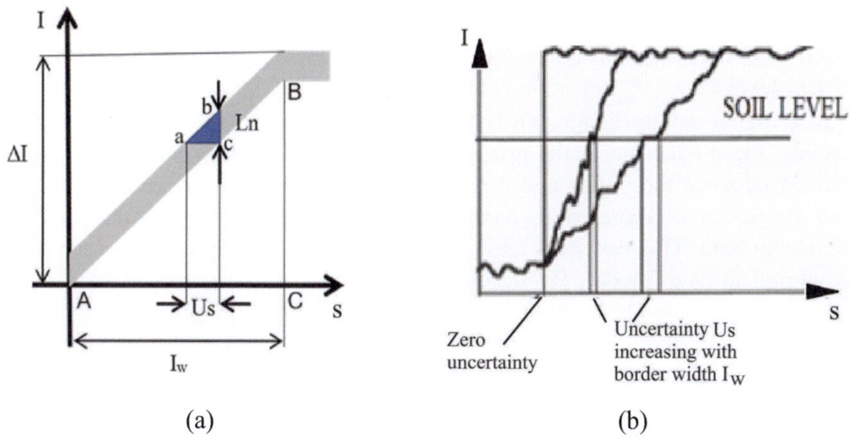

Fig. 4.32 Uncertainty in edge position detection

where l is the length of the measurement performed or of the field of view.

From this equation, it is clear that:

- If the edge width I_w is very small (approaching zero or one pixel), the impact of noise on uncertainty becomes negligible, and U_s tends toward zero. Therefore, thin edges are preferable for minimizing uncertainty.
- Conversely, if the edge contrast ΔI is zero, the uncertainty U_s becomes infinite. This highlights the importance of having edges with high contrast to ensure accurate edge detection.

An estimation of uncertainty for a typical medium level camera having n = 12 bit, taking $L_n = 1/(2^n) = 0{,}000244$, $I_w = 10$ pixels and $\Delta I = 2000$ (a very good edge contrast equal to about half camera grey level range) give us on a field view l = 1 m an uncertainty of $U_s = 1.2$ μm, an order of magnitude less than the previously estimated U_c uncertainty due to spatial sampling.

Anyway, the combined uncertainty U due to spatial sampling, vertical noise and discretization can be roughly estimated by the common way to combine uncorrelated uncertainty sources as:

$$U = \sqrt{U_s^2 + U_c^2} \tag{4.3}$$

And used for example to do a first estimation of the minimal spatial resolution (NxM pixels on the camera sensor) and number of bits n of the camera AD converter we need to perform measurements with a specified uncertainty level U. Assuming that no other noise sources are more relevant due to optical distortions, lighting, etc.

(2) *Setting a global soil level on the image histogram*

A grey-level histogram of an image is a graphical representation in the form of vertical bars, where each bar corresponds to a specific grey level. The height of each bar indicates the number of pixels in the image that possess that grey level. Figure 4.33 illustrates this concept using a simple, low-resolution image with only 16 grey levels.

In a higher-quality image, such as the one shown in Fig. 4.34, the histogram contains more bars due to the greater range of grey levels. When there is good contrast between the object and the background, the histogram can exhibits two distinct peaks, each representing the most frequent grey level of either the object or the background. This pattern is known as a bimodal distribution, where each peak is referred to as a 'mode'. By selecting a threshold value located between these two peaks, it is possible to convert the image into a binary format: assigning a grey level of 0 to all pixels above the threshold and a grey level of 1 to those below. This binary transformation define the object's borders, making it suitable for further image analysis and measurement.

In some cases, it may be useful to define more than one threshold level, allowing for a type of gray-level classification known as image segmentation [3]. Unlike

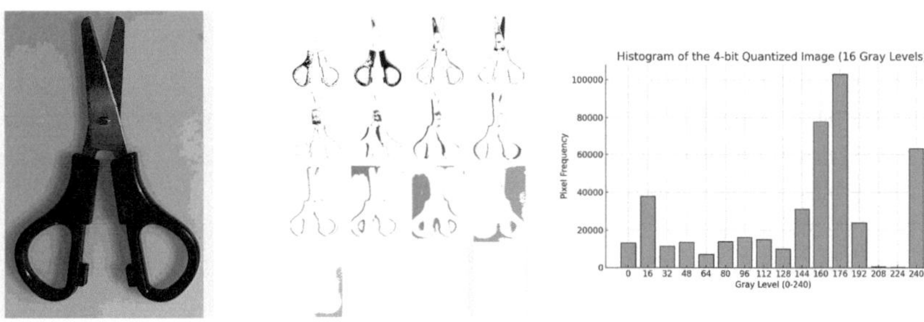

(a) (b) (c)

Fig. 4.33 A 4 bit grey level image (16 levels) (**a**), the 16 images of pixel having same grey level (**b**) and the histogram of number of pixels vs grey level (**c**)

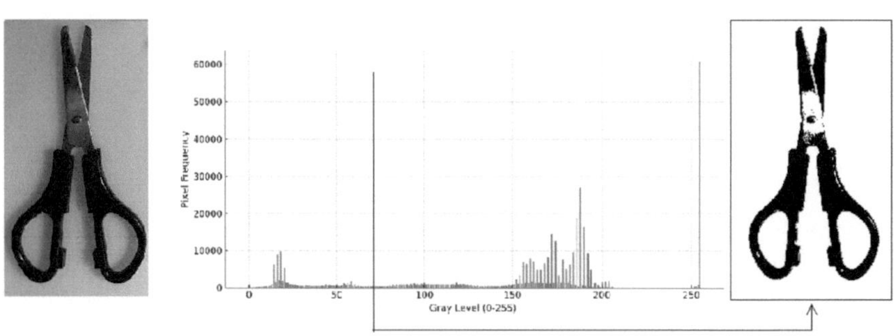

Fig. 4.34 Generating a binary image by a soil on its histogram of grey levels

binary images, which consist of only two levels, segmented images contain multiple grey levels (or different colour or other image parameters). This enables the identification and measurement of different features or objects within the same image. Image segmentation is commonly used in fields such as medical diagnostics, where it facilitates the extraction of anatomical structures such as organs or tissues from medical x-ray images. This kind of measurements supports a variety of applications, including diagnosis, treatment planning, and image-guided surgery.

(3) *Identifying local maxima and minima in the first derivative of the image*

As mentioned earlier, an edge in an image corresponds to a sudden and rapid change in grey level. This change can be detected by analyzing the rate at which the grey level changes by calculating image first derivative. As illustrated in Fig. 4.35, a positive edge is characterized by a transition in slope: for example, starting from 0°, increasing to 45°, reaching a maximum at 75° at the center of the transition (or 'step'), then decreasing back to 45°, and finally returning to 0°. The point where the slope reaches its maximum at 75°can be considered the location of the edge.

Similarly, the edge corresponding to a decreasing grey-level step can be identified by locating the minimum of the first derivative. By taking the absolute value of the first derivative, both positive and negative edges can be detected through their local maxima, allowing for the identification of all object boundaries relative to the background.

This concept can be extended to a two-dimensional image I(x, y), which can be considered a grey-level field, similar to the fields introduced in Chap. 1. In this

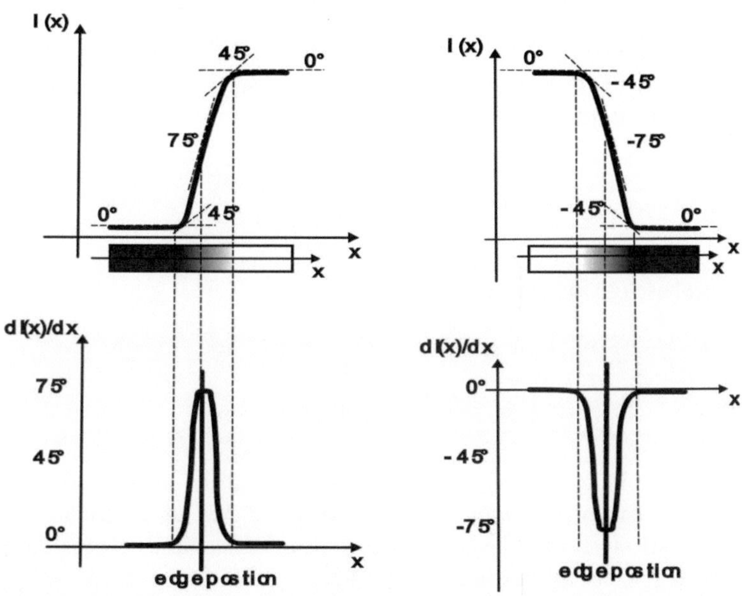

Fig. 4.35 1D edge detection by first derivative

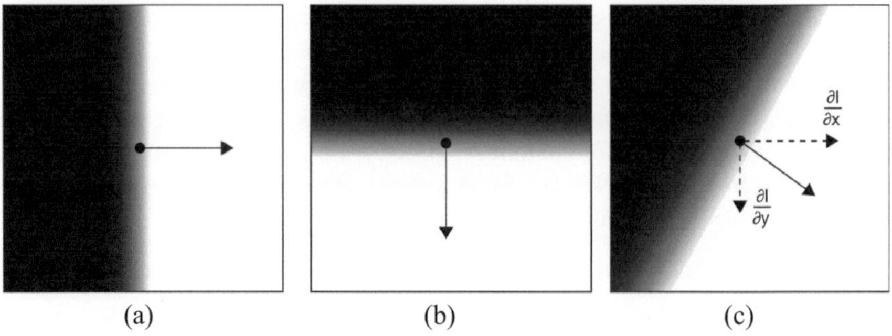

Fig. 4.36 2D edge detection by computing image gradient, **a** and **b** changing only in one direction, **c** changing in both directions

context, we compute partial derivatives along the x and y axes to obtain the two component of the image gradient vector. The gradient provides information on both the magnitude and direction of the most rapid change in grey level, as illustrated in Figure 4.36. Since the gradient is a vector, it can be represented either by its two components (along x and y) or by its magnitude and orientation (i.e. the phase). These representations result in two separate images that describe the spatial variation in intensity across the original image.

In Fig. 4.36 we have, for the tree images the tree gradients:

$$\nabla I = \left[\frac{\partial I}{\partial x}, 0\right] \quad \nabla I = \left[0, \frac{\partial I}{\partial y}\right] \quad \nabla I = \left[\frac{\partial I}{\partial x}, \frac{\partial I}{\partial y}\right] \tag{4.4}$$

From the two gradient vector components along x and y direction image gradient magnitude of the third image will be:

$$M = \sqrt{\left(\frac{\partial I}{\partial x}\right)^2 + \left(\frac{\partial I}{\partial y}\right)^2} \tag{4.5}$$

And its phase (giving edge orientation) is:

$$\theta = \tan^{-1}\left[\frac{\left(\frac{\partial I}{\partial y}\right)}{\left(\frac{\partial I}{\partial x}\right)}\right] \tag{4.6}$$

However, the image I(x, y) formed on the image plane by the camera lens is digitized by the camera sensor into a discrete matrix of grey levels, denoted as I(i, j). As a result, derivatives must be approximated using differences between the values of neighboring pixels, as illustrated in Fig. 4.37.

To compute differences, we need at least two values along each direction (x and y) which requires a minimum of four neighboring pixels, as shown in Fig. 4.37.

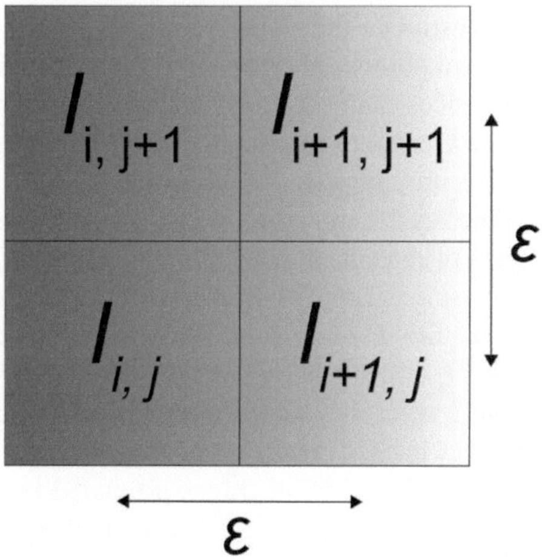

Fig. 4.37 Four pixels used to estimate derivatives in two directions

A basic approximation of the grey-level variation, corresponding to the image's first derivative or gradient, can then be calculated using these pixel values by finite differences, as:

$$\frac{\partial I}{\partial x} \approx \frac{1}{2\varepsilon}\left[\left(I_{i+1,j+1} - I_{i,j+1}\right) + \left(I_{i+1,j} - I_{i,j}\right)\right] \tag{4.7}$$

$$\frac{\partial I}{\partial y} \approx \frac{1}{2\varepsilon}\left[\left(I_{i+1,j+1} - I_{i+1,j}\right) + \left(I_{i,j+1} - I_{i,j}\right)\right] \tag{4.8}$$

where ε is the distance between pixels, i.e. pixels width. It can be easily seen that above formulas correspond to the definition given in Cap 3 of the convolution of the image matrix I on each of it's I(i, j) elements with the two matrix:

$$\mathrm{C_x} = \frac{1}{2\varepsilon}\begin{bmatrix} -1 & 1 \\ -1 & 1 \end{bmatrix} \quad \mathrm{C_y} = \frac{1}{2\varepsilon}\begin{bmatrix} -1 & 1 \\ -1 & 1 \end{bmatrix} \tag{4.9}$$

Adding a scale factor of $\frac{1}{2\varepsilon}$. So we can calculate the complete two first derivative matrix $\frac{\partial I(i,j)}{\partial x}$ and $\frac{\partial I(i,j)}{\partial y}$ along x and y directions of I(i,j) as:

$$\frac{\partial I(i,j)}{\partial x} \approx I(i,j) * C_x \quad \frac{\partial I(i,j)}{\partial y} \approx I(i,j) * C_y \tag{4.10}$$

For each image element I(i, j) thus obtaining the gradient vector field:

$$\nabla \mathrm{I} = \left[\frac{\partial I(i,j)}{\partial x}, \frac{\partial I(i,j)}{\partial y}\right] \tag{4.11}$$

To estimate the slopes (derivatives) around each pixel I(i, j) of our image more than two points can be used, so more complex formulations than a simple two pixel differences can be used, by taking other surrounding pixels to better estimate slopes. This corresponds to use larger matrix for making convolution. The most used matrix kernels are illustrated in Table 4.2.

The key difference between the four kernels matrices in Table 4.2 lies in their trade-offs between edge localization, noise sensitivity, and computational complexity. The simplest 2 × 2 convolution matrices offer more precise localization of edge positions and require less computational effort. However, they are more sensitive to noise. In contrast, larger matrices such as Prewitt or Sobel provide better noise suppression through averaging but come at the cost of increased computational effort and reduced accuracy in edge localization. Therefore, for high-contrast images, such as those with sharp black-and-white edges produced by backlighting, the use of simpler 2 × 2 matrices is generally more effective. On the other hand, for images with blurred gray levels, significant noise, or low spatial and amplitude resolution due to digitization, larger convolution kernels are preferable as they help average out noise and reduce uncertainty in edge position detection.

An example of image derivative and gradient calculation is shown in Fig. 4.38. The brighter areas indicate stronger variations in light intensity, thus highlighting edges and contours (such as people, the trail, rocks, and ridges). By analysing the first derivative of the image and applying a thresholding criterion, such as selecting the median value between the maximum and minimum slopes, or using a threshold based on the histogram of the gradient magnitude, it is possible to identify edge positions.

This process results in a binary image that highlights object boundaries, allowing for measurements and further analysis. In some cases, defining more than one threshold level can be useful to classify different types of edges or regions obtaining an image segmentation.

Table 4.2 Convolution matrix kernels used to estimate partial derivatives of images

Gradient by:	Roberts	Prewitt	Sobel (3 × 3)	Sobel (5 × 5)
$\frac{\partial I}{\partial x}$	0 1 −1 0	−1 0 1 −1 0 1 −1 0 1	−1 0 1 −2 0 2 −1 0 1	−1 −2 0 2 1 −2 −3 0 3 2 −3 −5 0 5 3 −2 −3 0 3 2 −1 −2 0 2 1
$\frac{\partial I}{\partial y}$	1 0 0 −1	1 1 1 0 0 0 −1 −1 −1	1 2 1 0 0 0 −1 −2 −1	1 2 3 2 1 2 3 5 3 2 0 0 0 0 0 −2 −3 −5 −3 −2 −1 −2 −3 −2 −1

Fig. 4.38 An image and its partial derivative computed by Sobel kernel

Unlike binary images, segmented images contain multiple (usually a small number of) grey levels, each representing a distinct region or feature within the image.

(4) *Finding zeroes on images second derivative*

Another method of identifying edges is to use the second derivatives. As illustrated in Fig 4.39 the second derivative is the slope of the first derivative of a function describing the grey level profile on the edge.

The zero points of the second derivative correspond to the local maxima or minima of the first derivative, and to inflection points where the function changes its concavity. Consequently, edges are located at the zero crossings of the second derivative. In the case of a two-dimensional image I(x,y), which can be interpreted as a grey-level field (as the other field described in Chap. 1), we can compute the Laplacian operator (i.e., the two-dimensional second derivative) to identify points where the Laplacian field (image) is zero. According to the definition provided in Chap. 1, the Laplacian of an image is given by:

$$\nabla^2 I = \frac{\partial^2 I}{\partial x^2} + \frac{\partial^2 I}{\partial y^2} \tag{4.12}$$

And it is a scalar field, so can be seen as a second image where zeroes points are the border lines, i.e. the edges.

As before seen our images are digital, i.e. matrix I(i, j) of numbers, one number for each pixel, and second derivatives can be estimated by using at least two surrounding points of each pixel, i.e. the tree elements of the matrix, so taking a minimum of nine contiguous pixels as illustrated in Fig 4.40.

A first possible approximation of the Laplacian two components by finite differences can be done numerically as follows:

$$\frac{\partial^2 I}{\partial x^2} \approx \frac{1}{\varepsilon^2}\left(I_{i-1,j} - 2I_{i,j} + I_{i+1,j}\right) \tag{4.13}$$

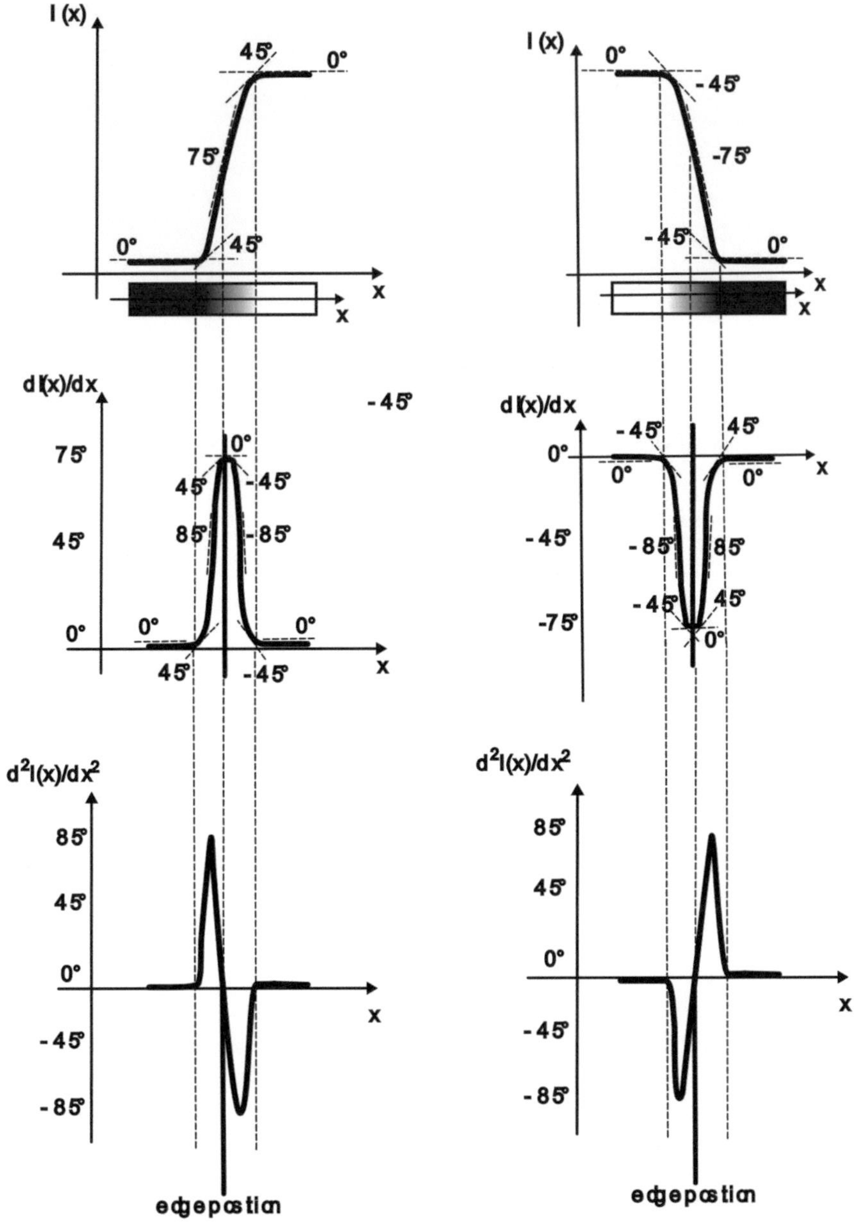

Fig. 4.39 Finding edges by second derivatives

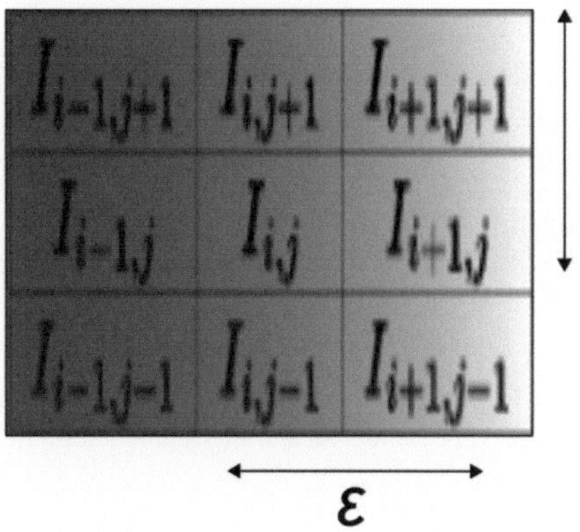

Fig. 4.40 Elements of the matrix used to estimate second derivative

$$\frac{\partial^2 I}{\partial y^2} \approx \frac{1}{\varepsilon^2}\left(I_{i,j-1} - 2I_{i,j} + I_{i,j+1}\right) \tag{4.14}$$

where ε is the distance between pixes, i.e. pixels width.

It can be easily seen that above formulas correspond to the definition given in Cap 3 of the convolution of the image matrix I aroud each of its I(i, j) elements with the matrix:

$$\mathrm{C_L} = \frac{1}{\varepsilon^2}\begin{bmatrix} 0 & 1 & 0 \\ 1 & -4 & 1 \\ 0 & 1 & 0 \end{bmatrix} \tag{4.15}$$

Adding a scale factor of $\frac{1}{\varepsilon^2}$. So, we can calculate the Laplacian of an image as:

$$\nabla^2 I \approx I(i,j) * C_L \tag{4.16}$$

Alternative formulations using more complex convolution matrix kernels are also possible, offering improved accuracy and reduced sensitivity to noise, as previously discussed in the context of gradient-based methods.

Figure 4.41 illustrates an example of edge detection based on identifying zero-crossings of the Laplacian.

The edge detection methods based on derivatives, as described above, are conceptually simple and straightforward. However, they do not account for noise, which is typically present in most images for various reasons discussed earlier in Chap. 3.

A typical example of a real grey-level step profile affected by noise is shown in Fig. 4.42 with its corresponding derivative below. As a result, maxima, minima, and zero-crossings appear not only at the actual edge location but also at many other points, due to the presence of noise identifying a lot of false step by the many maximum and minimum on first derivative and the many zero's on the second derivative.

(a) (b) (c)

Fig. 4.41 Detecting edge by Laplacian zero crossing: original image in (**a**), image of the Laplacian in (**b**), Laplacian black/white segmentation in (**c**)

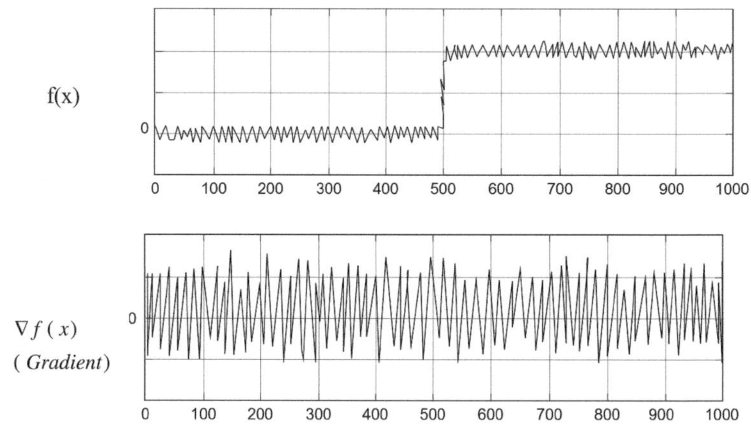

Fig. 4.42 Step of grey level at an edge with noise and its derivative

Therefore, a preliminary image processing step is required to smooth the grey-level distribution removing noise. As previously discussed, this can be achieved through image filtering (smoothing) in the frequency domain by multiplying the Fourier transform of the grey-level profile with a low-pass filter. A more efficient approach, however, is to perform a convolution in the spatial domain (equivalent to multiplication in the frequency domain) using a band-limited function, such as a Gaussian function as a filter.

The Gaussian is defined by its standard deviation σ, which determines its width in the spatial domain. As shown in Chap. 3, the corresponding spatial frequency bandwidth is approximately width $1/\sigma$; thus, performing a convolution with this Gaussian function, noise components with spatial frequency content above $1/\sigma$ are attenuated or removed. This image processing technique is illustrated in Fig4.43a.

Image processing for noise reduction can also be simplified by using the convolution properties discussed in Chap. 3. Specifically, the derivative of the convolution between a signal and a Gaussian function n_σ is equal to the convolution of the signal with the derivative of the Gaussian.

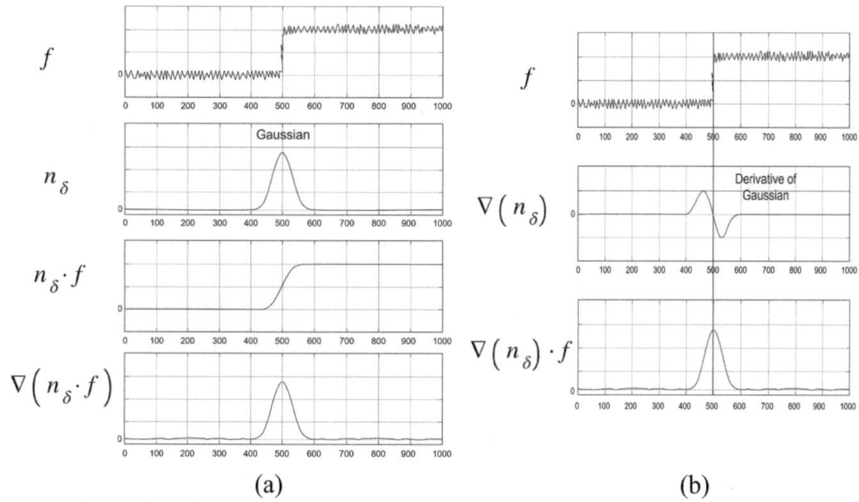

Fig. 4.43 Removing noise effects on derivatives calculation by gaussian convolution (**a**) or more efficiently making convolution with the derivative of the gaussian (**b**)

This means that one can skip a step, as illustrated in Fig. 4.43, and instead perform a single convolution using the analytically known gradient of the Gaussian function ∇n_σ, thereby streamlining the computation, so as:

$$\nabla(n_\sigma * I) = \nabla(n_\sigma) * I \tag{4.17}$$

For the second derivative approach to find edges, the same processing technique to reduce noise effects can be applied and the same simplification by making directly only one convolution with the Laplacian of the gaussian function, because also for the second derivatives, from the properties of convolution is true that:

$$\nabla^2(n_\sigma * I) = \nabla^2(n_\sigma) * I \tag{4.18}$$

The simplified algorithm for this spatial filtering noise removing image processing is illustrated in Fig. 4.44.

The image processing techniques described above for edge detection and noise reduction can be applied to 2D images. Therefore, the convolution must be performed using the 2D gradient and the 2D Laplacian of the Gaussian function, both illustrated in Fig. 4.45.

The advantages and disadvantages of the two alternative approaches for edge detection are summarized in Table 4.3.

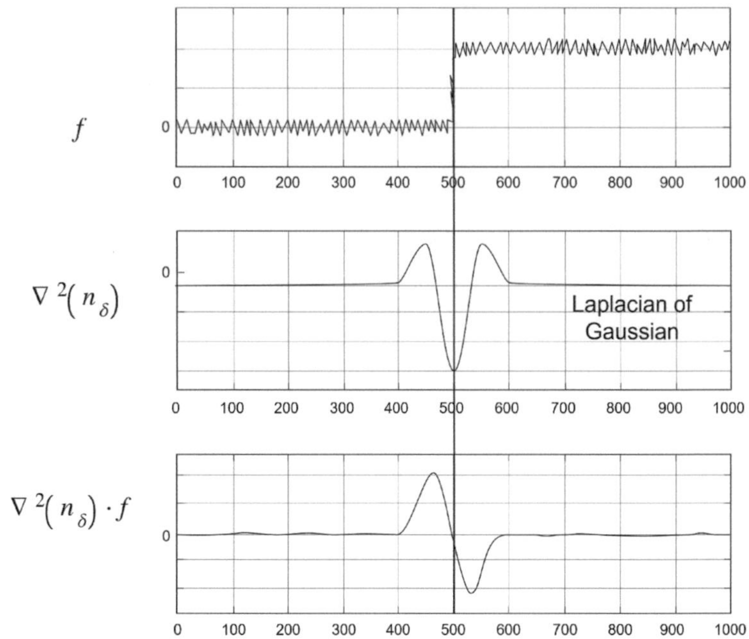

Fig. 4.44 Convolution with Laplacian of Gaussian to reduce noise effects and to find edge positions

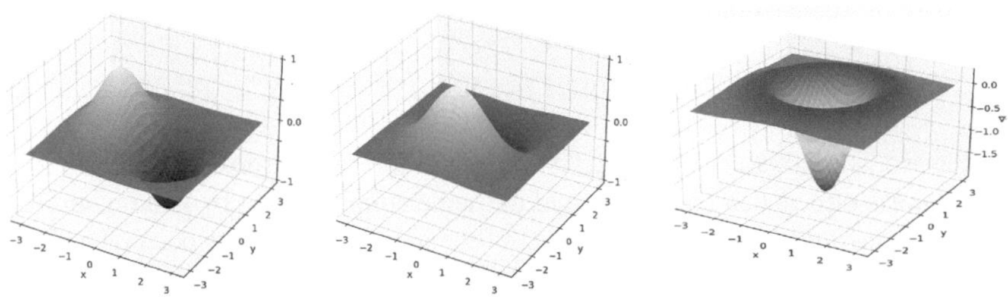

(a) $\nabla x\,(n_\sigma)$ (b) $\nabla y\,(n_\sigma)$ (c) $\nabla^2(n_\sigma)$

Fig. 4.45 Gradient components and Laplacian of the gaussian function

Table 4.3 Finding edges by image gradient or by Laplacian

Gradient	Laplacian
Provides location, magnitude and direction of the edge	Provides only location of the edge
Needs the definition of a solil level to identify relative maximum or minimum	Detection based simply on Zero-Crossing. No needs of level definition
Requires two convolutions with the two components of the gradient vector	Requires only one convolution

Building on the core concepts described above, numerous algorithms have been developed for edge detection in image processing and computer vision. These algorithms aim to identify boundaries or significant intensity transitions within an image and are widely used in tasks such as object detection, image segmentation, and feature extraction. The most commonly used edge detection algorithms include:

1. *Sobel Edge Detection*

Based on convolution with two 3×3 kernels (horizontal and vertical) to compute the gradient magnitude in the x and y directions. Produces edges that emphasize regions with high spatial frequency (e.g., object boundaries).

It is sensitive to noise; commonly is followed by a thresholding.

It is used mainly for basic edge detection tasks where speed is essential.

2. *Prewitt Edge Detection*

It is similar to Sobel, but uses simpler convolution kernels for gradient computation. Slightly less accurate than Sobel, but computationally more efficient. It is suitable for simple and lightweight edge detection tasks.

3. *Canny Edge Detection*

It is a multi-step algorithm that applies Gaussian smoothing. Uses Sobel-like operators to find intensity gradients. Highly robust to noise and provides thin, accurate edges. Well suited for general-purpose edge detection, especially in noisy images.

4. *Laplacian of Gaussian (LoG)*

It combines Gaussian smoothing and the Laplacian operator to detect edges.

It is sensitive to second-order intensity changes, useful for detecting fine details but prone to false edges if the Gaussian smoothing is inadequate. Very useful for edge detection in detailed or textured images.

5. *Roberts Cross Operator*

It uses 2×2 convolution kernels to calculate the gradient in diagonal directions. Lightweight and fast, but highly sensitive to noise. Useful for edge detection in low-noise images or real-time systems.

6. *Scharr Operator*

It is a variant of the Sobel operator with optimized kernels for improved rotational invariance. Produces more accurate gradient estimates than Sobel.

Useful for high-precision edge detection tasks.

7. *Hysteresis Thresholding*

Is often used as part of Canny Edge Detection. Ensures edge continuity by retaining edges that link strong and weak gradients. Useful for enhancing edge coherence in multi-step algorithms.

8. *Marr-Hildreth Algorithm*

Detects edges by locating zero-crossings in the second derivative of the image smoothed by a Gaussian. Is sensitive to noise but effective for detecting continuous edges. Useful and applied to medical imaging, such as detecting boundaries in X-rays.

9. *Sobel-Feldman*

A generalization of Sobel for multi-directional gradients. Useful for complex texture analysis.

10. *Phase Congruency*

Detects edges based on local phase information rather than intensity gradients. Useful for illumination- and contrast-invariant edge detection.

11. *Deep Learning-based Edge Detection*

Based on convolutional Neural Networks (CNNs) trained to detect edges, such as Holistically-Nested Edge Detection (HED). Useful for complex scenes, real-time edge detection in AI systems.

In the following table a comparison between main algorithms for edge detection.

Algorithm	Advantages	Disadvantages
Sobel	Simple, efficient	Noise-sensitive
Prewitt	Lightweight	Less accurate than Sobel
Canny	Accurate, robust to noise	Computationally intensive
Laplacian of Gaussian	Detects fine details	Sensitive to noise
Roberts cross	Fast, lightweight	Prone to noise and aliasing
Scharr	Precise gradient estimation	More computational than Sobel
Deep learning	Handles complex scenarios, adaptable	Requires training and large data

The choice of best algorithm depends on the specific application, computational constraints, and noise levels in the image. For general use, Canny Edge Detection is widely considered the gold standard due to its balance of accuracy and robustness, but for well formed and digitized images with high contrast using algorithms with a lower computational effort can be better.

4.7 Binary Images Processing

Once edges are defined, we have images with only two states, before and after the edge. So black and white images with only one bit resolution, only two grey levels 0 or 1, like that in Fig. 4.46.

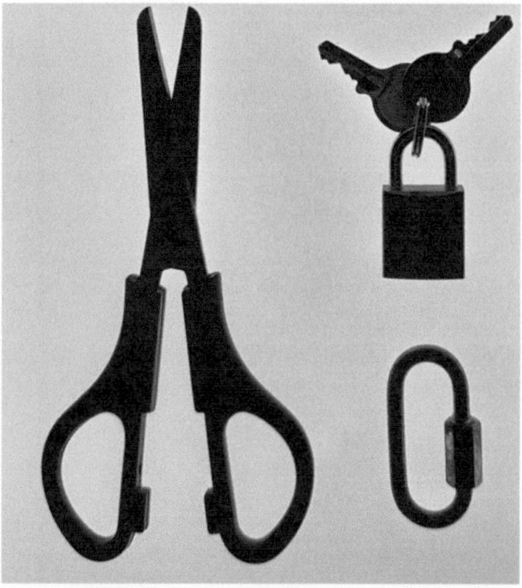

Fig. 4.46 Black and white images, binary two grey levels images

Binary two grey levels images can be obtained by edge detection or by establishing a threshold value with the image histogram and define what is below the threshold by giving it the value zero grey level and what is above the threshold by giving it the value one grey level or vice versa, as previously said to find edges. An example of this process is above illustrated in Fig. 4.34.

Another very efficient way to get binary images is that of analogic processing by back lighting. An example is given in Fig. 4.47.

From binary images, represented as a matrix I(i, j) containing values of 0 and 1, it is possible to extract various shape-related measurements [3, 4]. These include distances between 0 and 1 transitions along lines, profiles, or rays, as well as key parameters such as the object's area, centroid (first-order moment), and principal axes (first and second moments of inertia). These measurements can be computed using the following equations:

Area

$$A = \sum_{i=1}^{N} \sum_{j=1}^{M} I(i,j)\varepsilon^2 \tag{4.19}$$

Centroid coordinates

$$C_x = \frac{1}{A} \sum_{i=1}^{N} \sum_{j=1}^{M} i\varepsilon I(i,j) \quad C_y = \frac{1}{A} \sum_{i=1}^{N} \sum_{j=1}^{M} j\varepsilon I(i,j) \tag{4.20}$$

Or

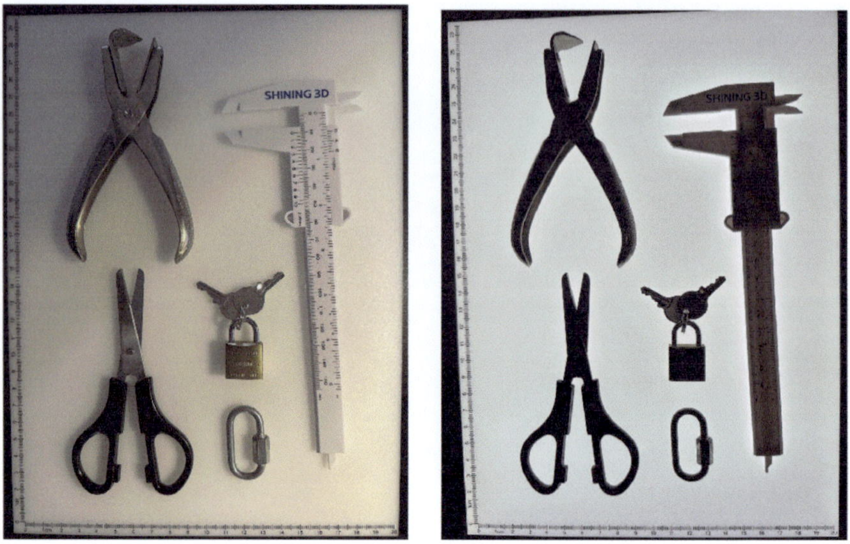

(a) (b)

Fig. 4.47 Binary images obtainded by using a back lighting led pad

$$C_x = \frac{1}{N}\sum_{i=1}^{N} x_i \quad C_y = \frac{1}{M}\sum_{j=1}^{M} y_j \tag{4.21}$$

The moment of inertia for a digital plane figure are calculated around the centroid. For a digital figure, the moment of inertia about an axis passing through the centroid is the sum of squared distances of the pixels from that axis, weighted by their mass (which, in this case, is usually uniform and can be taken as 1).

Moment of inertia respect x and y axis, as well known, can be computed as:

$$I_x = \sum_{j=1}^{M} (y_j - Cy)^2 \quad I_y = \sum_{i=1}^{N} (x_i - Cx)^2 \tag{4.22}$$

And the product of inertia as:

$$I_{xy} = \sum_{i=1}^{N} (x_i - Cx)(y_i - Cy) \tag{4.23}$$

So, the inertia tensor **I** is given by:

$$\boldsymbol{I} = \begin{bmatrix} I_x & I_{xy} \\ I_{xy} & I_y \end{bmatrix} \tag{4.24}$$

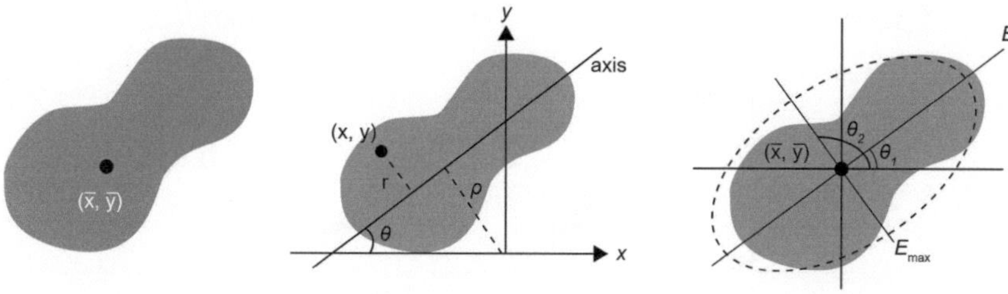

Fig. 4.48 Geometrical features definition on a binary 2D image

The principal moments of inertia I_1 and I_2 correspond to the eigenvalues of this tensor, and the directions of the principal axes are given by the corresponding eigenvectors. These can be computed, as well known by linear algebra, by solving the characteristic equation:

$$\det\left(\boldsymbol{I} - \lambda \boldsymbol{I}_{\text{identity}}\right) = 0 \tag{4.25}$$

where λ values are the eigenvalues. Solving for λ gives the principal moments of inertia, and solving for the eigenvectors gives the directions of the principal axes. The principal axes of inertia are the axes along which the moments of inertia are either maximal (E_{max}) or minimal (E_{min}). These axes are typically orthogonal to each other. The principal moments of inertia correspond to the moments about these axes, and they represent the "resistance" of the figure to rotation around each axis. Above parameters are illustrated in Fig. 4.48.

Roundness R_o of the plane figure can be defined as:

$$R_o = \frac{E_{min}}{E_{max}} \tag{4.26}$$

Estimation of these set of digitized and binarized figure parameters can be used to recognize, classify and measure characteristics of objects.

An example of application of this approach is given in Fig. 4.49.

Binary images can be used to identify objects or shapes based on patterns stored in memory, such as character recognition. By comparing the pixels of a stored reference image with those of the current image, after applying an appropriate scale factor, we can measure the differences. A perfect match will yield zero difference, while a sufficiently low difference may indicate a good enough match for reliable identification of objects, shapes, character or categories.

In the following paragraph of this chapter, we will introduce more general techniques for identifying objects, shapes, or features within images. These methods form the foundation for performing 2D and 3D measurement tasks by detecting and using reference points.

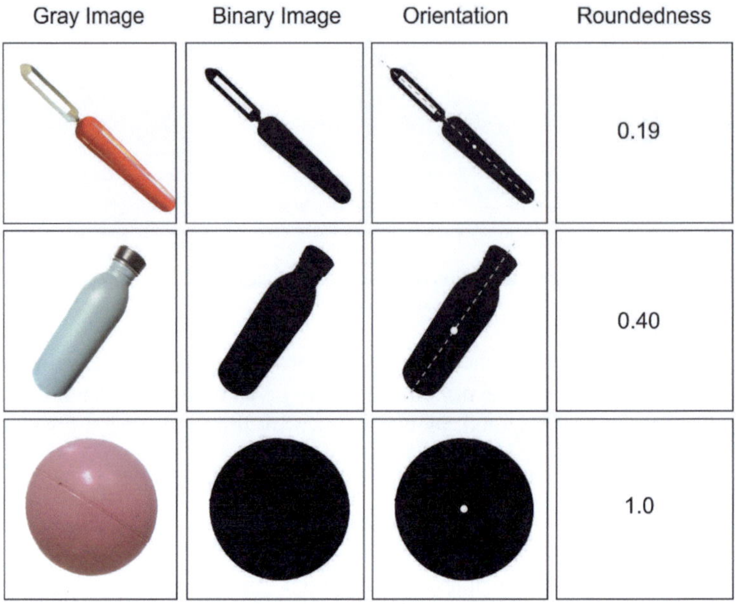

Fig. 4.49 Geometrical features on some binary images of objects

4.8 Point of Interest Detection

Many general image processing techniques exist to identify points of interest in complex images [2–4]. But what exactly are these points of interest? Consider the example shown in Fig. 4.50. Thanks to our eyes, connected to the remarkably powerful image processing system of the human brain, we can easily recognize the same object inside the red square in the two images, even if it has been altered by changes in exposure, lighting, scale, or rotation. Our brains effortlessly match the two instances. In fact, we can often match a shape in an image to a shape stored in our memory, such as a letter, a number, or a familiar object like an apple, a book, or a fork.

A single pixel cannot be uniquely identified across one or more images. As we have seen through image histograms, hundreds or even thousands of pixels may share the same grey-level value in both images. Therefore, relying on a single pixel is not sufficient for matching. To reliably recognize a point as the same across different parts of an image or between two images, we need to consider a surrounding region, an area with a distinctive and unique distribution of grey levels. This local pattern allows the point to stand out and be matched. Such a region may correspond to a specific visual feature: a hole, a letter, or a geometric shape like a triangle, square, or star, features that our minds are trained to recognize and associate. We are able to recognize this pattern, whether it's a distribution of grey levels or colors, using our eyes and brain, even when it is rotated, varies in brightness, or appears blurred, as shown in Fig. 4.50c.

(a) (b) (c)

Fig. 4.50 Point of interest definition

In contrast, it is not possible to identify the position of points, objects, or blocks of pixels in images like that of Fig. 4.51a along the stripes or anywhere in regions inside red squares with same grey level distribution, or in uniform colour images like that in Fig. 5.51b.

So... what exactly is a "blob" that we can visually identify in an image? To define it we need:

- Localized contrast: A blob is a region centered around a point that shows a noticeable variation in brightness or color (e.g., a bright or dark area compared to its surroundings).
- Distinct shape: It has a well-defined, recognizable shape that stands out from the background.
- Clear position: It occupies a specific, easily identifiable location, for instance, a white dot on a black background is an ideal example of a blob or point of interest.

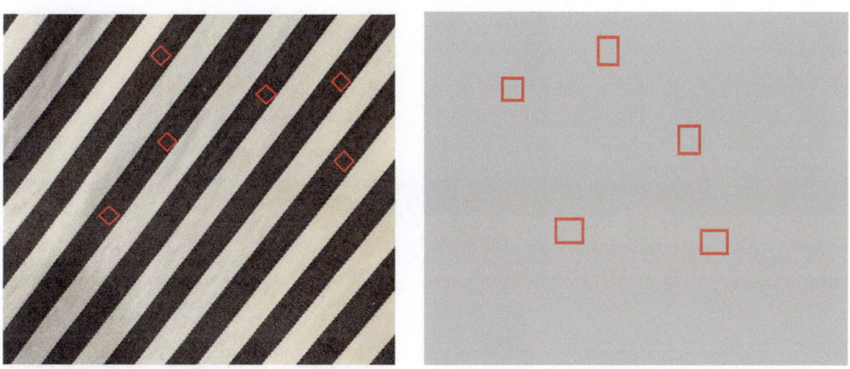

(a) (b)

Fig. 4.51 Images where is not possible to define positions of point of interest

– Invariant to scale and rotation: A good blob remains recognizable even if the image is scaled or rotated. For example, a black dot on a white background is still identifiable even if it becomes larger, smaller, or slightly lighter or darker.
– Robust to brightness changes: It can still be detected despite changes in overall brightness. For instance, a white dot on a black background should be recognizable regardless of how bright or dim the white appears.

Lines or borders are not considered points of interest because all the points along a line tend to have the same grey level distribution, making it hard to pinpoint a unique location just by examining local brightness patterns. This is illustrated in Fig. 4.52.

Instead, the little square with a statue and a window on Fig. 4.52 are clear points of interest because their intensity distribution are present only in a single specific positions in the image.

We can identify the central point of a blob, such as the black dot in Fig. 4.53a, similarly to how we did for binary images. For instance, we might select a threshold value from the blob's histogram (e.g., a "soil level") to determine the center by above Eq. 4.20. To estimate the blob's size, we can define a radius that includes all pixels above a certain grey level, or we can enclose the blob within a square and use the size of that square as a measure of the blob's dimensions.

But how can we automatically detect a point of interest (blob) and its previously defined characteristics using image processing techniques?

Fig. 4.52 Image with clear blog and undefined blog along lines on a picture of San Sabino Bari Chatedral in Italy

Fig. 4.53 Point of interest (blog) characteristics: central position black point in (**a**), dimension in (**b**), orientation in (**c**) and signature in (**d**)

We start by analyzing a simple one-dimensional grey level profile, such as the one shown in Fig. 4.54. The grey level profiles pulses in the figure can be considered one-dimensional blobs because each has a clearly identifiable central position on x axis, a measurable size (its width), no orientation (as it exists in only one dimension), and a characteristic shape that defines its signature.

Therefore, a 1D blob is something delimited by at least 2 edges, one positive and one negative, or viceversa, but we previously define image processing techniques useful to find edges so the same approach can be applied to find double edge, so blog position. As previously done for edge a possible image processing technique is making convolution of the grey level profile with the second derivative of a gaussian function to reduce the effects of noise.

Therefore, a one-dimensional blob can be defined as a region bounded by at least two edges, one positive and one negative, or vice versa. Since we have already introduced image processing techniques for edge detection, we can extend the same approach to detect pairs of edges, which in turn allows us to locate blobs.

As with edge detection, one effective technique is to convolve the grey level profile with the second derivative of a Gaussian function. This not only highlights regions corresponding to blobs but also helps suppress noise, improving the robustness of the detection.

By varying the width (σ) of the Gaussian function and performing convolution, we obtain the results shown in Fig. 4.55. It becomes clear that when the Gaussian's width matches the width of the blob, the convolution exhibits an extremum (maximum or minimum). This allows us to determine the position of the blob by sliding the Gaussian along the x-axis, and to estimate its size by identifying the value of σ that maximizes or minimize the convolution.

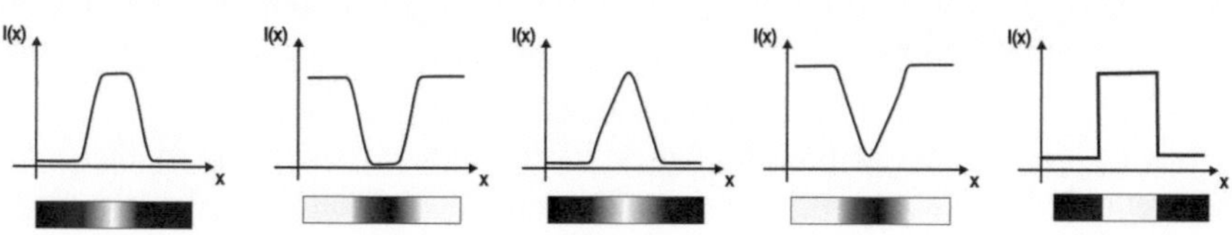

Fig. 4.54 Monodimensional points of interest (blogs)

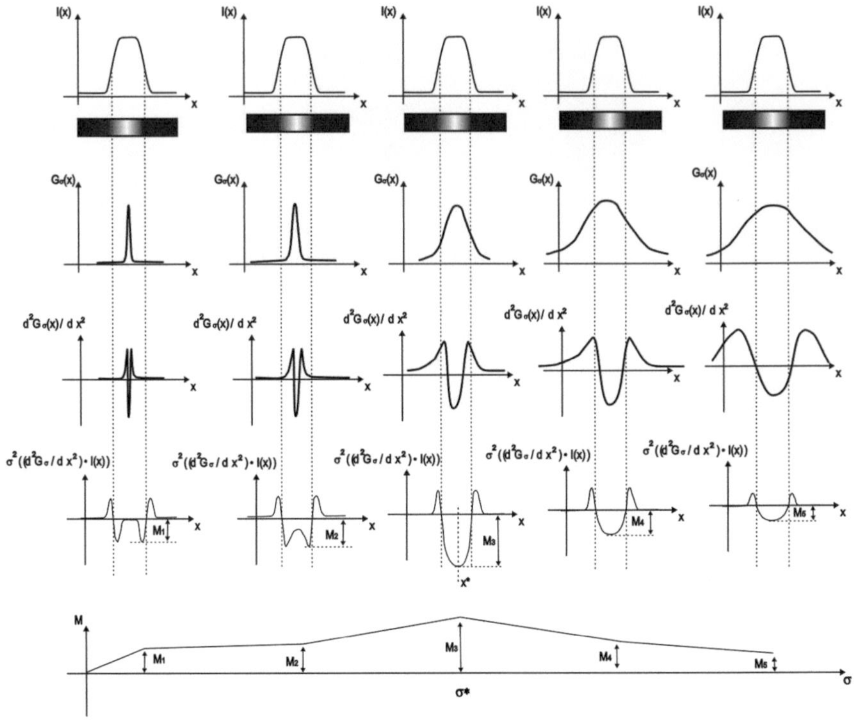

Fig. 4.55 Detecting 1D blog parameters (position and width) by convolution with gaussian second derivative scaled by σ^2 and finding its extreme value points coordinates (x*, σ *)

Thus, detecting a blob's position and scale becomes a classic mathematical problem of finding the extrema of a two-dimensional function (the convolution function of the pulse grey level profile with the Gaussian) with respect to the variables x and σ.

The step to do this grey levle profile processing are therefore:

(1) Define the monodimensional grey level profile I(x) where to find possible blog positions x_i*
(2) Compute $\sigma_i^2 \frac{\partial^2 G_\sigma}{\partial x^2} * I(x)$ at many scales ($\sigma_0, \sigma_1, \sigma_2, \ldots \sigma_k$)
(3) Find the two coordinates (x_i*, σ_i*) where the function $\left|\sigma_i^2 \frac{\partial^2 G_\sigma}{\partial x^2} * I(x)\right|$ have its maximum value.

Values x_i* will define blog central positions, σ_i* its scale (width).

Having for example a profile of grey level width tree blogs as in Fig. 4.56 this method will find tree x positions and tree σ values identifying the tree blogs parameters.

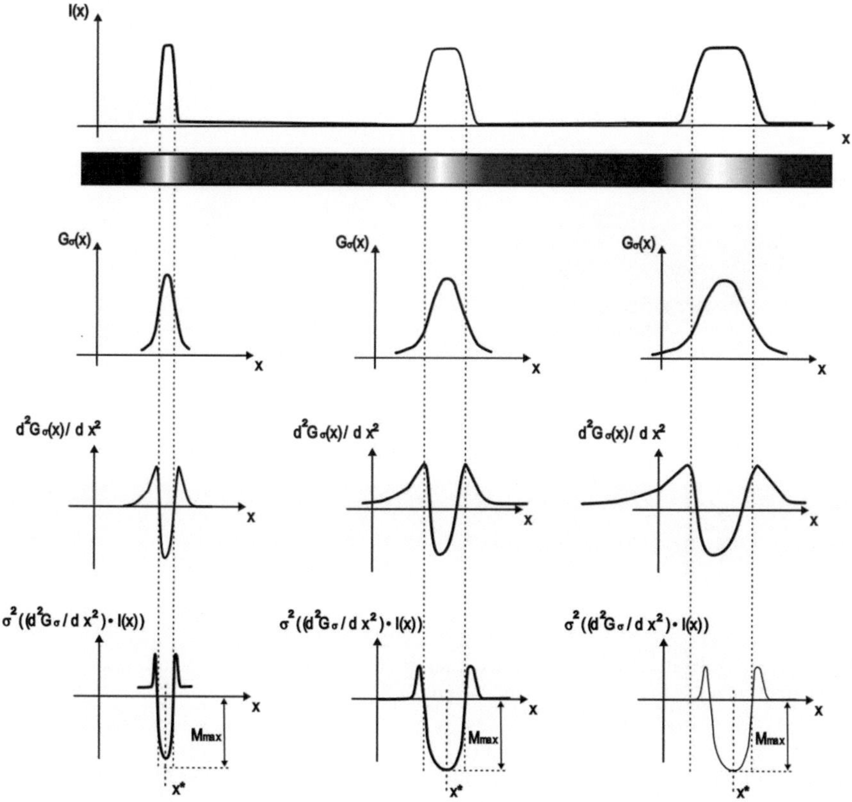

Fig. 4.56 Detecting 1D blog parameters by convolution with Gaussian second derivative scaled by σ^2 and finding its extreme value points coordinates (x^*, σ^*)

For 2D images the algorithm is the same, using Laplacian of the gaussian 3D function with different σ values instead of second derivatives, normalized by multiplication with σ^2 as illustrated in Fig. 4.57.

Convolution with a Gaussian or a Laplacian of Gaussian, according to the convolution theorem described in Chap. 3, corresponds to filtering the image using a filter whose Fourier transform is that of the Gaussian or the Laplacian of Gaussian. The bandwidth of this filter is inversely proportional to the standard deviation σ, approximately equal to 1/σ. Therefore, increasing the value of σ results in a narrower bandwidth, meaning the filter acts as a low-pass filter that retains only image frequencies below 1/σ. As a result, the image is increasingly blurred, as illustrated in Fig. 4.58.

To determine whether a blob is present around a given point in an image, we compute the scale-normalized Laplacian of Gaussian (LoG) response, i.e. the convolution of the image with LoG filters, at multiple scales (defined by different values of σ). A local maximum in the response across scale at a given point indicates the

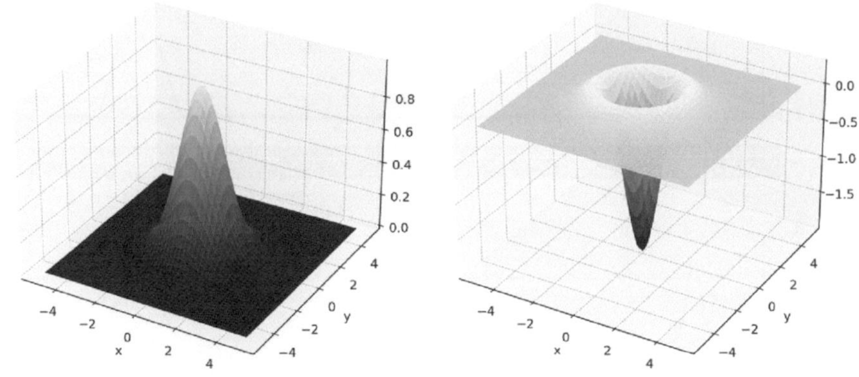

Fig. 4.57 Bidimensional Gaussian function and its Laplacian

Fig. 4.58 Convolving an image with increasing σ gaussian function (in pixel) means reducing image frequencies so blurring/defocusing the image

presence of a blob centered at that point. The scale σ* at which the maximum occurs corresponds to the blob's characteristic size.

This process is illustrated in Fig. 4.59a, which shows plots of the LoG response as a function of σ at two different image locations, (X_1, Y_1) and (X_2, Y_2). At (X_1, Y_1), the response curve exhibits a clear maximum, signifying the presence of a blob with radius proportional to σ*. Conversely, at (X_2, Y_2), the absence of a distinct maximum implies that no blob is detected at that location.

The algorithm for detecting blobs in 2D images proceeds as follows:

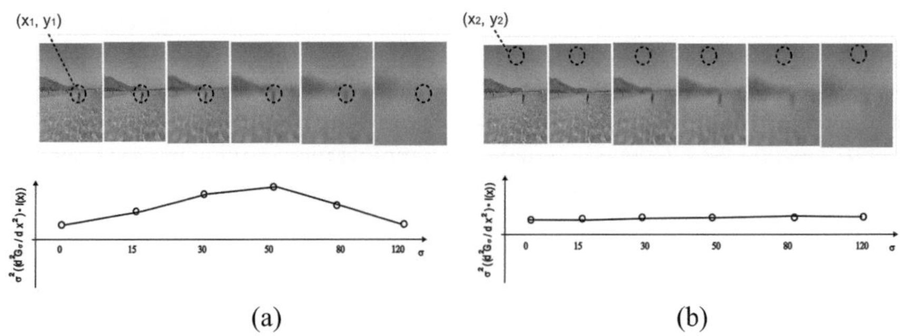

Fig. 4.59 Identifying a blog on an image in (**a**) not in (**b**)

1. Define the region of interest in the image, denoted as *I(x, y)*, where blob detection is to be performed.
2. Compute the scale-normalized Laplacian of Gaussian response $\sigma^2\nabla^2 G_\sigma * I(x, y)$ at multiple scales (σ_0, σ_1, σ_2, … σ_k), for each pixel location (x, y).
3. Identify extrema: Find the coordinates (x^*,y^*,σ^*) where the absolute value of the response $\left|\sigma^2\nabla^2 G_\sigma * I(x, y)\right|$ attains a local maximum across both spatial and scale dimensions.

The resulting coordinates (x^*, y^*) represent the centers of detected blobs, while σ^* defines its characteristic scale (i.e., the radius of the blob, proportional to σ^*).

It is possible to simplify the computation by approximating the scale-normalized Laplacian of Gaussian $(n - 1)\sigma^2\nabla^2 G_\sigma$ (NLoG) with the Difference of Gaussians (DoG), defined as $(G_{n\sigma} - G_\sigma)$. This approximation significantly reduces computational complexity while maintaining accuracy in blob detection tasks. As shown in Fig. 4.60, the two functions closely match, demonstrating that the DoG is a reliable substitute for the NLoG. This justifies the use of DoG in practical implementations of blob detection algorithms.

The general algorithm for blog detection by using DoG is therefore as follow. Input are.

- A grayscale image I(x, y)
- A base scale σ and a multiplicative factor k > 1
- Number of scales (levels) to be computed.

1. The first step is the construction of a scale space (illustrated in Fig. 4.61a) by generate a set of progressively blurred images by convolving the input image with Gaussian filters of increasing standard deviation, so the images

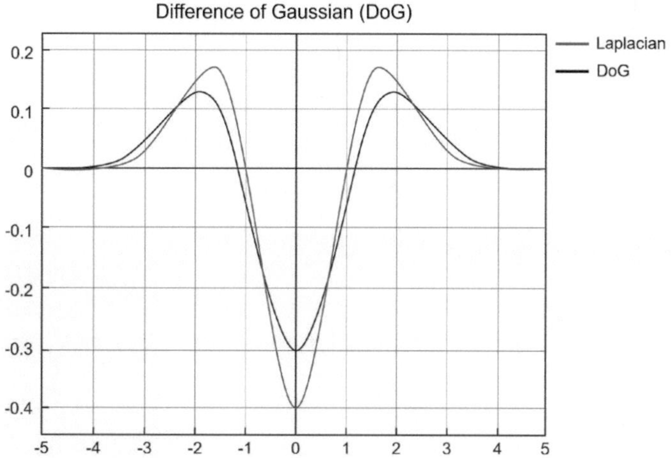

Fig. 4.60 Difference of plot between DoG and NLoG functions

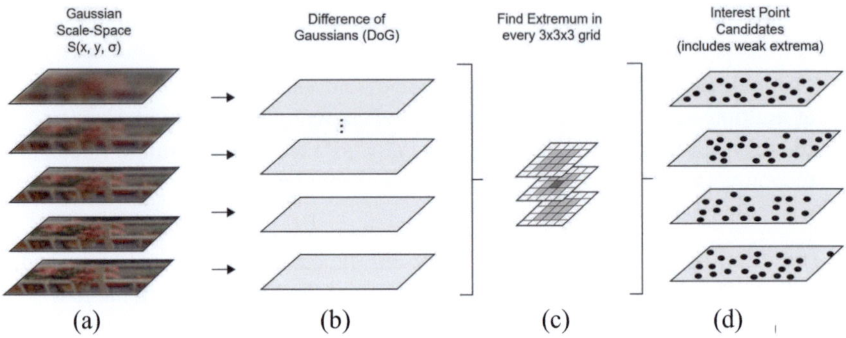

Fig. 4.61 Difference of gaussian image stack to find extremum of convolution

$$L(x, y, \sigma_i) = G_{\sigma i} * I(x, y)$$

where $\sigma i = k^i \cdot \sigma$, $i = 0,1,2,\ldots,n$

2. At each adjacent scale, compute the Difference of Gaussians:

$$D(x, y, \sigma i) = L(x, y, \sigma i + 1) - L(x, y, \sigma i)$$

This step approximates the scale-normalized Laplacian as illustrated in Fig. 4.61b.

3. For each pixel (x, y) in each DoG image, compare the value with its generally 26 neighbors in 3D (8 neighbors in the same scale, 9 in the scale above, and 9 in the scale below). The (x, y) point is considered a blob candidate if it is a local maximum or minimum in this 3D neighborhood
4. Sub-pixel and sub-scale refinement may be performed (e.g., using Taylor expansion around the extremum) and discard low-contrast points or unstable responses (e.g., along edges), as illustrated in Fig. 4.62.

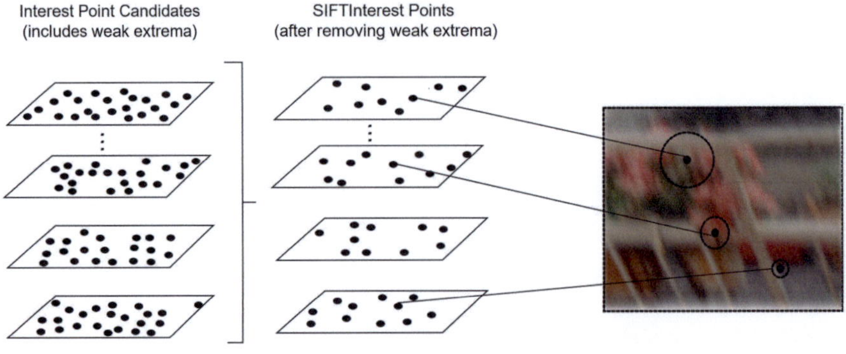

Fig. 4.62 Filtering and identifying on the image the points positions and scale

(a) (b)

Fig. 4.63 Example of finding interest points on a picture of penguins in Patagonia

Output is each detected blob represented by its position (x*, y*) and scale σ*, which is proportional to the radius of the blob, as illustrated in Fig. 4.62.

Many software are available to make this calculation, like Matlab, Pyton with OpenCV package of functions, etc. Using that routines on image of Fig. 4.63a many point of interest can be identified on it, highlighted by the circles in Fig. 4.63b.

Once the central position and scale of the blob have been identified, its orientation can be determined by selecting a square region of pixels that approximates the circle defined by the value of σ*, as shown in Fig. 4.64a.

For each pixel within this square, the gradient is estimated, and its orientation is then discretized into one of the eight principal directions: 0, 45, 90, 135, 180, 225, 270, and 315°. A histogram having 8 bars H(i) is then constructed by counting the number of pixels aligned with each of these 8 directions, as illustrated in Fig. 4.64b. The principal direction of the blog is that defined by the most frequent orientation within the square corresponding to the highest bar in the histogram, or equivalently, the maximum value in the vector H(i), denoted as H_{max}, as illustrated in Fig. 4.64c.

The final step is to define the blob signature in a way that is as independent as possible from lighting levels and conditions. To achieve this, the square region used

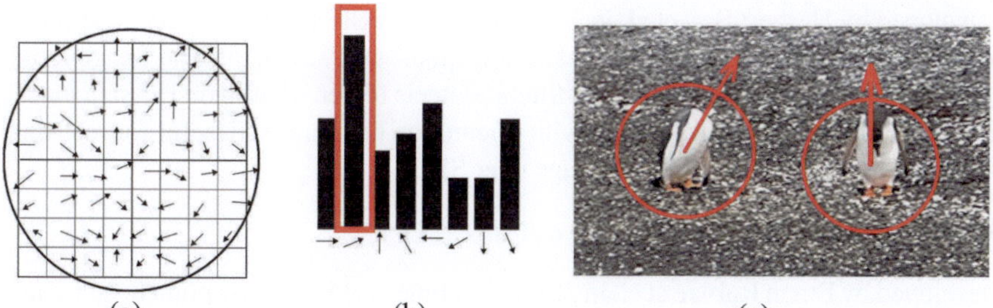

(a) (b) (c)

Fig. 4.64 Definition of blog orientation

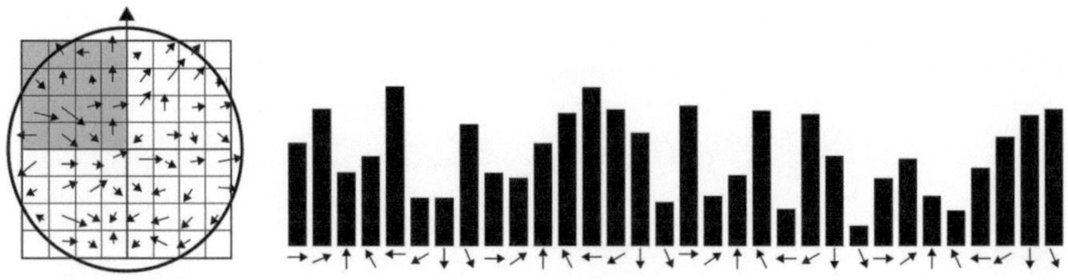

Fig. 4.65 The blog signature

to determine the blob's orientation is divided into four quadrants. For each quadrant, a histogram, similar to the one previously described, is computed. These four histograms are then concatenated to form a single histogram (and its corresponding vector H(i)) consisting of $8 \times 4 = 32$ elements. This vector is defined to be the block signature, as illustrated in Fig. 4.65.

What above described is the main workflow of the so called scale invariant feature transform (SIFT) algorithms proposed by David Lowe first time in 1999 [5] to identify point of interest and blobs. Many other algorithms has been subsequently developed in computer vision for identifying key points or features in images that remain stable under various transformations, such as scaling, rotation, and changes in illumination. These algorithms are fundamentals in applications like object and character recognition, image stitching, tracking and 3D stereo vision. The main are:

1. *SIFT (Scale-Invariant Feature Transform)*

Developed by David Lowe in 1999. Detects key points in scale-space by searching for local extrema in the Difference of Gaussian (DoG) pyramid. Computes a descriptor for each key point based on the gradient orientation histograms around the key point. Invariant to scale, rotation, and partially invariant to affine transformations. Used for image matching and object recognition. Works well in diverse lighting and viewpoint conditions.

2. *SURF (Speeded-Up Robust Features)*

Developed by Herbert Bay et al. in 2006. Inspired by SIFT but faster due to approximations in calculations (e.g., Haar wavelet responses). Uses an integral image representation for fast computation of box filters. Detects features based on Hessian matrix approximation and describes them with orientation histograms. Useful for real-time applications due to its speed advantage over SIFT.

3. *ORB (Oriented FAST and Rotated BRIEF)*

Developed by Ethan Rublee et al. in 2011. Combines the FAST key point detector and the BRIEF descriptor. Adds orientation invariance and scale invariance using multi-scale pyramids. Is lightweight and suitable for real-time applications, especially in resource-constrained environments. Useful for embedded systems, mobile vision.

4. *MSER (Maximally Stable Extremal Regions)*

Detects regions rather than points. Finds connected components of pixels that remain stable across a range of intensity thresholds. Invariant to affine transformations and scale changes. Useful for scene text recognition, object detection.

5. *Harris-Laplace and Hessian-Laplace*

Combines Harris corner detection with a Laplacian operator to achieve scale invariance. Detects corners at multiple scales using scale-space theory. Useful for image stitching, 3D modelling.

6. *ASIFT (Affine-SIFT)*

Developed by Jean-Michel Morel et al. in 2009. Extends SIFT to achieve full affine transformation invariance. Simulates viewpoint changes to account for affine distortions. Useful for complex object recognition with large viewpoint changes.

7. *KAZE and AKAZE*

Developed by Pablo F. Alcantarilla et al. in 2012.Dete cts features in a non-linear scale-space constructed using a Perona-Malik diffusion equation. AKAZE is a faster variant optimized for real-time applications. Useful for Image matching with enhanced detail sensitivity.

8. *BRISK (Binary Robust Invariant Scalable Keypoints)*

Developed by Stefan Leutenegger et al. in 2011. Based on binary descriptor that is scale- and rotation-invariant. Utilizes circular sampling patterns around key points. Useful for mobile vision and augmented reality.

In the following table key comparison criteria are given.

Algorithm	Invariant to	Speed	Use case
SIFT	Scale, rotation	Moderate	Robust object recognition
SURF	Scale, rotation	Fast	Real-time systems
ORB	Scale, rotation (limited)	Very fast	Embedded devices
MSER	Affine, scale (regions)	Moderate	Scene text, object segmentation
ASIFT	Full affine	Slow	Large viewpoint changes
KAZE/AKAZE	Scale	Moderate	Detail-sensitive applications
BRISK	Scale, rotation	Fast	Lightweight, mobile systems

Each algorithm has trade-offs between computational complexity, accuracy, and robustness, and the choice of algorithm depends on the specific application and constraints.

4.9 Finding Corresponding Point of Interest

Matching points of interest identified within the same image or across two different images can be used to detect identical features, either between images or between extracted image block features and those in a reference library, for tasks such as object or character recognition [3, 4]. This matching can be performed by comparing the corresponding blob signature vectors, H_1 and H_2, for example by computing the squared differences as follows:

$$d(H_1, H_2) = \sqrt{\sum_k (H_1(k) - H_2(k))^2} \tag{4.27}$$

The smaller the difference, the better the correspondence between the two blogs. An alternative approach involves computing a normalized difference, defined as:

$$d(H_1, H_2) = \frac{\sum_k \left[\left(H_1(k) - \overline{H}_1\right)\left(H_2(k) - \overline{H}_2\right)\right]}{\sqrt{\sum_k \left(H_1(k) - \overline{H}_1\right)^2} \sqrt{\sum_k \left(H_2(k) - \overline{H}_2\right)^2}} \tag{4.28}$$

In this case, a perfect match corresponds to a value of 1. In general, a threshold below 1 can be defined to indicate a sufficiently good matching between the two blogs.

By performing correspondence analysis between blobs in two images, a set of matching points can be identified, as illustrated in Fig. 4.66.

A classic application of identifying corresponding points between two images is image stitching, as illustrated by the composition of Fig. 4.67 using the two images scaled and rotated of Fig. 4.66.

Fig. 4.66 Finding same point of interest on two rotated pictures of Patagonia mountains of Cerro Torre chain

Fig. 4.67 Using corresponding points of pictures in previous Fig. 4.66 to make image sticking

Fig. 4.68 Using corresponding points to make traking by trackpy in python

Another highly valuable application of image-based measurement techniques is tracking the temporal position of the same object across two or more images. These tracked objects, such as moving animals, particles, geometric shapes, or markers, can be identified and followed over time, as illustrated in Fig. 4.68.

In the following chapters, we will explore several key measurement applications that rely on identifying corresponding points across two or more images. This fundamental process underpins a wide range of measurement techniques, including 3D shape reconstruction, displacement and deformation analysis, as well as velocity measurement methods in fluids and solid objects.

References

1. *The International System of Units (SI)*, vol. 2, 9th edn. (2022). ISBN 978-92-822-2272-0
2. B.G. Batchelor, *Machine Vision Handbook* (Springer, Berlin, 2012)
3. R. Zelinski, *Computer Vision Algorithms and Applications* (Springer, Berlin, 2010)
4. A. Kaehler, G. Bradski, *Learning OpenCV 4: Computer Vision with Python* (O'Reilly Media, 2019)
5. D.G. Lowe, Distinctive image features from scale-invariant keypoints. Int. J. Comput. Vis. (IJCV) **60**(2), 91–110 (2004)

Chapter 5
3D Shape Measurements by Images

Abstract In this chapter, we present the fundamental principles behind the most used classical and modern techniques for measuring the 3D shape of objects or surrounding scenes using 2D image processing and structured light. These principles form the basis of what are widely known as optical "scanners". Over the past few decades, these systems have significantly advanced in terms of measurement range (from small to very large objects), resolution, and measurement uncertainty. As a result, they have enabled a wide array of applications across various sectors. These include industrial use cases such as reverse engineering, quality assurance, and tolerance control of products and mechanical components. They are also applied to large-scale measurements, such as capturing the geometry of entire cars, ships, aircraft, and their parts; mapping buildings, bridges, and natural landscapes; and documenting cultural heritage through the 3D digitization of artworks, statues, monuments, and historical sites, for purposes of visualization, conservation, and restoration. The chapter discusses both active techniques (which use structured light) and passive techniques (which rely on natural illumination), alongside key image processing methods. These techniques enable the extraction of (x, y, z) coordinates for millions of surface points, collectively referred to as a point cloud. This point cloud forms the foundation for creating mathematical surface models and geometric primitives, measuring dimensional properties, comparing against CAD models, developing new models, or replicating objects through 3D printing, including scaled versions.

5.1 Introduction

Over the past decades, three-dimensional optical shape measurement techniques based on image analysis have gained increasing importance due to their expanding range of applications across diverse fields. These include industrial quality control, robotics, the entertainment industry, medical surgery, civil engineering, and the preservation and restoration of artworks. A comprehensive review of these techniques is available in [1], while an overview of their industrial applications is provided in [2]. Numerous other publications further explore the subject in detail. The 2D image

G. Rossi, *Mechanical and Thermal Measurements by Images and Waves*, Smart Sensors, Measurement and Instrumentation 51, https://doi.org/10.1007/978-3-032-03464-9_5

acquisition and processing methods used for 3D shape measurement can be broadly classified into two main categories:

(1) Passive techniques, which do not employ structured lighting.
(2) Active techniques, which rely on structured lighting, either spatially or temporally modulated.

Among passive techniques, some of the most widely adopted approaches include:

- Image defocus and focus variation methods [3];
- Stereovision [4, 5] and multi-stereovision (also known as photomodeling), which generate 3D point clouds representing surface geometry;
- Photometric stereo, which estimates surface geometry based on the distribution of surface normal vector.

More recent methods using multi-view 2D image acquisition like multi-stereovision but combined with advanced models such as Neural Radiance Fields (NeRF) and Gaussian Splatting. These techniques enrich traditional photomodeling by incorporating in the output model not only point position but also surface orientation, transparency, and view-dependent reflectance surface properties. The four main 2D image acquisition schemes commonly employed to reconstruct the 3D surface geometry of objects or scenes are illustrated in Fig. 5.1.

The fastest and generally most accurate 3D shape measurement techniques are those based on structured lighting, as illustrated in Fig. 5.2. These methods typically exhibit very low measurement uncertainty and are particularly effective for objects whose size is compatible with the limitations of the projection system. Structured light techniques are most used for small (centimeter-scale) or medium-sized (up to several meters) objects and surfaces, ranging from mechanical components to entire vehicles (such as cars or aircraft), and even indoor architectural environments (e.g., the interior of a room). For larger surfaces, the shape is often acquired in segments, which are then reconstructed into a complete model through post-processing. A key advantage of structured light systems is the simplification of image processing. These systems eliminate the need for computationally intensive algorithms such as SIFT (described in the previous Chap. 4), which involve searching for correspondences between image features and computing dense point clouds.

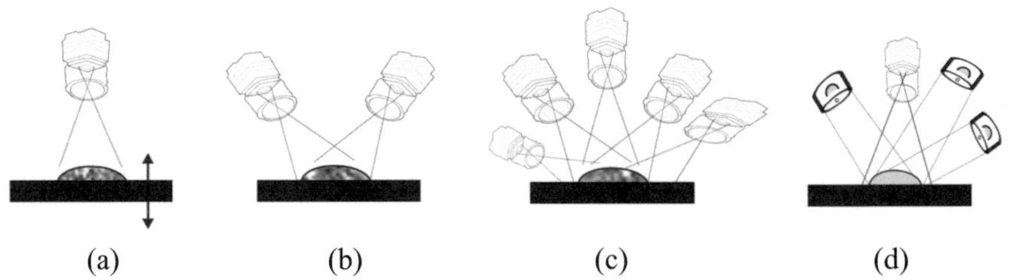

Fig. 5.1 Passive 3D optical shape measurement techniques based on: **a** focus variation, **b** stereo vision, **c** multistereovision, **d** photometric stereo

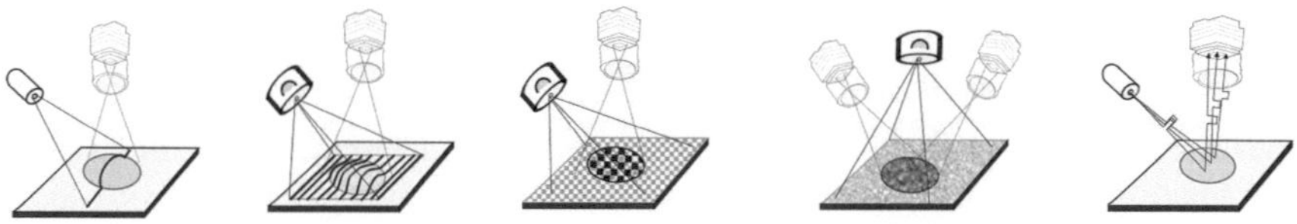

Fig. 5.2 Active 3D measurement techniques based on structured lighting

Structured lighting offers an alternative to more complex surface reconstruction approaches based on Neural Radiance Fields (NeRF) or Gaussian splatting. While those models are well-suited for photorealistic scene synthesis, structured lighting focuses on precise geometry acquisition. By projecting known patterns onto the surface, it effectively reduces the influence of variations in reflectivity, color, and texture, resulting in significantly greater speed, accuracy, and repeatability. This makes it the preferred choice in scenarios where metrological precision and efficiency are paramount.

The output of these measurement techniques varies depending on the method used. For instance, photometric stereo yields normal vectors that describe the orientation of surface elements, while structured light and stereovision-based methods produce millions of 3D coordinate points on the surface of an object or scene. This dense collection of spatial data is known as a point cloud. The point cloud serves as the foundation for further modeling and analysis. It can be used to:

- Generate a mathematical surface representation in the form of a triangular mesh, typically stored in stl standard format, which is widely used for 3D printing;
- Extract geometric primitives, such as lines, planes, circles, or cylinders, for metrological or reverse engineering purposes;
- Convert the data into more complex parametric models like Non-Uniform Rational B-Spline (NURBS) surfaces, enabling the creation of complete CAD representations of objects, environments, or entire scenes.

Many software are available to processing point clouds to obtain CAD models and mathematical definition of surfaces. These technologies are applied across a broad range of sectors, including:

- Industrial quality control and dimensional inspection of products and mechanical components;
- Architectural and civil engineering, for the measurement of indoor spaces and structural elements by generating CAD models;
- Urban space and city modelling;
- Cultural heritage, for the documentation, preservation, restoration, and virtual accessibility of artworks and historical artifacts.
- Medical applications, such as dental scanning and anatomical body part digitization for diagnostics or prosthetics.

5.2 Depth from Focus and from Defocusing

The first of these 3D shape measurement techniques determines the third spatial coordinate, the distance of surface points from the camera's optical center, by identifying which points appear in focus across a series of 2D images. As illustrated in Fig. 5.3, this method involves capturing a stack of images at varying distances r between the camera and the object's or scene's surface. For each point on the surface, the image in which it appears sharpest is used to estimate its depth. Assuming the camera objective can be approximated as a single thin lens with focal length f, the thin lens equation (Eq. 2.8) allows us to calculate the object distance r as:

$$r = \frac{if}{i - f} \tag{5.1}$$

where i is the distance between image sensor and lens (objective) optical centre, as previously explained in Chap. 2, with the scheme and parameters defined in Fig. 2.7.

So, from a stack of images like that of Fig. 5.4 we can make a "discretization" of distance values of points of object surfaces that are "in focus" at that distance.

By calculating the depth for each surface point, i.e. identifying at which camera-object distance r a point appears in focus, we can acquire a sequence of images taken at different focal distances. This sequence, known as a "focus stack", enables the estimation of depth and thus the reconstruction of the third spatial coordinate of the 3D model.

But how do we determine which points are actually in focus in an image, and thereby located at a specific distance r?

To better understand the concept of "focus" we refer to the diagram in Fig. 5.5. In this illustration, a point R on the object generates a sharp image point I on the image plane when it lies at the correct focal distance. However, when the object point is displaced to a new position R', located at a different distance r', it no longer forms a single point image. Instead, due to the change in geometry, the light ray c becomes c', and the resulting image is a blurred circle of diameter b centered around I. The

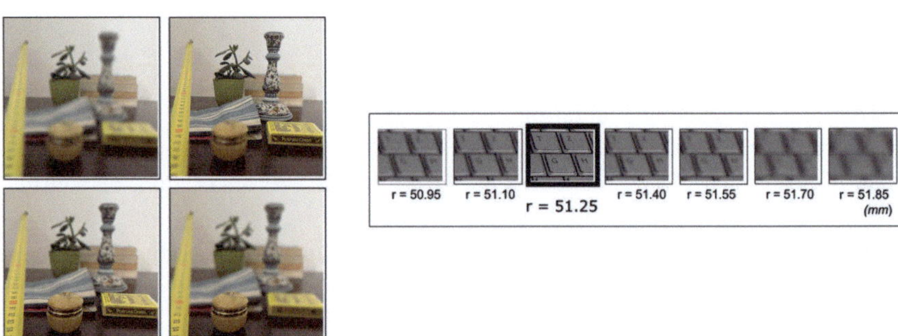

Fig. 5.3 Distance measurement of points in an image by focus position

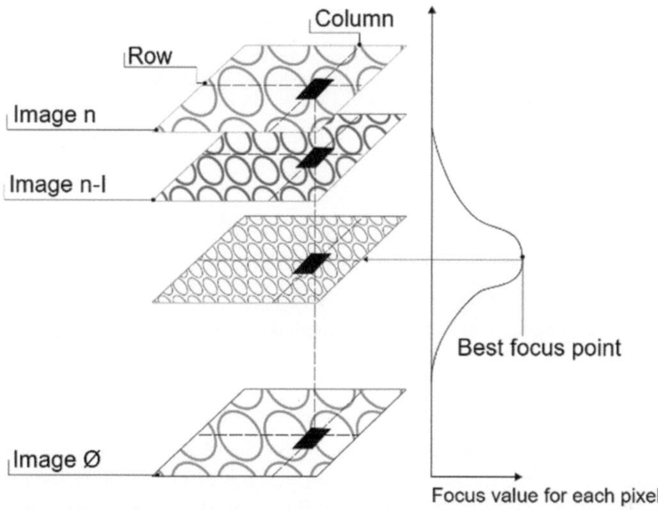

Fig. 5.4 Discretization of depth of a surface by focus variation

amount of blur increases with deviation from the ideal focus distance, following the principles of image formation by thin lenses, as discussed in Chap. 2.

From the similar triangles CC′F′ and RR′F′ of Fig. 5.5 is easy to see that the circle blurring diameter b can be estimated by:

$$b = \frac{r - r'}{r} f \tag{5.2}$$

This blurring effect due to defocusing is analogous to the application of a low-pass filter in image processing, as previously discussed in Chap. 3. Specifically, Figs. 3.33 and 3.35 demonstrate that a sharply focused point contains higher frequency content

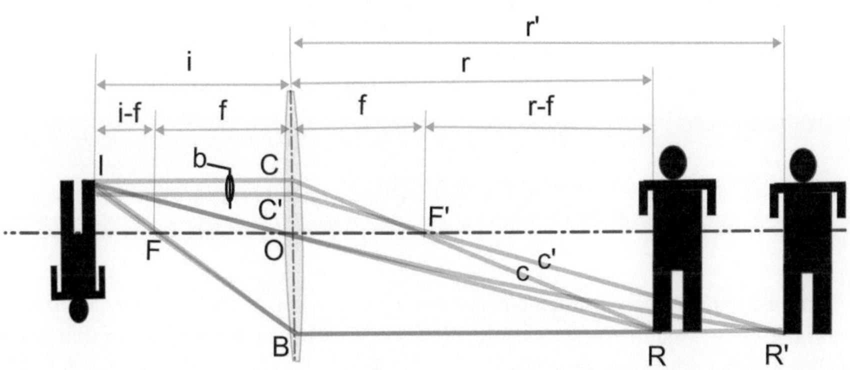

Fig. 5.5 Image blurring by defocusing: points become circles

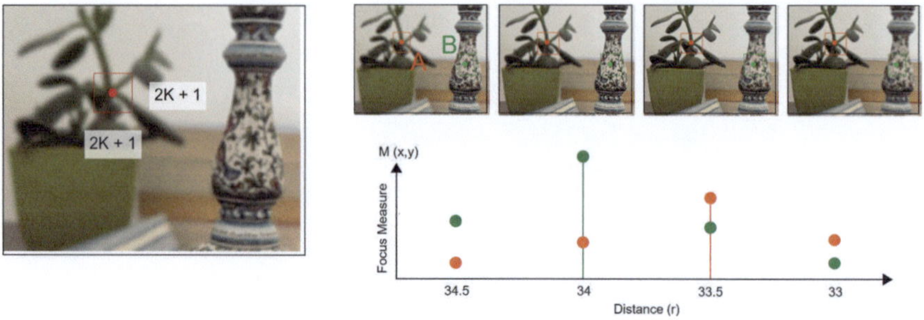

Fig. 5.6 Detection of points in focus by M maximum and therefore their distance r to the camera

compared to the same point when it becomes blurred and enlarged, such as when it appears as a circular spot due to defocus. This loss of high-frequency components results in diminished detail and visually mimics the effect of displacing the object away from the focal plane. According to the convolution theorem, this blurring process can also be interpreted as a convolution between the point (or a small region surrounding it) and a broader intensity distribution, typically modeled by a Gaussian function. This concept was previously illustrated in Fig. 3.39. Consequently, regions around points that are in focus within the image stack tend to retain higher spatial frequency content. These regions are less affected by blurring, exhibit sharper edges, and show greater local contrast, as illustrated in Fig. 5.6.

A practical way to quantify this frequency content is through the absolute value of the Laplacian. As a second-order derivative operator, the Laplacian responds strongly to areas of rapid intensity change, such as small features and thin edges, which are characteristic of high-frequency information. Therefore, the magnitude of the Laplacian can be used as an effective focus level measure, helping to identify in-focus regions within the image stack. We can therefore estimate the frequency content of the area around each point of an image by the function:

$$\nabla_M^2 f = \left| \frac{\partial^2 f}{\partial x^2} \right| + \left| \frac{\partial^2 f}{\partial y^2} \right| \tag{5.3}$$

And an estimate of the frequency content of the area around the image point (x, y) by the number:

$$M(x, y) = \sum_{i=x-K}^{x+K} \sum_{j=y-K}^{y+K} \nabla_M^2 f(i, j) \tag{5.4}$$

So calculating this parameter M(x, y) on each point of the image stack by its maximum it is possible to estimate the distance of focus, so the distance of that point to the camera optical centre.

This method effectively results in a discretization of object depth r, assigning a depth value corresponding to each image position in the focus stack. However, by applying interpolation techniques, it is possible to achieve sub-stack resolution, thereby increasing the accuracy of the 3D shape reconstruction, as illustrated in Fig. 5.7. It is important to note, however, that the effectiveness of this technique depends on the presence of texture or contrast in the surface being measured. Surfaces with uniform color and no texture cannot be measured using this approach. This is because Eq. 5.3 relies on detecting local maxima in frequency content to identify in-focus regions. In the case of a textureless surface, the grey-level distribution remains constant across all distances, yielding identical frequency content and preventing any meaningful focus detection. This limitation can be overcome by introducing artificial texture. If feasible, the surface can be painted with a random speckle pattern. Alternatively, a random speckle pattern can be projected onto the surface using structured light, enabling the necessary frequency variations for accurate focus-based depth estimation. Figure 5.8a, b shows a homemade 3D measurement system built using a standard digital camera. The setup was specifically designed to measure the fracture surfaces of broken specimens, which naturally exhibit a random speckle pattern suitable for focus-based shape reconstruction. The specimens are mounted on a motorized optical stage, which enables precise movement along the optical axis to capture the focus stack, a series of images taken at different distances. To ensure coaxial illumination and imaging, the system employs a beam splitter that aligns the lighting and the camera's optical path. Both image acquisition and stage movement are fully computer-controlled via an I/O interface board and dedicated software, allowing automated and synchronized operation of the measurement process. Typical results obtained are illustrated in Fig. 5.8c.

A focus variation shape measurement system can be realized using a standard optical microscope, as it already integrates all the necessary components: optical elements, illumination, and precise sample positioning along the optical axis to acquire images at different distances, as illustrated in Fig. 5.9a.

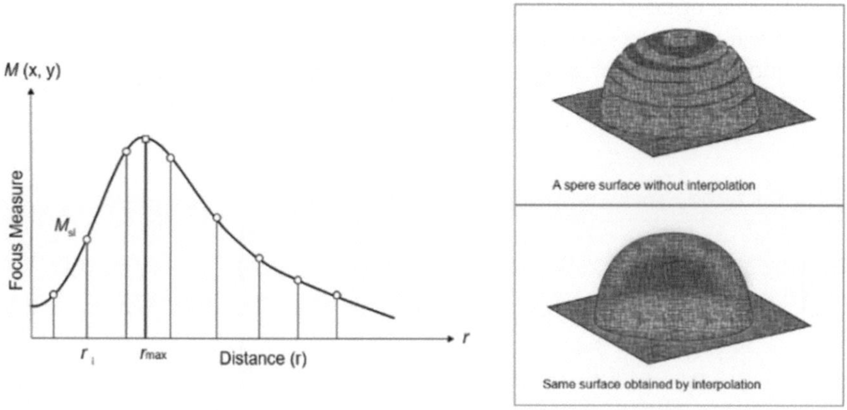

Fig. 5.7 Depth of focus and improving resolution by interpolation

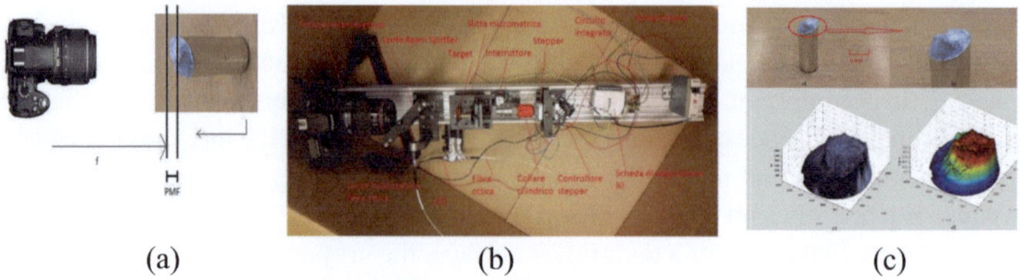

Fig. 5.8 A custom made focus variation measurement system

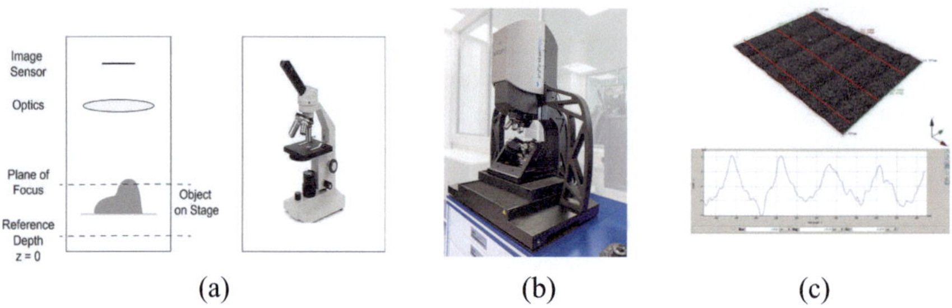

Fig. 5.9 Depth from focus measurement systems and results

High performance commercial systems, based on this measurement principle, such as the one shown in Fig. 5.9b, an InfiniteFocus G6 by Bruker Alicona, can achieve measurement uncertainties as low as few nm, over large measurement volumes up to 200 × 200 × 180 mm^2 [3], making them highly effective for high-resolution surface characterization.

As previously discussed in Chap. 2, image defocusing can be achieved not only by altering the distance r between the camera and the object, but also by modifying other optical parameters such as the numerical aperture (NA), the focal length f, or the position of the image sensor s relative to the optical center. These effects are illustrated in Fig. 5.10.

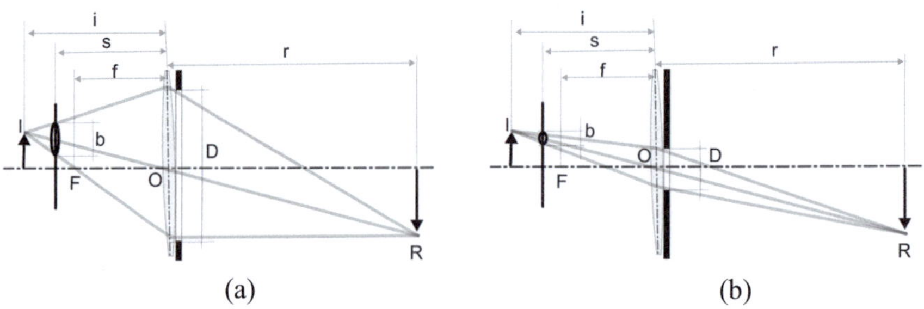

Fig. 5.10 Defocusing by changing image sensor position (**a**); or by changing diaphragm diameter (**b**)

Simplifying these concepts by the scheme of Fig. 5.10, from similar triangles we can say that:

$$\frac{b}{D} = \frac{i-s}{i} \quad so \quad b = \frac{i-s}{i} \tag{5.5}$$

There exists a simple relationship involving the optical diaphragm aperture D: decreasing D (for example, by increasing the f-number N_a) leads to a reduction in the diameter of the blurring circle b. Conversely, the blur diameter b increases when the image sensor is shifted from the ideal focal plane at distance i to a new position s, thereby increasing the defocus amount by increasing the difference i − s.

From similar triangles in Fig. 5.10 we can also see that $r = \frac{if}{i-f}$, but from (5.5) $i = \frac{Ds}{D-b}$ and finally we have:

$$r = \frac{sf}{s - f + b\left(\frac{f}{D}\right)} \tag{5.6}$$

A useful relation to calculate the distance r of the object from our camera if we know blurring circle diameter b and optical parameters s, f and D.

According to the above simplified geometrical optics model, a point in the real world is imaged as a uniformly grey circular spot with diameter b, since the rays converging toward point I form a cone. However, due to diffraction at the aperture edges, lens imperfections, slight misalignments of optical components, and various aberrations, the actual shape of the blur is better approximated by a Gaussian distribution. This Gaussian has a standard deviation approximately equal to b/2, as illustrated in Fig. 5.11b. Unlike the ideal step-like spot, the Gaussian blur exhibits reduced high-frequency content in its Fourier transform.

This Gaussian function expresses the transformation and blurring of each point on the image plane is called the gaussian point spread function expressed as follows:

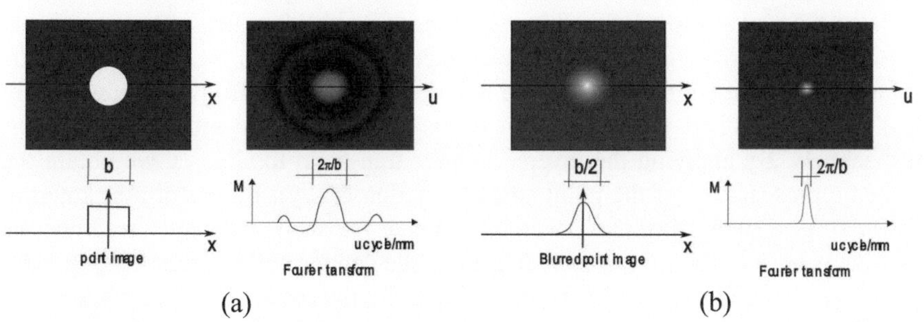

Fig. 5.11 **a** Geometric optics step function model for a point, a more realistic model of it by a Gaussian function

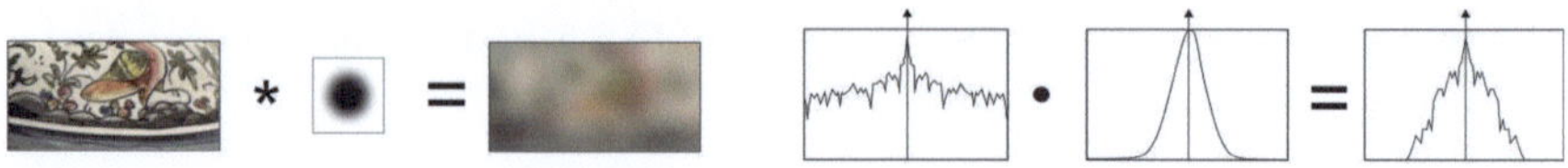

Fig. 5.12 Defocusing as image low pass filtering by convolution in space domain or multiplication in spatial frequency domain

$$g(x, y) = \frac{1}{2\pi\sigma^2} e^{\frac{-(x^2+y^2)}{2\sigma^2}} \tag{5.7}$$

Defocusing of each point in an image is a linear and shift-invariant process. As such, it can be mathematically represented as the convolution of the original image with a point spread function (PSF), typically modeled as a Gaussian function with standard deviation $\sigma = b/2$, where b is the blur diameter. This is illustrated in Fig. 5.12.

According to the convolution theorem (see Chap. 3), convolution in the spatial domain corresponds to multiplication in the spatial frequency domain. Therefore, the defocused image i(x, y), obtained by convolving the focused image $i_0(x, y)$ with the Gaussian PSF g(x, y), i.e. $i(x, y) = i_0(x, y) * g(x, y)$ can also be computed in the spatial frequency domain as:

$$I(u, v) = I_0(u, v) \cdot G(u, v)$$

where I(u, v), $I_0(u, v)$, and G(u, v) are the Fourier transforms of the defocused image, the focused image, and the Gaussian PSF respectively. Notably, the Fourier transform of a Gaussian is also a Gaussian, as shown in Fig. 5.12.

This formulation highlights that defocusing acts as a low-pass filter, attenuating high-frequency components of the image. The cut-off characteristics of this filter are determined by the standard deviation $\sigma = b/2$, and the blur diameter b is a function of the object's distance r from the camera, as described by Eq. (5.6).

If we have access to a focused image, or even a small focused sub-image around a point, and a corresponding defocused image (or sub-image), we can compute the frequency response of the point spread function as $G(u, v) = I(u, v)/I_0(u, v)$, where I(u, v) and $I_0(u, v)$ are the Fourier transforms of the defocused and focused images, respectively. By applying the inverse Fourier transform to G(u, v), we obtain the spatial representation of the point spread function g(x, y). By performing interpolation on g(x, y) and Eq. (5.4), we can estimate the shape parameter σ. Since the blur diameter $b = 2\sigma$, we can then determine the depth r of the corresponding point (sub-image around it) in the image using Eq. (5.3), thus estimating its distance from the camera.

However, in practical scenarios, the focused image $i_0(x, y)$ is usually not available. To overcome this limitation, we can instead use two defocused images $i_1(x, y)$ and $i_2(x, y)$, captured under different optical settings that produce different point spread functions $g_1(x, y)$ and $g_2(x, y)$. These can then be used in place of the unknown focused image to estimate depth through a comparative analysis.

From convolution theorem, in frequency domain, we have:

$$\frac{I_1(u, v)}{I_2(u, v)} = \frac{G_1(u, v)}{G_2(u, v)} \tag{5.8}$$

An equation and two unknown $G_1(u, v)$ and $G_2(u, v)$. Anyway, if our two spread functions $g_1(x, y)$ and $g_2(x, y)$ are generated by changing the optical parameter aperture D from D_1 to D_2, from Eq. (5.5) we have the two relationship:

$$b_1 = \frac{i - s}{i} D_1 \quad \text{and} \quad b_2 = \frac{i - s}{i} D_2.$$

From the ratio of these two equations, we have:

$$\frac{b_2}{b_1} = \frac{D_2}{D_1} \quad \text{and because } \sigma = \text{b}/2 \text{ also } \frac{\sigma 2}{\sigma 1} = \frac{D_2}{D_1} \tag{5.9}$$

A second equation which allows to solve the problem, using both Eq. (5.8) and Eq. (5.9), so finding blurring diameter b and therefore from the Eq. (5.5) the depth r.

Due to the presence of noise, this direct approach for estimating the depth r may suffer from numerical instabilities, making it difficult to apply in practice. However, more robust signal processing techniques that are less sensitive to noise can be employed, thereby making the method suitable for a wide range of applications.

An alternative way to obtain two or more defocused images is by varying the position of the image sensor i, as defined in Fig. 5.10, and capturing at least two images at positions i_1 and i_2. The image processing techniques and mathematical relationships involved remain essentially the same as previously described. A practical implementation of this method involves acquiring two images simultaneously using two sensors placed at positions i_1 and i_2 (or equivalently at distances D_1 and D_2), using for example a beam splitter which directs the same optical image onto two CCD or CMOS sensors.

This second approach, which estimates depth based on focus variation between two or more images, is generally less precise than methods based on a full focus stack. However, it is significantly faster and better suited for real-time applications.

It has been effectively applied in various domains, such as autonomous vehicles for depth estimation in navigation and obstacle avoidance; robotics for scene interpretation and object manipulation; medical imaging for 3D reconstruction of tissues or cells; photography, for post-capture digital automatic refocusing.

5.3 Stereovision Basics

Stereovision, also known as stereo imaging or stereo photography, is a passive technique used to measure three-dimensional shapes, based on the principle of triangulation. This principle allows us to determine the spatial position of a point by intersecting two known lines of sight in a common plane. It is essentially the same mechanism that our two eyes use to perceive depth. As discussed in Chap. 3, we introduced the basic mathematical relationship between a point in 3D space, defined by coordinates (x, y, z), and its projection onto the 2D image plane with coordinates (u, v), using the ideal pinhole camera model. According to this model, knowing the image coordinates (u, v) of a point only tells us that the point lies somewhere along a straight line extending from the camera's optical center through the (u, v) location on the image plane. This is known as the line of sight, illustrated in Fig. 5.13. To pinpoint the 3D position of the point, we need a second image taken from a different viewpoint. In this stereo setup, the same point in space will project onto two different locations, one in each image. Each of these projections defines a line of sight in 3D space. The intersection of these two lines identifies the exact 3D position of the point through triangulation, as also shown in Fig. 5.13.

If we use two cameras whose image sensors lie on the same plane and are horizontally aligned, shifted only along the u-axis by a fixed distance b (known as the *baseline*, i.e., the distance between the two optical centers), then, as shown in Fig. 5.13, the same point in space will appear at the same v-coordinate in both images but will have different u-coordinates. This horizontal difference between the two projections is called *disparity*.

This camera arrangement is referred to as a *calibrated stereo setup* because the relative position and orientation of the two cameras are known, and their internal parameters are also specified. In the simplest model, these intrinsic parameters

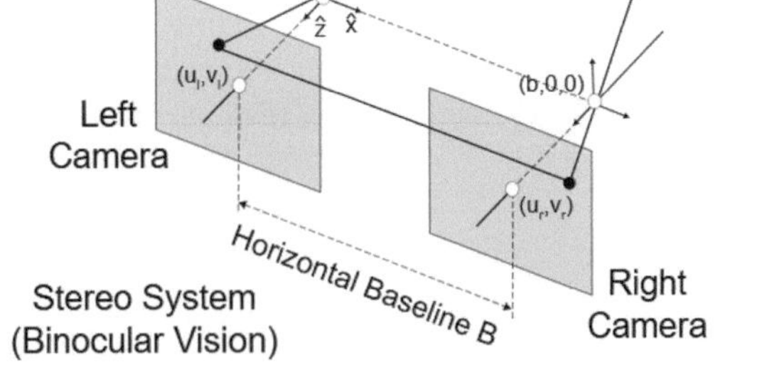

Fig. 5.13 Basic principle of stereovision

include the focal length f, and the half-height and half-width of the image sensor, assumed identical for both cameras.

Using the principles of perspective projection and the geometric configuration shown in Fig. 5.13, we can derive the depth of the 3D point from the disparity using the equations (2.26), so:

$$(u_l, v_l) = \left(f\frac{x}{z} + o_x, f\frac{y}{z} + o_y\right) \quad \text{and} \quad (u_r, v_r) = \left(f\frac{x-b}{z} + o_x, f\frac{y}{z} + o_y\right) \tag{5.10}$$

Solving the above four equations for (x,y,z) we can calculate the 3D coordinates of same point identified in the two images as:

$$x = \frac{b(u_l - o_x)}{(u_l - u_r)} \qquad z = \frac{bf}{(u_l - u_r)} \qquad z = \frac{bf}{(u_l - u_r)}$$

From the equations it is evident that the depth (z) of a point is inversely proportional to the *disparity*, the horizontal difference between the projections of the point on the left and right image planes, denoted as $(u_l–u_r)$. This means that as the disparity decreases, the depth increases. For points that are very far from the cameras, the disparity becomes so small that it may fall below the resolution limit of the image sensor or be lost in noise, making it difficult or even impossible to detect accurately.

As shown in Fig. 5.13 and supported by equations such as Eq. (5.10), disparity is directly proportional to the *baseline length* (b), the distance between the two camera optical centers. Increasing the baseline makes the disparity larger for a given depth, which improves the ability to detect faraway objects. A longer baseline also enhances the precision of depth estimation: when two lines of sight intersect at a very narrow angle (as with a short baseline), even slight errors or noise in direction can cause large deviations in the estimated 3D position. A larger baseline increases this angle, reducing sensitivity to such uncertainty and improving measurement accuracy.

These geometric principles allow us to compute the 3D coordinates (x, y, z) of a point, provided we can identify the corresponding image points in both views. In practical terms, once we select a point in one image (e.g., point L in the left image with coordinates (u_l, v_l)), we must locate the matching point R in the right image with coordinates (u_r, v_r). With both sets of coordinates, we can apply Eq. (5.11) to calculate the actual 3D position of the point in space.

In a stereo imaging system configured as in Fig. 5.13, with two identical cameras aligned horizontally (i.e., shifted only along the x-axis), the v-coordinate (vertical position) of corresponding points remains the same. This simplifies the correspondence search: to find the right image point R that corresponds to point L in the left image, we can restrict our search to a horizontal line at the same v-coordinate.

Practically, this is done by extracting a small image patch E_L around point L and searching for the most similar patch E_R along the corresponding scanline in the right image. This process, illustrated in Fig. 5.14, is commonly carried out using a

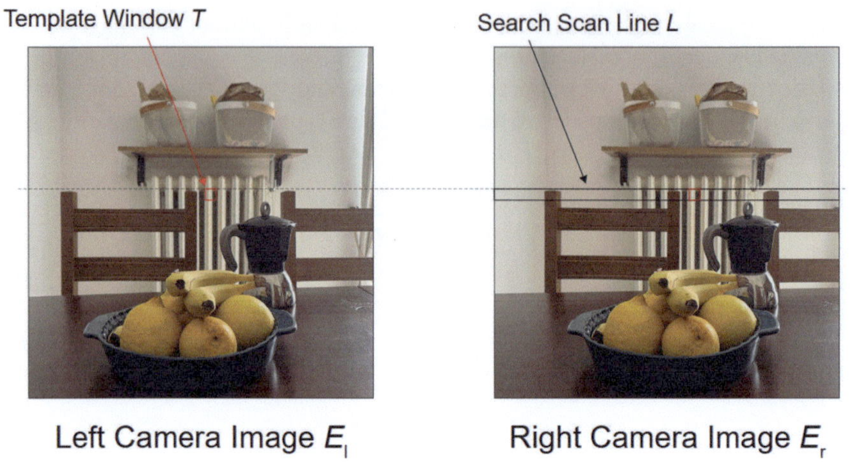

Fig. 5.14 Finding corresponding points between left and right image

technique called *template matching*, which compares pixel patterns to identify the best match.

Template matching in stereo vision can be performed using the same principles described in Chap. 4 for identifying corresponding points of interest, specifically, using the equations (4.27) and (4.28) based on the comparison of block signatures.

In this context, rather than comparing feature descriptors, the method involves directly comparing the raw pixel values of two small sub-images: E_L from the left image and E_R from the right image. The goal is to find the position where these patches best match.

This is typically done by evaluating similarity using either Sum of Absolute Differences (SAD) or Sum of Squared Differences (SSD) defined by:

$$SAD(k,l) = \sum_{i,j} |E_l(i,j) - E_r(i+k, j+l)| \; SSD(k,l) = \sum_{i,j} |E_l(i,j) - E_r(i+k, j+l)|^2 \quad (5.11)$$

The best match is found at the position that minimizes the SAD or SSD value, indicating the highest similarity between the two image regions. Both should be zero for a perfect match between the two sub images E_l and E_r, but because of noise, different lighting of the two images, different perspective projections, aberrations etc. this does not normally happen, so we will search for a minimum of these quantities. It is also possible to find a good matching by finding the maximum of the normalized correlation function NCC defined by:

$$NCC(k,l) = \frac{\sum_{i,j} E_l(i,j)E_r(i+k,j+l)}{\sqrt{\sum_{i,j} E_l(i,j)^2 \sum_{i,j} E_r(i+k,j+l)^2}} \tag{5.12}$$

This parameter is less sensitive to noise, variations in lighting conditions, or other optical effects that may introduce differences between the two sub-images E_l and Er.

It is essential that the surface being imaged has a non-repetitive, distinctive texture. If the surface contains repetitive patterns, the minimum in the difference metrics (such as SAD or SSD), or the maximum in normalized cross-correlation, may occur at multiple locations corresponding to each repetition of the pattern. This ambiguity makes it impossible to uniquely identify corresponding points, which is critical for accurate depth estimation. The choice of window size for the sub-images E_L and E_R plays a critical role in the accuracy and robustness of template matching. A small window, such as 5×5 pixels, offers higher spatial resolution and allows for more precise localization of corresponding points. However, it is also more susceptible to noise, illumination changes, geometric distortions, and lens aberrations. Conversely, a larger window provides greater robustness to such disturbances by averaging over more pixel data, making it less sensitive to local variations. The trade-off is a reduction in spatial resolution, which can lead to less precise depth estimation, especially in areas with fine details or sharp edges. Some calibrated stereovision systems and 3D commercial cameras are illustrated in Fig. 5.15.

A 3D measurement system can also be implemented using two standard cameras mounted on a fixed and properly aligned support. By performing a calibration procedure, using available software tools or libraries in environments such as MATLAB or Python, it is possible to determine both the intrinsic and extrinsic parameters of the stereo setup. This measurement system, illustrated in Fig. 5.16, enables accurate 3D measurements from stereo images.

This approach was applied in the 2011 thesis work of Luca Giulivi at the University of Perugia, with the help of Michele Moretti [7], demonstrating that it is possible to achieve high-precision measurements using high quality cameras in a stereo setup. The goal was to long time monitor the relative displacement between two parts of a cracked wall using a non-contact optical measurement technique.

To facilitate tracking, randomly shaped dots were painted on the test wall, each uniquely identified by a code. This allowed for precise localization and tracking of the wall sections over long time. One practical application of this technique is structural

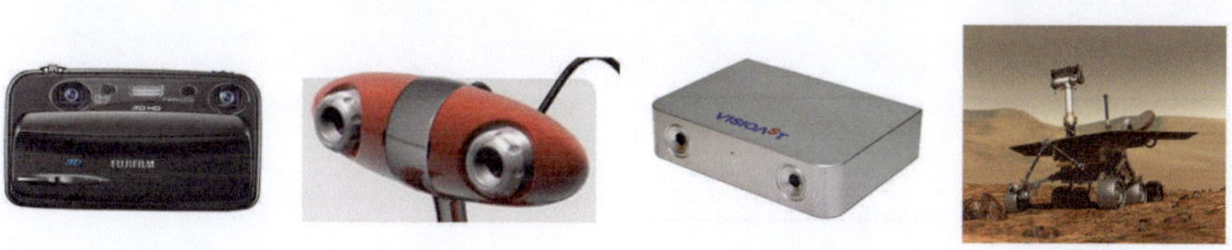

(a) (b) (c) (d)

Fig. 5.15 Commercial 3D cameras in (**a**) (**b**) and (**c**) and in (**d**) one stereocamera realized for a Mars spatial mission [6]

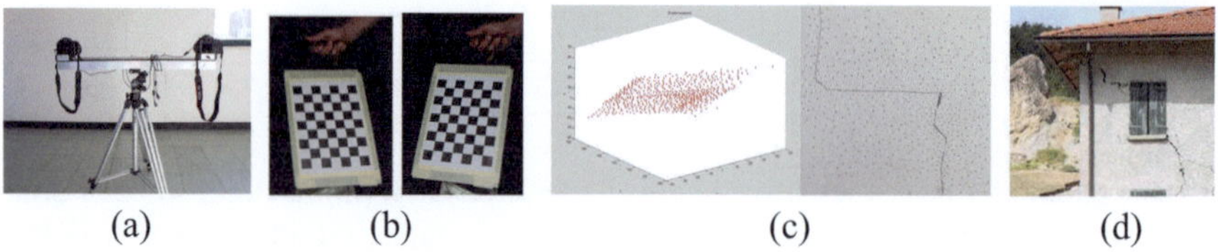

Fig. 5.16 A 3D custom made measurement system by two cameras mounted on an optical rail in (**a**); calibration of the system by a Matlab package using chessboard references in (**b**); using the system for measurement of positions of points painted on a wall in (**c**); and for monitoring building cracks evolution in (**d**)

health monitoring, for instance, tracking crack movement in buildings caused by seismic activity, as illustrated in Fig. 5.16d.

The stereo system calibration and measurement process was performed using MATLAB's Stereo Camera Calibrator toolbox, as shown in Fig. 5.17.

Stereovision, while powerful, presents several limitations and challenges that can impact measurement accuracy and reliability. These challenges primarily stem from the inherent differences in how a scene appears to each camera.

A major source of error arises from photometric distortions, where corresponding points in the two images may exhibit different brightness levels. These differences can result from variations in viewpoint, illumination, or disparities in the radiometric response of the two cameras (e.g., sensor sensitivity, exposure settings, or lens aberrations). In addition, noise and optical imperfections can further alter the appearance of the same point in each image, complicating the correspondence search.

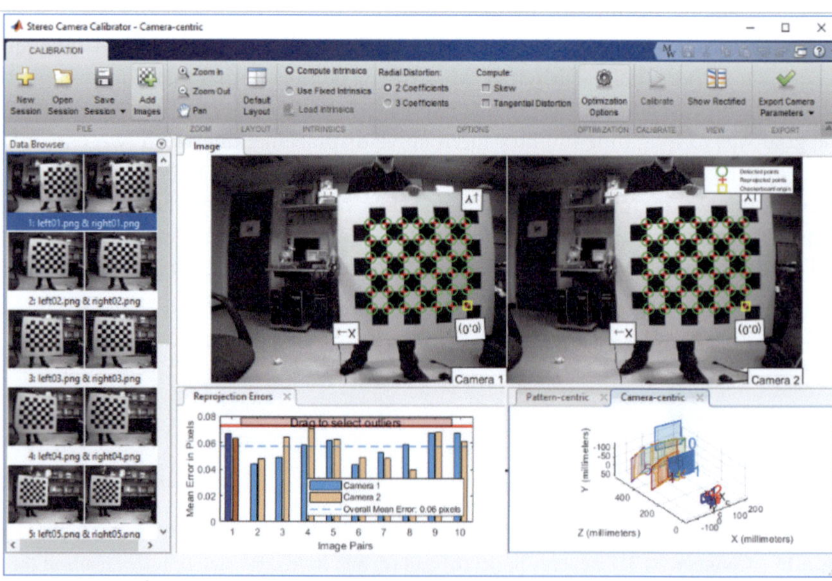

Fig. 5.17 Matlab camera calibration toolbox used for stereocamera system calibration

Perspective distortions also introduce challenges: the same physical area may appear with different sizes or shapes in each image due to the different viewing angles of the cameras. More critically, occlusions can prevent certain points in the scene from being visible to all cameras. If a point is visible in one image but not in another, it is impossible to compute its depth through triangulation.

Stereovision requires sufficient texture in the scene to identify unique and reliable correspondence points. In regions with low texture or repetitive patterns, it becomes difficult, or even impossible, to match points accurately. In such cases, artificial texture must be added using methods such as projected light patterns, painted markers, or printed textures, assuming it is practical and permissible for the application.

One of the fundamental disadvantages of stereovision systems is the computational cost of finding correspondences between images. This process is inherently intensive and can be too slow for real-time applications without specialized hardware or algorithmic optimization. Moreover, the accuracy of depth estimation is highly dependent on the quality and distinctiveness of image textures; uniform or featureless surfaces pose serious limitations.

The measurement range of a stereovision system is primarily determined by the optical configuration. With appropriate lenses, stereovision can be applied across scales, from microscopic measurements to large-scale 3D mapping, such as mountain profiling or complete cities mapping.

In terms of measurement precision, stereovision is generally less accurate than structured light techniques. However, high-quality systems can still achieve impressive results. For typical mechanical components in the range of tens of centimeters, sub-millimeter accuracy is attainable. A widely accepted benchmark for stereo system accuracy is approximately 0.1% of the measurement range [4, 5], although specialized systems for specific applications may exceed this performance.

5.4 Uncalibrated Stereovision

There are methods that enable 3D shape reconstruction even when the relative positions and orientations of the two cameras, or of a single camera capturing two images from different viewpoints, are not known in advance. This situation, illustrated in Fig. 5.18, assumes that we have a pair of images capturing the same object, with many common parts and easy to identify same points of interest on both images.

While the extrinsic parameters (i.e., position and orientation) of the cameras are unknown, it is assumed that the intrinsic parameters, such as focal length f and the coordinates of the principal point (o_x, o_y), are known for both cameras (or for the same camera). These internal calibration values are essential for interpreting the geometric relationships within each image, even in the absence of information about the cameras' spatial relationship.

As shown in the following sections, it is possible to estimate the relative position and orientation between two cameras by identifying corresponding points in the pair

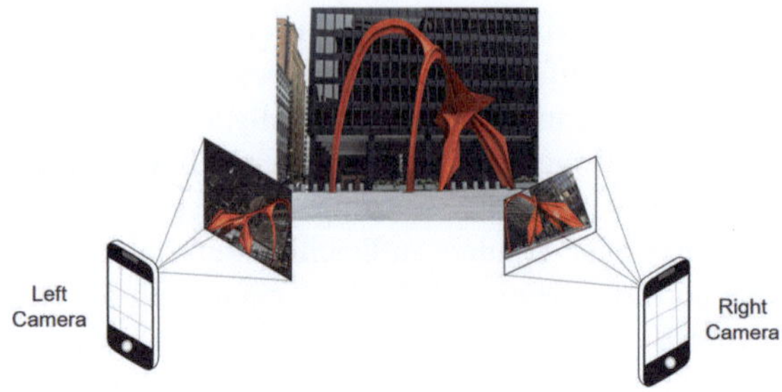

Fig. 5.18 Uncalibrated sterovision problem

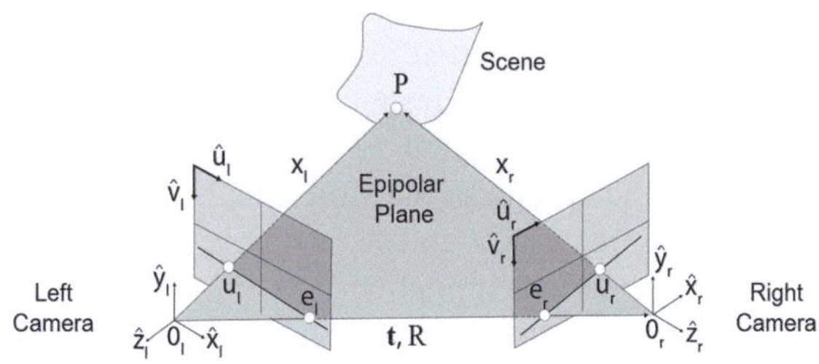

Fig. 5.19 Epipolar geometry

of images. These matched points serve as references for calculating the rotation and translation matrices that relate the two camera coordinate systems. This process is conceptually similar to what we discussed in Chap. 2, where the projection matrix was defined based on the transformation between the object and camera reference frames.

The first step in this approach involves the use of epipolar geometry. Epipolar geometry describes the intrinsic projective geometry between two views and is independent of the scene structure. It is based on the concept of epipoles, as illustrated in Fig. 5.19.

- The right epipole is the point in the right image where the optical center of the left camera is projected. In other words, it is the point from which the right camera "sees" the left camera's optical center.
- Similarly, the left epipole is the projection of the right camera's optical center into the left image.

Each pair of images defines a unique pair of epipoles. These points are fundamental to construct the epipolar lines, which constrain the possible locations of corresponding points across the two images. Leveraging these geometric constraints allows us to compute the essential or fundamental matrix and, from it, derive the relative camera pose.

The epipolar plane is defined as the plane that passes through the 3D point P and the optical centers (camera centers) of both cameras. This plane contains the projection rays from each camera to the point P, and it intersects each image plane in a line known as the epipolar line. For each 3D point P in the scene, there exists a unique epipolar plane, assuming a stereo setup with two distinct camera centers.

Within this framework, we also define a vector **n**, which is normal to the epipolar plane. This vector can be computed as the cross product between the translation vector **t**, representing the (currently unknown) position of the right camera relative to the left camera, and the position vector $\mathbf{x_l}$, which gives the coordinates of point P in the coordinate system of the left camera.

Mathematically, this is expressed as:

$$\boldsymbol{n} = \boldsymbol{t} \times \boldsymbol{x}_l \tag{5.13}$$

n is a vector perpendicular to the epipolar plane, as illustrated in Fig. 5.19. Because for definition **n** and $\mathbf{x_l}$ are perpendicular, we have that:

$$\boldsymbol{x}_l \cdot \boldsymbol{n} = 0$$

From the above two equations we can write the so-called epipolar constraint:

$$\boldsymbol{x}_l \cdot (\boldsymbol{t} \times \boldsymbol{x}_l) = 0$$

That can be written more explicitly according to the cross product definition as:

$$\left| x_l \; y_l \; z_l \right| \begin{bmatrix} t_y z_l - t_z y_l \\ t_z x_l - t_x z_l \\ t_x y_l - t_y x_l \end{bmatrix} = 0$$

Or equivalently in the matricial form:

$$\text{defining} \quad \boldsymbol{T} = \begin{bmatrix} 0 & -t_z & t_y \\ t_z & 0 & -t_x \\ -t_y & t_x & 0 \end{bmatrix} \text{as} \left| x_l \; y_l \; z_l \right| \begin{bmatrix} 0 & -t_z & t_y \\ t_z & 0 & -t_x \\ -t_y & t_x & 0 \end{bmatrix} \cdot \begin{bmatrix} x_l \\ y_l \\ z_l \end{bmatrix} = 0 \tag{5.14}$$

Coordinates of point P in left camera, given by vector $\mathbf{x_l}$, are related to that in the right camera, given by $\mathbf{x_r}$, by the rototraslation of the vectors between the two different coordinate systems, so that:

$$\boldsymbol{x}_l = \boldsymbol{R}\boldsymbol{x}_r + \mathbf{t}$$

where $\boldsymbol{R}$ is the rotation matrix of the right reference system respect to the left and **t** is the translation vector between the two origins of the two reference systems. In matrix form above equation can be written as:

$$\begin{bmatrix} x_l \\ y_l \\ z_l \end{bmatrix} = \begin{bmatrix} r_{11} \; r_{12} \; r_{13} \\ r_{21} \; r_{22} \; r_{23} \\ r_{31} \; r_{32} \; r_{33} \end{bmatrix} \begin{bmatrix} x_r \\ y_r \\ z_r \end{bmatrix} + \begin{bmatrix} t_x \\ t_y \\ t_z \end{bmatrix} \text{ where } \boldsymbol{R} = \begin{bmatrix} r_{11} \; r_{12} \; r_{13} \\ r_{21} \; r_{22} \; r_{23} \\ r_{31} \; r_{32} \; r_{33} \end{bmatrix}.$$

Substituting into the epipolar constraint (Eq. 5.14) we obtain:

$$\left|x_l\ y_l\ z_l\right|\left(\begin{bmatrix} 0 & -t_z & t_y \\ t_z & 0 & -t_x \\ -t_y & t_x & 0 \end{bmatrix}\cdot\begin{bmatrix} r_{11}\ r_{12}\ r_{13} \\ r_{21}\ r_{22}\ r_{23} \\ r_{31}\ r_{32}\ r_{33} \end{bmatrix}\begin{bmatrix} x_r \\ y_r \\ z_r \end{bmatrix}+\begin{bmatrix} 0 & -t_z & t_y \\ t_z & 0 & -t_x \\ -t_y & t_x & 0 \end{bmatrix}\begin{bmatrix} t_x \\ t_y \\ t_z \end{bmatrix}\right)=0$$

But the second term of the sum inside parenthesis is the vector product $\mathbf{t} \times \mathbf{t}$ that is equal to zero for each vector, so above equation simplify to:

$$\left|x_l\ y_l\ z_l\right|\begin{bmatrix} 0 & -t_z & t_y \\ t_z & 0 & -t_x \\ -t_y & t_x & 0 \end{bmatrix}\cdot\begin{bmatrix} r_{11}\ r_{12}\ r_{13} \\ r_{21}\ r_{22}\ r_{23} \\ r_{31}\ r_{32}\ r_{33} \end{bmatrix}\begin{bmatrix} x_r \\ y_r \\ z_r \end{bmatrix}=0$$

Defining the essential matrix **E** as:

$$\boldsymbol{E} = \boldsymbol{TR} \tag{5.15}$$

The above equation simplifies to:

$$\left|x_l\ y_l\ z_l\right|\boldsymbol{E}\begin{bmatrix} x_r \\ y_r \\ z_r \end{bmatrix}=0\text{or}\left|x_l\ y_l\ z_l\right|\begin{bmatrix} e_{11}\ e_{12}\ e_{13} \\ e_{21}\ e_{22}\ e_{23} \\ e_{31}\ e_{32}\ e_{33} \end{bmatrix}\begin{bmatrix} x_r \\ y_r \\ z_r \end{bmatrix}=0 \tag{5.16}$$

That gives the transformation of coordinates from the left image to the right image. The matrix **T** is skew-symmetric, **R** is orthonormal as already seen in Chap. 2, and we know from linear algebra that matrix **E** can be splitted into its two components, one orthonormal and the other symmetric that are **T** and **R**.

Relation (5.16) is between coordinates (x, y, z) of the point in 3D but unfortunately, we only know the corresponding points in 2D image plane coordinates (u, v). Even if we don't know the 3D coordinates, we can use the relationship between 2 and 3D coordinates, the perspective projection Eq. (2.26), defined in Chap. 2. From that we can write:

$$u_l = f_l\frac{x_l}{z_l} + o_{xl};\ \ v_l = f_l\frac{y_l}{z_l} + o_{yl}$$

And also, multiplying by z_l:

$$u_l z_l = f_l x_l + o_{xl} z_l;\ \ v_l z_l = f_l y_l + o_{yl} z_l$$

In matrix form, using homogeneous 2D coordinates, as done in Chap. 2, above relationship can be written as:

$$z_l\begin{bmatrix} u_l \\ v_l \\ 1 \end{bmatrix}=\begin{bmatrix} z_l u_l \\ z_l v_l \\ z_l \end{bmatrix}=\begin{bmatrix} f_l x_l + o_{xl} z_l \\ f_l y_l + o_{yl} z_l \\ z_l \end{bmatrix}=\begin{bmatrix} f_l & 0 & o_{xl} \\ 0 & f_l & o_{yl} \\ 0 & 0 & 1 \end{bmatrix}\begin{bmatrix} x_l \\ y_l \\ z_l \end{bmatrix}$$

where the matrix of intrinsic parameters f_l, o_{xl} and o_{yl} is known. For both the left and right camera we therefore have:

$$z_l \begin{bmatrix} u_l \\ v_l \\ 1 \end{bmatrix} = \begin{bmatrix} f_l & 0 & o_{xl} \\ 0 & f_l & o_{yl} \\ 0 & 0 & 1 \end{bmatrix} \begin{bmatrix} x_l \\ y_l \\ z_l \end{bmatrix} \text{ and } z_r \begin{bmatrix} u_r \\ v_r \\ 1 \end{bmatrix} = \begin{bmatrix} f_r & 0 & o_{xr} \\ 0 & f_r & o_{yr} \\ 0 & 0 & 1 \end{bmatrix} \begin{bmatrix} x_r \\ y_r \\ z_r \end{bmatrix} \tag{5.17}$$

And defining the two known internal parameter matrix of left and right camera:

$$\boldsymbol{K_l} = \begin{bmatrix} f_l & 0 & o_{xl} \\ 0 & f_l & o_{yl} \\ 0 & 0 & 1 \end{bmatrix} \text{ and } \boldsymbol{K_r} = \begin{bmatrix} f_r & 0 & o_{xr} \\ 0 & f_r & o_{yr} \\ 0 & 0 & 1 \end{bmatrix}$$

We can write in compact form Eqs. (5.16) the inverted relation between 3D (x, y, z) coordinates from their 2D images as:

$$\boldsymbol{x}_l^T = \left| u_l \ v_l \ 1 \right| z_l \boldsymbol{K}_l^{-1^T} \ \boldsymbol{x}_r = z_r \boldsymbol{K}_r^{-1} \begin{bmatrix} u_r \\ v_r \\ 1 \end{bmatrix} \tag{5.18}$$

Substituting the above Eq. (5.18) relations in epipolar costrain Eq. (5.16) we get:

$$\left| u_l \ v_l \ 1 \right| z_l \boldsymbol{K}_l^{-1^T} \begin{bmatrix} e_{11} & e_{12} & e_{13} \\ e_{21} & e_{22} & e_{23} \\ e_{31} & e_{32} & e_{33} \end{bmatrix} z_r \boldsymbol{K}_r^{-1} \begin{bmatrix} u_r \\ v_r \\ 1 \end{bmatrix} = 0$$

In this equation the two numbers z_l and z_r are the distances of the real point from left and right camera that cannot be 0, so we can remove these two numbers from the above equations, simplifying it to:

$$\left| u_l \ v_l \ 1 \right| \boldsymbol{K}_l^{-1^T} \begin{bmatrix} e_{11} & e_{12} & e_{13} \\ e_{21} & e_{22} & e_{23} \\ e_{31} & e_{32} & e_{33} \end{bmatrix} \boldsymbol{K}_r^{-1} \begin{bmatrix} u_r \\ v_r \\ 1 \end{bmatrix} = 0$$

Defining the new three matrix product **F,** named the fundamental matrix, as:

$$\boldsymbol{F} = \boldsymbol{K}_l^{-1^T} \begin{bmatrix} e_{11} & e_{12} & e_{13} \\ e_{21} & e_{22} & e_{23} \\ e_{31} & e_{32} & e_{33} \end{bmatrix} \boldsymbol{K}_r^{-1} \tag{5.19}$$

We have finally a fundamental relation between 2D coordinates of points in left and right image and matrix F, that can be written as:

$$\left| u_l \; v_l \; 1 \right| \begin{bmatrix} f_{11} \, f_{12} \, f_{13} \\ f_{21} \, f_{22} \, f_{23} \\ f_{31} \, f_{32} \, f_{33} \end{bmatrix} \begin{bmatrix} u_r \\ v_r \\ 1 \end{bmatrix} = 0 \tag{5.20}$$

In the equation involving the fundamental matrix $\mathbf{F} = [f_{ij}]$, there are nine unknown coefficients to determine. However, for each pair of corresponding points, specifically, the image coordinates (u_l, v_l) in the left image and (u_r, v_r) in the right image, we obtain one equation that relates them through the fundamental matrix.

To solve for the nine unknowns, we need at least eight independent point correspondences. These 8 points can be identified either manually (by selecting matching points in the two images) or automatically using feature-matching algorithms such as SIFT (Scale-Invariant Feature Transform) illustrated previously in Chap. 4.

For a generic *i*-th pair of corresponding points, we can write:

$$\left| u_l^i \; v_l^i \; 1 \right| \begin{bmatrix} f_{11} \, f_{12} \, f_{13} \\ f_{21} \, f_{22} \, f_{23} \\ f_{31} \, f_{32} \, f_{33} \end{bmatrix} \begin{bmatrix} u_r^i \\ v_r^i \\ 1 \end{bmatrix} = 0$$

And from this, making the matrix products we get a linear equation related to the point i:

$$\left(f_{11} u_r^i + f_{12} v_r^i + f_{13}\right) u_l^i + \left(f_{21} u_r^i + f_{22} v_r^i + f_{23}\right) v_l^i + f_{31} u_r^i + f_{32} v_r^i + f_{33} = 0$$

And rearranging in matrix form by taking m points we can write the following linear system of equations:

$$\begin{bmatrix} u_l^1 u_r^1 & u_l^1 v_r^1 & u_l^1 & v_l^1 u_r^1 & v_l^1 v_r^1 & v_l^1 & u_r^1 & v_r^1 & 1 \\ \vdots & \vdots & \vdots & \vdots & \vdots & \vdots & \vdots & \vdots & \vdots \\ u_l^i u_r^i & u_l^i v_r^i & u_l^i & v_l^i u_r^i & v_l^i v_r^i & v_l^i & u_r^i & v_r^i & 1 \\ \vdots & \vdots & \vdots & \vdots & \vdots & \vdots & \vdots & \vdots & \vdots \\ u_l^m u_r^m & u_l^m v_r^m & u_l^m & v_l^m u_r^m & v_l^m v_r^m & v_l^m & u_r^m & v_r^m & 1 \end{bmatrix} \begin{bmatrix} f_{11} \\ f_{21} \\ f_{31} \\ f_{21} \\ f_{22} \\ f_{23} \\ f_{31} \\ f_{32} \\ f_{33} \end{bmatrix} = \begin{bmatrix} 0 \\ 0 \\ 0 \\ 0 \\ 0 \\ 0 \\ 0 \\ 0 \\ 0 \end{bmatrix} \text{ or in compact form:}$$

$$\mathbf{Af} = 0.$$

Where all the coefficients of matrix A are known values of the 2D coordinates of corresponding points founded in the two images. Anyway an indetermination still remain because if **f** is a fundamental vector in above equation (or its matrix form **F** in previous equations) also k **f** is a fundamental vector because keeps the solution of all the above equations. i.e. matrix **F** and k**F** describe the same epipolar geometry scaled by a k factor. So, this linear system of equations can be solved by classical least square methods of linear algebra (minimizing $\|\boldsymbol{Af}\|^2$) and using the constraint $\|\boldsymbol{f}\|^2 = 1$ to set the scale k at an arbitrary value.

Once the above system is solved, we can compute essential matrix **E** from known left and right intrinsic camera matrices and the fundamental matrix **F**, by using Eq. (5.19), or in compact form, by:

$$\boldsymbol{E} = \boldsymbol{K}_l^T \boldsymbol{F} \boldsymbol{K}_r \tag{5.21}$$

And finally, because of properties of rototranslation matrices, we can extract rotational **R** and translational ***T*** matrices from **E** by the singular value decomposition techniques of linear algebra, so that:

$$\boldsymbol{E} = \boldsymbol{T}\mathbf{R} \tag{5.22}$$

After completing this procedure, we obtain a fully calibrated stereo camera system. This means we have determined both the intrinsic parameters of each camera and the extrinsic parameters, namely, the distance and relative pose (rotation and translation) between the two camera coordinate systems. Additionally, all the elements of the epipolar geometry, including the fundamental or essential matrix, are now known, enabling 3D shape measurements.

If we define epipolar lines as the intersections of the epipolar plane with the left and right image planes (as illustrated in Fig. 5.18), then for any point in one image, the corresponding point in the other image must lie along its associated epipolar line. This geometric constraint significantly simplifies the correspondence search: instead of searching in two dimensions across the entire image, the search is reduced to a one-dimensional search along a line.

As a result, for each point in one image, we can search for its match along the corresponding epipolar line in the other image, just as we did in the case of a simple, calibrated stereo system described in the previous section.

This allows us to calculate the 3D position potentially of all the image points, thus performing the so called dense triangulation for a complete shape measurement (Fig. 5.20).

Epipolar line equation for a known point (u_l, v_l) in left image can be defined on right image from epipolar constrain Eq. (5.20) in terms of the two variables on left plane (u_r, v_r) by making the matrix product of Eq. (5.20), we obtain:

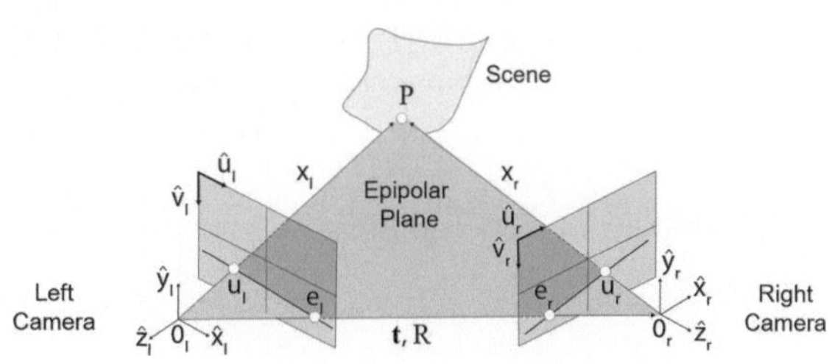

Fig. 5.20 Searching corresponding points along epipolar lines for dense triangulation

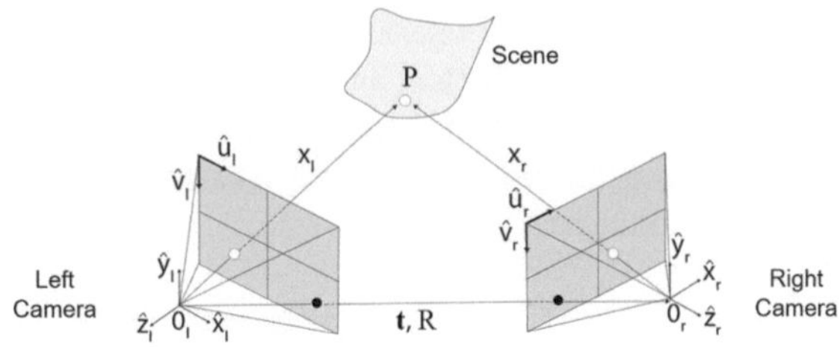

Fig. 5.21 Computing 3D coordinates of dense corresponding points cloud

$$(f_{11}u_l + f_{12}v_l + f_{13})u_r + (f_{12}u_l + f_{22}v_l + f_{32})v_r + f_{13}u_l + f_{23}v_l + f_{33} = 0$$

Or simply:

$$a_l u_r + b_l v_r + c_l = 0 \tag{5.23}$$

And along this line we should search the corresponding point of that on the left image.

As a last step we can calculate the coordinates of all the points we have found corresponding in the left and right images. For each one of this point the scheme is that of Fig. 5.21.

Exactly as we did before in calibrated stereovision, we can calculate the depth of a point on the scene by knowing its 2D coordinates on the left and right image.

Giving the intrinsic parameters, the perspective projections of each one of these points in the two images in homogeneous coordinates are given by:

$$\begin{bmatrix} u_l \\ v_l \\ 1 \end{bmatrix} = \begin{bmatrix} f_l & 0 & o_{xl} & 0 \\ 0 & f_l & o_{yl} & 0 \\ 0 & 0 & 1 & 0 \end{bmatrix} \begin{bmatrix} x_l \\ y_l \\ z_l \\ 1 \end{bmatrix} \quad \text{and} \quad \begin{bmatrix} u_r \\ v_r \\ 1 \end{bmatrix} = \begin{bmatrix} f_r & 0 & o_{xr} & 0 \\ 0 & f_r & o_{yr} & 0 \\ 0 & 0 & 1 & 0 \end{bmatrix} \begin{bmatrix} x_r \\ y_r \\ z_r \\ 1 \end{bmatrix}$$

$$\text{or in compact form} : \tilde{\boldsymbol{u}}_r = \boldsymbol{M}_{intr} \tilde{\boldsymbol{x}}_r \tag{5.24}$$

And we know now by calibration the relationship of 3D coordinates between left and right camera, in homogeneous coordinates:

$$\begin{bmatrix} x_l \\ y_l \\ z_l \\ 1 \end{bmatrix} = \begin{bmatrix} r_{11} & r_{12} & r_{13} & t_x \\ r_{21} & r_{22} & r_{23} & t_y \\ r_{31} & r_{32} & r_{33} & t_z \\ 0 & 0 & 0 & 1 \end{bmatrix} \begin{bmatrix} x_r \\ y_r \\ z_r \\ 1 \end{bmatrix}$$

Substituting the left coordinate vector expression on the previous, for the left camera we get:

$$\begin{bmatrix} u_l \\ v_l \\ 1 \end{bmatrix} = \begin{bmatrix} f_l & 0 & o_{xl} & 0 \\ 0 & f_l & o_{yl} & 0 \\ 0 & 0 & 1 & 0 \end{bmatrix} \begin{bmatrix} r_{11} & r_{12} & r_{13} & t_x \\ r_{21} & r_{22} & r_{23} & t_y \\ r_{31} & r_{32} & r_{33} & t_z \\ 0 & 0 & 0 & 1 \end{bmatrix} \begin{bmatrix} x_r \\ y_r \\ z_r \\ 1 \end{bmatrix}$$

Making the product of the intrinsic matrix with the rototranslation matrix in the above expression, giving the projection matrix $\mathbf{P}_l$, in compact form we can write the previous equation as:

$$\widetilde{\boldsymbol{u}_l} = \boldsymbol{P}_l \widetilde{\boldsymbol{x}_r} \tag{5.25}$$

So taking the two above equations $\widetilde{\boldsymbol{u}_r} = \boldsymbol{M}_{intr} \widetilde{\boldsymbol{x}_r}$ and $\widetilde{\boldsymbol{u}_l} = \boldsymbol{P}_l \widetilde{\boldsymbol{x}_r}$ in general matrix form as:

$$\begin{bmatrix} u_r \\ v_r \\ 1 \end{bmatrix} = \begin{bmatrix} m_{11} & m_{12} & m_{13} & m_{14} \\ m_{21} & m_{22} & m_{23} & m_{24} \\ m_{31} & m_{32} & m_{33} & m_{34} \end{bmatrix} \begin{bmatrix} x_r \\ y_r \\ z_r \\ 1 \end{bmatrix} \quad \text{and} \quad \begin{bmatrix} u_l \\ v_l \\ 1 \end{bmatrix} = \begin{bmatrix} p_{11} & p_{12} & p_{13} & p_{14} \\ p_{21} & p_{22} & p_{23} & p_{24} \\ p_{31} & p_{32} & p_{33} & p_{34} \end{bmatrix} \begin{bmatrix} x_r \\ y_r \\ z_r \\ 1 \end{bmatrix}$$

Since the only unknowns are the three coordinates (x_r, y_r, z_r) of the 3D point, we can rearrange the terms from the two projection equations to write a linear system of the form:

$$\begin{bmatrix} u_r m_{31} - m_{11} & u_r m_{32} - m_{12} & u_r m_{33} - m_{13} \\ v_r m_{31} - m_{21} & v_r m_{32} - m_{22} & v_r m_{33} - m_{23} \\ u_l p_{31} - p_{11} & u_l p_{32} - p_{12} & u_l p_{33} - p_{13} \\ v_l p_{31} - p_{21} & v_l p_{32} - p_{22} & v_l p_{33} - p_{23} \end{bmatrix} \begin{bmatrix} x_r \\ y_r \\ z_r \end{bmatrix} = \begin{bmatrix} m_{14} - m_{34} \\ m_{24} - m_{34} \\ p_{14} - p_{34} \\ p_{24} - p_{34} \end{bmatrix} \tag{5.26}$$

So, we have four equations with three unknowns, an overdetermined linear system of equations that can be solved by the linear algebra methods.

Summarizing, the steps of stereo uncalibrated measurement technique are:

(1) Taking at least two images where it is possible to identify at least 8 corresponding points.
(2) By using epipolar geometry estimating the fundamental matrix, finding relative positions of an image respect to the other, using coordinates in the image planes of these points.
(3) By using epipolar geometry again defining dense correspondences between all the pixel on one image that can be founded on the other, along the epipolar lines
(4) Computing 3D coordinates for each of the above correspondences founded, i.e. a dense point cloud allowing to define the object shape

What expressed above is referred to a couple of images but it is possible to extend this method to more couples of images, so to a set of images. The conditions are anyway that:

(a) It is possible to identify at least 8 corresponding points between all images couples
(b) It is possible to identify dense correspondences between all images couples

To ensure reliable correspondence matching, the captured images must have sufficient overlap and contain textured regions that allow the identification of common features. Algorithms such as SIFT or similar feature detection methods can then be used to match corresponding points between the two views. In cases where natural texture is lacking, artificial texture can be added to the surface using techniques such as attaching markers, painting distinctive patterns, or projecting a random structured light pattern onto the object. These methods enhance feature detectability and improve the accuracy of stereo matching. In recent years, stereovision techniques have seen significant advancements, with the development of innovative research methods and robust algorithms designed to overcome the limitations discussed earlier. These advances have also made it possible to optimize performance for specific application domains. A comprehensive overview of these algorithms and developments can be found in reference [8].

5.5 Photomodeling

The uncalibrated stereovision technique previously described can be applied to a sequence of overlapping images taken from different viewpoints around an object in order to reconstruct its 3D shape. This method is commonly known as photomodeling or Structure from Motion (SfM). The core idea involves capturing multiple 2D images of the object from various positions and orientations. By identifying common reference points (keypoints) across pairs of these images, it is possible to estimate the relative positions and orientations (poses) of the camera for each shot. Using these correspondences and applying the previously explained geometric techniques, such as epipolar geometry, the spatial relationships between image planes can be determined with respect to an arbitrary but consistent reference frame.

This constitutes the first stage of data processing. In the next phase, dense correspondence matching is performed across the image set to identify as many common points as possible. These matched points are then used to compute their 3D coordinates, resulting in a dense point cloud that represents the surface geometry of the object. An overview of this measurement technique is illustrated in Fig. 5.22.

Information about point coordinates in the images can only be obtained in regions where surface texture is present. This texture is essential for the matching algorithms previously described, as it enables the identification of corresponding points between pairs of 2D images. Consequently, for objects like the one shown in Fig. 5.23, areas with uniform color, surfaces without texture, such as the uniformly white regions, may result in gaps or “holes” in the reconstructed data, due to the absence of distinguishable features.

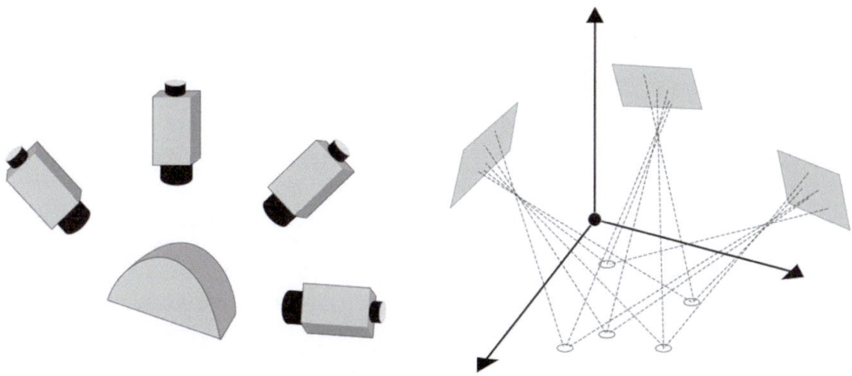

Fig. 5.22 Taking pictures for photomodeling

Fig. 5.23 Photomodeling indetermination in white texture missing areas

This limitation can be addressed by artificially adding texture to the surface. Possible solutions include illuminating the object with a random pattern, applying markers, or temporarily painting the surface. For example, easily removable spray coating, such as water-based powders, can be used. As shown in Fig. 5.23, a practical approach involves dusting the figurine with wet curry powder to create sufficient texture for feature detection.

Photomodeling can be used to reconstruct the shape of an object, a portion of it, or an entire surrounding environment, such as the interior of a room. This technique is particularly valuable in applications like autonomous navigation and robotic vision, where accurate spatial awareness is essential.

To achieve high-quality 3D reconstruction, it is crucial to ensure adequate overlap between images. Various image acquisition strategies are employed for this purpose, some of which are illustrated in Fig. 5.24. Image capture can be performed using handheld cameras, drones flying over buildings or natural landscapes such as mountains, or by extracting frames from a video recording.

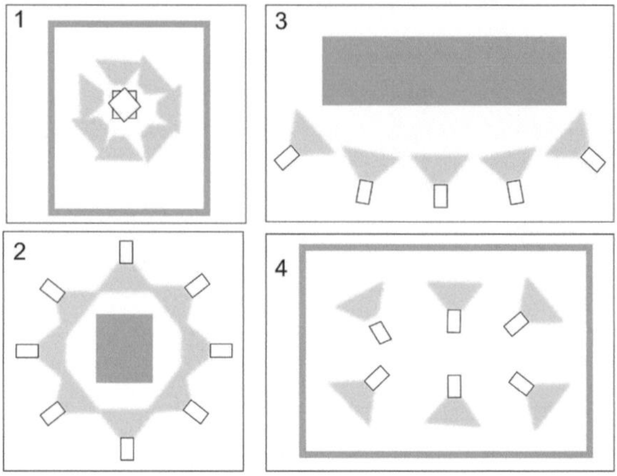

Fig. 5.24 Scheme to take pictures for measurement 3D objects shape or 3D surrounding shapes

A wide range of software solutions is available for processing images in photomodeling workflows. Popular commercial tools include Agisoft Metashape, RealityCapture, Pix4Dmapper, Autodesk ReCap Pro, and 3DF Zephyr. There are also powerful free and open-source alternatives, such as COLMAP, OpenMVG, Meshroom, VisualSFM, and Regard3D. In addition, dedicated libraries in Python and MATLAB are available for developing custom photogrammetry pipelines.

The typical photomodeling process involves the following steps:

1. Importing the images or extracting frames from a video captured filming around the object.
2. Preprocessing the dataset, such as filtering out unsuitable or redundant images.
3. Initial data processing (camera calibration and alignment): detecting and matching keypoints across images to estimate the relative camera poses and camera intrinsics parameters.
4. Dense point cloud generation, estimating the 3D coordinates of thousands (or millions) of matched points across the image set.

The output of this process is typically a point cloud file containing the 3D coordinates of all identified points. This point cloud can then be converted into a 3D mesh, representing the object's surface geometry. The smallest element of this mesh is a triangle, defined by three points.

The resulting triangular mesh is usually exported in the standard STL format. To enhance visual realism, portions of the original images can be projected onto each triangular surface to generate a textured 3D model that includes colour and material appearance. An example of such a reconstruction is shown in Fig. 5.25.

A comparison of results between some of the software available to perform photomodeling, from [9], is illustrated in Fig. 5.26.

Fig. 5.25 Common outcomes of photomodeling applied to works of art

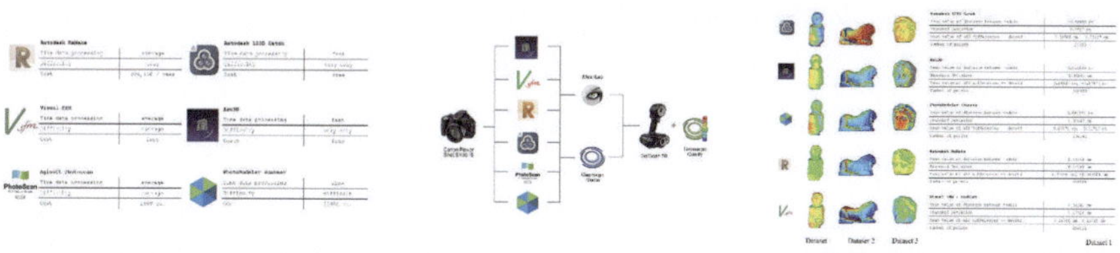

Fig. 5.26 Comparison between photomodeling data processing softwares for art digitalization

5.6 Marker Based Techniques

Photomodeling cannot be applied on uniform color object surface like a car's exterior, but placing on it physical markers the above illustrated technique allows the measurement of position of each marker, and normally with low uncertainty, useful for many industrial purposes, as illustrated in Fig. 5.27.

Markers can be square, rectangular, circular shaped, or coded accordingly with a predefined black and white distribution in circular or matrix shape. Marker in images

Fig. 5.27 Marker based multistereovision for uniform colour surfaces

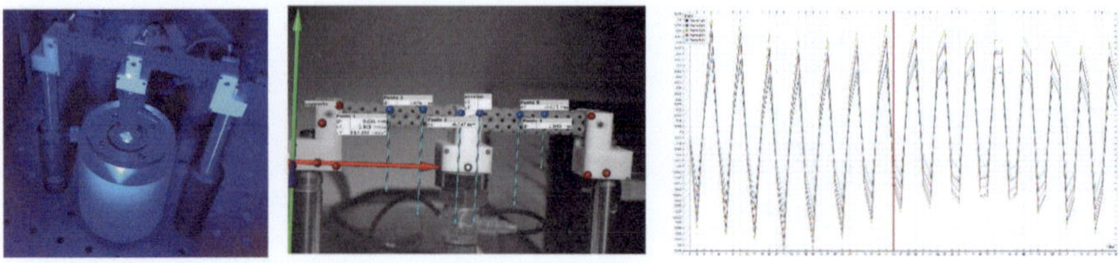

(a) (b) (c)

Fig. 5.28 Marker based dynamic displacement measurements on a beam obtained by a GOM measurement systems—ZEISS INSPECT Optical 3D software

can be identified by template matching, SIFT, SURF or other algorithms. They can be adhesive or magnetic.

This type of measurement is widely used for the evaluation of the deformation of mechanical components subjected to static or dynamic loadings. By measuring before and after the application of the load we can obtain a displacement vector for each marker identified in the two images, or by making a film of the dynamically loaded component, as the beam in Fig. 5.28a loaded by an electrodynamic exciter giving a sinusoidal force at the beam center, it is possible to measure the amplitude of the displacement of each marker placed on the beam as illustrated in Fig. 5.28b and also its time history as illustrated in Fig. 5.28c.

5.7 Photometric Stereo

Photometric stereo is a computer vision technique used to estimate the field of surface normals of an object surface by analyzing multiple images captured under varying lighting conditions. This method enables the reconstruction of surface geometry by determining the direction of the surface normal at each point. It was first introduced by Woodham in 1980 [10].

The fundamental principle behind photometric stereo is that the amount of light reflected by a surface depends on the surface's orientation with respect to both the light source and the observer [11]. By capturing how lighting variations affect pixel intensities, it becomes possible to infer the underlying surface orientation.

A special case of this technique, known as shape-from-shading, involves estimating surface shape using only a single image. This approach was thoroughly analyzed by B. K. P. Horn in 1989 [12].

Over the past decades, photometric stereo has been generalized beyond its original formulation. Extensions have been developed to accommodate non-Lambertian reflectance properties, extended (non-point) light sources, and more complex materials [13, 14]. Moreover, alternative configurations have been proposed, such as using a fixed light source with multiple viewing angles instead of varying illumination [15, 16].

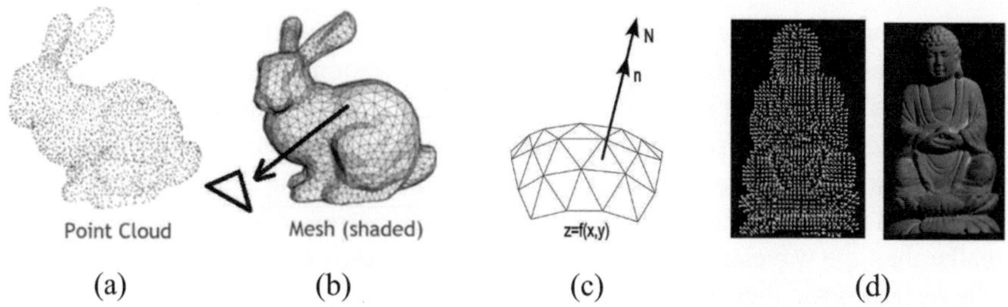

Fig. 5.29 Three different ways to represent a surface: point cloud (**a**), triangular mesh (**b**) and from the normal to the surface vector on each triangle (**c**) the complete normal vector field (**d**)

Objects surfaces can be represented in various formats, such as point clouds, triangular meshes, or surface normal vector fields, examples of which are illustrated in Fig. 5.29.

Given a surface defined by a mathematical function z = f(x, y), discretized through a triangular mesh obtained by interpolating a measured point cloud (x, y, z), we can compute its surface gradient vector as follows:

$$\left(-\frac{\partial z}{\partial x}, -\frac{\partial z}{\partial y}\right) = (p, q)$$

And from this gradient vector we can define the surface normal vector as:

$$\mathbf{N} = \left(-\frac{\partial z}{\partial x}, -\frac{\partial z}{\partial y}, 1\right) = (p, q, 1)$$

And the unit length normal vector **n** as:

$$\boldsymbol{n} = \frac{\boldsymbol{N}}{\|\boldsymbol{N}\|} = \frac{(p, q, 1)}{\sqrt{p^2 + q^2 + 1}}$$

Conversely, if we measure the normal vector field, given by **N** or **n** on all the object surface points, we know the two gradients on each point of the surface $\left(\frac{\partial z}{\partial x}, \frac{\partial z}{\partial y}\right)$ and from these it is possible to estimate surface shape simply by integration as:

$$z(x, y) = \int \frac{\partial z}{\partial x} dx + \int \frac{\partial z}{\partial y} dy$$

To reliably estimate this integral, it is necessary to apply numerical methods that reduce the impact of noise, as derivatives are particularly sensitive to it. Common approaches include pre-processing the data through smoothing techniques, applying FFT-based filtering, using least squares optimization, or performing integration along carefully selected paths. Additionally, assumptions about boundary conditions are essential, for example, setting z = 0 along the edges or at designated reference points. When representing the surface in a 3D coordinate system, a gradient plane can be defined (as illustrated in Fig. 5.30) by setting z = 1. In this context, each point (p, q) in the gradient space corresponds to a unique orientation of the surface normal.

As previously discussed in Chap. 1, when a light beam strikes a surface, it may be reflected specularly, if the surface behaves like a mirror, meaning its roughness is smaller than the wavelength of the light, or scattered diffusely in multiple directions if the surface is rougher. This behavior is illustrated in Fig. 1.34. In the case of diffuse reflection, the intensity of the incident light is distributed across many directions.

A general model for this phenomenon is presented in Fig. 5.31a: the reflectance map R(p, q), which describes the intensity I of light reflected by the surface in the direction defined by the vector (p, q,1). This model captures the relationship between surface orientation and the resulting light intensity in a compact mathematical form described by the function I = R(p, q).

Simple cases are a uniform diffusing surface model, so:

$$\mathrm{R(p, q) = K = constant}$$

As illustrated in Fig. 5.31b.

The most widely used and meaningful model for many types of surfaces is the Lambertian model, which assumes that light is diffusely reflected with maximum intensity in the direction of the surface normal and zero intensity at angles of 90°, relative to the normal, as illustrated in Fig. 5.31c.

This model is given by the expression:

$$I = \mathrm{K}\cos\theta_i = K(\boldsymbol{n}\cdot\boldsymbol{s}) \tag{5.27}$$

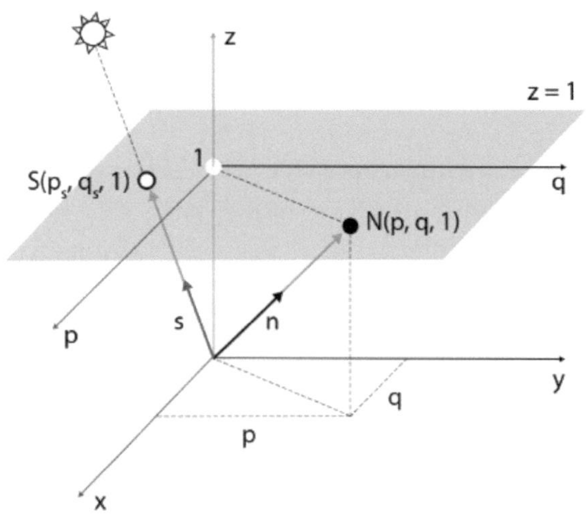

Fig. 5.30 Gradient space plane

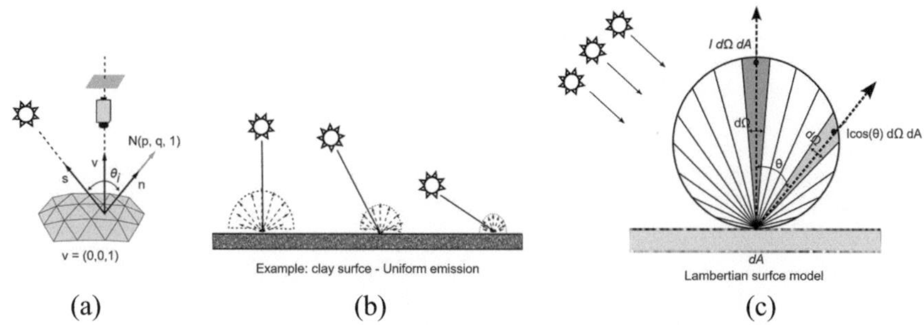

Fig. 5.31 Radiance model definition (**a**) for a perfect diffuser in (**b**) and for a Lambertian reflectance model in (**c**)

and taking $K = 1$ the classical Lambertian model can be simply described by the expression:

$$I = \cos\theta_i = (\boldsymbol{n} \cdot \boldsymbol{s})$$

So in gradient space the Lambertian model is given by the function:

$$I = (\boldsymbol{n} \cdot \boldsymbol{s}) = R(p, q) = \frac{pp_s + qq_s + 1}{\sqrt{p^2 + q^2 + 1}\sqrt{p_s^2 + q_s^2 + 1}} \tag{5.28}$$

Reflectance maps generated by plotting the previous equation are shown in Fig. 5.32a, where each point (p, q) represents a reflectance value R(p, q). When considering a fixed lighting direction **s**, all viewing directions that form the same angle θ with **s** will exhibit identical intensity under the Lambertian model, as the reflectance depends solely on the angle θ. These directions correspond to iso-brightness contours in the (p, q) plane. Since the angle θ defines a cone of equal inclination around **s**, the iso-brightness lines are given by the intersection of this cone with the (p, q) plane, as depicted in Fig. 5.32c.

The intersection of a cone with a plane produces classical conic curves, namely ellipses, parabolas, or hyperbolas, depending on the relative orientation. In the case of a Lambertian surface viewed perpendicularly, the intensity I is zero when the

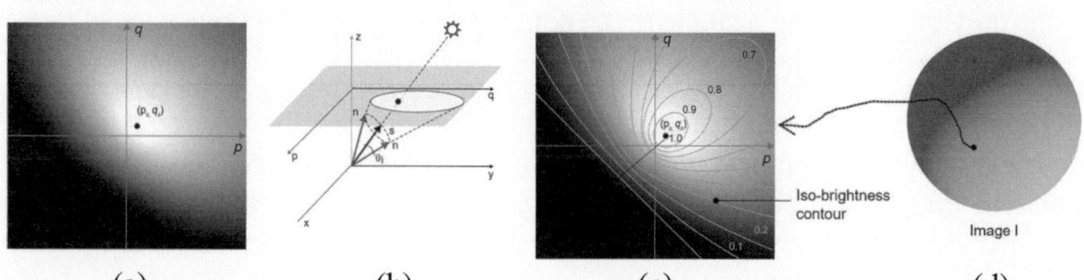

Fig. 5.32 Reflectance maps and iso surfaces

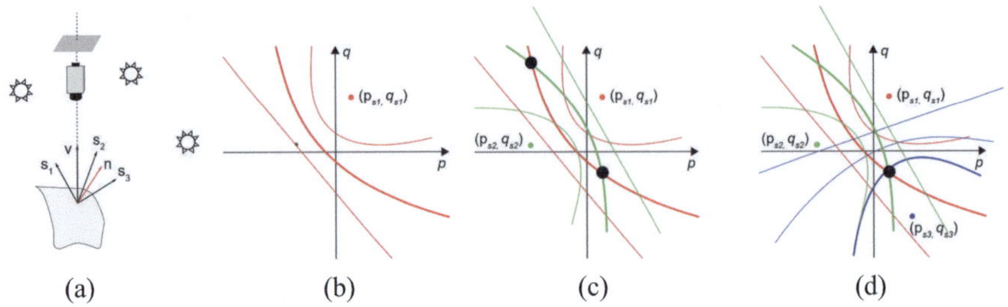

Fig. 5.33 Photometric stereo measurement principle

numerator of the reflectance function R(p, q) equals zero. This condition, expressed as $pp_s + qq_s + 1 = 0$, defines a straight line in the (p, q) plane.

For any given intensity value *I*, there exist infinite surface orientations (p, q) lying along the corresponding iso-brightness conic curve in the (p, q) plane, as shown in Fig. 5.32d. Therefore, when analyzing a single image where the intensity at each point is known only from one illumination direction, it is not possible to determine a unique surface normal. Instead, each point corresponds to an entire iso-brightness curve of possible normal directions.

This leads to the core concept of the photometric stereo method: by illuminating the surface from multiple known directions $\mathbf{s_i}$, as illustrated in Fig. 5.33a, we obtain multiple iso-brightness curves as illustrated in Fig. 5.33b that can be used to compute the normal to the surface **n**.

Considering multiple lighting directions $\mathbf{s}_i(p_{si}, q_{si})$ as in Fig. 5.33a. We will have the reflectance maps for each ith light source $R_i(p, q)$ and the intensity recorded by our camera on same **v** direction for the ith lighting source will be $I_i(x, y)$.

- Illuminating along $\mathbf{s}_1$ direction we record by the camera the I_1 intensity defining the R_1 iso-brightness line of Fig. 5.33b were all the possible orientation of our normal to the surface can be.
- Illuminating along $\mathbf{s}_2$ direction we record by the camera the I_2 intensity defining the R_2 iso-brightness line of Fig. 5.33c were all the possible orientation of our normal to the surface can be. This curve has the two intersection with the previous, in both that point our normal to the surface can stay.
- Illuminating along $\mathbf{s}_3$ direction we record by the camera the I_3 intensity defining the R_3 iso-brightness line of Fig. 5.33d. This curve has only one intersection with the previous, defining a single (p, q) points, so our founded unique normal to the surface **n** vector in point (x, y).

Therefore, the photometric stereo technique, based on the Lambertian reflectance model, requires a minimum of three distinct lighting directions and three corresponding images to compute the surface normals. However, the method can be extended to handle more complex, non-Lambertian reflection models by employing a greater number of lighting directions and images.

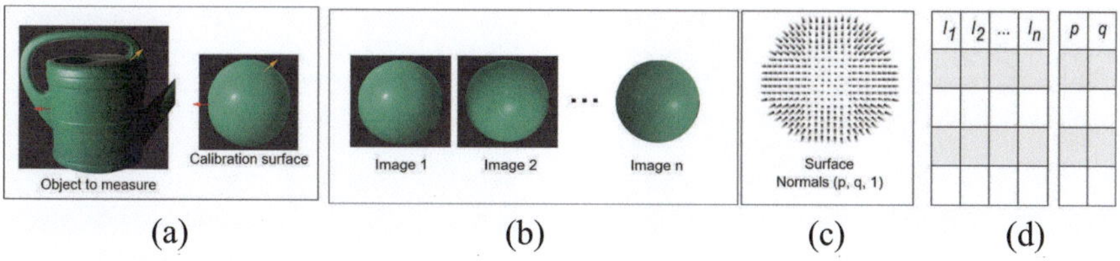

Fig. 5.34 Calibrated photometric stereo: calibration technique

It is important to note that this technique is not suitable for perfectly specular (mirror-like) surfaces, as such surfaces would theoretically require an infinite number of lighting directions to reconstruct their normals.

Photometric stereo algorithms are readily available in various software tools for image processing, including the Python OpenCV library, as well as other specialized image analysis frameworks.

An alternative approach to surface shape measurement is calibrated photometric stereo. The core idea is to use a reference object, typically the double curvature surface of a sphere, that shares the same optical properties and reflectance function R(p, q) as the target object to be measured. Since both objects exhibit the same surface reflectance, illuminating them from the same direction will produce identical intensity responses, as shown in Fig. 5.34a. When the lighting direction changes, the intensity values on both objects will change accordingly, but the correspondence between intensity and lighting direction remains the same. This creates a direct relationship between the intensity captured by the imaging system and the direction of illumination. So, lighting the reference sphere under n different lighting directions we get the images illustrated in Fig. 5.34b and for a sphere having a known radius we can compute the surface normal field (p, q) at every visible point of its images. These correspondences are then stored in a calibration table, such as the one illustrated in Fig. 5.34d.

Because the target object shares the same reflectance function, we can use this calibration table as a lookup tool: for any measured intensity on the target object surface point, under a known lighting direction, we can infer the corresponding surface normal in that point by referencing the precomputed data from the sphere, obtaining the results illustrated in Fig. 5.35. This enables accurate surface normal estimation for a generic shape having same optical properties of the calibration sphere, without explicitly solving the inverse reflectance problem.

5.8 Neural Radiance Fields (NeRF)

A recent alternative [15] to the classical deterministic approaches based on point clouds, surfaces, normal vectors, etc., previously illustrated, is the artificial intelligence-based method known as Neural Radiance Fields (NeRF) [17, 18]. This

Fig. 5.35 Calibrated photometric stereo: measurement of normal vector field of an object surface

new approach emerged in the fields of computer graphics and computer vision around 2020, to model 3D scenes and objects in videogames and computer graphics. Instead of representing the geometry using point coordinates, NeRF uses the weights of a neural network that learns the shape and surface optical properties of the scene or of an object.

Anyway, like photomodeling, NeRF reconstructs the 3D shape of a scene or object by using a set of 2D images. However, the shape is not represented by explicit 3D coordinates, but rather implicitly encoded in the network's weights, trained by these images.

The NeRF model, illustrated in Fig. 5.36, supports downstream applications such as novel view synthesis, 3D geometry reconstruction, and estimation of reflectance properties of each scene or object point in all directions, thus providing richer information than just the 3D point clods coordinates. This method is particularly effective for tasks such as photorealistic rendering and 3D scene reconstruction. NeRF learns how light interacts with a scene and can generate new views from arbitrary angles, often using only a sparse set of 2D images as input. Once trained the NerF model, it is also possible to derive traditional models, such as point clouds or surface meshes.

NeRF modelling starts from a set of images captured from multiple viewpoints around the scene or object, like photomodeling. These images serve as the training data for learning the implicit 3D representation, with each viewpoint providing

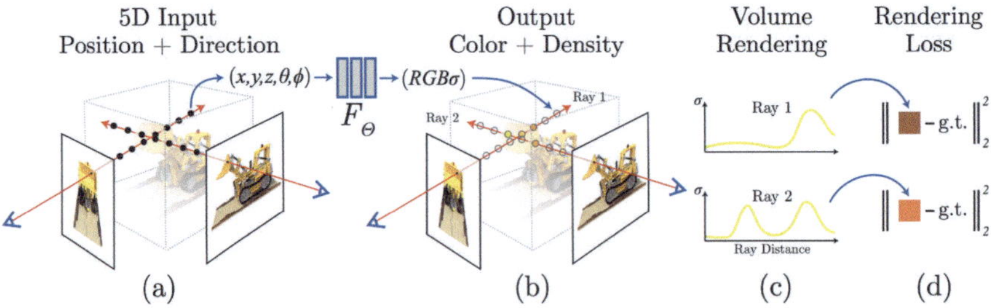

Fig. 5.36 A NeRF model [15]: Sample points and viewing directions along camera rays and pass them through the network (**a**); network output is a color vector and density value for each sample (**b**); combine the outputs of the network via volumetric rendering to go from discrete samples in 3D space to a 2D image (**c**); compute the loss and update network gradients via backpropagation to represent scene (**d**)

unique geometric and photometric information that is essential for accurate reconstruction of the scene. To effectively use these images, precise camera calibration and pose estimation are necessary. Each input image must be associated with an accurate camera pose, including its position and orientation in three-dimensional space. This geometric calibration is fundamental for establishing the correspondence between the 2D pixels in the images and the 3D coordinates in the scene. Furthermore, the camera's intrinsic parameters, such as focal length, principal point, and lens distortion are encoded in the intrinsic matrix, which defines the projection relationship between the 3D world coordinates and the 2D image plane. Before feeding the images into the neural network, pre-processing operations are applied to ensure consistency and efficiency. The input images are resized to standardized dimensions, typically 256 by 256 or 512 by 512 pixels, to optimize computational performance and maintain uniformity in the network input. Additionally, the pixel intensity values are normalized, commonly scaled to ranges such as [0, 1] or [–1, 1], which facilitates stable gradient flow during the training process and guarantees consistent scaling across images with varying lighting conditions.

For each pixel in the input images, a ray is generated that originates from the camera centre and passes through the pixel into the 3D scene. Along these rays, multiple sample points are distributed at varying depths using stratified sampling to uniformly cover the viewing volume while maintaining computational efficiency. More advanced implementations utilize hierarchical sampling, where an initial set of coarse samples guides the placement of finer samples in regions of higher density variation, thus optimizing the sampling distribution and improving rendering quality.

Neural network architecture and processing is based on multilayer perceptron (MLP) receiving two primary inputs for each sampled point. The first input is the spatial coordinates, represented as a 3D vector

$$\boldsymbol{x} = (x, y, z)$$

which specifies the precise location of the point in the 3D space. The second input is the viewing direction, expressed as a unit vector

$$\boldsymbol{d} = (\theta, \phi)$$

which encodes the direction of the camera ray passing through that point. This viewing direction allows the network to model view-dependent effects, such as changes in appearance based on the observer's perspective. To enable the network to capture fine geometric details, the input coordinates undergo a high-frequency positional encoding transformation. This encoding applies sinusoidal functions to the coordinates, effectively mapping them into a higher-dimensional space. This process compensates for the MLP's natural bias towards learning low-frequency functions and enhances its ability to represent intricate spatial variations and sharp details within the scene. For each spatial point, the neural network produces two fundamental outputs. The first is the radiance, given as an RGB color vector

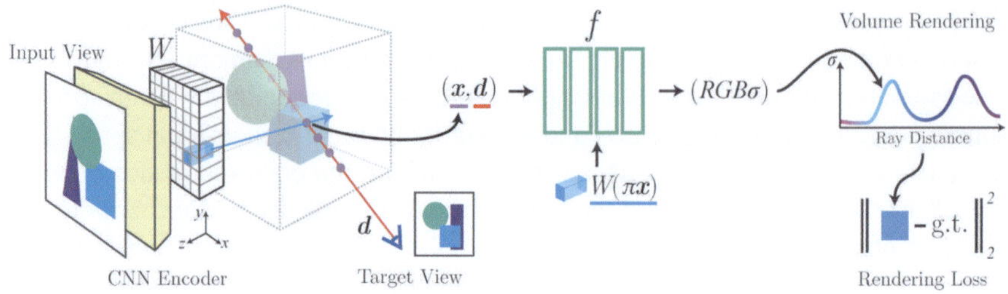

Fig. 5.37 NERF neural network architecture [15]

$$c = (\mathrm{R},\ \mathrm{G},\ \mathrm{B})$$

which represents the light emitted or reflected from the point in the specified viewing direction. The second output is a scalar density value σ(x), which represents the probability that light traveling through that point is absorbed or scattered. This density essentially models the material properties at the location, indicating how opaque or transparent the sampled point is within the volumetric scene (Fig. 5.37).

Volume rendering and image synthesis, as illustrated in Fig. 5.38, is given by the final color of each pixel, calculated through volumetric integration, which involves numerically integrating the sampled radiance and density values along the corresponding camera ray. This process follows the volume rendering equation:

$$C(r) = \int T(t) \cdot \sigma(r(t)) \cdot c(r(t), d)\, dt \tag{5.29}$$

In this equation, T(t) represents the transmittance, which quantifies the fraction of light that reaches each sample point along the ray without being absorbed or blocked by the medium. By integrating these values, the rendering process synthesizes the pixel color by accounting for how light interacts with the volume at different depths. A key feature of this rendering approach is its differentiability, meaning that the entire pipeline supports gradient-based optimization. This property enables the neural network to be trained by minimizing the photometric error between the synthesized images produced by the model and the actual captured images. Through this continuous feedback, the network iteratively improves its internal representation of the scene, leading to increasingly accurate reconstructions and novel view syntheses.

Once the neural network is trained, it can be used to produce a volumetric density field that implicitly encodes the geometry of the scene or object. This implicit representation allows for further processing, such as surface reconstruction, where algorithms like Marching Cubes extract an explicit polygonal mesh from the volumetric density data. This mesh can then be used for detailed geometric analysis. From the volumetric density, high-density regions can be sampled to generate point clouds, which are useful for a variety of applications.

These point clouds can be processed and analysed with specialized tools such as Open3D or CloudCompare. Similarly, the extracted meshes can be examined using computational geometry libraries like Trimesh or PyMesh, which enable the calculation of important morphometric features including surface area, volume, and curvature. The resulting 3D data is highly versatile and can be integrated into various platforms and workflows.

Deep learning frameworks such as PyTorch and TensorFlow provide comprehensive support for both the training of NeRF models and the post-processing of their outputs.

Visualization tools like CloudCompare offer interactive environments for exploring and measuring the reconstructed 3D data. Furthermore, the polygonal meshes can be exported to computer-aided design (CAD) software, facilitating high-precision engineering analysis and measurement tasks.

5.9 Gaussian Splatting

Gaussian splatting is a technique for processing 2D images captured around an object or a scene, enabling direct rendering of volumetric data without converting it into surface, point, or line primitives. Initially introduced as *splatting* by Lee Westover in the early 1990s [20], recent developments and comparisons with Neural Radiance Fields (NeRF) have been presented in [21].

This method is especially useful for efficiently representing complex or dense point clouds, volumetric data, or neural 3D representations. The core concept is to associate a Gaussian function with each point in the scene, defining a region of influence around it. The rendering is then performed by projecting these Gaussians onto a 2D image plane.

From a Gaussian splatting model, it is possible to extract either a point cloud or a surface model, which can be leveraged for accurate 3D measurements, offering benefits in various applications. Each Gaussian is parameterized by its position, orientation, and shape in 3D space, defining how it affects its surroundings. During rendering, the Gaussians are projected onto the camera's view plane, where their overlap and blending produce a smooth, continuous surface representation that enhances rendering quality beyond simple point sets (Fig. 5.39).

The input to the model consists of a set of images of a static scene along with camera positions, represented as a sparse point cloud. Each point is associated with a 3D Gaussian function described by its mean $\mu = [x, y, z]$, covariance matrix Σ controlling its shape and orientation, and opacity α. The view-dependent appearance is modelled through spherical harmonics.

Model parameters are optimized by stochastic gradient descent to minimize a combined loss function incorporating L1 loss and D-SSIM, as illustrated in the Plenoxels framework [22]. Gaussian splatting has also been used to render NeRFs by converting volumetric neural fields into point-based Gaussians, enabling much faster rendering. Its efficiency makes it ideal for displaying dense scientific or medical data and for real-time applications.

Gaussian splatting stands out by offering faster rendering and more memory-efficient representations than traditional volumetric NeRFs, while still producing smooth, high-quality visuals even with sparse data.

This technique is not limited to neural rendering but also excels in volumetric data visualization from sources such as Lidar or 3D scanning. In scientific and medical fields, Gaussian splatting smooths transitions between data points in CT or MRI scans, improving the visualization of biological structures like neurons or organs. It is also employed in visualizing complex climate or weather datasets. Each Gaussian's defining parameters include:

- $\mu = [x, y, z]$: the Gaussian's centre in 3D space
- Σ: the covariance matrix that controls its spread and orientation
- C: color vector (RGB values)
- α: opacity for blending.

To render these Gaussians, they are projected onto the image plane according to the camera's position and orientation. A splatting kernel determines how much each Gaussian contributes to each pixel, blending their influence to produce the final rendered image.

Several software tools facilitate Gaussian splatting visualization and processing. Open3D offers libraries for point cloud manipulation that can be extended to support Gaussian splatting. Similarly, the Point Cloud Library (PCL) provides point cloud handling with integration possibilities for splatting techniques. Deep learning frameworks such as PyTorch and TensorFlow are widely used to train and optimize neural networks or representations that rely on Gaussian Splatting.

5.10 Triangulation and Structured Lighting Basics

Shape measurement systems based on the principle of triangulation can be implemented not only with two cameras, using stereovision by intersecting their lines of sight, but also with a single camera line of sight in combination with a laser beam pointer line. In this setup, the triangulation occurs by intersecting the camera's line of sight with the path of the laser beam, as shown in Fig. 5.40.

When a point (x, y, z) on the surface is illuminated by a high-intensity, collimated laser beam of a pointer for example, originating from a known position (x_0, y_0, z_0), at a baseline distance b, and traveling along a known direction vector $\mathbf{r}$, the camera captures the image of the illuminated point P, where the laser beam intersects the surface. The observed position of P in the camera's image corresponds to the intersection between the camera's viewing ray $\mathbf{s}$ and the laser beam path. From this intersection the 3D coordinates of point P can be easily calculated. The baseline and the laser ray define a plane in which both the illuminated point P on the object and its image projection lie. Therefore, a line-scan (or line array) 1D camera aligned with this plane can be effectively used for measurement instead of a 2D camera.

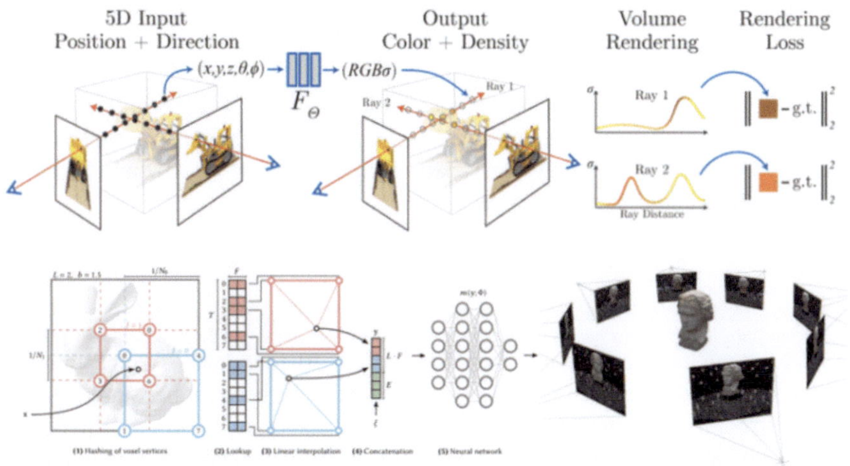

Fig. 5.38 Volume rendering process [15]

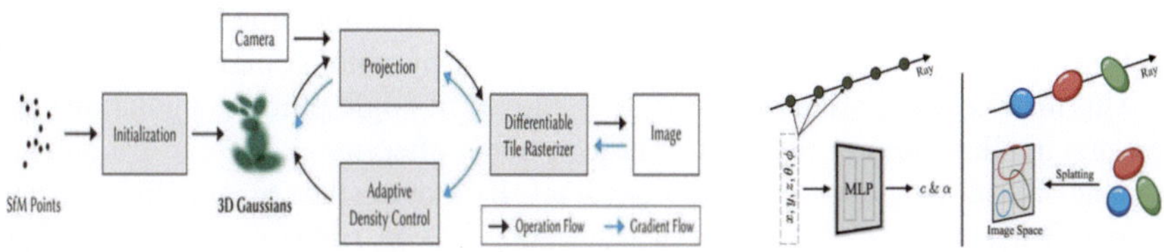

(a) (b)

Fig. 5.39 **a** Gaussian splatting model from [21] and comparison between NeRF and GS

A major advantage of this method over stereovision is that locating the illuminated point P does not require complex correspondence searches along epipolar lines or advanced feature-matching algorithms such as SIFT. Instead, the process is greatly simplified: by capturing two images of the same surface, one with the laser illumination and one without, and computing their difference, the illuminated point is immediately identified. This allows for a simple direct extraction of the 2D coordinates (x_p, y_p) of the laser spot, the only thing in the difference image shown in Fig. 5.41.

By the (x_p, y_p) coordinates in the image plane and the equation of camera ray of sight, that from the projection Eq. 2.25 will be given by:

$$x = x_p \frac{z}{f} \qquad y = y_p \frac{z}{f} \tag{5.30}$$

where f is the focal length of the receiving optics.

Lighting ray of the pointer can be described by the equation:

$$\frac{x - x_0}{a} = \frac{y - y_0}{b} = \frac{z - z_0}{c} \tag{5.31}$$

where (x_0, y_0, z_0) is the starting laser point coordinates and (a, b, c) parameters giving its orientation, all known. The coordinates (x, y, z) of the lighted point can be easily found by intersection of the two lines expressed by the Eqs. (5.30) and (5.31), we obtain for the z coordinate the expression:

$$z = \frac{cx_0 - az_0}{cx_p \frac{1}{f} - a} \tag{5.32}$$

And we can calculate x and y from the above 5.30 equations. By repeating this process for each illuminated point, so scanning the laser beam, it becomes possible to reconstruct a complete 3D point cloud [23] of the surface. However, achieving a dense point cloud, such as one with a resolution of 640 × 480 points, would require acquiring 640 × 480 × 2 = 614,400 images (accounting for both the illuminated and non-illuminated frames for each point). With a camera operating at 30 frames per second, this would take approximately 614,400/30 = 20,480 s, or just under 6 h. This highlights a significant limitation in terms of acquisition time when aiming for high-resolution 3D reconstructions using this point-by-point scanning method.

This measurement principle is widely employed in single-point triangulation laser sensors, as illustrated in Fig. 5.40. These systems often use a line camera and are capable of achieving sub-micrometer accuracy. By moving the sensor or the object in a controlled manner, a process known as scanning, it is possible to capture the surface shape with very high precision. However, this approach is generally not suitable for fast measurements, as it involves sequential data acquisition point by point or line by line, as illustrated in Fig. 5.41, resulting in longer acquisition times.

One possible improvement to the above measurement techniques is to replace the single laser point with a laser sheet, using a structured light approach to project an

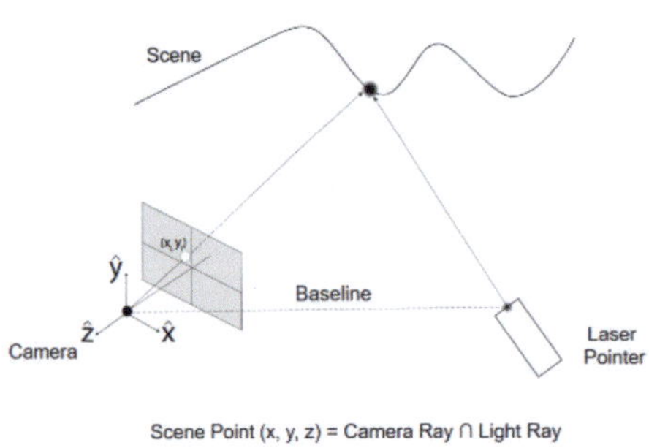

(a)

(b)

Fig. 5.40 Laser line structured lighting for triangulation (**a**), and an application for measurement of circular shape of a rotating shaft

Fig. 5.41 Finding the 2D coordinates of the lighted point by images subtraction

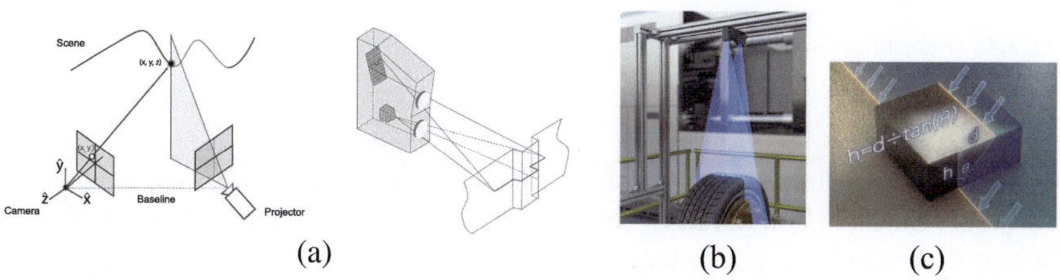

Fig. 5.42 Laser sheet structured lighting principle [24]

entire line onto the surface, as shown in Fig. 5.42. The projected light is a plane in 3D space, typically described by the equation $Ax + By + Cz + D = 0$, where the parameter A, B, C and D are known from its position and orientation.

From intersection of this lighting plane equation with each line of sight on the points on the lighted profile, given by above Eq. 5.30, it is possible to get the z coordinate of each point as:

$$z = \frac{-Df}{Ax_p + By_p + Cf} \tag{5.33}$$

And from Eq. 5.30 also their x and y coordinates. This for each point (x_p, y_p) of the lighted profile giving its complete 3D definition.

This measurement principle is largely used to perform 2D profile measurement by laser sheet sensors, illustrated in Fig. 5.42b, with typically sub-micrometer accuracy.

An application developed within a research project funded by the Italian Ministry in 2019, aimed at verifying the geometry of 3D-printed trabecular structured specimens, is illustrated in Fig. 5.43. By translating the specimen along a direction perpendicular to the laser sheet, multiple cross-sectional profiles can be captured. These profiles are then combined to reconstruct an accurate 3D shape model of the object.

By using laser sheet illumination, the number of required images for reconstructing a surface with a resolution of 640 × 480 points is significantly reduced: only 640 × 2 = 1,280 images are needed. With a camera operating at 30 frames

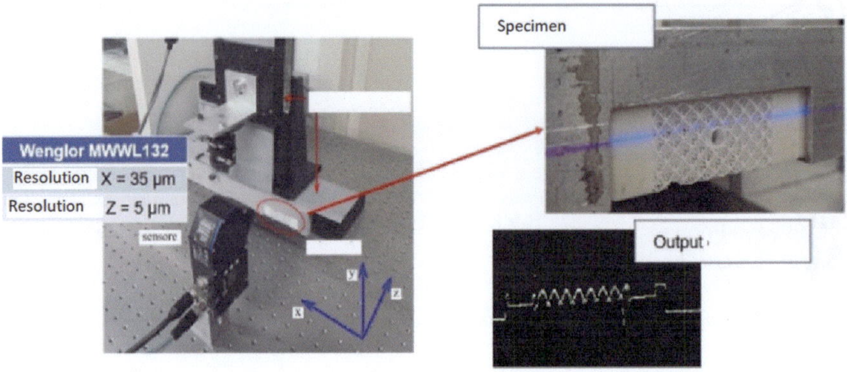

Fig. 5.43 Measurement geometry of a trabecular 3D printed specimen by scanning a laser sheet [25]

per second, this results in an acquisition time of approximately 42.6 s, a much more acceptable duration compared to point-by-point scanning. While still not sufficient for real-time applications, further improvements are possible. For instance, using multiple laser sheets simultaneously, as illustrated in Fig. 5.44, or projecting structured light patterns such as black-and-white stripe sequences, can reduce acquisition time proportionally to the number of planes used. With a sufficiently large number of projection planes, e.g., ten or more, it becomes feasible to reduce total acquisition time to below one second.

For complex scene or object, the problem of lighting with many planes can be ambiguous making difficult the determination of what plane generate a line, as in Fig. 5.45.

This limitation can be addressed by encoding the structured light pattern, as illustrated in Fig. 5.46. Instead of capturing a single image, multiple images are acquired, each corresponding to a different combination of active and inactive lighting planes or projected strips. For example, using three separate images with independent control

Fig. 5.44 Structured lighting with more light sheets in non-ambiguous (**a**) and ambiguous (**b**) conditions

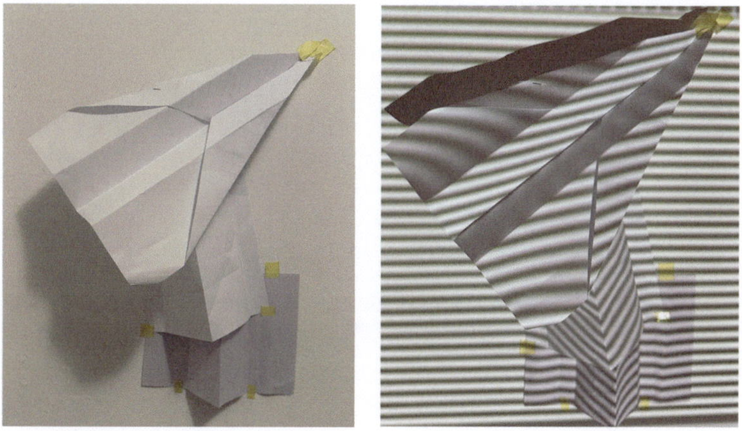

Fig. 5.45 Ambiguity in "following a lighting line on a complex surface realized by folded sheet of papers with planes at different orientation angles and distance from the projector

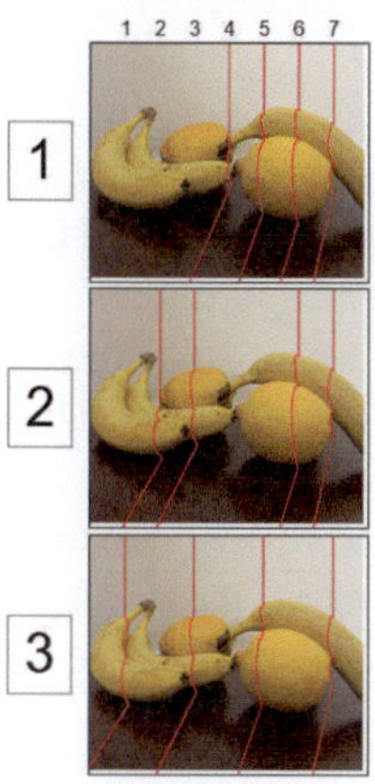

Image	Projection Pattern						
1 Bit1	0	0	0	1	1	1	1
2 Bit1	0	1	1	0	0	1	1
3 Bit1	1	0	1	0	1	0	1
(Binary)	(001)	(010)	(011)	(100)	(101)	(110)	(111)
Stripe Numbers	1	2	3	4	5	6	7

Fig. 5.46 Structured light coding of 7 combinations in 3 images

over each of the three lighting planes, we can create $2^3 = 8$ possible lighting combinations. Excluding the case where all planes are off, which provides no useful information, we obtain 7 effective lighting patterns. In general, using n images, it is possible to encode up to $2^n - 1$ distinct and useful structured light configurations.

To identify the illuminated planes or strips in structured light systems, it is necessary to detect transitions between illuminated and non-illuminated regions in the captured images. However, this image processing step can be sensitive to noise and other disturbances. To minimize the number of transitions, and thus improve robustness, Gray code encoding can be used instead of standard binary encoding. Gray code reduces the likelihood of errors by ensuring that only one bit changes between consecutive codes, thereby minimizing black-to-white transitions. This can

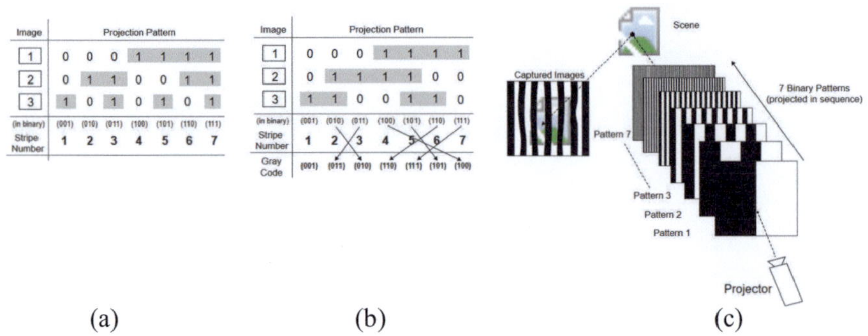

Fig. 5.47 Reducing transitions from 10 in (**a**) to 6 in (**b**) on 7 lighting combinations in 3 images; in (**c**) a general Grey coding of structural lighting

be achieved by simply reordering the sequence of binary-coded values, as illustrated in Fig. 5.47.

An alternative to binary (on–off) encoding is to use k distinct grey levels of illumination. In this case, it becomes possible to encode up to $k^n - 1$ unique patterns using only n images, thereby reducing the number of required acquisitions. However, this approach introduces a new challenge: accurately distinguishing between adjacent gray levels with lower contrast becomes more susceptible to errors caused by noise, optical aberrations, and variations in the surface's reflective properties.

Another approach to structured light projection is the use of invisible infrared illumination. This technique was adopted by early low-cost consumer 3D sensing systems, such as PrimeSense, used in the first generation of Microsoft Kinect, which projected a fixed pattern of infrared dots to reconstruct a dense 3D image. Subsequent versions of the Kinect transitioned to a more advanced method based on time-of-flight (ToF) technology, where depth information is derived from the time delay of reflected infrared light.

5.11 Phase Shifting Methods

Phase-shifting methods rely on the continuous modulation of projected light patterns. The basic concept, illustrated in Fig. 5.48, involves illuminating the scene with a pattern whose intensity varies smoothly, such as a linear gradient from right to left L_1, superimposed on a constant illumination level L_2.

Along the x-direction, perpendicular to the projector's lighting planes, it is possible to compute the ratio L_1/L_2. For a generic point P on the surface, the camera receives two intensity values: $I_1 = \alpha L_1$ and $I_2 = \alpha L_2$ where α accounts or the surface's reflectance and other optical properties. Since this factor is the same for both measurements, the intensity ratio simplifies to: $(I_1/I_2) = (\alpha L_1/\alpha L_2) = (L_1/L_2)$.

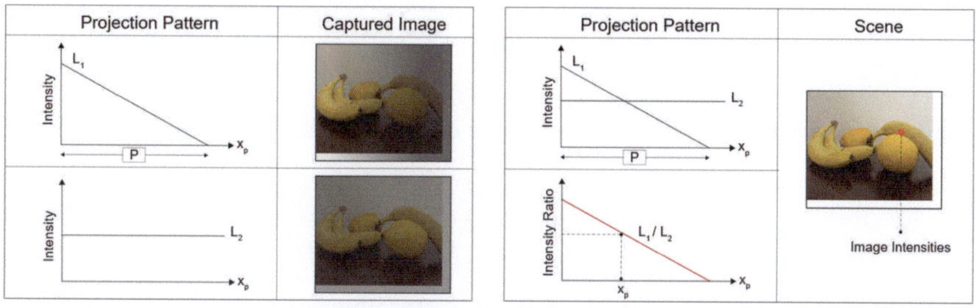

Fig. 5.48 Phase shifting basics for shape measurement

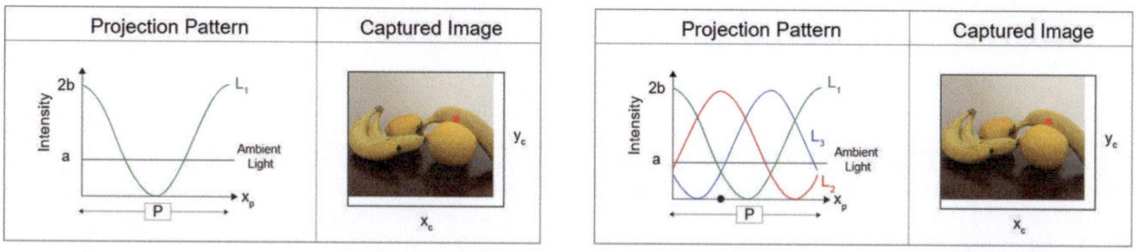

Fig. 5.49 Phase shifting using cosine lighting patterns

This ratio allows us to determine the specific projector column (and thus the corresponding light plane) illuminating point P. Once the equation of the light plane is known, intersecting it with the camera's line of sight passing through point P yields the full 3D coordinates of that point, analogous to the principle used in laser sheet triangulation.

A second approach is to use a cosine lighting pattern as illustrated in Fig. 5.49. Lighting level in point P will be:

$$L_1(x_p) = a + b\cos\left(\frac{2\pi x_p}{P}\right)$$

where a is the ambient light, b the mean amplitude value of the cosine light pattern. This produces on the image of point P the intensity:

$$I_1(x_c, y_c) = \alpha a + \alpha b\cos\left(\frac{2\pi x_p}{P}\right)$$

where α is the coefficient giving the back reflected intensity on image from point P, and αa, αb and x_p are the unknown. Taking other two images with the shifted lighting intensity distribution of $\pm 2\pi/3$ we can record the other two intensities:

$$I_2(x_c, y_c) = \alpha a + \alpha b\cos\left(\frac{2\pi x_p}{P} - \frac{2\pi}{3}\right)$$

$$I_3(x_c, y_c) = \alpha a + \alpha b \cos\left(\frac{2\pi x_p}{P} + \frac{2\pi}{3}\right)$$

So, a system of three unknown and three equation who's solution is:

$$x_p = \frac{P}{2\pi}\tan^{-1}\left(\sqrt{3}\frac{I_2 - I_3}{2I_1 - I_2 - I_3}\right) \quad (5.34)$$

Once again x_p gives the column of the projector illuminating that point and therefore the light plane equation from where, performing the intersection of this plane with the line of sight of that point, known from its 2D coordinates in the image plane, it is possible to determine the three coordinates of that point.

5.12 Multicamera Structured Light Optical Scanners

The highest measurement accuracy in modern optical scanning systems is currently achieved using multi-camera setups integrated with structured light projection. These advanced optical scanners are based on the measurement principles previously described, but enhanced by proprietary software algorithms that combine stereovision, reference markers, and structured lighting techniques.

Such systems commonly achieve dimensional measurement accuracies in the range of few to 50 μm over measurement volumes of 0.1–0.5 m, making them highly suitable for verifying mechanical tolerances in a wide range of industrial applications. This capability is critical in modern manufacturing and globalized production chains, driving both technological advancement and the growing market demand for these systems.

The general architecture of a high-performance optical scanner is illustrated in Fig. 5.50. It typically includes at least one structured light projector or a laser sheet projection system, along with two or more cameras to enable stereoscopic vision. The use of multiple cameras helps reduce occlusions caused by complex geometries, such as holes, recesses, or shadowed regions on the object's surface.

The primary advantage of structured lighting in industrial optical measurement systems lies in its ability to minimize the influence of the surface's optical properties. While features such as color, gloss, or texture may be important for realistic surface representation in applications like video games or cultural heritage preservation, they are not relevant for dimensional inspection of mechanical components. A second key benefit of structured lighting is its ability to directly highlight measurement points through high-contrast illumination, such as with a laser line or structured light pattern, eliminating the need for complex point correspondence searches across images. As discussed earlier, such searches typically require significant computational resources. In some systems, a third camera is integrated specifically for capturing color and texture information, allowing the simultaneous acquisition of geometric and visual data when needed (Fig. 5.51).

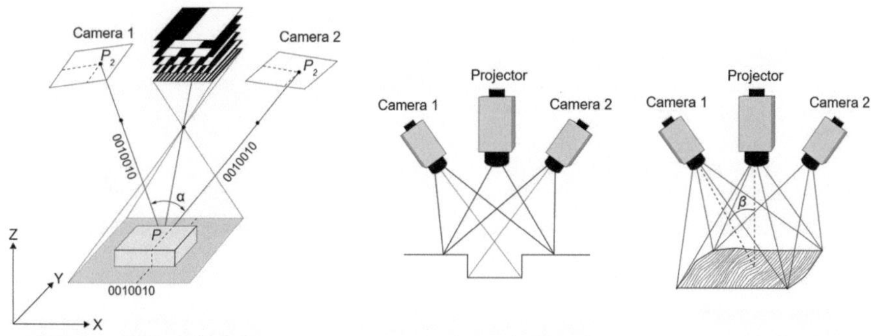

Fig. 5.50 General architecture of a high metrological performances optical scanner

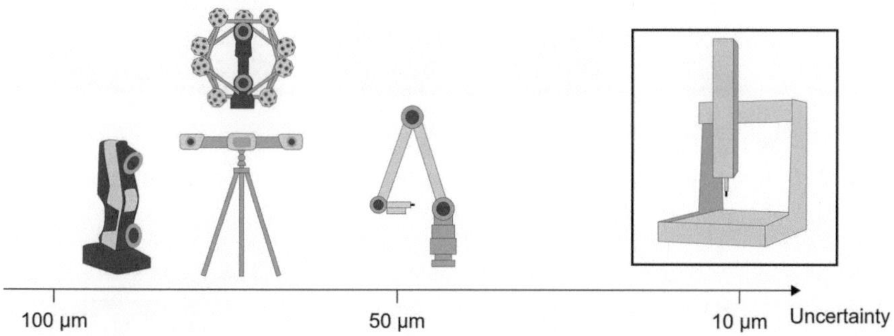

Fig. 5.51 Performance levels of optical scanners

Structured lighting systems commonly use white light projection patterns or laser-based illumination, including single, multiple, or crossed laser sheets, as illustrated in Fig. 5.52. In recent years, high-performance systems have increasingly adopted blue laser sheets, either in parallel or crossed configurations, as a replacement for traditional red lasers. This shift is due to the shorter wavelength of blue light, which helps to reduce unwanted reflections and scattering on specular or glossy surfaces, thereby improving measurement accuracy and surface detail capture performances.

Optical scanners are available in both portable and fixed configurations, each suited to different measurement scenarios, as illustrated in Fig. 5.53. In fixed-head systems, a rotary table is often used to position and rotate the object, enabling full surface acquisition without moving the scanner. In contrast, portable scanners can be manually operated or mounted on robotic arms, enabling automated inspection workflows directly on production lines. These systems are capable of rapidly capturing large volumes of 3D data, acquiring millions of points in a single scan. Regardless of the scanner's position relative to the object, each scan captures a dense point cloud of the visible surface region. To reconstruct the complete geometry of the object, multiple partial scans are aligned and merged. This alignment process can be guided

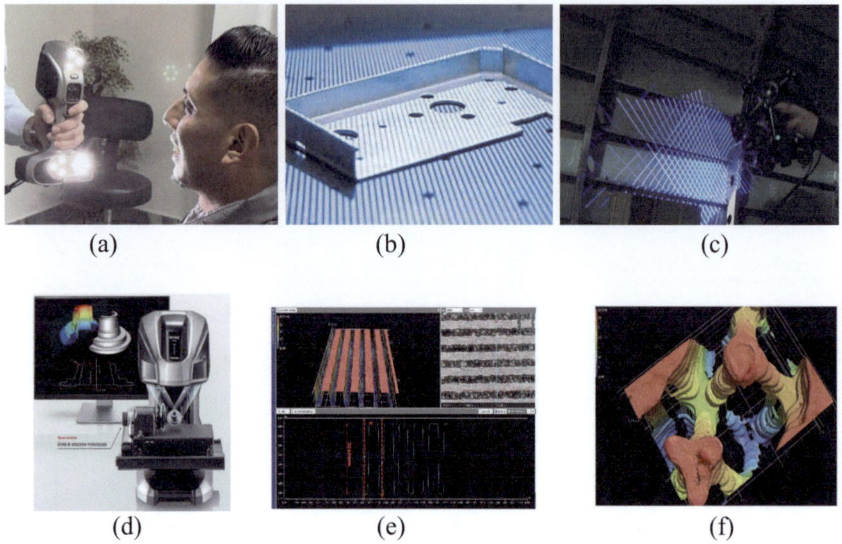

Fig. 5.52 Different kinds of structured light projection optical scanners: white light (**a**), blue laser (**b**), crossed sheets (**c**) [24, 26]

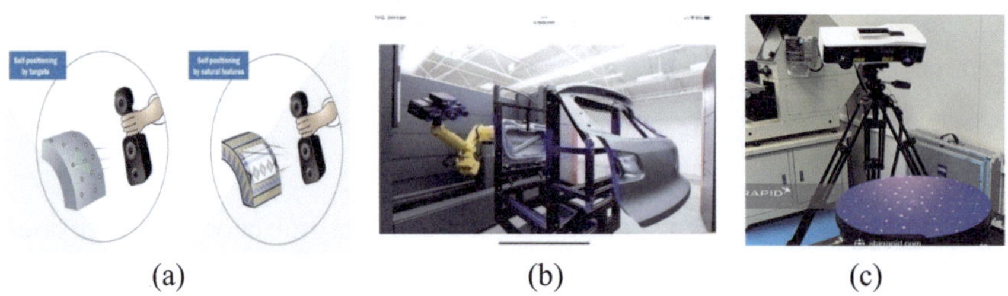

Fig. 5.53 Moving the optical scanner manually (**a**), by a robotic arm around the object (**b**) or using a rotating table (**c**)

by natural reference features on the object's surface or by artificial markers intentionally placed for this purpose. The data registration and stitching techniques used are similar to those described in Chap. 3.

An example of complex surface measurement using a portable scanner is shown in Fig. 5.54. In this study, the shape of a boat component was measured using a reference Creaform scanner (ScanGo30), which employs white light and offers a certified accuracy of 0.1 mm. The same part was also measured using photogrammetry techniques, with images captured by a Fujifilm X Series X-A2 16.3 MP camera. The image sets were processed using software such as Agisoft PhotoScan, Autodesk ReMake, and VisualSfM.

The comparison between the two methods revealed maximum deviations of approximately 10 mm. These findings can assist boat maintenance companies in

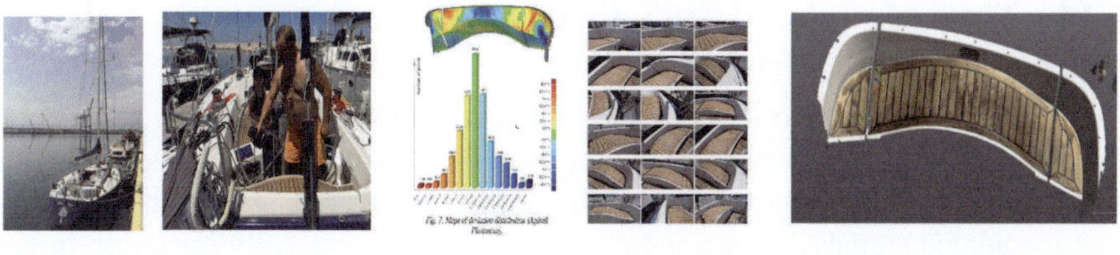

Fig. 5.54 Shape measurement of boat part by white light scanning and photomodeling, from [17]

selecting the most suitable measurement technique based on the required level of accuracy and the intended use of the 3D data.

5.13 Reference Image Generation for Uncertainty Analysis of Data Processing Techniques

The quality of images used for 3D measurements and the associated uncertainty can be assessed through a comprehensive simulation of the image generation process. This simulation can be implemented using a mathematical model that incorporates lighting conditions, surface properties, and both camera and scanner characteristics. Software environments like Blender, an open-source 3D graphics platform, enable such simulations effectively, as illustrated in Fig. 5.55.

Using these models, it is possible to generate a series of images for performing photomodeling or structured light shape measurements. Since the simulated images are free from uncontrolled noise sources, such as digitization artifacts, optical imperfections, or random variations in lighting, they can serve as a well-controlled reference for image data processing performance analysis. These simulations allow for precise performance evaluation and uncertainty estimation in shape reconstruction, as the reference geometry is exactly known within the 3D CAD model. Additionally, they enable analysis of how specific factors, such as lighting conditions, camera parameters, or surface optical properties, affect algorithmic performance.

Fig. 5.55 Image virtual generation by models of lighting, surface optical properties, camera parameters and object shape

5.14 Time Modulated Structured Lighting

An alternative approach to structured lighting for obtaining 3D shape measurements of objects or scenes is to modulate the light in time rather than in space. The core concept involves emitting pulses of light, such as short flashes or high-frequency sine-modulated signals, produced by fast LED sources, typically operating in the infrared spectrum as described in Chap. 3. By measuring the time Δt it takes for the light to travel from the source to the reflecting surface and back to the detector (a round-trip path equal to twice the distance *d* between the system and the surface), the 3D shape can be reconstructed. This time-of-flight method is commonly known as Imaging Light Detection and Ranging (i-LIDAR).

Techniques to detect time delay along the path from light source to reflecting point on the surface and back to a receiver are that illustrated in paragraph 1.13 of Chap. 1. Basically the two approach are used:

1. Direct time of flight (dTOF) detecting the time Δt that a pulse need, travelling at light speed c, to go from source to target and back to a receiver, that gives the distance $d = c\ \Delta t/2$
2. Indirect time of flight (iTOF) detecting the phase shift φ of a modulated light signal at frequency f, travelling at velocity c. From this phase shift we have the distance $d = c\ \varphi/(4\pi f)$.

Indirect time of flight approach is that more used because of the best signal to noise ratio and lower lighting power needs. But iTOF presents several limitations compared to dTOF:

- Range ambiguity: The phase measurement is periodic, creating ambiguity in distance measurement. For a modulation frequency f, the unambiguous range is limited to c/(2f). Beyond this range, the phase wraps around, making it impossible to determine the correct distance without additional information or multiple frequency measurements.
- Accuracy vs. range trade-off**:** Higher modulation frequencies provide better depth resolution but reduce the unambiguous measurement range. Lower frequencies extend the range but compromise precision. This fundamental trade-off requires careful system design based on application requirements.
- Motion artifacts: Moving objects during the integration time can cause phase errors, leading to distance measurement inaccuracies. The continuous wave nature of iTOF makes it more susceptible to motion blur compared to the instantaneous nature of dTOF pulses.
- Multipath interference: Reflections from multiple surfaces can interfere with the primary signal, causing phase errors. This is particularly problematic in indoor environments or scenes with complex geometry where secondary reflections are significant.
- Temperature sensitivity**:** The phase detection circuitry is sensitive to temperature variations, requiring calibration and compensation for accurate measurements in varying environmental conditions.

(a) (b)

Fig. 5.56 Time of flight sensor arrays: iTOF sensors from Hamamatsu (**a**), other dTOF typical sensors with Single Photon Avalanche Detector (SPAD) having 168 × 597 pixels,10 μm pixel size, 15 cm resolution on 300 m range (**b**)

LIDAR system sensors are composed of pixels that detect not light intensity, but the time delay, according to the techniques described above. These sensors produce images representing either the time delay or the phase shift, both of which are directly proportional to distance. As a result, they generate 3D distance maps directly, without the need for additional image processing to extract surface shapes. Typical sensors of this kind commercially available are illustrated in Fig. 5.56.

Currently, CMOS 3D image sensor technology based on photonic mixer devices (PMDs) is available with resolutions up to 640 × 480 pixels and measurement uncertainty in the range of 1–2 mm. These sensors are increasingly used in a wide range of applications, including smartphones, vehicles, and robotics for autonomous navigation, as well as in cars to enhance safety through driver assistance systems and enable autonomous driving (Fig. 5.57).

The main advantage of time-of-flight (ToF) 3D cameras lies in their ability to directly measure the phase shift, and thus depth, at each pixel, eliminating the need for complex image processing. This makes them significantly faster than other technologies described previously. However, they are nowadays not as accurate at micrometer (μm) scale resolution compared to methods based on focus variation or spatially structured light.

Limitations of ToF cameras include nowadays lower spatial resolution, which restricts the amount of detail they can capture about scene geometry and surface

Fig. 5.57 Lidar sensor for people counting from Hamamatsu

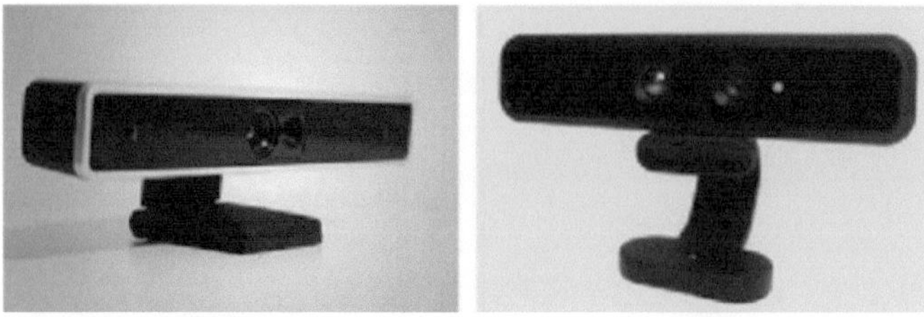

(a) (b)

Fig. 5.58 Two typical kind of commercial Lidar 3D cameras

characteristics. Performance may also be affected by poorly reflective materials (such as fabric) and by incident angles that deviate significantly from the surface normal.

Figure 5.58 illustrates two commercially available 3D LIDAR cameras based on this technology

References

1. Z. Wang, C. Zhang, Z. Wang, Y. Shuang, Y. Li, Real-time three-dimensional measurement techniques: a review. Optik—Int. J. Light Electron Opt. **202** (2020)
2. Z. Liu, H. Ukida, P. Ramuhalli, K. Niel, *Integrated Imaging and Vision Techniques for Industrial Inspection* (Springer, Berlin, 2015)
3. R. Danzl, F. Helmli, S. Scherer, Focus variation—a new technology for high resolution optical 3d surface metrology, in *The 10th International Conference of the Slovenian Society for Non-Destructive Testing "Application of Contemporary Non-destructive Testing in Engineering"* (Ljubljana, Slovenia, 2009)
4. J.J. Aguilar, F. Torres, M.A. Lope, Stereo vision for 3D measurement: accuracy analysis, calibration and industrial applications. Measurement **18**(4), 193–200 (1996)
5. W. Sankowski, M. Włodarczyk, D. Kacperski, K. Grabowski, Estimation of measurement uncertainty in stereo vision system. Image Vis. Comput. **61** (2017)
6. NASA, National Aeronautics and Space Administration, Available online: https://www.nasa.gov/. Accessed: data di accesso
7. L. Giulivi Sviluppo di metodologie di misura di campi di spostamento attraverso analisi di immagini, Thesis advisor Gianluca Rossi, co Advisors Michele Moretti and Matteo Becchetti, University of Perugia, Academic year 2010–2011
8. A. Bhat, Advances in theory and applications of stereo vision, Copyright © 2011 InTech, Open Access, ISBN 978-953-307-516-7
9. F. Bianconi1, S. Catalucci, M. Filippucci, R. Marsili, M. Moretti, G. Rossi, E. Speranzini, Comparison between two non-contact techniques for art digitalization. J. Phys. Conf. Ser. **882**, 012005 (2017)
10. R.J. Woodham, Photometric method for determining surface orientation from multiple images. Opt. Eng. 19(I), 139–144 (1980)
11. Y. Wu, Radiometry, *BRDF and Photometric Stereo (PDF)* (Northwestern University). Accessed 25 March 2025
12. B.K.P. Horn, Obtaining shape from shading information, in *Shape from Shading*. ed. by B.K.P. Horn, M.J. Brooks (MIT Press, 1989), pp.121–171

13. S. Barsky, M. Petrou, The 4-source photometric stereo technique for 3-dimensional surfaces in the presence of highlights and shadows. IEEE Trans. Pattern Anal. Mach. Intell. **25**(10), 1239–1252. IEEE
14. S. Ikehata, D. Wipf, Y. Matsushita, K. Aizawa, Robust photometric stereo using sparse regression, in *IEEE Conference on Computer Vision and Pattern Recognition (CVPR)* (2012), pp. 318–325
15. O. Drbohlav, R. Sara, Specularities reduce ambiguity of uncalibrated photometric stereo, in *European Conference on Computer Vision (ECCV)* (2002), pp. 46–62.
16. Y. Quéau, R. Mecca, J.-D. Durou, Unambiguous photometric stereo using two images, in *International Conference on Image Analysis and Processing (ICIAP)* (2015), pp. 286–297
17. B. Mildenhall, P.P. Srinivasan, M. Tancik, J.T. Barron, R. Ramamoorthi, R. Ng, Nerf: Representing scenes as neural radiance fields for view synthesis, in *European Conference on Computer Vision* (Springer, Berlin, 2020), pp. 405–421
18. K. Gao, Y. Gao, H. He, D. Lu, L. Xu, J. Li, NeRF: Neural Radiance Field in 3D Vision, Introduction and Review. https://doi.org/10.48550/arXiv.2210.00379
19. A. Yu et al., Pixelnerf: Neural radiance fields from one or few images, in *Proceedings of the IEEE/CVF Conference on Computer Vision and Pattern Recognition* (2021)
20. L.A. Westover, *SPLATTING: A Parallel, Feed-Forward Volume Rendering Algorithm* (1991)
21. B. Kerbl, G. Kopanas, T. Leimkühler, G. Drettakis, 3D Gaussian Splatting for Real-Time Radiance Field Rendering (2023). arXiv:2308.04079, https://doi.org/10.48550/arXiv.2308.04079
22. S. Fridovich-Keil et al., Plenoxels: Radiance fields without neural networks, in *Proceedings of the IEEE/CVF Conference on Computer Vision and Pattern Recognition* (2022)
23. S. Catalucci, R. Marsili, M. Moretti, G. Rossi, Comparison between point cloud processing techniques. Meas. J. Int. Meas. Confederation **127**, 221–226 (2018)
24. S.P.A. Keyence Italia, Soluzioni di automazione e strumenti di misura per l'industria [Online]. Available: https://www.keyence.it. Accessed 25 May 2025
25. G. Allevi, F. Docchio, S. Pasinetti, G. Rossi, P. Castellini, R. Marsili, A. Quattrocchi, G. Sansoni, P. Chiariotti, R. Montanini, R. Rossetti, E.P. Tomasini, Qualification of additive manufactured trabecular structures using a multi-instrumental approach, in *IEEE I2MTC* (Auckland, New Zealand, 2019)
26. Creaform Inc., 3D Scanners, CMM Scanners and Metrology Measurement Solutions [Online]. Available: https://www.creaform3d.com/en. Accessed 25 May 2025

Chapter 6
3D Shape Measurements By Scanning Techniques

Abstract In this chapter, the principles of 3D shape measurement are presented, focusing on techniques that utilize various types of wave-based scanning. These methods operate by analyzing how waves, such as electromagnetic, ultrasonic, light or X-rays, are reflected or transmitted through objects. Today, such techniques are widely employed to reconstruct the three-dimensional geometry of objects and environments. Additionally, tomographic imaging using mainly but not only X-rays enables the non-contact acquisition of 3D volumetric data from within an object. All of these methods rely on the fundamental physical interactions between waves and matter, as introduced in Chap. 1. This chapter provides a deeper description of the key components involved in measurement systems and the data processing algorithms used. The aim is to enhance the reader's understanding of the core principles, capabilities, and limitations of these 3D scanning techniques.

6.1 Introduction

Shape measurement techniques using scanning beams are widely used in various applications, including industrial quality control, 3D modeling, and biomedical imaging. The waves used for capturing the geometrical properties can be of two different natures: electromagnetic or mechanical. When employing an electromagnetic wave, different techniques can be used based on the wavelength of the electromagnetic emission that is being used. Depending on this parameter, one can distinguish between:

(1) *Radio wave scanning*—to determine the shape, distance, position and velocity of a backreflecting object, discovered for military purposes during the Second World War. These techniques are generally called radio detection and ranging (RADAR). Are used also to detect void and different materials beow grownd (GeoRADAR).
(2) *Visible light wave scanning*—to determine the shape, distance, position and velocity of an object by light reflections from its surface. These techniques are generally called light detection and ranging (LIDAR). Are also used in

G. Rossi, *Mechanical and Thermal Measurements by Images and Waves*, Smart Sensors, Measurement and Instrumentation 51, https://doi.org/10.1007/978-3-032-03464-9_6

Table 6.1 Beam scanning techniques to measure shape of objects

Technique	Medium used	Resolution	Depth measurement	Main applications
Ultrasonic	Sound waves	Medium	Internal and surface	NDT, medical imaging, underwater
Radar	Radio waves	Low to medium	Surface and internal	Defense, air traffic control, weather, automotive, terrain maps, underground analysis
Lidar	Laser pulses	High	Surface and internal	Autonomous driving systems, mapping, NDT, weather
X-Ray	X-ray beams	Very high	Internal	Medical, industrial, security

metereology to detect presence and concentration of atmosphere water and ice particles, gases, etc.

(3) *X-ray scanning*—that uses high-energy electromagnetic waves that most materials do not reflect but transmit and attenuate; from this attenuation, the shape and volumetric density distribution can be measured, scanning the measurand from many directions.

As for the technique that leverages mechanical waves:

(4) *Ultrasonic scanning*—high-frequency mechanical waves reflected by the surface and transmitted to the internal structure of an object. The waves transmitted into the object, and the reflected echoes are analyzed to measure shape, material properties and to obtain acoustic images. In water, for marine applications, are generally named sound detection and ranging techniques (SONAR), and for medical purposes are also named Echography.

Their performances and applications can be summarized in Table 6.1

Clearly, the performance of the measurement technique (in terms its range, resolution and uncertainty), strictly depends on the interaction between the wave used and the measurand, as well as its internal physical characteristics. The physical principles governing these interactions are illustrated in Cap 1 and will depend on the characteristic of the surface, materials, and the kind of wave used.

6.2 Measuring Shapes By Mechanical Scanning

The reference method for high-precision 3D shape measurement does not rely on traveling waves, beam scanning, or image processing. Instead, the industry standard for such measurements is set nowadays by Coordinate Measuring Machines (CMMs). These systems [1] operate based on direct mechanical contact: a probe physically touches various points on an object's surface to record their spatial coordinates. This

(a) Cartesian bridge CMM (b) anthropomorphic arm (c) touching probe

Fig. 6.1 Coordinate measuring machines and touching probes

is the fundamental principle underlying traditional dimensional metrology tools used in mechanical manufacturing, such as calipers, micrometers, and dial comparators.

CMMs are widely employed in industrial settings to verify the geometric features of complex mechanical components, ensuring they meet design specifications and can be correctly assembled into larger products, machines, or structures. In a typical Cartesian coordinate CMM, the probe is moved along three orthogonal axes (X, Y, and Z) using precision-guided linear slides. Each axis is equipped with a high-resolution position sensor, allowing for accurate determination of the probe's location when contact with the surface is made.

To enhance measurement accuracy and minimize thermal expansion effects, the machine frame is usually constructed from dimensionally stable materials such as granite or ceramics. When the probe touches the object's surface, the position sensors simultaneously register the contact point's three-dimensional coordinates.

In addition to Cartesian systems, there are also articulated-arm CMMs and other rotational designs that use angular displacements instead of linear translations.

These systems detect the probe's position based on rotational joints, typically capturing three angular values at each contact point. While these designs are more compact, lightweight, and easier to use they generally offer lower measurement accuracy compared to cartesian CMM.

Typical examples of CMM and tactile probes are shown in Fig. 6.1.

The scanning processes needed for measuring the shape can be of two kinds:

- Using the probe as touch-trigger, recording a single point when the probe makes contact; this method gives the best possible measurement accuracy.
- Moving the probe continuously along the surface, collecting multiple points for a higher resolution; in this way the scanning is faster best measurement speed at the expense of accuracy.

For soft surfaces or other solutions, non-contact sensors can be used instead of touch probe to detect positions. Laser triangulation sensors, illustrated in previous Chap. 5, are largely used for this purpose. The main application of CMMs is geometric inspection, tolerances verification, comparing measurements with CAD models of mechanical components, but they are also used for general dimensional inspections or for identifying unknown geometries. The applications are many in all industrial sectors where a component is realized or checked to be used for an assembly. Nowadays they are generally the highest performing measurement system of objects shapes, with accuracy lower than 1 μm in ranges of meters, but with high costs

Table 6.2 Main characteristics of CMM

CMM type	Accuracy	Measuring range	Speed	Portability	Applications
Bridge CMM	Very high (<1–5 μm)	Medium	Moderate	Low	Small to medium parts, precision industries
Cantilever CMM	High (2–6 μm)	Small	High	Low	Small parts, electronics, and precise work
Gantry CMM	Moderate to High (3–10 μm)	Large (up to 10 m or more)	Moderate	Low	Large parts, automotive, aerospace
Horizontal arm CMM	Moderate (5–15 μm)	Very large	Moderate to High	Low	Sheet metal, automotive body panels
Portable CMM	Moderate (10–30 μm)	Small to Medium (varies)	High	Very high	Field inspections, flexible environments
Optical CMM	High (2–10 μm)	Small to Medium	Very High	High	Delicate, soft, or small parts
Laser tracker	Moderate (5–20 μm)	Very large (up to 80m)	High	Very high	Large components, alignment, aerospace

(roughly in the range 50 K€–1 M€). Main characteristics of CMM are given in Table 6.2.

In Fig. 6.2, from [2], is illustrated the comparison between the shapes of a human tooth replica realized with zirconium, measured with CMM and with intra oral scanners, typically used in dentistry applications.

The goal is that of characterize by comparison intra-oral scanner performance using a reference object, a reference measurement technique based on CMM and a specific numerical method on purpose developed to estimate a global uncertaintyof intra-oral scanners.

6.3 Measure By Scanning Electromagnetic Beams: Radar

RADAR is an acronym of "RAdio Detection and Ranging": a measurement technique mainly developed for defense purposes during the second world war to detect long distance airplanes or ships that were approaching. Since all these vehicles are made of steel, they reflect a collimated radio beam generated by an antenna. By detecting

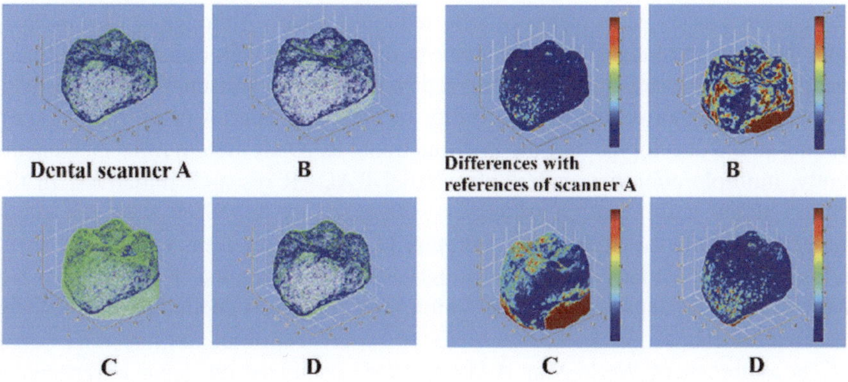

Fig. 6.2 A reference tooth measured by CMM (blu surface) for four kind of intra-oral scanner performance comparisons

the time of flight, i.e., when the wave comes back to the emitter, it is possible to measure the distance and position of the reflecting target [3]. Usually, the antenna is moving between two scanning angle: the result is the position of the object in polar coordinates, as illustrated in Fig. 6.3.

The development of RADAR is rooted in a series of scientific and engineering milestones. The theoretical basis for RADAR can be traced to the electromagnetic field equations formulated by James Clerk Maxwell in the 1860s (see Chap. 1), which mathematically predicted the existence and behavior of electromagnetic waves. This theoretical framework was experimentally validated by Heinrich Hertz between 1886 and 1888, who demonstrated that radio waves could propagate through space and undergo reflection and refraction, phenomena essential to the concept of electromagnetic wave-based detection.

An early practical application emerged in 1904, when German inventor Christian Hülsmeyer patented the "Telemobiloscope," a device capable of detecting metallic

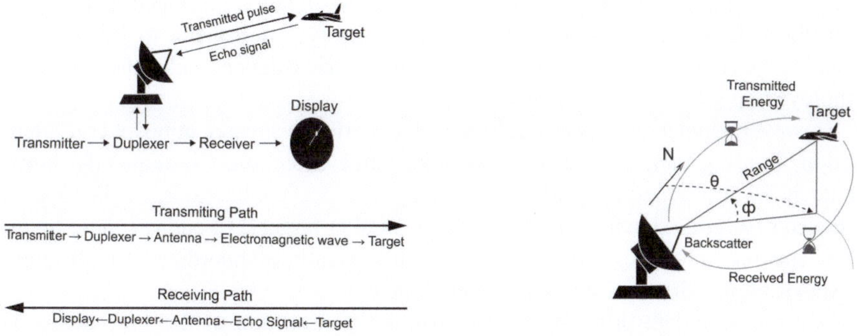

Fig. 6.3 RADAR measurement principle and its polar coordinates definition

objects, such as ships, by means of reflected radio waves. Although limited in function, this system represented a precursor to modern RADAR. Substantial advancements occurred during the 1930s. British physicist Robert Watson-Watt and his team successfully demonstrated the feasibility of detecting aircraft using radio waves, leading to the development of the first functional RADAR system in 1935. Concurrently, multiple countries undertook parallel RADAR research initiatives. In the United States, significant contributions were made by the Naval Research Laboratory (NRL) and the Radiation Laboratory at the Massachusetts Institute of Technology (MIT). Germany developed military RADAR systems such as the Freya and Würzburg, while Japan, the Soviet Union, and Italy also conducted independent research efforts.

The outbreak of World War II (1939–1945) catalyzed the rapid advancement and deployment of RADAR technology. A notable achievement was the British "Chain Home" network, the first operational long-range RADAR defense system, which played a critical role during the Battle of Britain. Central to this development was the cavity magnetron, a high-frequency microwave generator that significantly enhanced RADAR resolution and range. The magnetron was later shared with the United States via the Tizard Mission, accelerating transatlantic RADAR collaboration. During the war, RADAR found diverse applications, including aircraft and ship detection, artillery fire control, and anti-aircraft defense. Following the war, RADAR technology expanded into civilian domains such as air traffic control, maritime navigation, meteorology, automotive safety (e.g., collision avoidance systems), planetary and space exploration, and subsurface imaging via ground-penetrating radar (GPR). Modern RADAR systems employ advanced digital signal processing, phased array antennas, and high-speed computing to achieve precise object detection, tracking, and classification across vast distances. Operating across a broad frequency range, from approximately 10 MHz to 300 GHz, RADAR has evolved into a cornerstone of contemporary sensing and measurement technologies.

The interaction of electromagnetic waves with obstacles has been discussed in Chap. 1. As shown in Fig. 1.32, efficient reflection occurs primarily when the dimensions of the obstacle are comparable to or larger than the wavelength of the incident wave. Within the RADAR frequency spectrum classification presented in Table 6.3 and Fig. 6.4, wavelengths range from approximately 1 millimeter to 100 m. This broad range enables the detection of various object sizes, depending on the frequency employed.

However, higher-frequency radio waves experience greater atmospheric attenuation, limiting their effectiveness for long-range detection. Consequently, lower frequencies are generally preferred for long-distance applications, at the cost of reduced resolution, i.e. the ability to discern smaller targets. To address these trade-offs, specific frequency bands have been historically standardized for different RADAR applications, as summarized in Table 6.3.

The measurement principle of RADAR is the detection of the time delay Δt between the instant the radio wave is generated by the transmitter antenna, and instant when the wave reflected hits the receiver antenna [4].

Table 6.3 RADAR bandwidth for different applications

Band designation	Frequency range	Usage
HF	3–30 MHz	OTH (Over-the-Horizon) surveillance
VHF	30–300 MHz	Very-long-range surveillance
UHF	300–1,000 MHz	Very-long-range surveillance
L	1–2 GHz	Long-range surveillance
S	2–4 GHz	En route traffic controlModerate-range surveillance
C	4–8 GHz	Terminal traffic controlLong-range tracking
X	8–12 GHz	Airborne tracking detectionShort-range trackingMissile guidance
Ku	12–18 GHz	Mapping, marine radarAirborne intercept
Ka	26.5–40 GHz	High-resolution mappingSatellite altimetry
K	18–27 GHz	Little use (water vapor)
Millimeter	40–100 + GHz	Very-high-resolution mapping, airport surveillance, experimental

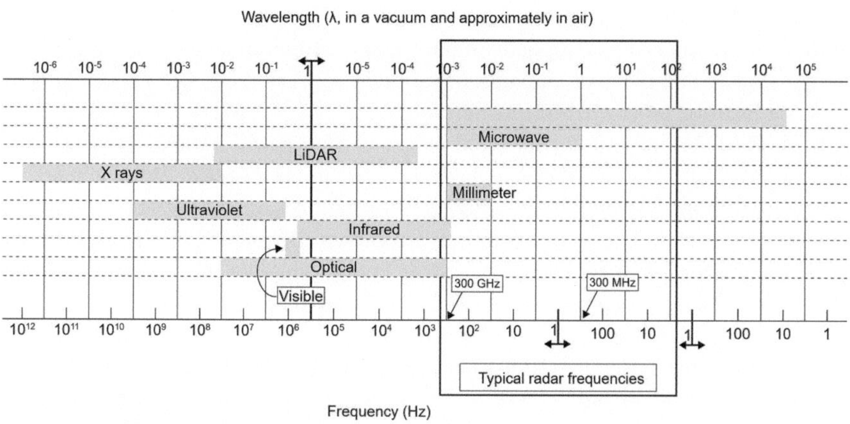

Fig. 6.4 Frequency range of RADAR's, LIDAR's and X ray

The time delay Δt can be measured using the two LIDAR approaches, previously described in Cap 5:

(1) *Direct time of flight (dTOF)*: the distance *R* between the target and the emitter is calculated directly by measuring the Δt, which is then multiplied by the speed of light *c*.

$$2R = c \cdot \Delta t \tag{6.1}$$

(2) *Indirect time of flight (iTOF)*: the distance d between the target and the emitter is calculated by measuring the phase shift φ of a modulated radio signal at frequency f, travelling at the speed of light c along the path. From this phase shift we have the distance $d = c\,\varphi/(4\pi f)$.

$$2R = c \cdot \frac{\varphi}{4\pi f} \tag{6.2}$$

Given the three polar coordinates (R, θ, ϕ) measured by a RADAR system, the cartesian coordinates (x, y, z) of the position of the target can be estimated using the following classical equation for coordinate transformation:

$$(x, y, z) = (R \cdot \cos(\theta) \cdot \sin(\varphi), R \cdot \sin(\theta) \cdot \cos(\varphi), R \cdot \cos(\theta)) \tag{6.3}$$

where R is estimated by the time delay; θ and ϕ are the two known orientation angles of the radio beam on the horizontal and vertical plane respectively.

Two basic schemes are available: the bistatic or the monostatic RADAR, both illustrated in Fig. 6.5. In bistatic systems an antenna is used to send a radio beam and the signal reflected by the target is received by a second antenna, placed at a different location in space. While in a monostatic system a single antenna behaves as both the emitter and the receiver. Also, quasi-monostatic configurations are available, where the transmitter and receiver antennas are slightly separated but still appear to be at the same location as viewed from a far target.

If the RADAR system is kept fixed in space, then it will be able to detect only the of a target along a line; these are called non imaging RADAR. But the antennas can be continuously rotated in order to get information along multiple directions, as described in Fig. 6.6. These configurations are known as “imaging” RADAR systems, where images are taken not by a sensor array but by scanning (i.e. moving, rotating) the sending and receiving line of the radio signal and making a polar plot of received echoes generated by the reflecting objects [5].

The radio signals used in RADARs are generated by powerful transmitters, while the received should be highly sensitive, due to the low amplitude of the reflected

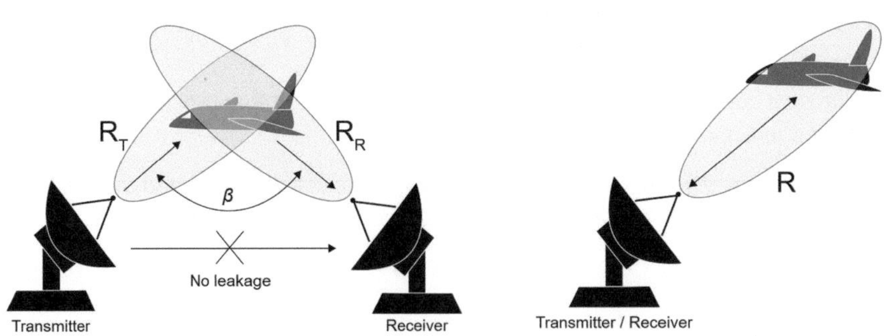

Fig. 6.5 Bistatic and monostatic RADAR

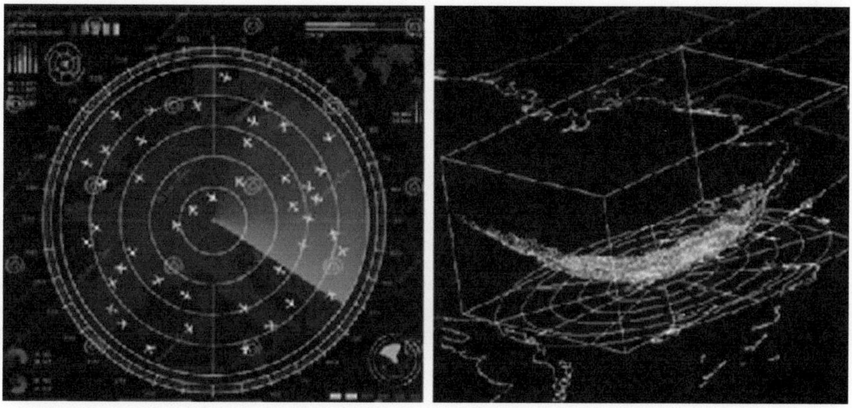

Fig. 6.6 2D and 3D typical imaging radar systems output

signals, because reflecting target spread the wave power in all directions and its back detection, only in one direction, is strongly attenuated. Special data processing techniques are required to detect these low level signals received.

The main components of a radar system are illustrated in Fig. 6.7 and are the following.

(1) *The transmitter*

The transmitter produces the short-time and high-power RF electrical pulses that are radiated into space by the antenna. The transmitter must have the following technical and operational characteristics:

- It must be able to generate the required average RF power and the required peak power.
- It must have adequate RF bandwidth.
- It must have high RF stability to meet the signal processing requirements.

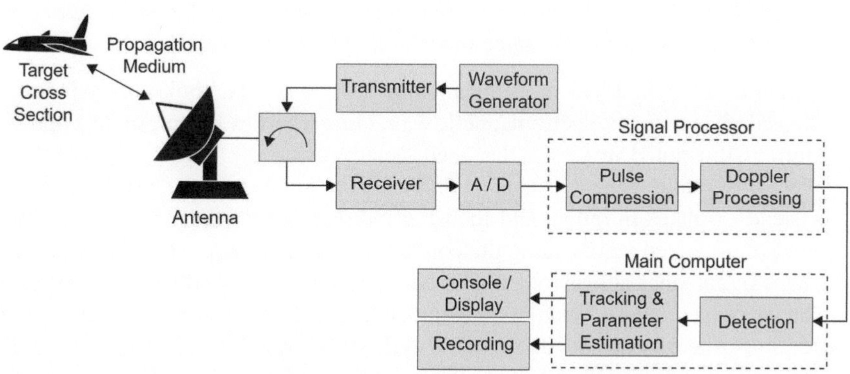

Fig. 6.7 Main components of a RARAR system

- It must be easily modulated to meet waveform design requirements.
- It must be efficient, reliable, and easy to maintain.

Generally, the principle types of transmitters are:

(a) Power Oscillator Transmitter (POT): High power oscillators where a tube, usually a magneton, produces the RF pulse. The oscillator tube is synchronized with a high-power continuous energy pulse, generated by a separate unit.
(b) Power Amplifier Transmitters (PAT): used in many newly developed radar systems. In this system the transmission pulse is generated by a waveform generator and then brought to the necessary power with a subsequent amplifier (Amplitron, Klystron or Solid-State Amplifier).

(1) *The antenna*

The antenna transmits and/or receives electromagnetic waves. It is a transducer with the function of converting electromagnetic waves into electric currents and vice versa. The very basic principle is illustrated in Chap. 1, in Fig. 1.9, is a simple wire with moving charge inside. Typically, the mechanical dimensions of the antenna are fractions of the radio wavelength used. There are antennas with a length of over one hundred meters that use frequencies in the long wave range and antennas with a length of a few millimeters for microwave radio waves.

The antenna performs the following essential functions:

- It ensures that the signal has the required pattern in space, sufficiently focused.
- It shall provide the required frequency.
- It must measure the angular position with a low uncertainty.

The basic radar performance is proportional to the product of the antenna area or aperture and the transmitted average power. The investment in the antenna therefore leads to direct results in terms of system performance. Antennas, illustrated in Fig. 6.8, are crucial components that strongly determine the radar's performance in terms of range, resolution, and accuracy. The simplest and low-cost dipole and monopole antennas consisting of a conductive rod or wire, generating omnidirectional radiation, that can be used for short range radar or for Ground Penetrating Radar (GPR) applications. The three main kind of RADAR antenna are:

1. Parabolic or cylindrical reflector antennas that use a parabolic or cylindrical shaped plate to focus electromagnetic waves into a narrow beam. Are like optical mirrors illustrated in Cap.2, designed usually using ray tracing techniques or electromagnetic finite element modelling. Antennas can have a metallic grid instead of plates to reflect and focusing radio waves, in order to generate and receive polarized waves along the conductive rod direction. Also, the parabolic shape of antennas allows for a higher gain, narrower beamwidth and detection at longer ranges. These kinds of antennas, illustrated in Fig. 6.8, are largely used for air traffic control, weather radar, and military surveillance.
2. Phased array antennas operate based on the principle of beamforming, where

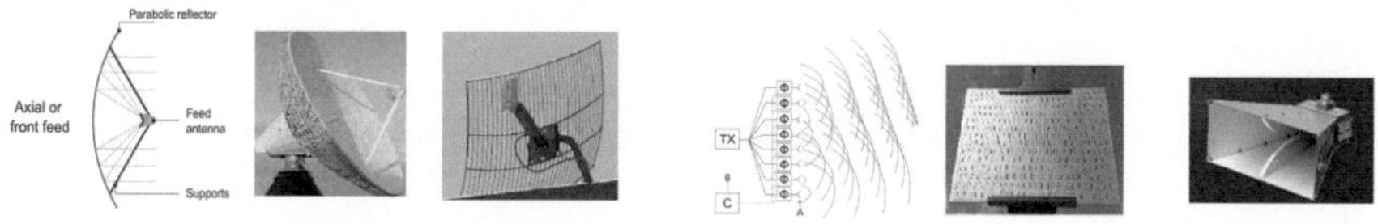

(a) Parabolic linear and vertically polarized (b) phased array (c) horn

Fig. 6.8 RADAR antennas main kinds

multiple radiating elements emit signals with controlled time delays. This technique is an application of the Huygens principle, as discussed in Chap. 2. By varying the relative delays between the emitted signals, the direction of the resulting electromagnetic beam can be steered electronically, allowing for rapid scanning without any mechanical movement of the antenna. An example of such a system is illustrated in Fig. 6.8b.

3. Horn antennas focus radio waves by utilizing reflections within a metallic, flared waveguide structure. This configuration effectively directs the electromagnetic energy, resulting in high directivity. The principal advantages of horn antennas include low signal loss, wide operational bandwidth, and strong directional performance. They are commonly used in microwave RADAR systems and other high-frequency applications. An example is shown in Fig. 6.8c.

The *gain* of the antenna is defined as the ratio between the amount of energy propagated in one direction, and the amount that would be propagated in all directions if the antenna was isotropic. Figure 6.9a illustrates the concept of *gain* of the antenna.

The gain depends on the wavelength of the radio wave, the effective area of the antenna, the opening, as well as its the efficiency. The antenna gain G is the radiation intensity relative to a lossless isotropic reference, given by [3–5]:

$$G = \frac{4\pi A\epsilon}{\lambda^2} \tag{6.4}$$

where A is the antenna area, ε antenna efficiency (a number between 0 and 1), $\lambda = c/f$ is the wavelength of transmitted and received radio signal. Increasing the antenna size relative to the wavelength enhances gain and narrows the beamwidth, improving detection range and angular resolution. A typical plot of energy radiation

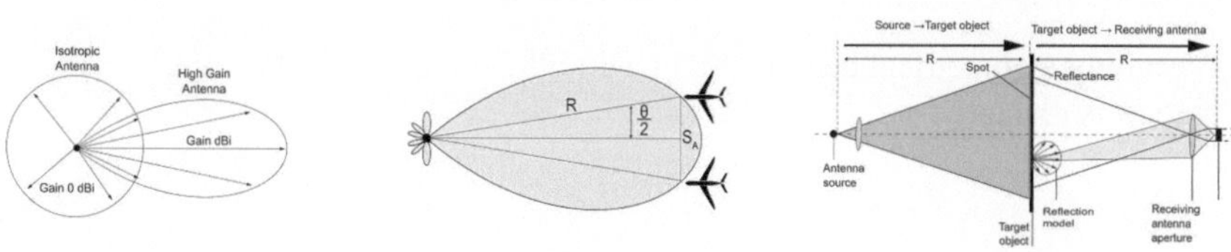

(a) (b) (c)

Fig. 6.9 Antenna transmission gain (**a**), resolution (**b**) and back reflected power model (**c**)

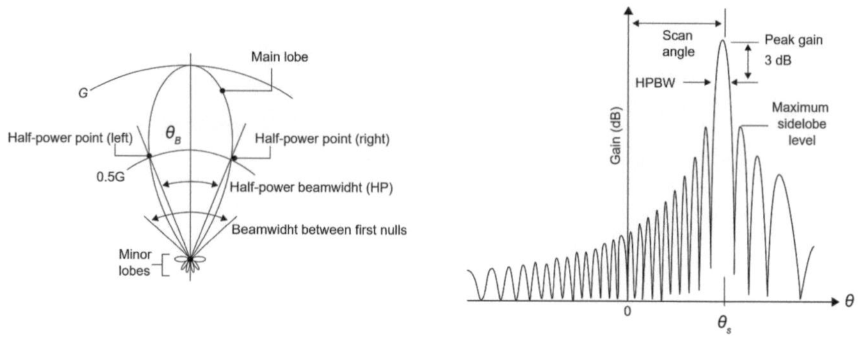

Fig. 6.10 Antenna directional performances in polar and cartesian diagrams

distribution with the angle of an antenna is illustrated in Fig. 6.10, both in polar and cartesian coordinates. The antenna logarithmic gain is defined as:

$$G(dB) = 10log\left(\frac{I}{I_o}\right) \tag{6.5}$$

Hence, a drop of 3 dB in gain corresponds to a 50% reduction in the radiation intensity.

The angular positions that correspond to the radial directions where the Gain decreases to 50% are the so called Half Power BeamWidth apertures (HPBW) θ_B, illustrated in Fig. 6.10. This parameter directly impacts the tangential (cross-range) resolution Δs_t, approximated as:

$$\Delta s_t = 2Rsin\left(\frac{\theta_B}{2}\right) \tag{6.6}$$

Another important parameter is the range resolution Δs_r, which determines the radar's ability to distinguish between closely spaced targets, as illustrated in Fig. 6.9b. It depends on the pulse width τ or related bandwidth B_w [4]. Therefore, the theoretical range resolution of a radar system can be estimated from the following formula:

$$\Delta s_r = \frac{c\tau}{2} = \frac{c}{2B_w} \tag{6.7}$$

Assuming symmetrical beamwidth in azimuth and elevation, the overall volume resolution is given by:

$$\Delta s_r \times (\Delta s_t)^2 \tag{6.8}$$

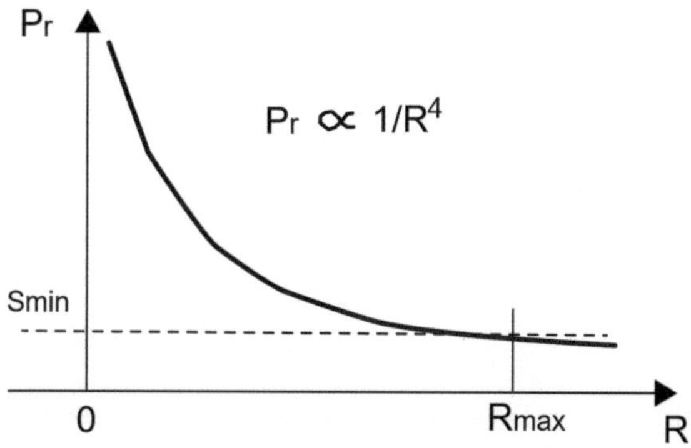

Fig. 6.11 RADAR received signal power versus distance

The characteristics of the antenna define the main RADAR performances, and its measurement range. The received radio wave signal power P_r will be a small fraction of the whole transmitted power P_t, mainly due to the reduced viewing angle of the receiving antenna when the target is far away (Fig. 6.9c), as well as the not perfect alignment of all emitted rays. The received power P_r can be calculated knowing the distance d from the reflecting target and the transmitting power P_t [3–5]:

$$P_r = \frac{P_t G_t \sigma A_{er}}{\left(4\pi d^2\right)^2} = \frac{P_t G_t G_r \sigma \lambda^2}{(4\pi)^3 d^4} \tag{6.9}$$

where A_{er} is the effective aperture area of the receiving antenna, σ is the RADAR cross section, G_r is the receiving antenna gain, G_t is the transmitting antenna gain. Eq. (6.9) underlines how the power of the signal reflected by the target decreases with the forth power of the its distance, as illustrated in Fig. 6.11.

The minimum received power that the radar receiver can capture, S_{min}, as well as the maximum detection range, R_{max}, i.e. the measurement range of the RADAR, can be calculated with the following equation:

$$P_{min} = \frac{P_t G_t G_r \sigma \lambda^2}{(4\pi)^3 d^4} \quad \text{and therefore} \quad R_{max} = \sqrt[4]{\frac{P_t G_t G_r \sigma \lambda^2}{(4\pi)^3 S_{min}}} \tag{6.10}$$

Here, the effect of atmospheric attenuation on the power of the signal is not considered. For RADAR frequencies below 10 GHz ($\lambda > 3$ cm), atmospheric attenuation is neglectable (lower than 0.01 dB/km). The reason lies in the physical phenomena detailed in Chap. 1, for which long wavelength signals are not reflected, diffused or affected by small obstacles, in this case the particles in the atmosphere.

The lower limit of a RADAR's measurement range is given by the pulse width τ, or its corresponding bandwidth, and by other delays, τ_e, introduced by the processing electronics (mainly the duplexer):

$$R_{min} = \frac{(\tau + \tau_e)c}{2} \tag{6.11}$$

Range limitations are also given:

(a) For continuous modulated RADAR, by the modulation frequency f_m, given by Eq. (1.82), or using dual frequency modulation technique by Eq. (1.83):

$$\mathrm{R}_{max} = \frac{\mathrm{c}}{2\mathrm{f_m}} \tag{6.12}$$

(b) The RADAR must receive the echo before sending out the next pulse. Otherwise, it won't know whether a received echo corresponds to the current pulse or a previous one, leading to range ambiguity. Hence, for pulsed RADAR with pulse repetition frequency f_s, the maximum measurable distance is given by:

$$R_{max} = \frac{2c}{f_s} \tag{6.13}$$

(3) *The receiver and visualizing RADAR components*

In some RADAR systems, the target's velocity can also be determined using the Doppler effect, as illustrated in Chap. 1. This is achieved by detecting the frequency shift of the back-reflected wave, a physical phenomenon caused by the relative motion between the emitting source and the target.

Other target characteristics can be inferred from single reflections. For example, size, shape, and material properties can be assessed based on signal strength, polarization, and other signal features, which are utilized in certain advanced radar systems.

The time-delayed back-reflected signal pulses have traditionally been visualized using three types of 2D plots, known as A, B, and C displays, as shown in Fig. 6.12. Originally displayed using cathode-ray tubes (CRTs) in oscilloscopes, capable of visualizing only one signal component at a time, these visualization methods are still employed today on modern computer screens.

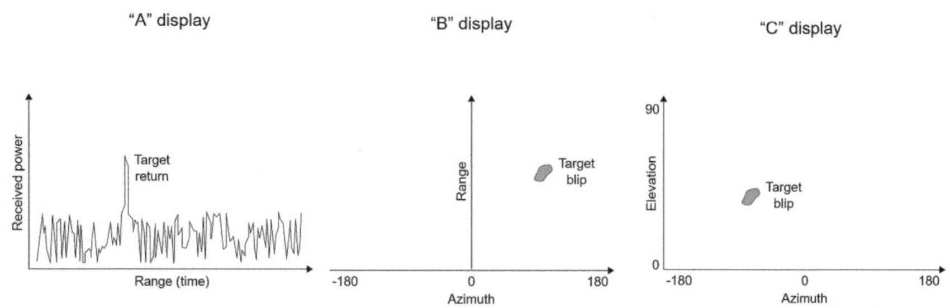

Fig. 6.12 Output of a RADAR system using A, B and C type of display

- *A-display*: *The* amplitude of the radio signal received is plotted against distance (or the corresponding delay time). Peaks in the signal correspond to the time it takes for the radio wave to travel to the target and back. This time is proportional to twice the distance to the reflecting target or surface.
- *B-display*: The received signal amplitude is represented as screen brightness, plotted over time against the azimuth angle. A peak in brightness (indicating high signal amplitude) reveals the target's position in 2D polar coordinates, distance d and azimuth angle θ.
- *C-display*: The received signal amplitude is shown as screen brightness plotted against elevation angle φ and azimuth angle θ. A bright spot in this 2D display indicates a reflection from a target, providing its position by two angular coordinates.

RADAR measurement techniques are also used to measure inside solid materials through Ground-Penetrating Radar (GPR), or Geo-RADAR, which is used in various media such as rock, soil, ice, water, pavements, and civil structures. GPR operates with high-frequency, typically polarized, radio waves, usually in the range of 10 MHz to 2.6 GHz. The antennas are generally placed in contact with the ground for optimal signal transmission, although air-launched GPR antennas can also be used for non-contact measurements. One of the primary applications of GPR is the detection and location of underground utilities such as pipes, cables, and voids. However, it is also widely used in geological surveys, archaeology, mining, and numerous other detection and mapping tasks.

6.4 Measure By Light Beam Scanning: Lidar

About 30 years after the introduction of radio wave reflection measurement techniques (RADAR), laser light began to be used for measuring distances and other characteristics of the crossed media, following the same basic principle: time-of-flight delay between a transmitted and a received pulse or delay in amplitude-modulated waves [6]. This measurement technique is known as Light Detection and Ranging (LiDAR). Lasers used in LiDAR operate within atmospheric transmission windows, typically in the visible range (0.4–0.7 μm), the near-infrared range (0.7–1.5 μm), the mid-infrared range (3–5 μm), and the far-infrared range (9–13 μm), as illustrated in Fig. 6.4. The technology began with the invention of the laser in 1960, as discussed in Chap. 1. Early LiDAR systems, similar to RADAR, were developed for military and aerospace applications, particularly for range finding and target position detection.

The first airborne LiDAR systems were tested by NASA and the U.S. military to measure terrain elevation and shape. These systems were soon adopted for topographic mapping. A typical output of such measurements is shown in Fig. 6.13.

Initially, LiDAR technology was bulky, expensive, and limited to governmental and research institutions. In the 1990s, commercialization began, and LiDAR applications expanded into forestry, agriculture, and urban planning. A significant cost

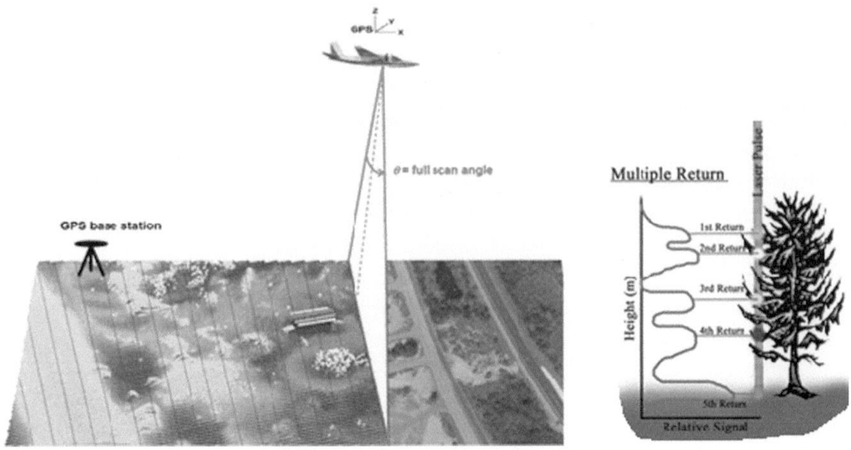

Fig. 6.13 Airborn LiDAR for high resolution mapping

reduction occurred in the 2000s thanks to advances in semiconductor and solid-state laser diode technology, making terrestrial, mobile, and airborne LiDAR more accessible. LiDAR also became integral to autonomous vehicle research, with early experiments conducted during the DARPA Grand Challenges of 2004 and 2005.

In the 2010s, LiDAR became a widely adopted measurement technology for self-driving cars. Companies such as Velodyne and Luminar developed high-resolution, long-range sensors. The development of solid-state and MEMS-based LiDAR systems further reduced costs and increased durability. In 2020, Apple introduced LiDAR in consumer devices like the iPad Pro and iPhone 12 Pro, bringing the technology into mainstream consumer electronics.

Today, LiDAR is extensively used in robotics, smart cities, environmental monitoring, and safety applications. Artificial intelligence-powered data processing has significantly enhanced object recognition and scene interpretation. Current research focuses on new developments such as solid-state LiDAR, flash LiDAR, and frequency-modulated continuous wave (FMCW) LiDAR for improved real-time performance. Quantum LiDAR (and RADAR) also represents a promising frontier in measurement technologies.

To detect time delays Δt, the same two approaches described in Chap. 1 for i-LiDAR and RADAR are used:

1. *Direct Time of Flight (dTOF)*: Measures the time delay Δt that a light pulse takes, traveling at the speed of light c, to go from the emitting laser (typically a laser diode) to the reflecting spot and back to the photodiode. For example, from Eq. 6.1, a delay of $\Delta t = 0.67\ \mu s$ corresponds to a distance of 100 m, or approximately 6.7 ns/m of distance.

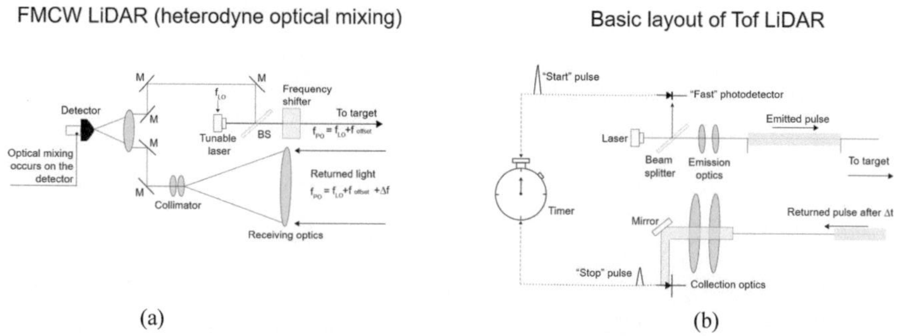

Fig. 6.14 iTOF heterodyne optical mixing LiDAR (**a**) compared to simple dTOF (**b**)

2. *Indirect Time of Flight (iTOF)*: Measures the phase shift φ of a modulated light signal at frequency f, traveling at speed c over a path of length $2R$. The distance is given by Eq. 6.2.

The advantages of iTOF techniques, including frequency-modulated continuous wave (FMCW) LiDAR, which uses heterodyne optical mixing (as explained in Chap. 1), include photon shot-noise-limited detection, low sensitivity to ambient light, and the ability to extract both distance and velocity information in the frequency domain. Additionally, they require lower-bandwidth electronics, making them less expensive. However, iTOF systems typically have shorter measuring ranges than dTOF systems, with the limit depending on the modulation frequency.

A typical configuration of an iTOF LiDAR system using heterodyne optical mixing, compared to a simpler dTOF setup, is illustrated in Fig. 6.14.

The position of the reflecting laser spot is measured using three polar coordinates (d, θ, φ), as RADAR, where θ and φ define the direction of the transmitted and received laser beam, and d is determined as described above. Scanning the laser beam enables the measurement of many points position on a surface.

In LiDAR systems, atmospheric attenuation is more significant than in RADAR systems, where it can often be neglected due to longer wavelengths. In LiDAR, the received optical power P_r from the laser spot is typically estimated using a formulation that includes an exponential attenuation factor to account for atmospheric losses. A typical expresson is [7]:

$$P_r = P_0 \rho \frac{A_0}{\pi d^2} \eta_r e^{-2\kappa d} \tag{6.14}$$

where P_0 is the laser source power transmitted, ρ the target reflectivity, A_0 the aperture area of the receiver, η_r the receiving optic transmission coefficient, κ the atmosphere absorbing coefficient and d the target or reflecting point distance. This assuming normal incidence, Lambertian reflection, flat beam profile and negligible divergence, laser spot smaller than the target, and κ independent of d.

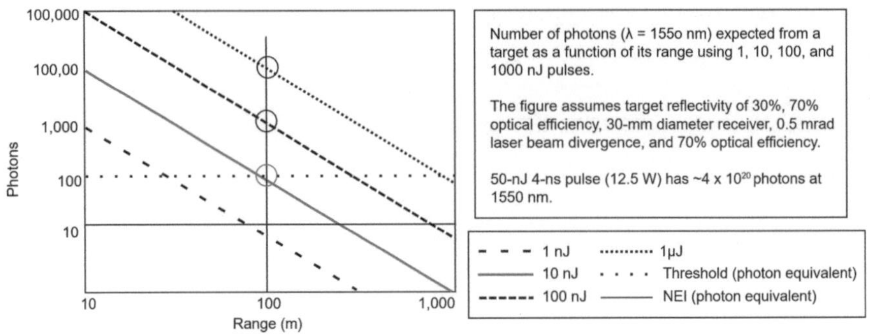

Fig. 6.15 Number of photons received for each generated laser pulse

From above Eq. (6.14) it is possible to build a diagram, illustrated in Fig. 6.15, of the received number of photons (at $\lambda = 1550$ nm) expected from a target as a function of its range using 1, 10, 100, and 1000 nJ pulses. Here a LiDAR laser diode source at 1550 nm is assumed, able to generate light pulses of 50 nJ energy and 4 ns pulse widht (12.5 W) so having ~4×10^{20} photons @ 1550 nm. The figure assumes target reflectivity of 30%, 70% optical efficiency, 30 mm diameter receiver, 0.5 mrad laser beam divergence, and 70% optical efficiency in Eq. (6.14).

From the values shown in Fig. 6.15 it is clear that the main challenge of a dTOF LiDAR system is the need of very low-noise detectors capable of detecting backscattered signals consisting of few photons.

Typical laser diodes used in these systems operate at either 1550 nm or 905 nm. The 1550 nm wavelength offers better eye safety and lower background noise, but requires infrared (non-silicon) photodetectors. In contrast, 905 nm lasers provide better atmospheric transmission and allow the use of silicon-based photodetectors.

Photodetectors for LiDAR applications must feature high photosensitivity and gain, low noise, and minimal timing jitter. Jitter refers to the random fluctuations in the timing of received pulses caused by the detector, leading to variations in transit time. A typical jitter of 100 ps results in a depth uncertainty of approximately 1.5 cm. Avalanche photodiodes (APDs), discussed in Chap. 3, are the most widely used detectors due to their high gain (about 100) and high quantum efficiency, despite their relatively high noise levels. Other types of detectors are also employed.

When the laser spot is small compared to the target, distance uncertainty is mainly due to detector jitter, represented as $\delta\Delta t$. In this case, the range uncertainty can be estimated as:

$$\delta R = c\,\frac{\delta\Delta t}{2} \tag{6.15}$$

When the laser spot is large compared to the target, the distance uncertainty depends on the time duration τ of the pulse and its propagation depth, so can be estimated by:

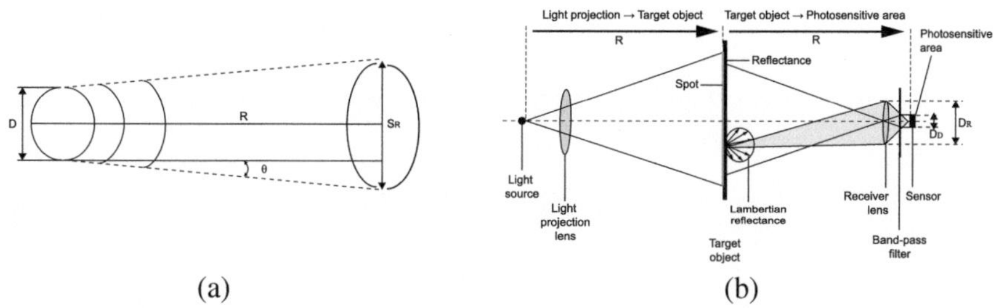

Fig. 6.16 Increasing spot diameter with distance (**a**) and power received reduction scheme (**b**)

$$\delta R = c\,\tau/2 \tag{6.16}$$

Beam divergence is determined by the optical focusing systems, both in laser and radio beams. Even with perfect collimation, diffraction effects, discussed in Chap. 1, introduce a divergence that can be estimated by:

$$\theta = \frac{1.22\lambda}{D} \tag{6.17}$$

where λ is the wavelength and D is the beam diameter. As illustrated in Fig. 6.16a the resulting spot diameter S_R at distance R becomes:

$$S_R = D + 2R\,\tan\theta \approx D + 2R\theta \approx D + 2.44\,R\,\frac{\lambda}{D} \tag{6.18}$$

For a radar system operating at 77 GHz ($\lambda \approx 3.9$ mm) with an aperture diameter of 15 cm, equation (6.17) gives a divergence angle of about 1.8°, and equation (6.18) results in a spot diameter of approximately 6.5 m at a distance of 100 m.

For a LiDAR system using a 905 nm infrared laser with a beam aperture of 10 mm, the divergence angle calculated from Eq. (6.17) is approximately 0.006°, resulting in a spot diameter of about 3.2 cm at 100 meters, as possible to estimate by Eq. (6.18).

This make clear the fundamental physical reason why LiDAR offers significantly higher spatial resolution than RADAR. However, due to its longer wavelength, RADAR can detect objects at greater distances and through fog or clouds. Its lateral resolution is limited by antenna size, with typical values in the order of several meters at 100 m. LiDAR, by contrast, provides a compact and highly accurate solution for 3D mapping.

For this reason, LiDAR is commonly used in topographic mapping, while RADAR remains the preferred choice for applications such as aircraft collision avoidance, air traffic control, and astronomy.

A widely used example of a single-point LiDAR measurement system largely used in civil engineering is the Disto measurement system, the first handheld laser distance meter in the world introduced by Leica in 1993, illustrated in Fig. 6.17a. When combined with two-axis rotation, similarly to a RADAR antenna, it can acquire

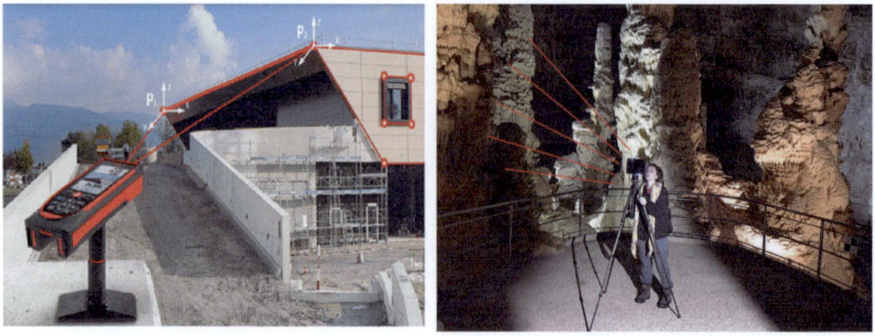

a) Single point LIDAR b) Scanning LIDAR

Fig. 6.17 A single point Leica disto (**a**) and a mechanical scanning LiDAR system (**b**).

the coordinates of multiple surface points. Typical scanning systems based on this principle are illustrated in Fig. 6.17b.

Single-point LiDAR systems measure distance and determine the two angles relative to the north direction (using a compass sensor) and to the vertical direction (using an inclinometer sensor), providing the three polar coordinates of the laser spot with respect to the instrument's position. Typical measurement ranges extend up to 300 m, with an uncertainty of around 1 mm. Mechanical scanning LiDAR systems include internal sensors that precisely measure their rotations around two axes, giving the laser spot's polar coordinates. These systems often integrate a video camera to capture color and texture information by making pictures of the scanned surface. Since laser collimation is not perfect, spatial resolution depends on the spot diameter S_R, which increases with distance R due to the beam divergence angle θ, as specified by the manufacturer. The spot diameter at distance R of a laser beam starting with a diameter D can be estimated by the above Eq. (6.18), simply as $D+2R\,\theta$. Table 6.4 summarizes the typical specifications of several commercially available time-of-flight (TOF) and phase-shift (PS) LiDAR scanning systems (Table 6.4).

The resulting scanned surface will have an uncertainty and level of detail that will be dependent on the angular uncertainty and resolution of the LiDAR used, as well as on the orientation angles θ and φ and distance of the target, as illustrated in Fig. 6.18a.

From the polar-to-Cartesian transformation described in Eq. (6.3), the uncertainties in Cartesian coordinates δx, δy, and δz due to uncertainties in the measured distance δr, and angles $\delta\theta$ (azimuth) and $\delta\varphi$ (elevation), can be estimated using a first-order approximation error propagation model by the expression [8]:

$$\delta x = |\cos(\varphi)\cdot\cos(\theta)\cdot\delta R| + |R\cdot\cos(\varphi)\cdot\sin(\theta)\cdot\delta\theta| + |R\cdot\sin(\varphi)\cdot\cos(\theta)\cdot\delta\varphi|$$

$$\delta y = |\cos(\varphi)\cdot\sin(\theta)\cdot\delta R| + |R\cdot\cos(\varphi)\cdot\cos(\theta)\cdot\delta\theta| + |R\cdot\sin(\varphi)\cdot\sin(\theta)\cdot\delta\varphi| \tag{6.19}$$

Table 6.4 Typical metrological performances of topographic Lidar scanning systems

Model	Measuring principle	Max FOV	Field of view	Min/Max range [m]	Angular accuracy	Accuracy in Distance	Camera
Leica ScanStation P2	TOF	18.5 kg	270° × 360°	0.3–1000	0.003°	4 mm @ 50 m	Integrated, 1 Megapixel
Leica C10	TOF	13 kg	270° × 360°	0.3–300	0.0025°	4 mm @ 50 m	Integrated, 4 Megapixel
Leica HDS6200	PS	14 kg	360° × 310°	1.5–79	0.006°	2 mm @ 25 m	Add-on optional
Riegl VZ-400	TOF	9.8 kg	100° × 360°	1.5–600	0.0005°	5 mm @ 100 m	Add-on optional
Riegl VZ-1000	TOF	9.8 kg	100° × 360°	1.5–1400	0.0005°	5 mm @ 100 m	Add-on optional
Optech ILRIS	TOF	14 kg	40° × 360°	3–1200	0.0064°	7 mm @ 100 m	Integrated, 3.1 Megapixel
Optech ILRIS HD-ER	TOF	14 kg	40° × 360°	3–3000	0.0064°	7 mm @ 100 m	Integrated, 3.1 Megapixel
Optech ILRIS	TOF	14 kg	40° × 360°	3–1200	0.0064°	7 mm @ 100 m	Integrated, 3.1 Megapixel

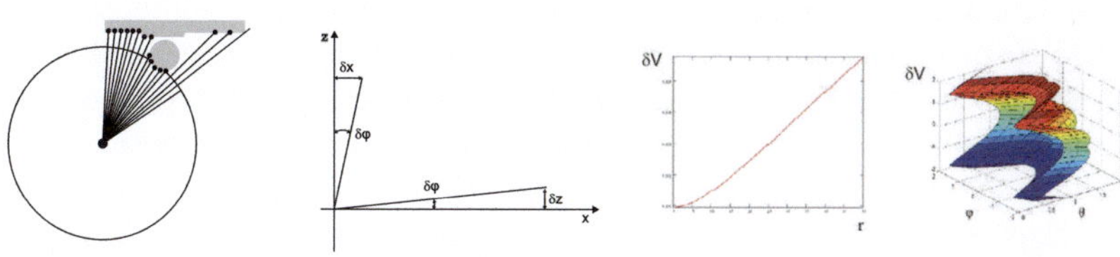

(a) (b) (c) (d)

Fig. 6.18 Uncertainty estimation in position measurement by scanning technique

$\delta z = |\sin(\varphi) \cdot \delta R| + |R \cdot \cos(\varphi) \cdot \delta\varphi|$ and so globally $\delta V = \sqrt{\delta x^2 + \delta y^2 + \delta z^2}$

These equations clearly express the contribution of each uncertainty component to the resulting Cartesian coordinates. Here, R is the measured radial distance, θ is the azimuth angle (typically measured from the north), and φ is the elevation angle (measured from the horizontal plane). The uncertainties δR, $\delta\theta$, and $\delta\varphi$ are assumed to be small and represent the standard deviations (or maximum expected uncertainty) of the corresponding distance and angular measurements.

Plotting the global uncertainty evaluation δ V respect to distance R gives a typical result illustrated in Fig. 6.18 c), an almost liner trend, but the effect on uncertainty given by viewing angle θ and φ, plotted in Fig. 6.18c are not linear and can be very

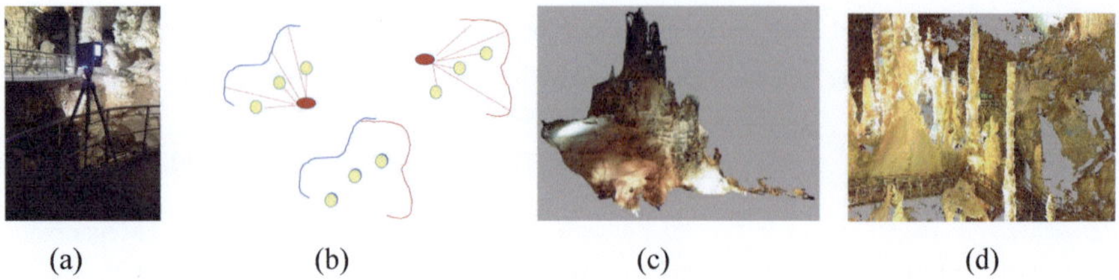

Fig. 6.19 Shape measurement of a big cave by iTOF LiDAR.

big for point having low values of viewing angle θ and φ for a "leverage" effect, i.e. for example a big uncertainty in height measurement of a far object by detecting its angular aperture from bottom to top point of it.

An example of a real-world application of such measurement technique is shown in Fig. 6.19, which illustrates the shape of the large karstic cave named "Abisso Ancona" [8] located within the Frasassi cave complex in the municipality of Genga, Ancona, Italy. The cave has approximate dimensions of 120 × 180 m and a height of 250 m. The survey, conducted in 2018 by the author and colleagues [8] utilized a Faro Focus3D X130 LiDAR scanner. The scanner features a range of up to 100 m, a distance measurement accuracy of ±2 mm at 25 m, and an angular resolution of approximately 50 µrad. It employs an indirect time-of-flight (iTOF) measurement principle and provides a field of view of 360° horizontally and 300° vertically, with a maximum scan rate of up to 976,000 points per second.

In very complex shaped and large environments like this, multiple scanning positions are required. In this case, 24 laser scans were performed from different tripod-mounted positions, capturing 24 separate portions of the cave surface in about half a day. Reference spheres were placed in the environment to serve as common alignment markers, allowing the separate scans to be accurately registered, positioned and merged, as shown in Fig. 6.19b. The complete model was obtained by combining all the scans, resulting in an aligned point cloud consisting of approximately 0.95×10^9 measurement points.

Similar techniques are widely used for the documentation and analysis of monuments, theaters, churches, and other indoor or outdoor environments for a variety of purposes [9].

To acquire "depth images" within fractions of a second needed for other LiDAR applications, such as autonomous navigation, surveillance, safety, and collision avoidance, alternative scanning techniques have been developed. These methods, illustrated in Fig. 6.20, are generally faster but less accurate than mechanical scanning LiDAR systems used in topographic applications. This evolution has been largely driven by the automotive industry in support of autonomous driving technologies.

Solid-state LiDAR systems offer the potential to significantly reduce costs, by up to an order of magnitude, while improving robustness, as they contain no moving parts and are highly compact. A key enabler of this advancement is the use of photonic integrated circuits (PICs), which are increasingly being adopted in LiDAR development.

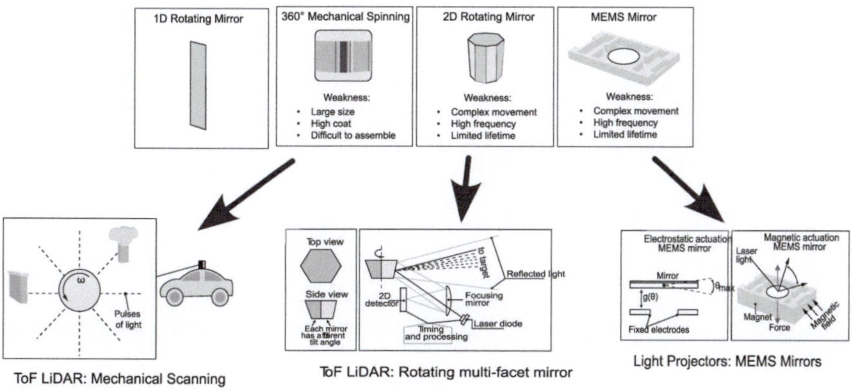

Fig. 6.20 Main scanning techniques of LiDAR systems

PICs allow the integration of numerous optical components, such as lasers, modulators, photodetectors, and filters, on a single chip, analogous to how electronic integrated circuits (EICs) integrate transistors, capacitors, and resistors. In particular, it is now possible to integrate indium phosphide (InP) lasers with silicon photonic circuits, enabling the creation of compact, reliable, and scalable LiDAR systems.

PIC technology is already being applied across fields like telecommunications, sensing, and LiDAR, and it is accessible to a wide range of users through multi-project wafer runs and open-access foundries. These platforms facilitate the integration of pulsed or frequency-modulated laser sources directly on-chip. One example is Blackmore, which has developed a frequency-modulated continuous-wave (FMCW) LiDAR system based on PICs. FMCW LiDAR provides both range and velocity information by measuring the Doppler effect.

Figure 6.21 illustrates the technological evolution and cost reduction of LiDAR systems, from early mechanical devices to the emerging generation of PIC-based solutions. Also shown are several state-of-the-art sensors used in autonomous driving, robotic navigation, safety monitoring, and other advanced applications.

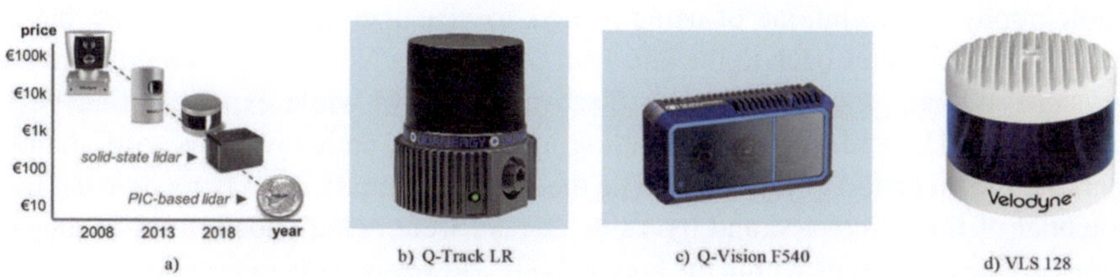

Fig. 6.21 Evolution of LiDAR technologies from the beginning systems to the actual PIC's two Quanergy sensors: **b** Q-Track a direct TOF scanning system (1.3 Mp/s, range 210 m) and **c** Q-Vision a solid state iTOF system (32 Mp/s, range 6 m) **d** A Velodyne VLS128 sensor (range 225 m, 2,3 Mp/s, accuracy + 3 cm)

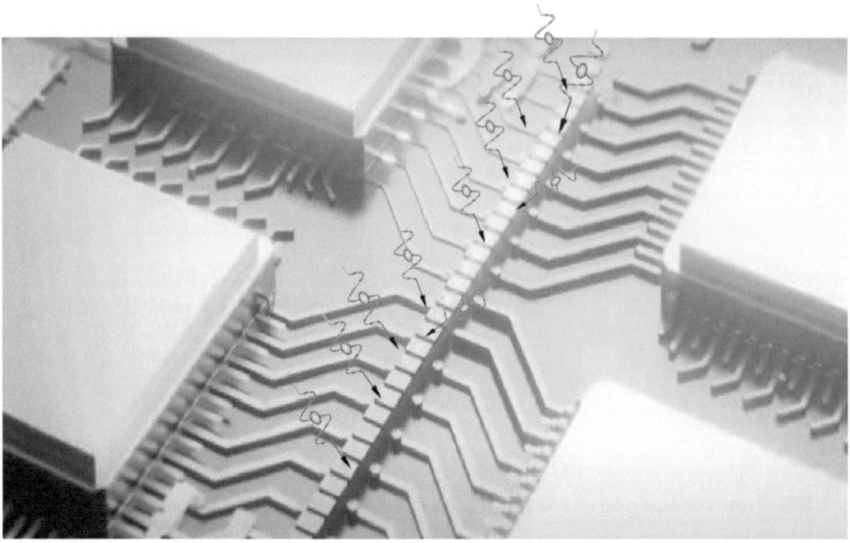

Fig. 6.22 Latest LiDAR sensor technologies

Among the latest innovations, Hesai's 2024 development of an ASIC-based LiDAR sensor marks a significant milestone. This technology, illustrated in Fig. 6.22 incorporates a large number of laser and optical components on a single chip, utilizing highly sensitive single-photon detectors capable of generating reliable signals from minimal photon returns. This ensures consistent performance even under challenging environmental conditions. The AT series by Hesai employs a rotating mirror design with substantially larger optical apertures than those of traditional MEMS mirrors, offering superior performance in adverse weather such as rain or fog.

6.5 Measure By X-Ray: Radiography and Tomography

Radiography, the technique of using X-rays to visualize internal structures, has a rich history that dates back to the late 19th century. On November 8, 1895, German physicist Wilhelm Conrad Röntgen discovered X-rays while experimenting with cathode rays. He produced the first radiographic image, an X-ray of his wife's hand, marking the inception of radiology. The medical community quickly recognized the potential of this discovery, and by 1896, X-rays were being widely used to image bones, tissue and organs within the human body. However, early radiographic images suffered from low resolution and required prolonged exposure times, often resulting in radiation damages.

A significant advancement came in 1913 with the invention of the Coolidge tube by William Coolidge. This vacuum tube allowed precise control over X-ray generation, leading to clearer images and enhanced safety. The introduction of fluorescent intensifying screens further reduced exposure times and enabled real-time imaging.

In 1972, computed radiography (CR) replaced traditional film with photostimulable phosphor plates, enabling digital storage and processing of X-ray images. By the 1990s and early 2000s, digital radiography (DR) emerged, eliminating the need for physical plates altogether and allowing for immediate image acquisition and viewing on digital platforms.

The mathematical foundations of tomography were developed in the 1960s, most notably by Allan Cormack, who formulated algorithms essential for computed tomography (CT). Independently, in 1967, Godfrey Hounsfield at EMI Laboratories further advanced these principles by applying them to the reconstruction of cross-sectional images from X-ray projection data. Their combined efforts culminated in the development of the first clinical CT scanner, built by Hounsfield in 1971, a breakthrough that revolutionized medical imaging by offering detailed cross-sectional views of internal anatomy.

Technological progress continued into the 1990s with the advent of Multi-Slice CT (MSCT), which incorporated multiple detector rows to simultaneously acquire several slices per rotation. This advancement dramatically improved imaging speed and spatial resolution. By the late 1990s, 64-slice CT scanners became the standard, enabling near-instantaneous imaging of the entire body.

Today, radiography and tomography [10] play critical roles across medicine, industry, and scientific research. Their continued evolution drives early diagnosis, precision treatment, and ongoing technological innovation (Table 6.5).

As illustrated in Fig. 6.4, X-rays occupy the frequency range of approximately 10^{15} to 10^{20} Hz, corresponding to wavelengths from 0.1 μm to 0.001 nm.

A typical X-ray measurement system comprises the following main components:

(1) *The X-ray source*
(2) *The detector*
(3) *Mechanical scanning systems for acquiring multiple projections*
(4) *Data processing algorithms.*

1. *The X-ray Source*

X-ray measurement systems are conceptually similar to optical imaging systems in that they require a source of electromagnetic radiation. However, instead of visible

Table 6.5 Main development in radiography and computer tomography

Period	Radiography	Tomography
1895	Discovery of X-rays by Röntgen	–
1913	Coolidge X-ray tube	–
1920s	Contrast media introduced	Early planar tomography
1930–1950s	Intensifying screens, fluoroscopy	First mechanical tomography
1960s	Digital radiography research	Mathematical foundations of CT
1971	Computed radiography (CR)	First CT scanner (Hounsfield)
1980s	Digital fluoroscopy	Spiral CT introduced
1990s	Digital radiography (DR)	Multi-slice CT (MSCT)
2000s-Present	AI-enhanced X-ray analysis	Dual-energy CT, AI-based imaging

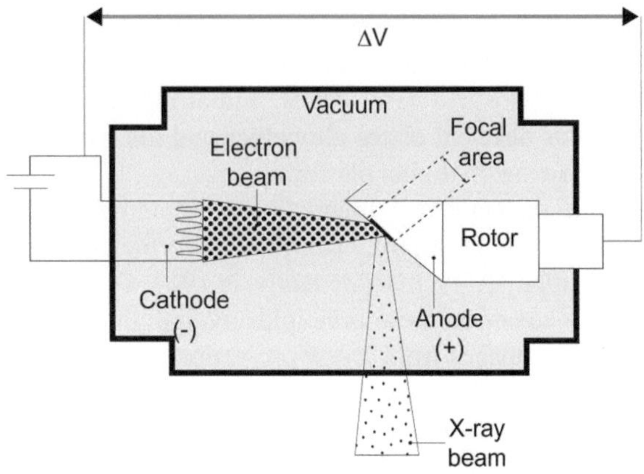

Fig. 6.23 The X-ray source

light, an X-ray source is used to "illuminate" the object. Unlike light, X-rays are not reflected by the object; instead, they penetrate it, as for an optical transparent material, and the transmitted intensity, attenuated by internal structures, is detected to form a transmission image.

The X-ray source, illustrated in Fig. 6.23, performs the same fundamental role as a light source in conventional imaging systems. It can emit the radiation in various geometries: a beam, a sheet, or a cone, depending on the application and design of the system.

X-rays are generated through interactions between high-energy electrons and atoms. When a focused electron beam strikes the target atoms, the electrons in those atoms may be displaced or excited, leading to the emission of X-rays as they return to lower energy states. This process occurs within an X-ray tube, a specialized device designed to produce X-ray radiation.

An X-ray tube consists of a sealed glass envelope, typically enclosed in a lead casing for radiation shielding, in which a high vacuum is maintained. Inside the tube, a cathode and anode are positioned at opposite ends. The cathode comprises a tungsten filament, which is heated by an electric current to emit electrons via thermionic emission. These electrons are then accelerated towards the anode by a high voltage potential difference. The anode, usually made of a heavy metal such as tungsten or molybdenum, can be either stationary or rotating. Rotating anodes are commonly used to increase heat dissipation and improve the efficiency and lifespan of the X-ray source.

In general, only 5% of the energy of the electrons is converted to X-rays, the rest is converted into heat and this can cause deterioration or even melting of the anode. For this reason, especially for diagnostics in the medical field, pipes with rotating anodes are usually used in such a way as to distribute the heat over a larger area and a larger mass. The affected area is generally called the target and is not perpendicular to the incident beam, but rather inclined by an angle of about 20° so that the electrons interact with a relatively large rectangular area, while the photons exit from a square area much smaller, the focal spot.

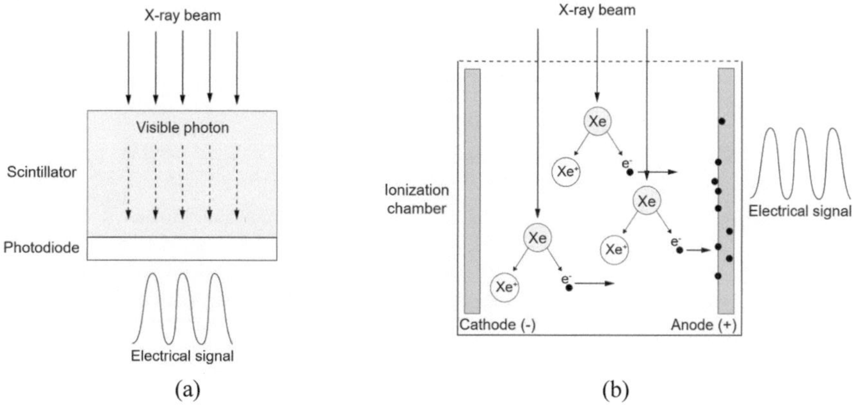

Fig. 6.24 X-Ray detectors

The x-ray tube is realized by a heat-resistant boron glass container inside which there is a vacuum to make the electron beam propagate and to be able to apply high potential differences to allow energy level jumps. The tube is wrapped around an oil filled sleeve to allow cooling and inside are mounted cathode and anode supported by a rotor to make it rotate and reduce the temperature.

(2) *X-ray Detectors and Attenuation Principles*

Early X-ray detectors consisted of radiographic films coated with X-ray-sensitive materials that changed their gray level in response to incident radiation intensity and exposure duration. Over time, more sophisticated detection systems have been developed, with two principal categories emerging: solid-state detectors and gas-based detectors (see Fig. 6.24a, b, respectively).

Solid-state detectors are the most widely used in modern applications. They are typically composed of scintillator crystals, such as bismuth germanate (BGO), sodium iodide (NaI), or calcium fluoride (CaF_2), coupled to photodetectors, often photomultipliers or photodiodes. The detection process involves a two-stage energy conversion mechanism:

- The incident X-ray photon is absorbed by the scintillator crystal, which then emits a number of visible-light photons (scintillation phenomenon).
- These visible photons are detected by a photodiode (similar to those discussed in Chap. 3), which converts the light into an electrical signal proportional to the original X-ray intensity.

This indirect detection process enables the transformation of X-ray energy into measurable electrical signals.

The advantages of a solid-state detector are:

- High efficiency.
- Operation at room temperature.

Fig. 6.25 Commercial solid state X-ray detector arrays (linear and 2D) and X-rays sources (Hamamatsu)

- Excellent sensitivity, with nearly 100% photon capture efficiency.

Recent developments have led to solid-state X-ray detector arrays integrated with CMOS technology. These detectors utilize multi-layer structures to enhance the interaction between X-rays and the scintillator material, resulting in increased energy resolution and spatial resolution. Such arrays are illustrated in Fig. 6.25, and are characterized by parameters such as physical dimensions, pixel size, and pixel count.

Gas-filled detectors, particularly ionization chambers using high-pressure xenon gas, represent another category of X-ray detection. Within these chambers, a strong electric field is applied. When an incident X-ray photon ionizes a xenon atom, it ejects an electron, producing a positive ion. The electron and ion are accelerated in opposite directions by the electric field and collected at the respective electrodes (anode and cathode), generating a current signal proportional to the photon intensity.

The advantages of these Xenon ionization chambers are:

- Uniform and linear response
- High temporal stability
- Compact size
- Lower production costs.

The disadvantage however is relatively low photon capture efficiency (60–93%).

The solid-state solution makes it possible to create X-ray CMOS sensors in which there are multiple layers to create interactions between X-rays and photons with ever increasing energy.

By using a single X-ray source and a 2D X-Ray detector as in Fig. 6.26, it is possible to obtain images of radiation absorption. The X-ray image is nothing more than an attenuation image of the X radiation.

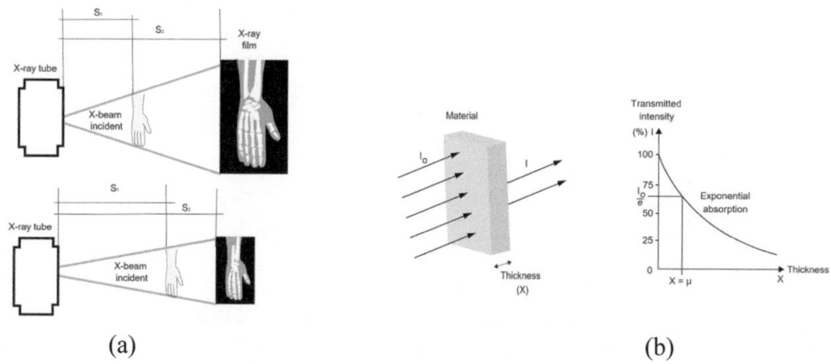

Fig. 6.26 X-Rays imaging (**a**) and attenuation parameters (**b**)

X-rays travel in straight line from the source to the imaging plane, so the only way to change the dimension of the image and/or its magnifying factor for a given aperture angle of the x-ray source, is to change the distance S_1 between the source and the object, as well as the distance S_2 between the source and the imaging plane, as illustrated in Fig. 6.26a. In this case the magnification M will be obviously:

$$M = \frac{S_2}{S_1} \tag{6.20}$$

The attenuation of an X-ray beam of initial intensity I_0, crossing a layer of length x of a material having an absorbing coefficient μ is given by [10]:

$$I(x) = I_0 e^{-x\mu} \tag{6.21}$$

The higher is the energy of the x-ray photon, the lower will be the attenuation coefficient of a given material, as illustrated in Fig. 6.27.

(3) *Mechanical scanning systems for multiple projections*

A full 3D scanning is achievable by mechanically moving an x-ray source and/or the detector around an object. This is the basis of the so called Computed Axial Tomography technique (CAT), initially developed for medical purposes and recently widely used in industrial applications, especially for measuring mechanical components and quality controls. The technique is known as Industrial Computed Tomography (ICT).

Tomographic measurements can also be performed using other types of waves beyond X-rays, such as light waves on transparent solid materials or fluids, radiofrequency microwaves, and acoustic or ultrasonic waves.

These alternative methods are viable when the object under investigation is partially transparent to the specific wave type. For example, light-based tomography can be used in semi-transparent media like water, glass, plexiglass, radiofrequency waves can penetrate certain solid materials, and acoustic or ultrasonic waves are

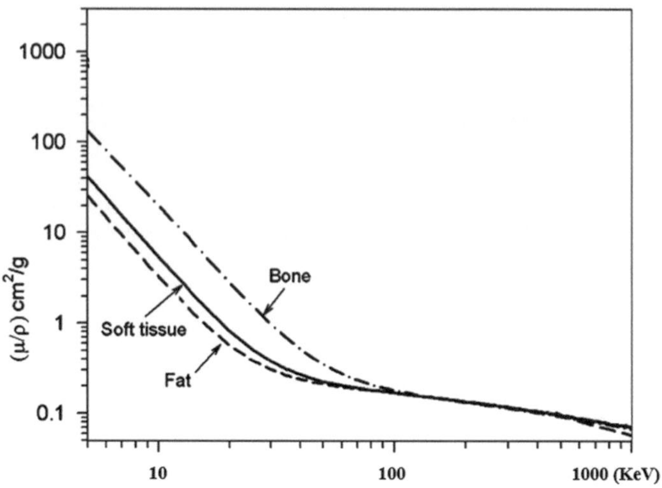

Fig. 6.27 Attenuation coefficients of different human tissue

widely applied in biological tissues and industrial components to analyze its internal structure.

The internal volume of the object being imaged is discretized into *voxels* (volumetric pixels), which represent the three-dimensional equivalent of 2D pixels, as illustrated in Fig. 6.28a. The final tomographic output is a three-dimensional voxel matrix *T(i, j, k)*, where each element corresponds to a discrete volume within the object.

This 3D matrix is typically reconstructed *slice by slice* through the processing of *projections*, data sets obtained by scanning the object with a narrow beam or a radiation sheet in a defined plane, as shown in Fig. 6.28b, c.

These projections are then mathematically combined using tomographic reconstruction algorithms to generate cross-sectional images, ultimately forming the full volumetric representation.

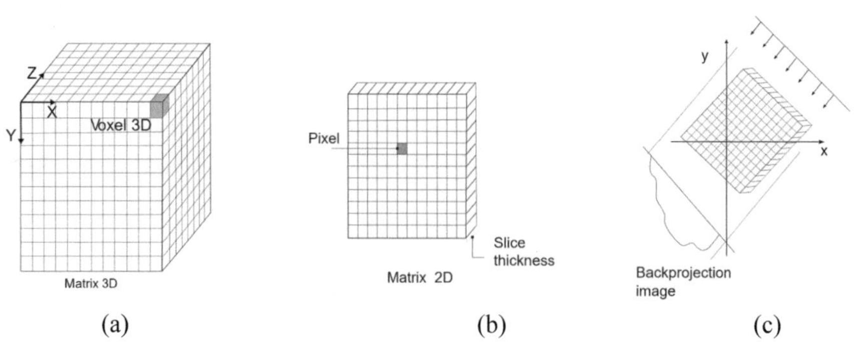

Fig. 6.28 Tomographic voxel model and projection

Given an object having lenght l with a non-uniform attenuation coefficient μ(x), the intensity of the attenuated ray can not be calculated using Eq. 6.21, but will be more in general given by:

$$I(x) = I_0 e^{-\int_0^l \mu(x)dx} \tag{6.22}$$

If we consider each voxel as a cube of dimension δ having a constant attenuation coefficient μ_i, then the attenuated ray that goes through the voxel will have the output intensity I_{ui}:

$$I_{ui} = I_0 e^{-\delta\mu_i} \tag{6.23}$$

So, crossing a series of N voxels, the total intensity measured by the detector will be:

$$I_u = I_0 \sum_{i=0}^{N} e^{-\delta\mu_i} \tag{6.24}$$

With just one measure of the output intensity I_u of a line of crossed voxel, even by knowing the initial intensity of the ray source I_0, it is impossible to retrieve information on the N unknown μ_i that constitute the attenuation profile, hence allowing the identification of the different material layers crossed by the ray. This issue is overcome by analyzing multiple images, scanning the object from different directions, as illustrated in Fig. 6.29 for a simple 4 pixel slice.

This simple slice of 4 voxel is crossed by X-ray along two directions. From Eq. 6.24 applied to all the 4 line of crossing rays, the following four equations can be written:

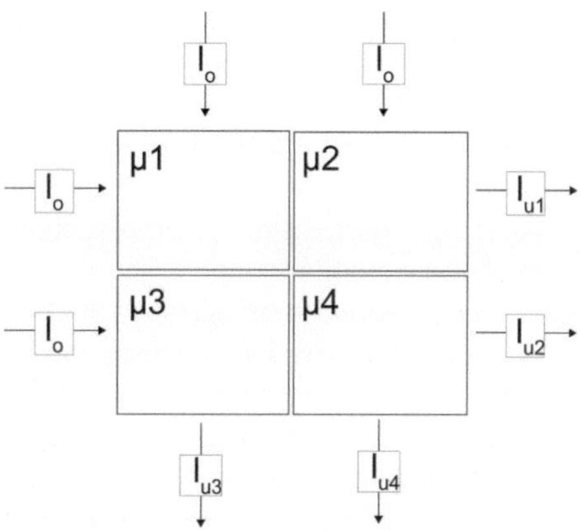

Fig. 6.29 Tomograpy basic principle

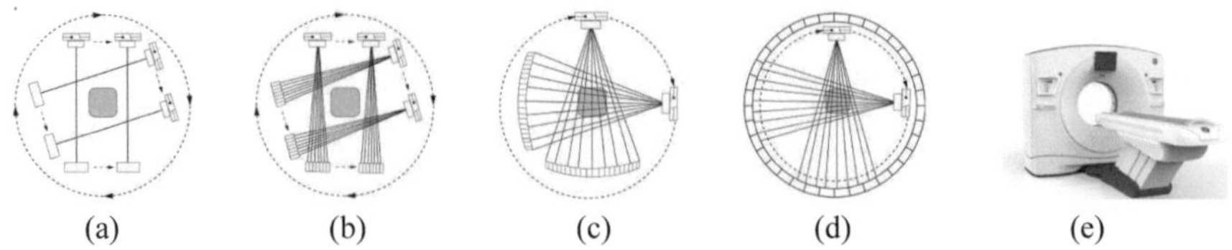

Fig. 6.30 Evolution of tomographic systems in medicine

$$\begin{cases} I_0e^{-\delta\mu_1} + I_0e^{-\delta\mu_2} = I_{u1} \\ I_0e^{-\delta\mu_3} + I_0e^{-\delta\mu_4} = I_{u2} \\ I_0e^{-\delta\mu_1} + I_0e^{-\delta\mu_3} = I_{u3} \\ I_0e^{-\delta\mu_2} + I_0e^{-\delta\mu_4} = I_{u4} \end{cases} \tag{6.25}$$

So, a system of four equations with four unknown that can be linearized by taking the logarithm of all the terms; giving a simple linear system. Hence, the unknowns $\mu_1, \mu_2, \mu_3, \mu_4$ can be obtained from the measurements of the attenuated intensities $I_{u1}, I_{u2}, I_{u3}, I_{u4}$.

Making a better discretization of the slice with more than 4 voxel, measuring more values of attenuated profiles of intensity and solving bigger system of more equations, makes it possible to get higher resolution information on intensity attenuation coefficients of a slice of the scanned object. Making the same for all the slice it is possible to obtain a complete 3D voxel cubic model. This is the basic principle of tomography.

Scan geometries used in medical tomographic systems have evolved significantly over time to improve acquisition speed and image quality. Four generations of systems can be described in tomography evolution, illustrated in Fig. 6.30.

- *First-generation systems* (Fig. 6.30a) featured a single X-ray source and a single detector. The system operated by translating the source-detector pair linearly across the object, followed by incremental angular rotations. Due to the sequential nature of data acquisition, scan times were relatively long.
- *Second-generation systems* (Fig. 6.30b) introduced a linear array of multiple detectors (n detectors) positioned opposite the X-ray source. Although the angular position of the system remained fixed during each acquisition step, all projections could be acquired simultaneously, significantly reducing scan time compared to the first-generation setup.
- *Third-generation systems* (Fig. 6.30c) improved upon this design by arranging the detectors in a curved arc opposite the X-ray source. Both the source and detector array rotate synchronously around the patient, enabling faster and more efficient data collection, while maintaining high image quality.
- *Fourth-generation systems* (Fig. 6.30d) employed a fixed ring of detectors surrounding the object, with only the X-ray source rotating around the central axis. This configuration eliminates the need for moving the detectors, further enhancing mechanical stability and acquisition speed.

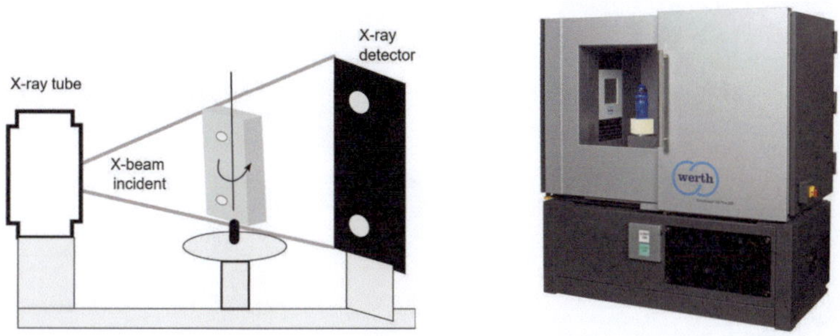

(a) (b)

Fig. 6.31 Industrial computed tomographic systems

These developments in scan geometries have been critical in enabling faster imaging, reducing patient exposure, and improving spatial resolution in clinical tomography.

Today they represent different possible configurations with advantages/disadvantages for different applications also in non-medical fields.

In industrial computer tomography (ICT) applications to solid and rigid objects it is possible to rotate the object, instead X-ray source and sensor array, as illustrated in Fig. 6.31a. The X-ray source and sensor array are therefore fixed, simplifying a lot the mechanics of the system and improving accuracy. A commercial systems of this kind is illustrated in Fig. 6.31b.

Performances of the most recent industrial tomographic systems are for example that of TomoScope XS Plus from Wert ^{R} illustrated in Fig. 6.31c allowing measurements on object dimensions L up to 530x725 mm with a length measurement uncertainy of 4.5 + L/75 μm, according to ISO 10360 and VDI/VDE 2617/13 standards.

For industrial quality control applications also 2D X-ray systems with linear arrays are used as illustrated in Fig. 6.32. These systems have typical field of view up to 1024 × 1280 mm (1024×1280 px @ 1.0 mm sampling) with 12-bit (4096-level) gray-scale

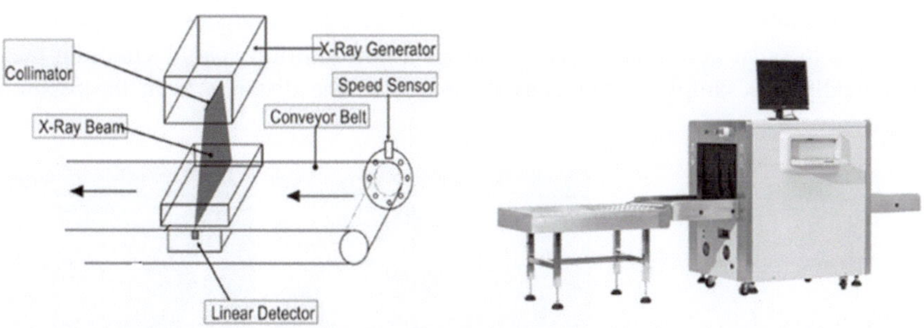

Fig. 6.32 In line 2D X-ray scanning systems and security applications for airport baggage scannin. A typical system with 1 mm spatial resolution, anode voltage 140 KV, image digital resolution of 1024 × 1280 pixel; 4096 grey levels

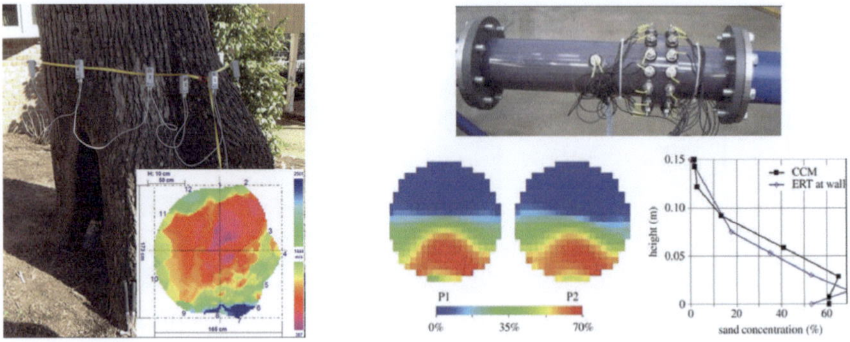

Fig. 6.33 Acoustic and impedance tomography for three and flows

imaging depth;140 kV sealed-oil X-ray source (100% duty, upward beam) enabling ~39 mm steel penetration; typical measurement uncertainty ~1 mm.

Tomographic measurement techniques are also used with other kind of rays, like light, microwave, ultrasound etc. [11]. Typical systems and applications are illustrated in Fig. 6.33.

A new field of application for high performances ICT systems are the analysis of mechanical components tolerances, defects and comparing with CAD model geometry or reverse engineering, i.e. realizing CAD models from ICT measurements. A typical example of quality control of reed relay sensors internal structure by X-ray and tomography is given in Fig. 6.34

This kind of technique turns out to be useful also for the quality assessment of objects obtained from 3D printing processes. In Fig. 6.35 an example of tomographic analysis of slices of lattice-structure samples, realized by additive manufacturing.

(4) *Data processing algorithms*

The main issue with tomographic reconstruction using multiple projections is the mathematical reconstruction process itself. Simply combining projections taken from different angles, as previously explained, can lead to a blurring effect, as shown in Fig. 6.36.

This occurs because acquiring projections at various angles corresponds to sampling in polar coordinates, which causes a non-uniform spatial sampling rate. Specifically, the sampling frequency decreases as the distance from the rotation

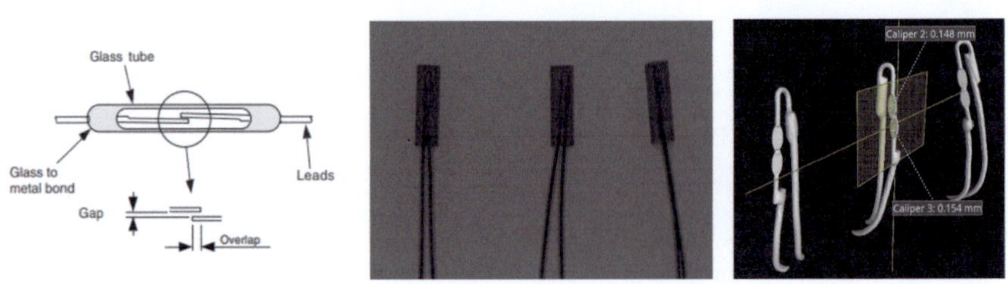

Fig. 6.34 Radiography and tomograpy of reed relay switch sensors

Fig. 6.35 Tomography of trabecular structure specimen realized by additive manufacturing

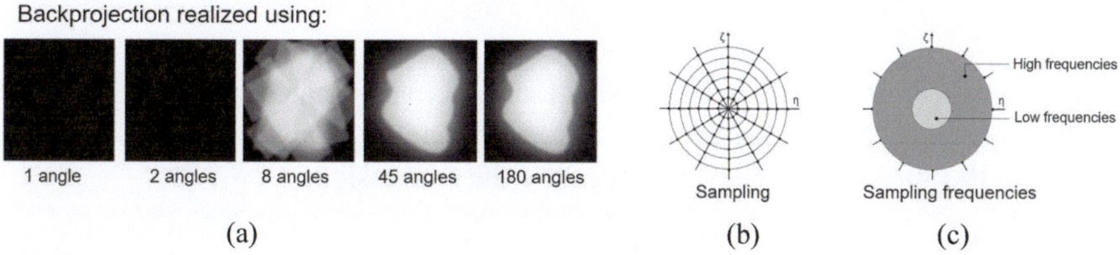

Fig. 6.36 Blurring effect given by projections along different orientation

center increases, leading to reduced resolution at the periphery of the object, as illustrated in Fig. 6.36c. This is analogous to the loss of detail seen in digital images sampled at low spatial frequencies, as discussed in Chap. 3.

To mitigate this issue, a widely used reconstruction method is the Filtered Backprojection (FBP), commonly applied in computed tomography (CT) and other imaging techniques based on beam transmission through materials [10]. FBP consists of two main steps: filtering and backprojection, which together reconstruct a cross-sectional image from multiple projections taken at different angles θ.

A projection profile p at each angle θ can be described by the function $p(s, \theta)$, representing the integral of attenuation along rays crossing the object with a projection angle θ. This function can be visualized either as a set of 1D projections for different θ values or as a 2D image called a sinogram, shown in Fig. 6.37b for a simple two-hole plate case.

Mathematically, this projection, no matter the nature of the wave used, is described by the Radon transform [12]:

$$p(s, \theta) = \int_{-\infty}^{\infty} I(x, y)\delta(x\cos\theta + y\sin\theta - s)dxdy \tag{6.26}$$

The goal of tomographic reconstruction is to invert this transform to retrieve slice image of attenuation $I(x, y)$ from many of its projections $p(s, \theta)$ measured at different angles θ.

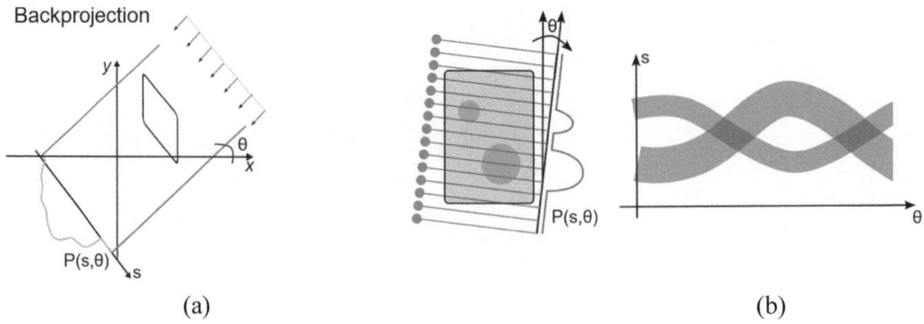

Fig. 6.37 The Radon projection transformation (**a**) and a sinogram (**b**)

To avoid the blurring effect previously described, each projection $p(s, \theta)$ is filtered to compensate for the loss of high-frequency components. A ramp filter $F(\omega)$ is used, defined in the spatial frequency domain ω by the expression:

$$F(\omega) = |\omega| \tag{6.27}$$

This means that each frequency component ω of a projection is multiplied by its absolute value, enhancing high frequencies to counteract the blurring introduced by the backprojection process. In the spatial domain, the ramp filter corresponds to the inverse Fourier transform of $|\omega|$. Anyway this filter can cause amplification of noise at high frequencies, so in practical applications it's often combined with a window function (e.g., Hamming, Hann, Shepp-Logan) to limit the frequency range, resulting in a modified ramp filter, typically by the model:

$$F(\omega) = |\omega| \cdot W(\omega)F(\omega) \tag{6.28}$$

where *W(ω)* is the chosen window function that smooths the filter and reduces noise. Fig. 38 a) graphically illustrates the effects of the ramp filter.

So taking the Fourier transform $P(\omega, \theta)$ of $p(s, \theta)$ as defined in Chap. 3 by Eq. (3.32) and multiplying by this ramp filter (in general modified using a window function like Hamming, Hann or others) we get:

$$P_f(\omega, q) = P(\omega, q)|\omega| \tag{6.29}$$

So, making the inverse Fourier transform of $P_f(\omega, \theta)$ by the Eq. (3.33) we obtain the filtered projection $P_f(s, \theta)$. Same result can be obtained using the convolution theorem (Eqs. 3.37 and 3.38), so performing a convolution in the spatial domain expressed by:

$$P_f(s, \theta) = P(s, \theta) * f(s) \tag{6.30}$$

where f(s) is the inverse Fourier transform of the ramp filter F(ω) = |ω|.

Using this filtered projection it is possible to make the backpropagation to obtain the 2D slice $f(x, y)$ by calculating the integral [12]:

$$\mathrm{f(x, y)} = \int_0^{\pi} \mathrm{P}_f(\mathrm{s}, \theta)\mathrm{d}\theta \tag{6.31}$$

where $P_f(s, \theta)$ is the filtered projection data and s coordinate is given by:

$$s = x\cos(\theta) + y\sin(\theta) \tag{6.32}$$

The effect of filtering is illustrated in Fig. 6.38b.

Filtered Backprojection (FBP) is implemented in several software packages, including Python's scikit-image library (with radon and iradon functions), MATLAB through the ASTRA tomography toolbox, and many other platforms. FBP is a fast and computationally efficient direct reconstruction method that does not require iterative approximations. It provides good image quality when a sufficient number of projections is available and performs well in applications like medical CT, where the signal-to-noise ratio is high. However, its main limitation is a high sensitivity to noise, since the ramp filter used in the frequency domain amplifies high-frequency components, contrary to typical image enhancement methods (like low-pass filtering) described in Chap. 3. For optimal results, FBP requires a large number of projections (Fig. 6.38).

Although FBP is widely used, alternative reconstruction techniques include:

- Iterative Reconstruction (IR)—More robust to noise, but computationally intensive.
- Algebraic Reconstruction Techniques (ART, SIRT)—Use optimization methods to solve the inverse problem.
- Deep Learning-Based Reconstruction—Leverage neural networks to improve image quality and reduce artifacts.

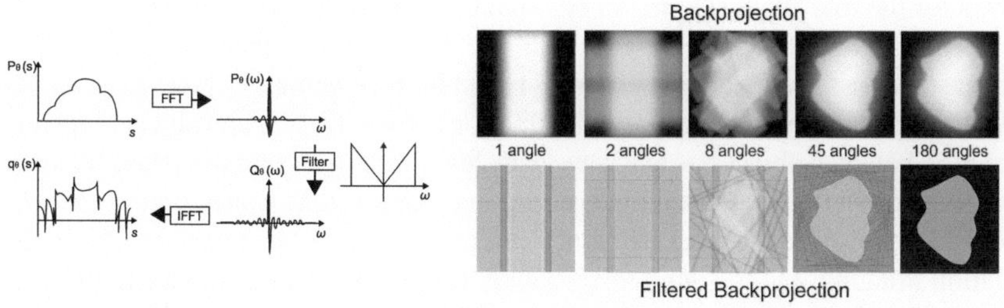

Fig. 6.38 Filtering benefits in filtered backpropagation

6.6 Measure By Scanning Acoustic Wave Beams

Acoustic wave beam scanning technology—originally developed for military use under the name SONAR (Sound Navigation and Ranging), analogous to RADAR, employs sound waves to detect and locate objects underwater. This technology has evolved to play critical roles in navigation, defense, oceanography, medicine (e.g., echography), and industrial non destructive control.

The origins of underwater acoustics can be traced back to 1826, when Jean-Daniel Colladon and Charles Sturm measured the speed of sound in Lake Geneva. A key scientific milestone enabling modern ultrasonic transducers was the discovery of the piezoelectric effect by Pierre and Jacques Curie in 1880. The tragic sinking of the Titanic in 1912 heightened awareness of the need for iceberg detection and hull inspection, accelerating interest in ultrasonic applications. By World War I (1915–1917), physicist Paul Langevin and engineer Constantin Chilowski developed the first practical sonar using quartz-based piezoelectric transducers, later known in Britain as ASDIC, to detect submarines. During World War II, sonar technology advanced with the development of both active (pulse emission) and passive (listening) systems. The Cold War further accelerated progress in underwater surveillance, with the U.S. deploying the SOSUS system for monitoring Soviet submarines. In 1928, Sergei Sokolov proposed using ultrasound to detect material defects, laying the groundwork for nondestructive testing (NDT). This field advanced in the 1940s when Floyd Firestone patented the "Supersonic Reflectoscope" to detect internal flaws in metals, establishing the basis of industrial ultrasonic testing. By the 1950s, ultrasonic inspection became widespread in aerospace and construction industries for weld and material integrity assessments.

Medical applications of ultrasound began in the 1940s. Karl Dussik attempted brain imaging via ultrasound in 1942. In 1949, George Ludwig used ultrasound to detect gallstones in animal tissues. In 1958, Ian Donald and Tom Brown developed the first practical fetal ultrasound scanner, revolutionizing prenatal care. Inge Edler and Carl Hellmuth Hertz pioneered M-mode echocardiography in 1954, producing the first heart images. Advancements in the 1960s transitioned imaging from A-mode to B-mode, providing two-dimensional images. In the 1980s, phased array technology allowed for flexible beam steering in both medical and industrial applications. Real-time scanning and Doppler echography in the 1980s–90s further enabled detailed cardiovascular assessments by measuring blood flow velocities.

As discussed in Chap. 1, the type of interaction between waves and objects depends on the ratio between object size and wave wavelength. To detect small surface features via reflection, shorter wavelengths are required. Given that wavelength $\lambda = v/f$ (with v as wave speed and f as frequency), high frequencies are necessary for high-resolution imaging. In air, sound travels at ~330 m/s, resulting in wavelengths from centimeters to meters at frequencies between 20 Hz and 20 kHz, suitable only for detecting large objects.

For higher resolution, ultrasound uses frequencies typically from 1 to 15 MHz. In solids like steel ($v \approx 5900$ m/s) or aluminum ($v \approx 6300$ m/s), this results in wavelengths from about 6 mm to 0.6 mm, enabling detection of small defects.

The propagation of ultrasonic waves at material interfaces is governed by acoustic impedance Z, defined as $Z = \rho v$, where ρ is the material density and v is the wave speed. The proportion of reflected (P_r) and transmitted (P_t) wave amplitudes at an interface between two materials with impedances Z_1 and Z_2 is given by:

$$\frac{P_r}{P_i} = \frac{Z_2 - Z_1}{Z_2 + Z_1} \quad \text{and} \quad \frac{P_t}{P_i} = \frac{2Z_2}{Z_2 + Z_1} \tag{6.33}$$

Significant back-reflection occurs only when the difference between the acoustic impedances of the two materials is large. For instance, a wave moving from air (very low impedance of $Z = 0.0004 \times 10^6$ kg/(m^2·s)) to steel (very high impedance of $Z = 45 \times 10^6$ kg/(m^2·s)) will experience nearly total reflection. This principle is critical in both industrial non-destructive testing and medical imaging applications.

The table below (Table 6.6) shows typical values of density and acoustic impedance for common materials encountered in ultrasonic measurement applications.

Intensity I of an ultrasonic beam is related to propagating medium impedance Z and density ρ; wave pressure amplitude P and velocity v by the relation [10]:

$$I = \frac{P^2}{2Z} = \frac{P^2}{2\rho v} \tag{6.34}$$

Reflection coefficient R, the ratio between reflected power (or intensity) I_r and the incident one I_i [10]:

Table 6.6 Density and acoustic impedance of common materials

Medium	Density $\left(\frac{kg}{m^3}\right)$	Acoustic impedance $\left(\frac{kg}{m^2 \cdot s}\right) \cdot 10^6$
Air	1.02	0.0004
Water	1000	1.480
Blood	1060	1.660
Bone	1620	7.800
Brain tissue	1030	1.550
Adipose tissue	920	1.380
Renal parenchyma	1040	1.620
Liver parenchyma	1060	1.640
Muscular tissue	1070	1.700
PZT (Lead Zirconate Titanate)	7650	29.000
Steel	7850	45.000
Aluminum	2700	17.300

$$R = \frac{I_r}{I_i} = \left(\frac{Z_2 - Z_1}{Z_2 + Z_1}\right)^2 \tag{6.35}$$

On the other hand, the transmission coefficient T, the ratio between the transmitted power (or intensity) I_t and the incident power I_i is also related to impedances values of the two materials by the relation [10]:

$$T = \frac{I_t}{I_i} = \frac{4Z_2Z_1}{(Z_2 + Z_1)^2} \tag{6.36}$$

For ultrasonic waves propagating through a continuous and homogeneous material, an exponential attenuation take places, as for the x-ray. Hence, if propagation start at point $x = 0$ with an initial amplitude P_0, after a pathlength x the amplitude will be [12]:

$$P(x) = P_0 e^{-\alpha x} \tag{6.37}$$

where α is the attenuation coefficient of ultasonic waves of that material. Same attenuation takes place for the intensity of the wave [12]:

$$I(x) = I_0 e^{-\mu x} \tag{6.38}$$

where μ is the attenuation coefficient of ultrasonic wave intensity, and from the above relation between intensity and pressure of the ultrasonic wave is easy to see that:

$$\mu = 2\alpha \tag{6.39}$$

Attenuation of wave intensity within materials imposes important limitations on measurement techniques based on beam scanning and the detection of reflections. In materials with high attenuation coefficients, wave intensity diminishes significantly with distance. Consequently, long propagation paths can result in reflected signals that fall below the noise threshold of the detector, rendering them undetectable.

The directions of incident, reflected, and refracted ultrasonic beams can be determined using Snell's Law (previously discussed in Chap. 1 for light waves), which also applies to mechanical waves. Specifically, the angle of incidence equals the angle of reflection with respect to the normal of the surface, and the angle of refraction depends on the ratio of propagation velocities in the two media, as described by Eq. (1.74).

Ultrasound waves are typically generated using piezoelectric materials [13], which are some natural crystals or can be synthetically manufactured. These materials produce electric charges when mechanically deformed, a phenomenon known as the direct piezoelectric effect. Conversely, when an electric field is applied, these materials undergo mechanical deformation—known as the inverse piezoelectric effect. Both effects are exploited in various technologies.

The piezoelectric principle is fundamental for many sensors used to measure mechanical quantities such as strain, length, vibration, ultrasound, acceleration,

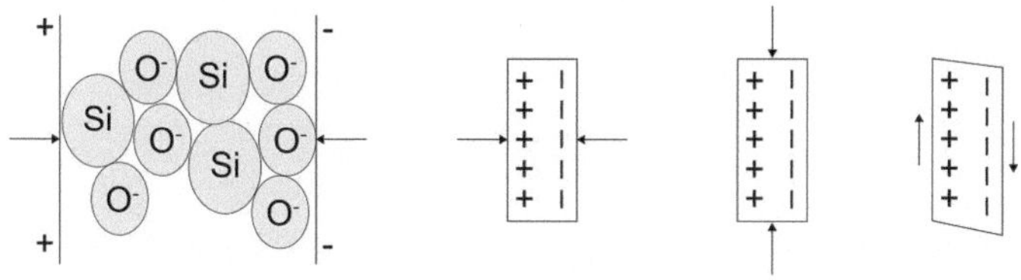

Fig. 6.39 Piezoelectric effects on a quartz crystal

force, and pressure. It also enables the generation of mechanical motion and ultrasound for applications in medical imaging, industrial inspection, and sensing technologies.

Common naturally occurring piezoelectric crystals include quartz, rock salt, and Rochelle salts. Among synthetic materials, lithium sulfate, ammonium dihydrogen phosphate, and especially ferroelectric ceramics like barium titanate and lead zirconate titanate (PZT) are widely used. In addition, polymeric materials such as polyvinylidene fluoride (PVDF) exhibit piezoelectric properties and are available as thin, flexible plastic films.

The piezoelectric effect can be activated through different types of mechanical deformation, including axial stretching, transverse deformation, and shear strain, as illustrated in Fig. 6.39, which refers to the behavior and phenomena of a quartz crystal under mechanical loads.

The piezoelectric sensitive element can be viewed as a capacitor, formed by two metallic plates enclosing the piezoelectric crystal, which also serves as the dielectric of the capacitor, as illustrated in Fig. 6.39. On the two faces where charges are generated, two thin metallic layers are deposited to collect them. The behaviour of the piezoelectric crystal can generally be assumed to be linear, meaning that the charge produced by the piezoelectric material varies linearly with the displacement x_i (the axial deformation of the crystal), and vice-versa, according to the following linear relations:

$$Q = K_q x_i \text{ and } x_i = \frac{1}{K_q} Q \tag{6.40}$$

where K_q is the piezoelectric coefficient. Because of the mechanical linearity of the crystal, the force generated is expressed as $F = Kx_i$, where K is the mechanical stiffness of the piezoelectric crystal. The piezoelectric transducer can be represented as a circuit consisting of a current generator, a resistor R_a (equal to the internal resistance of the crystal, which is generally very high due to the insulating nature of the material), and a capacitor with capacitance C_a, connected in parallel as in Fig. 6.40.

For a capacitor, its capacitance C_a is equal to the ratio of the charge on the plates Q to the voltage e_0 between them:

$$Ca = \frac{Q}{e_0} \tag{6.41}$$

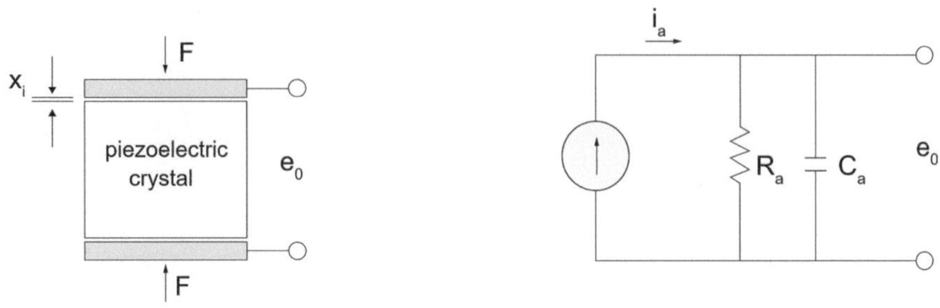

Fig. 6.40 Electromechanical model of a piezoelectric element

Therefore, the piezoelectric material generates a voltage e_0 equal to:

$$e_0 = \frac{K_q x_i}{C_a} \tag{6.42}$$

Conversely, if an electrical voltage signal $e_0(t)$ is applied, a mechanical vibration (an ultrasound if at high frequency) of the crystal is generated, given by the following equation:

$$x_i(t) = \frac{C_a}{K_q} e_0(t) \tag{6.43}$$

Or receive a signal:

$$e_0(t) = \frac{K_q}{C_a} x_i(t) \tag{6.44}$$

The sensing element can be composed of a single generator and/or receiver of ultrasonic waves as in Fig. 6.41a, or can it be built with multiple piezoelectric crystals arranged in arrays, to generate or receive complex beam with many rays of ultrasonic waves like that illustrated in Fig. 6.41b.

Here the scanning and focusing of ultrasonic beam is possible by analysing the phase shift between emitting elements of 1D or 2D arrays, as explained by the Huygens principle of Chap. 1.

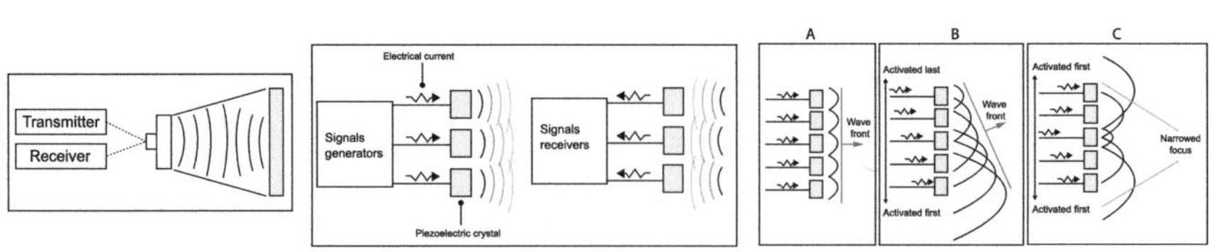

Fig. 6.41 Single and arrays ultrasonic transducers

Different operational modes are possible to detect defects, inclusions or voids inside materials:

(1) *Reflection technique (Pulse-eco)*: a single transducer is used both as emitter and receiver of waves. It bases its operation on the reflection property of the beam generated by any discontinuity surfaces. If there are no defects, the beam is reflected from the back wall (background echo). If there is a defect, an echo is delayed whose amplitude depends on the amount of energy that the transducer receives due to reflection. Reflected energy is a function of the size, inclination, and distance of the source.

(2) *Transmission technique*: Two transducers are used, one for transmitting and one for receiving the waves. They are divided into:

- *Transmission*: The measurand is interposed between the two transducers aligning the axes and the beam emitted with the receiver. If the part in analysis has defects, the receiving transducer will receive a lower energy than that emitted and will produce an electrical signal proportional to it. This technique does not allow to know depth and shape of the defect (Fig. 6.42a).
- *Reflection*: Two transducers, transmitter and receiver, are placed on the same surface of the workpiece and coupled in such a way that their beams converge at a point. This allows even portions of material very close to the surface to be analysed (Fig. 6.42b).

At the interface between air and a solid object, ultrasonic probes are subject to reflection and refraction phenomena due to mismatches in acoustic impedance. To minimize these effects, suitable acoustic coupling materials, typically gels, are applied between the probe and the target surface. These coupling agents are chosen to closely match the acoustic impedance of the transducer and the target material, thereby reducing unwanted reflections, improving signal transmission and signal to noise ratio.

A common approach in non-destructive testing is the immersion method, in which the ultrasound travels through a water medium before reaching the test specimen.

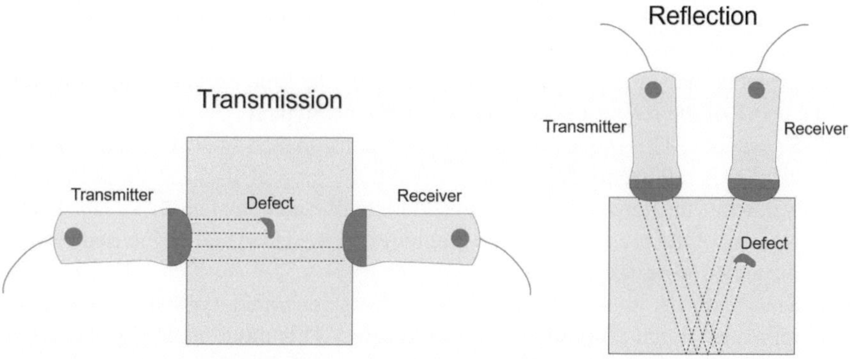

Fig. 6.42 Transmission and reflection operational modes

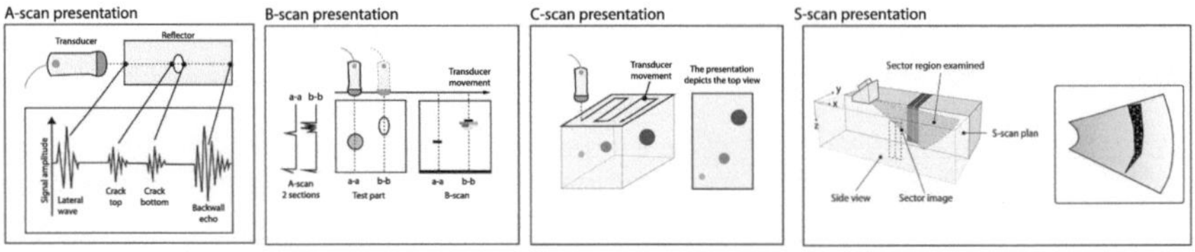

Fig. 6.43 Ultrasonic scanning techniques

This method provides a uniform acoustic coupling between the probe and the object, resulting in improved signal clarity and measurement consistency.

When analyzing the received signal, the first observed peak corresponds to the emission pulse of the transducer. A secondary, lower-amplitude peak results from partial reflection at the entry surface of the test object. The ultrasonic beam then continues through the material, producing additional reflections at any internal discontinuities, until it reaches the far surface, where the final echo is generated, as illustrated in Fig. 6.43a.

To detect time delays (Δt), the same principles used in LIDAR and RADAR systems, as introduced in Chap. 1, are employed:

1. Direct Time-of-Flight (dTOF): This technique measures the time Δt required for an ultrasonic pulse to travel from the emitting transducer to the target and back. For instance, in water where v = 1450 m/, a delay of $\Delta t = 13$ μs, from Eq. 6.1, corresponds to a distance R of 1 cm.
2. Indirect Time-of-Flight (iTOF): This method involves detecting the phase shift φ of a modulated ultrasonic signal of frequency f. The signal travels at velocity v along a round-trip path of length $2R$, and the distance is derived from Eq. 6.2.

As with RADAR-based systems, various imaging techniques can be used to reconstruct echo profiles and internal structures from the received ultrasonic signals.

Typical ultrasonic scanning modes are illustrated in Fig. 6.43.

Ultrasonic imaging employs several scan modes, each suited to different inspection needs. The primary scanning modes are described as follows:

- A-Scan Mode: This is a one-dimensional representation where the ultrasonic signal is visualized along a single line. Discontinuities within the material appear as peaks in the signal, and their positions along the time axis are proportional to the depth of the reflecting surfaces within the beam path.
- B-Scan Mode: This mode provides a two-dimensional cross-sectional image. The vertical axis represents depth (derived from the time-of-flight of the ultrasound pulse), while the horizontal axis corresponds to the scanning length, i.e., the lateral movement of the probe. This mode is commonly used to visualize internal profiles of the tested material.
- C-Scan Mode: A two-dimensional plan view is generated, typically color-coded to represent signal amplitude or other features. This mode is used primarily in automated or semi-automated ultrasonic inspections. The horizontal axis shows the probe movement (tracked by an encoder), while the vertical axis, known as the index axis, represents the orthogonal scanning direction.

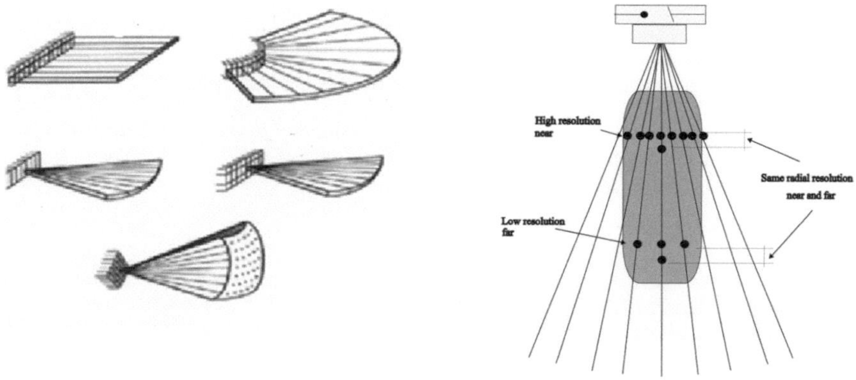

(a) (b)

Fig. 6.44 Axial and lateral resolution of complex beam ultrasonic probes

- S-Scan Mode (Sector Scan): This mode displays a sectorial cross-section of the examined component. It is constructed from a series of A-scans acquired at varying angles using a sector-scanning transducer. Each A-scan provides intensity information over time, which can be mapped to pixel positions in an image array. This allows for visualizing echo positions relative to the probe angle and depth, offering an angular view of internal features.

In systems employing complex one-dimensional or two-dimensional beamforming probes, such as those used in medical echography, the axial resolution is determined by the accuracy in measuring time delays between emitted and reflected ultrasonic signals. This resolution remains constant along the beam axis. However, the lateral resolution, which depends on the beam divergence, reduces with increasing distance from the transducer, as illustrated in Fig. 6.44b.

The wave returning from a reflection, if the interface between two acoustic impedances is in motion, undergoes the Doppler effect, as explained in Cap 1. A wave reflected by a moving structure has a lower or higher frequency than that of the incident wave depending on whether the structure approaches or moves away from the probe.

The frequency variation is linearly dependent on the speed of the material reflecting it, as illustrated in Chap. 1. So, we have a frequency shift given by:

$$\Delta f = f1 - f0 \approx \frac{2v_e f_0}{c} \cos(\Phi) \tag{6.45}$$

where Δf is called doppler shift, f_0 is the frequency of the incident beam, f_1 is the frequency of the reflected beam, c the propagation speed of waves in the medium, v_e is the velocity of the target along the wave direction and Φ is the angle between the velocity vector and the wave propagation direction (Fig. 6.45).

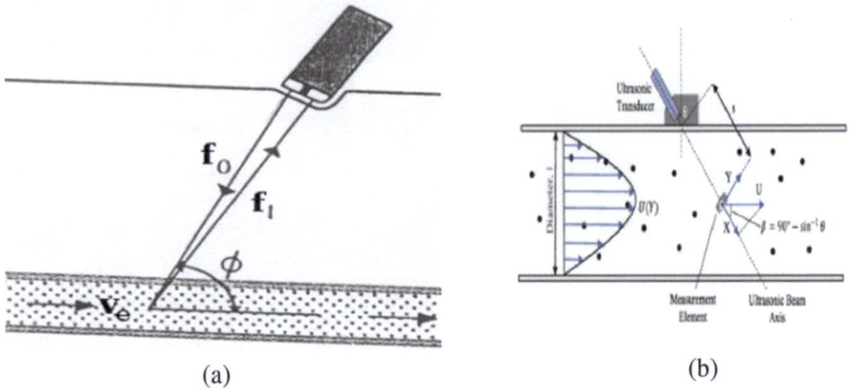

Fig. 6.45 Echo Doppler ultrasonic measurement systems and a velocity profile measurement from MET-FLOW SA company

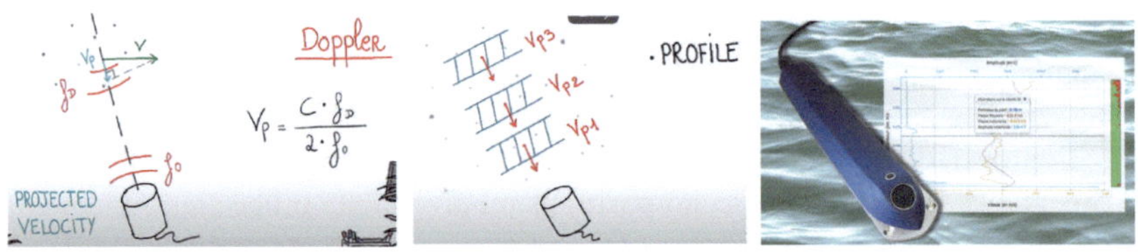

Fig. 6.46 Ultrasonic velocity profiler (UVP) measurement systems from UBERTONE company

Ultrasonic velocity profile measurements were first employed for blood detection and afterwards was used in other industrial applications, involving the quantification of fluid dynamic fields [14]. The measurement principle and the commercially available systems are illustrated in Fig. 6.46.

The angle Φ between the ultrasound beam and the direction of motion plays a critical role in Doppler measurements. The factor of 2 in the Doppler shift equation arises because the Doppler effect occurs twice in echo-based measurements: first, the moving erythrocytes experience a frequency shift due to their motion relative to the transmitted wave; then, the reflected signal undergoes a second shift when received by the stationary probe.

Two primary Doppler ultrasound techniques are commonly employed:

- *Continuous Wave Doppler (CW-D)*: CW Doppler utilizes two separate piezoelectric elements mounted on the same transducer—one dedicated to continuous transmission and the other to continuous reception. This method allows the detection of very high velocities, as it is not limited by pulse timing constraints. However, CW Doppler lacks depth selectivity, meaning it cannot distinguish between signals originating from different depths along the beam axis.
- *Pulsed Wave Doppler (PW-D)*: PW Doppler employs a single piezoelectric crystal that alternates between transmission and reception modes. This configuration

enables depth resolution along the ultrasound beam, as echoes can be time-gated to correspond to specific depths. Signals originating from shallower vessels arrive sooner than those from deeper structures, allowing for selective analysis of Doppler signals based on their arrival time.

To integrate functional Doppler information from PW-D with structural anatomical data from standard B-mode imaging, two main approaches are used. The first, known as Duplex Scanning or Echo-Doppler, overlays the sample volume used for PW-D onto the ultrasound image, providing spatial context for the flow measurements.

While CW-D lacks depth resolution, PW-D introduces its own limitations—particularly in measuring high flow velocities. According to the Doppler shift relationship, the measurable velocity depends on the transmitted frequency f. However, this frequency is bounded by the Pulse Repetition Frequency (PRF): a new pulse cannot be emitted until the echo from the previous pulse has been received. This constraint becomes more severe as imaging depth increases, reducing the maximum allowable PRF.

$$PRF_{max} = \frac{c}{2R} \tag{6.46}$$

With c being the propagation velocity of the wave in the medium and R the distance of the target. Based on Nyquist's Theorem, the highest Doppler frequency that can be unambiguously measured is half the PRF.

$$\Delta f_{max} = \frac{PRF_{max}}{2} = \frac{c}{4R} \tag{6.47}$$

Substituting this into the Doppler shift formula sets the upper limit for measurable velocity.

$$v_{max} = \frac{c}{8 \cdot R \cdot f_0 \cdot \cos(\Phi)} \tag{6.48}$$

To maximize the measurable velocity range and minimize aliasing artifacts, several adjustments can be made:

- Increase the PRF (within system constraints),
- Reduce the imaging depth (if clinically feasible),
- Use a lower transmitted frequency (accepting a trade-off in spatial resolution and echo strength),
- Position the transducer to maintain an optimal Doppler angle Φ, avoiding excessively small angles.

If these conditions are not met, aliasing occurs, causing ambiguity in velocity interpretation and potential diagnostic errors.

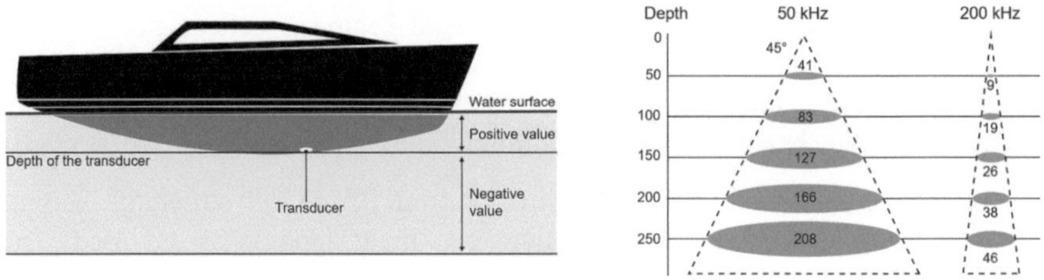

Fig. 6.47 Typical ultrasonic applications systems for sea depth measurement and fishing

Beyond the pulse-echo and Doppler techniques commonly used in ultrasonic diagnostics, SONAR systems represent a broader class of acoustic technologies optimized for long-range sensing and imaging in fluid environments, particularly underwater. SONAR (Sound Navigation and Ranging), illustrated in Fig. 6.47, operates on similar physical principles, wave propagation, reflection, and Doppler shift, but is adapted to overcome the specific challenges posed by large-scale aquatic domains. Active SONAR systems emit controlled acoustic pulses and analyze the returning echoes to determine target range, direction, and relative velocity, leveraging beamforming and time-delay processing for spatial resolution. In contrast, passive SONAR systems detect and analyze the acoustic emissions naturally produced by underwater sources, offering stealth capabilities in scenarios where signal transmission is undesirable. SONAR frequencies are generally lower than those used in medical or industrial ultrasonics, balancing resolution with attenuation constraints inherent in water. Modern SONAR architectures, such as Synthetic Aperture SONAR (SAS), exploit coherent motion compensation and aperture synthesis to generate high-resolution, wide-area imagery, even under conditions of significant turbidity or signal scattering. The effectiveness of SONAR systems depends not only on hardware design, transducer arrays, aperture geometry, and power handling, but also on sophisticated signal processing, including matched filtering, adaptive beamforming, and clutter suppression. Environmental variables such as salinity, temperature gradients, pressure, and seabed topography introduce nonlinear propagation effects that must be modeled to maintain localization accuracy. In metrological terms, the achievable range resolution is fundamentally limited by the signal bandwidth and pulse duration, while angular resolution depends on aperture size and wavelength, often reaching submeter precision under optimal conditions. Increasingly, AI and machine learning methods are being integrated into SONAR platforms to enhance target detection, classification, and anomaly identification, especially in autonomous or remotely operated systems. Applications span from naval operations and underwater navigation to marine biology, offshore engineering, and geophysical surveying, where optical or electromagnetic methods are infeasible due to absorption and scattering.

References

1. R. J. Hocken, P. H. Pereira, *Coordinate Measuring Machines and Systems*, 2nd edn. (Taylor & Francis, 2012)
2. S. Pagano, M. Moretti, R. Marsili, A. Ricci, Gi. Barraco, S. Cianetti, Evaluation of the accuracy of four digital methods by linear and volumetric analysis of dental impressions. Materials **2019**, 12, (1958). https://doi.org/10.3390/ma12121958
3. S. Kingsley, S. Quegan, *Understanding Radar Systems* (Scitech Publishing, Inc., 1999). ISBN 978-1891121050
4. F.E. Nathanson, *Radar Design Principles* (McGraw-Hill Book Company 1969), p. 7. ISBN: 978-1891121098
5. S.A. Hovanessian, *Radar System Design and Analysis* (Artech House 1984). ISBN: 0-89006-147-5
6. R.G. Torroméa, S. Barzanjeh Advances in quantum radar and quantum LiDAR, in *Progress in Quantum Electronics* (Elsevier, 2024)
7. P.F. McManamon, *LiDAR Technologies and Systems* (SPIE Digital library, 2019)
8. F. Radicioni, G. Rossi, G. Tosi, R. Marsili, Non contact shape an dimension measurements by LIDAR techniques of one of the biggest Italian cavern. J. Phys. Conf. Ser. **1249** (2019). IOP Publishing
9. W.P. Megarry, B.A. Davenport, D.C. Comer, Emerging applications of LiDAR/airborne laser. Conserv Manag Archaeol Sites **18**(4), 393–410 (2017)
10. W.A. Kalender, *Computed Tomography: Fundamantals, Systems Technology, Images Quality, Applications* (Wiley, 2006)
11. F. Le Chevalier, *Principles of Rdar and Sonar Signal Processing* (Artech House, 2002)
12. M. Ainslie, *Principles of Sonar Performance Modelling* (Springer, 2010)
13. L.J. Ziomek, *An Introduction to Sonar Systems Engineering* (Taylor & Francis, 2023)
14. Y. Takeda, *Ultrasonic Doppler Velocity Profiler for Fluid Flow* (Springer Nature, 2012)

Chapter 7
Displacement, Velocity and Deformation Field Measurements By Image Analysis

Abstract The fundamentals and applications of image generation and processing to measure displacement, velocity and deformation field on the surfaces of rigid and elastic solid bodies are here described. Those measurement techniques are widely applied in fields such as experimental mechanics, robotics, structural components optimization, qualification and analysis, as well as in material and component testing. Many of these techniques are based on bidimensional image cross-correlation and feature traking, known as Digital Image Correlation (DIC), are based on optical flow analysis, or on the use of fiducial markers placed on the object's surface where displacement and velocity can be measured. Many of these techniques, such as Digital Image Correlation (DIC), are based on bidimensional image cross-correlation and feature tracking, on optical flow analysis, or on the use of fiducial markers placed on the object's surface, where displacement and velocity can be measured. Illumination with polarized light of birefringent materials, such as mechanical components, specimens, or coating layers subjected to stress and deformation, can also be used to measure stress and strain fields through image analysis of polarized light. This technique is known as Photoelasticity. Interference patterns generated by laser illumination of a surface and its reflection combined with a reference light beam form the basis of highly sensitive surface deformation measurement techniques, such as Electronic Speckle Pattern Interferometry (ESPI), Shearography, and Speckle Interferometry.

7.1 Displacements and Deformations in Elastic Solid Bodies

In this chapter, the main quantity we focus on is the displacement field, which describes how each point in a body moves when the body is deformed. To understand deformation, we compare the shape of the body before and after it has been deformed, as shown in Fig. 7.1 [1]. In the simplest case, deformation in one direction (for small displacements) can be described as the relative change in length, $\Delta l/l$, so that:

$$\varepsilon = \frac{\Delta l}{l} \tag{7.1}$$

G. Rossi, *Mechanical and Thermal Measurements by Images and Waves*, Smart Sensors, Measurement and Instrumentation 51, https://doi.org/10.1007/978-3-032-03464-9_7

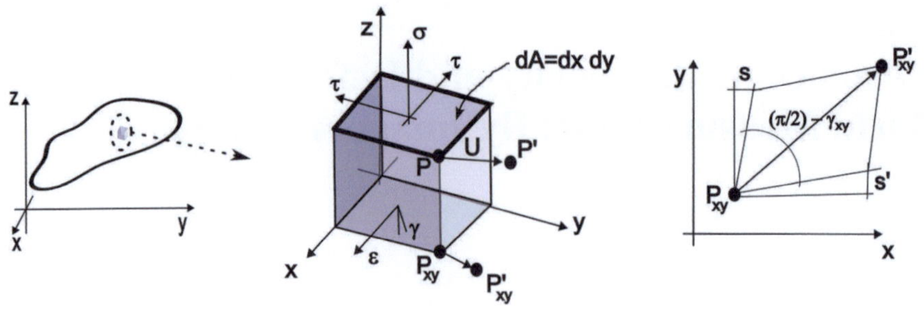

Fig. 7.1 Displacement of a linear elastic solid body: Definition of deformation and stress tensor components.

When we analyze deformation in three dimensions (along the x, y, and z directions), we introduce the displacement components u, v and w, which represent how much a material point has moved along the x, y, and z axes, respectively. The displacement vector field $\mathbf{U}(\mathrm{u},\ \mathrm{v},\ \mathrm{w})$ is defined as the difference between the position of a point in the deformed configuration, $\mathrm{P}'(\mathrm{x}',\ \mathrm{y}',\ \mathrm{z}')$, and its corresponding position in the undeformed configuration, $\mathrm{P}(\mathrm{x},\ \mathrm{y},\ \mathrm{z})$. In formula:

$$\mathbf{U}(\mathrm{u},\ \mathrm{v},\ \mathrm{w}) = \mathrm{P}'(\mathrm{x}',\ \mathrm{y}',\ \mathrm{z}') - \mathrm{P}(\mathrm{x},\ \mathrm{y},\ \mathrm{z}) \tag{7.2}$$

Let's look at two points in the material that are very close along the x-direction:

- The first point is at position x
- The second point is at position $x + \delta x$.

In the undeformed state, the distance between them is δx. In the deformed state, their displacements are:

- First point: $u(x)$
- Second point: $u(x + \delta x)$.

The difference in displacement between these two points is:

$$\delta u = u(x + \delta x) - u(x)$$

So, the relative stretch per unit length is:

$$\frac{\delta u}{\delta x} \approx \frac{\partial u}{\partial x}$$

This quantity represents the normal strain in the x-direction, and we write:

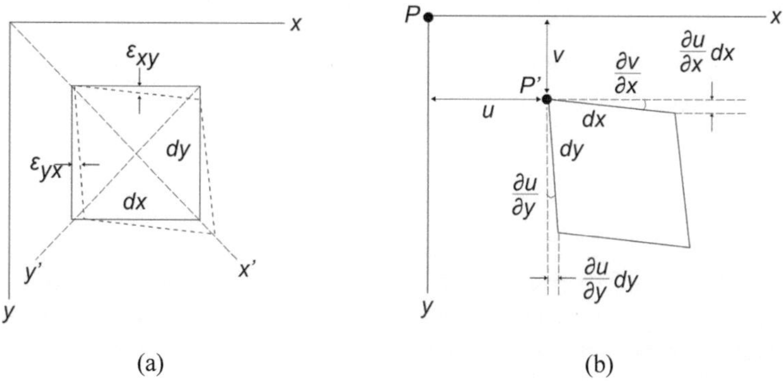

Fig. 7.2 Shear strain in an infinitesimal element: deformation changes the angle between sides in (**a**); total shear strain is the sum of shear effects along x and y directions in (**b**)

$$\varepsilon_x = \frac{\partial u}{\partial x}$$

This tells us how much a small element has stretched (or compressed) along x. Similarly:

$$\varepsilon_y = \frac{\partial v}{\partial y} \quad \varepsilon_z = \frac{\partial w}{\partial z} \tag{7.3}$$

These are the normal strains along y and z directions.

Building on this foundation, we can now describe another key aspect of deformation: shear strain. In addition to elongations or compressions along the principal directions, deformation in an elastic solid body also involves shearing strains, which correspond to changes in the angles between initially perpendicular lines in a small cubic element of material. As shown in Fig. 7.2, when a small cube of material is subjected to shear, the right angles between its sides change, they are no longer 90°. This means that the element is not just stretched but also distorted in shape.

This shear deformation can be considered as the result of two simple shearing actions applied on adjacent sides of the infinitesimal element. That can happen when the top and bottom faces slide relative to each other (along x), or when the left and right faces slide (along y). The total angular change is due to the combined effect of both these shearing actions. So, we describe the total shear strain in the xy-plane as:

$$\gamma_{xy} = \varepsilon_{xy} + \varepsilon_{yx}$$

We write in this way because each of the two components ε_{xy} and ε_{yx} contributes equally to the change in angle between sides. According to the sign convention, a positive shear strain corresponds to a displacement of a face in the positive direction of a coordinate axis. For example, if the top face moves rightward (positive x) and the right face moves upward (positive y), this results in a positive shear strain. Now, since ε_{xy} and ε_{yx} contribute equally to the total angular change, and because of the symmetry of the strain tensor (explained later), we define:

$$\varepsilon_{xy} = \varepsilon_{yx} = \frac{1}{2}\gamma_{xy}$$

This formulation [1] is used since in an elastic material under small deformations, these off-diagonal strain components are equal by definition: they measure the same physical effect from two different coordinate perspectives. So, to avoid double-counting, each is taken as half the total angular deformation. This highlights that the total angular distortion γ_{xy} is the sum of two symmetric components, which form part of the strain tensor.

As shown in Fig. 7.2b, an element that is only subjected to shear deformations shows displacement gradients in directions that are orthogonal to the displacement itself.

For instance:

- $\partial u/\partial y$ tells us how much displacement in the x-direction (u) changes along y,
- $\partial v/\partial x$ tells us how much displacement in the y-direction (v) changes along x.

Together, these describe how the square becomes a parallelogram. For small deformations, shear strains can be written in terms of displacement gradients [1] such as:

$$\begin{aligned} \gamma_{xy} &= \frac{\partial u}{\partial y} + \frac{\partial v}{\partial x} \\ \gamma_{yz} &= \frac{\partial u}{\partial z} + \frac{\partial w}{\partial y} \\ \gamma_{zx} &= \frac{\partial w}{\partial x} + \frac{\partial u}{\partial z} \end{aligned} \tag{7.4}$$

These derivatives describe how much one part of the material moves compared to a nearby part, in directions that are not aligned, which is exactly what shear is: a change in shape due to relative displacements in orthogonal directions.

These shear strain components also contribute to the differential displacements of a point in the material. In other words, they help determine how the material locally deforms due to shear.

Specifically [1], the shear-induced parts of the differential displacements *du*, *dv*, *dw* are:

$$\begin{aligned} (du)_{shear} &= \varepsilon_{xy}dy + \varepsilon_{xz}dz = \frac{1}{2}\left(\frac{\partial u}{\partial y} + \frac{\partial v}{\partial x}\right)dy + \frac{1}{2}\left(\frac{\partial u}{\partial z} + \frac{\partial w}{\partial x}\right)dz \\ (dv)_{shear} &= \varepsilon_{yx}dx + \varepsilon_{yz}dz = \frac{1}{2}\left(\frac{\partial v}{\partial x} + \frac{\partial u}{\partial y}\right)dx + \frac{1}{2}\left(\frac{\partial v}{\partial z} + \frac{\partial w}{\partial y}\right)dz \end{aligned} \tag{7.5}$$

$$(\mathrm{dw})_{\text{shear}} = \varepsilon_{zx}\mathrm{dx} + \varepsilon_{zy}\mathrm{dy} = \frac{1}{2}\left(\frac{\partial w}{\partial x} + \frac{\partial u}{\partial z}\right)\mathrm{dx} + \frac{1}{2}\left(\frac{\partial w}{\partial y} + \frac{\partial v}{\partial z}\right)\mathrm{dy}$$

These expressions provide a direct link between the measurable displacement field (e.g., through digital image correlation or optical flow techniques) and the strain state of the material. The complete strain state at a point in the material is described by the following strain tensor:

$$\begin{bmatrix} \varepsilon_x & \varepsilon_{xy} & \varepsilon_{xz} \\ \varepsilon_{yx} & \varepsilon_y & \varepsilon_{yz} \\ \varepsilon_{zx} & \varepsilon_{zy} & \varepsilon_z \end{bmatrix} \text{ or equivalently } \begin{bmatrix} \varepsilon_{11} & \varepsilon_{12} & \varepsilon_{13} \\ \varepsilon_{21} & \varepsilon_{22} & \varepsilon_{23} \\ \varepsilon_{31} & \varepsilon_{32} & \varepsilon_{33} \end{bmatrix} \tag{7.6}$$

This 3×3 tensor is symmetric ($\varepsilon_{xy} = \varepsilon_{yx} \approx \frac{1}{2}\gamma_{xy}$) because physical deformations in an elastic material under small strain do not depend on the direction; the shear in one direction is the same as in the other. Therefore, the strain tensor is symmetric, and we only need 6 independent values to fully describe the state of deformation at any point.

While strain quantifies deformation, stress quantifies the internal forces developed within a material due to external loading. In its simplest form [1], stress is defined as:

$$\sigma = \frac{F}{A} \tag{7.7}$$

where F is the internal force and A the area on which it acts. In uniaxial tension, stress and strain are related [1] by Hooke's Law:

$$\sigma = E\varepsilon \tag{7.8}$$

where E is the Young's modulus, a material property. In uniaxial tension a transverse strain $\varepsilon_t = -\nu\varepsilon$ is also generated, where ν is the Poisson's ratio [1]. In the general 3D case, we must also consider shear stress, denoted as τ, and define a stress tensor like the strain tensor:

$$\begin{bmatrix} \sigma_x & \tau_{xy} & \tau_{xz} \\ \tau_{yx} & \sigma_y & \tau_{yz} \\ \tau_{zx} & \tau_{zy} & \sigma_z \end{bmatrix} \text{ or } \begin{bmatrix} \sigma_{11} & \sigma_{12} & \sigma_{13} \\ \sigma_{21} & \sigma_{22} & \sigma_{23} \\ \sigma_{31} & \sigma_{32} & \sigma_{33} \end{bmatrix} \tag{7.9}$$

This tensor is also symmetric due to rotational equilibrium and includes all internal stress components acting on the six faces of a cubic element. The general constitutive law for linear, isotropic, and homogeneous materials (Hooke's law in 3D) connects [1] stress and strain tensors:

$$\sigma_{ij} = \frac{E}{(1+\nu)(1-2\nu)}\left[(1-\nu)\varepsilon_{ij} + \nu\,\varepsilon_{kk}\,\delta_{ij}\right] \tag{7.10}$$

where $\varepsilon_{kk} = \varepsilon_x + \varepsilon_y + \varepsilon_z$ is the volumetric strain, and δij is the Kronecker delta.

For practical applications, especially in plane stress conditions (e.g., thin plates with negligible σ_z), the 3D formulation simplifies into the Voigt notation [1], yielding a matrix form of the classical linear elastic materials constitutive law:

$$\begin{bmatrix} \sigma_{xx} \\ \sigma_{yy} \\ \tau_{xy} \end{bmatrix} = \frac{E}{1-\nu^2} \begin{bmatrix} 1 & \nu & 0 \\ \nu & 1 & 0 \\ 0 & 0 & \frac{1-\nu}{2} \end{bmatrix} \begin{bmatrix} \varepsilon_{xx} \\ \varepsilon_{yy} \\ \gamma_{xy} \end{bmatrix} \tag{7.11}$$

This representation, known as the Lamé equations, is widely used in finite element analysis (FEA) and experimental mechanics to efficiently relate stress and strain fields in two-dimensional problems for linear elastic materials.

7.2 Digital Image Correlation

7.2.1 Digital Image Correlation: Historical Background

Digital Image Correlation (DIC) is currently one of the most used full-field measurement techniques in experimental mechanics. Before the development of DIC, several optical methods were employed to analyze deformation fields. Among the earliest was Moiré Interferometry, introduced in the 1950s and 1960s. This technique used periodic grid patterns to visualize and measure material deformation with high sensitivity. However, it required complex optical setups and accurate alignment, which limited its practical application.

In the following decades, other optical techniques were developed, laying the groundwork for modern DIC. Speckle Photography, introduced in the 1960s and 1970s, involved illuminating the specimen surfaces with coherent laser light to generate interference speckle patterns. The displacement of these patterns, caused by surface deformation, was analyzed either manually or using early image processing techniques. Around the same period, Holographic Interferometry emerged as another laser-based method, capable of acquiring minute deformations through interference fringe analysis. Although both techniques offered high sensitivity and precision, they were constrained by the need for costly equipment, vibration isolation systems, complex calibration procedures, and significant manual effort.

The modern concept of DIC was pioneered in the early 1980s by researchers at the University of South Carolina, particularly by Peters et al. (1982). They developed algorithms to compare digital images captured before and after deformation, calculating displacements by tracking intensity variations within the images. Between 1990 and 2000, the development of subset-based DIC algorithms significantly improved accuracy, enabling displacement measurements at subpixel resolution. The technique was further extended to three-dimensional measurements using stereo-camera systems. Over the past two decades, advancements in ultra-high-resolution and high-speed cameras have allowed DIC to be applied to dynamic events, such as impact testing and micro- and nano-scale deformation analysis.

Compared to earlier techniques, DIC offers several key advantages. It does not require complex laser-based optical setups, instead relying on standard digital cameras and relatively simple equipment. This makes it suitable for in-field applications and full-field strain measurements. Additional benefits include good spatial and strain resolution, minimal sample preparation, long working distances, and the absence of laser safety concerns. Unlike other optical techniques such as ESPI or Moiré methods, DIC is relatively insensitive to vibrations and rigid body motions, broadening its range of applicability. Although computationally intensive, the simplicity of the experimental setup has facilitated widespread adoption. Today, DIC is used in a lot of disciplines including experimental mechanics, biomechanics, aerospace, and civil engineering. Several commercial high-performance systems are available, such as ARAMIS (from GOM-ZEISS), VIC-3D (from Correlated Solutions), Q-400 (from Dantec and LaVision), alongside open-source software solutions developed in Python and MATLAB for image data processing.

7.2.2 *Digital Image Correlation: Fundamentals*

Digital Image Correlation (DIC) is based on the processing of at least two images of a surface, one in the undeformed state and one after deformation. Since it is not possible to find correspondences between single pixels, as discussed in Chap. 2, the analysis is based on comparing small image subsets around each pixel or detecting features, blobs, or regions of interest to identify the same corresponding points on the two images. To facilitate this process, the surface must exhibit a distinctive texture. The best solution is to apply a random, high-contrast speckle pattern, typically black dots on a white background or vice-versa. However, naturally "texture-rich" surfaces, such as concrete or irregular stone walls, may also provide sufficient detail to identify matching points, as shown in Fig. 7.3 and previously discussed in Chap. 2. When an artificial pattern is required, several methods can be used to apply speckles to the surface, including airbrushing, aerosol spraying, rubber stamping, or using transferable adhesive patterns.

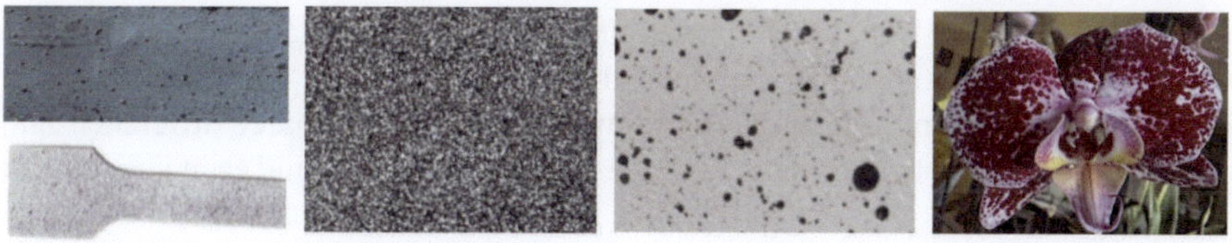

(a) (b) (c) (d)

Fig. 7.3 Examples of surfaces with speckle: concrete surface in (**a**), artificial speckle in (**b**), sprayed speckle in (**c**), natural speckle of a flower in (**d**).

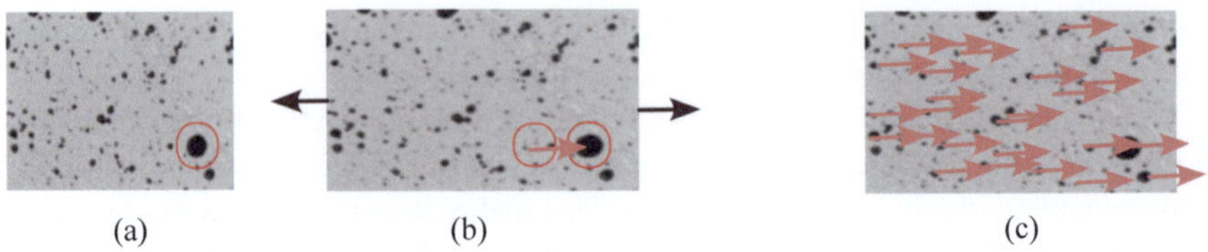

Fig. 7.4 Feature-based tracking DIC

Once the undeformed and deformed images of the object's surface are correctly acquired and digitized, as discussed in Chaps. 2 and 3, two main classes of image processing techniques can be applied to measure the displacement field:

1. Tracking-based DIC: This approach identifies corresponding features (or blobs) in both the undeformed and deformed images using feature detection techniques previously described in Chap. 2. Once a pair of corresponding features is found, the displacement is calculated as the vector connecting their estimated central points.
2. Cross-correlation based DIC: This method involves extracting a sub-image around each pixel in the undeformed image and searching for its most similar location in the deformed image using cross-correlation algorithms. Alternatively, the entire image can be divided into a regular grid of sub-images, and the same correlation process is applied to each sub-region.

An example of the tracking approach applied to a speckle-patterned surface is shown in Fig. 7.4a, b. Once a corresponding feature in the two images is matched, the displacement vector is determined by taking the difference between the two positions of their respective central points.

Using the tracking-based approach, displacement vectors illustrated in Fig. 7.4c can be estimated only in the locations where corresponding feature pairs are identified. As a result, the spatial sampling of the displacement field is non-uniform and depends on the distribution of detectable features.

An alternative approach involves defining how much a small sub-image, centered on a given point in the undeformed image, is moved in the deformed image. This method, based on sub-images comparison, provides a more uniform sampling of the displacement field and is illustrated in Fig. 7.5.

Given two square sub-images of size $N \times N$ pixels, one from the undeformed image $I(i, j)$ and one from the deformed image $I^*(i, j)$, a template matching can be performed by evaluating how much they are equal, as done for blobs in Chap. 4 by Eq. (4.2). So, matching can be measured by calculating the squared difference sum for each possible displacement (u, v) of the two sub images around each image point (x, y) as:

$$C^2(u, v) = \sum_{i,j=1}^{N} \left[I(i, j) - I^*(i + u,\ j + v)\right]^2 \tag{7.12}$$

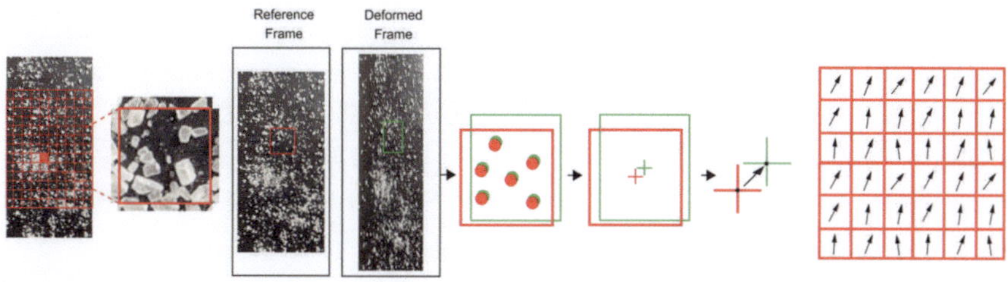

Fig. 7.5 DIC by subset correlation

If the two sub-images are perfectly equal, $I(i, j) = I^*(i + u, j + v)$, all the differences will be 0, so $C^2 = 0$. C^2 is a sum of square values so is always positive and increases with the differences between the two sub-images. In general the displacement values (u', v') that minimizes C^2 corresponds to the best match and therefore represents the estimated local displacement between the undeformed I and deformed I* images.

To illustrate this concept, consider a simple cross-shaped pattern image of 9x9 pixels illustrated in Fig. 7.6a. Each pixel is represented by a two levels grayscale value, with 0 corresponding to black and 100 to white. By giving a displacement of one pixel up and one pixel right we obtain the image of Fig. 7.6b.

Now, let us extract a 5×5 sub-image around the image center and compute here C^2. Taking as possible candidate displacement of this central point the vector $(u, v) = (-2, -2)$, by summing the differences squared of the numbers in the gray squares of Fig. 7.7b, according to Eq. (7.12), we obtain the number 18000, a high value indicating a poor match. Doing the same for $(u, v) = (1, 1)$ using the numbers in the gray squares of Fig. 7.7c, that are all equal, we obtain zero, which implies a perfect match between the two. Because C is always positive, the value of 0 is the minimum possible, so the displacement of the central point is given by the vector (1, 1).

In real images noise, lighting variations, and optical distortions introduce other grey level differences between the two sub-images, making the C^2 never exactly zero. Nevertheless, we can search for the minimum value of C^2 or, more robustly,

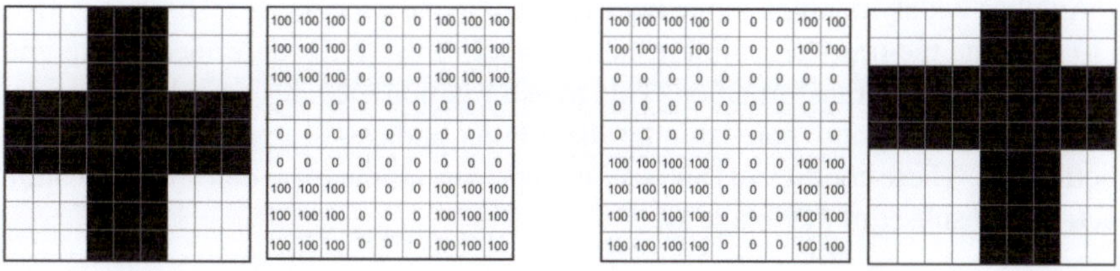

100	100	100	0	0	0	100	100	100
100	100	100	0	0	0	100	100	100
100	100	100	0	0	0	100	100	100
0	0	0	0	0	0	0	0	0
0	0	0	0	0	0	0	0	0
0	0	0	0	0	0	0	0	0
100	100	100	0	0	0	100	100	100
100	100	100	0	0	0	100	100	100
100	100	100	0	0	0	100	100	100

(a)

100	100	100	100	0	0	0	100	100
100	100	100	100	0	0	0	100	100
0	0	0	0	0	0	0	0	0
0	0	0	0	0	0	0	0	0
0	0	0	0	0	0	0	0	0
100	100	100	100	0	0	0	100	100
100	100	100	100	0	0	0	100	100
100	100	100	100	0	0	0	100	100
100	100	100	100	0	0	0	100	100

(b)

Fig. 7.6 A simple cross image (**a**) than displaced one pixel up and one right (**b**)

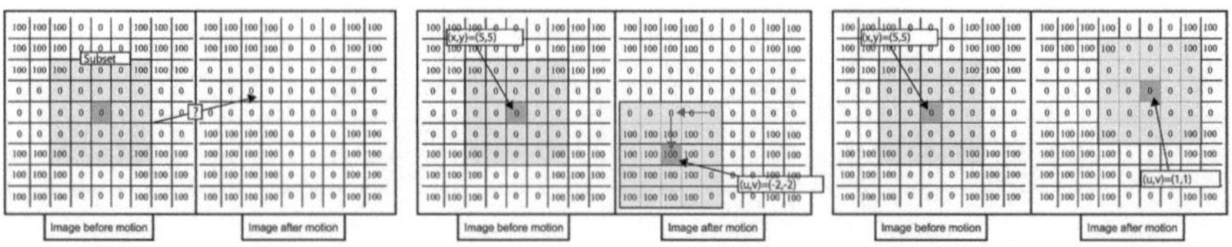

Fig. 7.7 Sub-images template matching to find displacement of its central point.

the maximum of a normalized cross-correlation function, which provides a more reliable estimate of displacement under non-ideal conditions.

As done in Cap 4 for comparing two blog signatures by Eq. (4.28), the normalized two-dimensional cross-correlation function between two images f and g, from [2], can be defined as:

$$C(u, v) = \frac{\sum_{i=1}^{m} \sum_{j=1}^{m} \left[f(x_i, y_j) - \bar{f}\right]\left[g\left(x_i', y_j'\right) - \bar{g}\right]}{\sqrt{\sum_{i=1}^{m} \sum_{j=1}^{m} \left[f(x_i, y_j) - \bar{f}\right]^2} \sqrt{\sum_{i=1}^{m} \sum_{j=1}^{m} \left[g\left(x_i', y_j'\right) - \bar{g}\right]^2}} \tag{7.13}$$

where the transformed coordinates $\left(x_i{\prime}, y_j{\prime}\right)$ account for sub-pixel displacements and possible gradients in the displacement field by the transformation:

$$\begin{aligned} x' &= x + u + \frac{\partial u}{\partial x}dx + \frac{\partial u}{\partial y}dy \\ y' &= y + v + \frac{\partial u}{\partial x}dx + \frac{\partial v}{\partial y}dy \end{aligned} \tag{7.14}$$

By calculating this C(u, v) function of two sub-images the displacement vector (u, v) corresponding to the maximum of C(u, v) function indicates the best local match between these two sub-images and thus provides the estimated displacement vector for the sub-image central point, as illustrated in Fig. 7.8. This method is widely used in DIC for its robustness to variations in illumination and noise level.

Another widely adopted modeling assumption and processing technique in DIC is to describe the transformation from the undeformed to the deformed image grid as a pseudo-affine transformation. This transformation includes rigid body translation, stretch, shear, and more complex distortions, and can be expressed using eight coefficients. These coefficients allow us to approximate the local deformation within each square subset of the image as:

$$\begin{aligned} x_t(a_0, a_1, a_2, a_3, x, y) &= a_0 + a_1 x + a_2 y + a_3 xy \\ y_t(a_4, a_5, a_6, a_7, x, y) &= a_4 + a_5 x + a_6 y + a_7 xy \end{aligned} \tag{7.15}$$

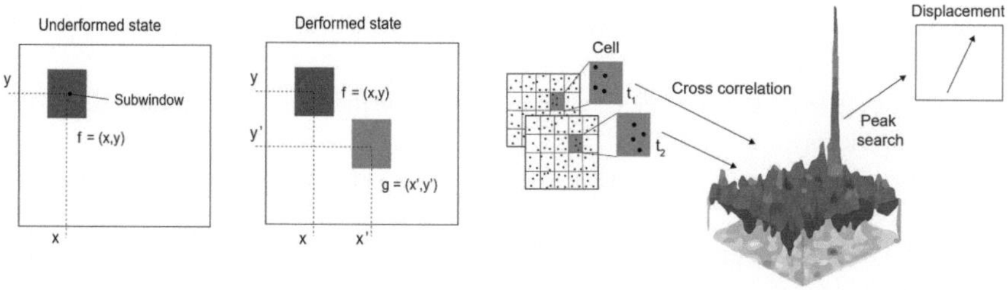

Fig. 7.8 Local displacement (u, v) from normalized cross-correlation, determined by the peak position of the correlation function

Fig. 7.9 Deformation model for DIC subsets

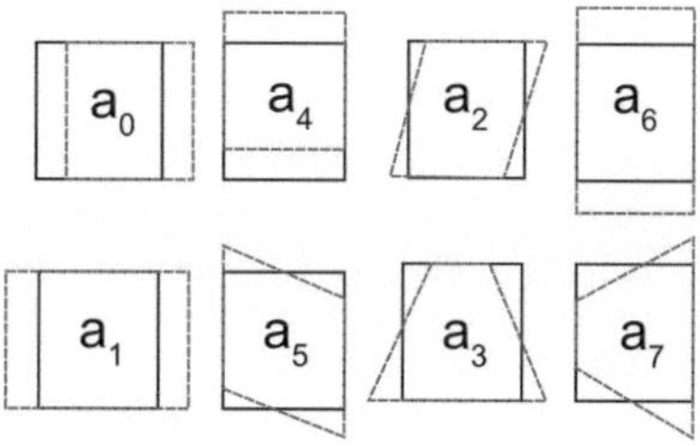

The objective becomes finding the optimal set of coefficients $a_0, \ldots, a_7$, illustrated in Fig. 7.9, along with gray level correction parameters g_0 and g_1, such that the difference between the reference square sub-image $G_1(x, y)$ and the transformed, gray-level corrected deformed image $G_t(x, y)$ is minimized, so solving the problem:

$$\begin{gathered}\min_{a_0,\ldots,a_1,\, g_0 g_1} \sum\nolimits_{x,y} \|G_1(x, y) - G_t(x, y)\|, \\ \text{with}\, G_t(x, y) = g_0 + g_1 G_2\big(x_t(x, y), y_t(x, y)\big)\end{gathered} \tag{7.16}$$

where G_1 is grey levels on the reference square sub-image, G_2 represent the gray levels of the deformed image, and the correction accounts for differences in illumination (g_0) and contrast (g_1).

7.2.3 *Digital Image Correlation: Implementation and Applications*

Commercial DIC software such as VIC-3D (Correlated Solutions), Istra4D (Dantec), LaVision, and LIMESS implement these techniques to extract full-field deformation measurements from images. The general procedure of a DIC analysis can be summarized in the following steps:

1. Apply an appropriate speckle pattern on the surface, with proper average speckle dimensions, if necessary.
2. Acquire at least two images, one in the undeformed configuration, and one or more in deformed configuration.
3. Perform image processing using cross-correlation, defining subset dimensions, or by tracking.
4. Define the displacement field by filtering and/or interpolating the displacement vectors.
5. Compute the strain field from the partial derivatives of the displacement field using the relations previously introduced, e.g. Eqs. (7.3) and (7.4).

It is important to note that, in bidimensional DIC, measuring deformation does not require knowing the exact physical scale (e.g., pixels to mm scale factor), since strain is a dimensionless quantity, expressed as a ratio of relative displacements $\Delta L/L$, and thus remains the same ratio whether measured in pixels or millimeters. To ensure accurate and reliable DIC measurements, several practical considerations must be considered when setting up a test bench:

- Proper and uniform lighting conditions
- Speckle pattern quality and size
- Camera resolution, lens distortion, and field of view
- Required displacement and strain resolution
- Correlation algorithm performance and settings
- Processing power of the computer
- Mechanical stability of the setup during testing, loading and acquisition.

The choice of speckle size respect pixel and subset window size plays a critical role: speckle patterns that are too small may move inside the pixel not changing its grey level, as in Fig. 7.10a, whereas excessively large speckles reduce spatial resolution due to insufficient image texture. According to the Nyquist criteria, described in Chap. 2, an appropriate sampling resolution is required, ideally at least double the expected minimum displacement, to capture deformation accurately. A practical rule of thumb is to ensure that each speckle dot covers at least 3×3 pixels and that each subset contains at least 3×3 speckles, as illustrated in Fig. 7.10b.

The subset dimension directly affects the spatial resolution and strain accuracy of the measurement. Smaller subsets yield higher spatial resolution but lower strain accuracy due to limited number of pixel correlated, while larger subsets improve accuracy at the expense of spatial resolution. An estimation of this trade-off is summarized in Table 7.1.

Moreover, DIC is not limited to 2D applications. The technique extends to 3D surface measurements using stereo camera setups, as described in Chap. 3. By calibrating a stereo system and acquiring paired images of the object before and after deformation, it is possible to compute the full 3D displacement field. Differentiation of this field yields the 3D strain field tensor, as illustrated in Fig. 7.11.

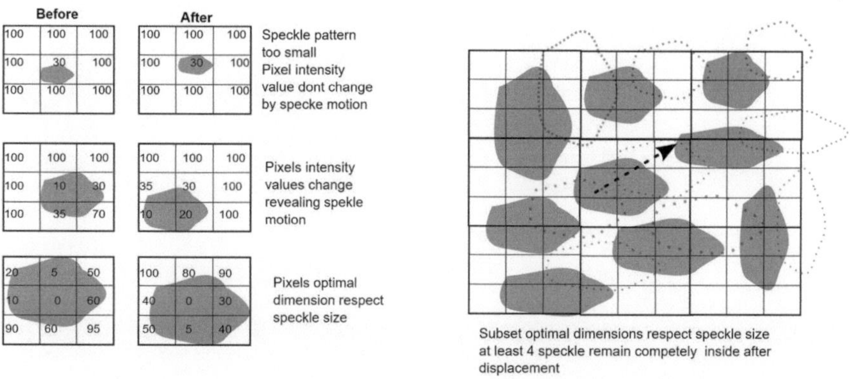

(a) (b)

Fig. 7.10 Choosing proper speckle pattern in relation to pixel and subset dimension

Table 7.1 Subset dimension and accuracy

Subset dimension	Displacement measure accuracy	Strain values accuracy (50% overlap) (%)
128 × 128	Da 0.01 a 0.03 pixel	0.094
64 × 64	Da 0.02 a 0.05 pixel	0.3
32 × 32	Da 0.05 a 0.2 pixel	1.25
16 × 16	Da 0.1 a 0.3 pixel	5

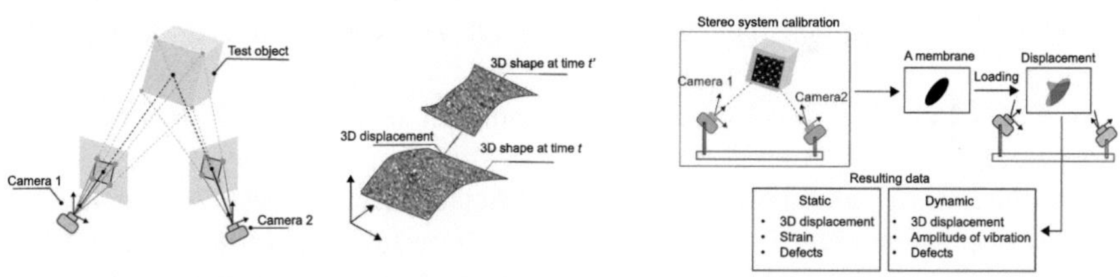

Fig. 7.11 3D deformation field measurements by DIC

In addition to surface measurements, DIC can be applied to volumetric data acquired through tomographic imaging systems (e.g., X-ray CT described in previous Chap. 6). This approach enables the analysis of internal deformation fields in composite and heterogeneous materials, such as concrete, rubber with steel reinforcements in tires, or fiber-reinforced polymers. An example of volumetric DIC is presented in Fig. 7.12.

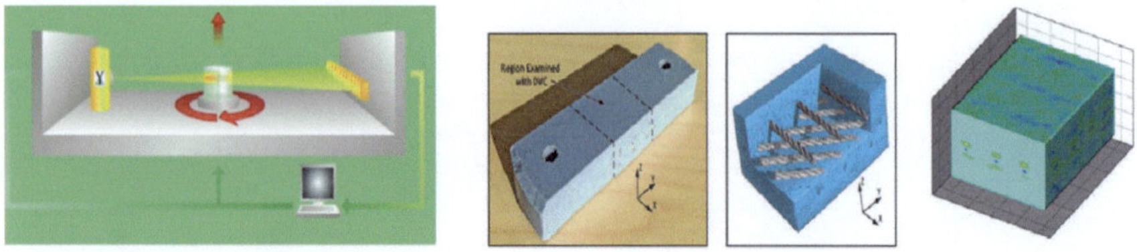

Fig. 7.12 Volumetric DIC—From LaVision depliant and website

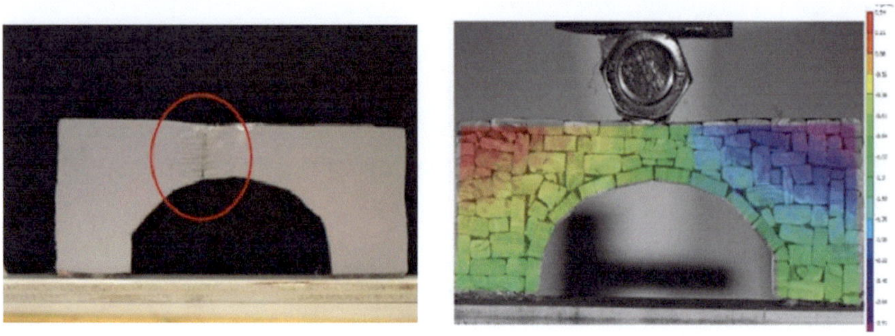

Fig. 7.13 Strain measurements by DIC on an arch model made by irregular bricks

Further examples demonstrate the versatility of DIC across a range of materials and applications. Figure 7.13 shows the strain distribution in a small-scale arch structure little model composed of irregular bricks glued onto a paper substrate and loaded with a steel weight. The deformation is clearly concentrated near the paper's crack line.

In Fig. 7.14a the classical case of an aluminum specimen with a central hole under axial loading is depicted, highlighting the strain concentration around the hole. Due to the uniform aluminum surface, an artificial speckle pattern was applied. In contrast, specimen in Fig. 7.14b, a trabecular titanium structure produced via additive manufacturing, benefits from its natural surface roughness, providing a good speckle for DIC without additional treatment.

Finally, Fig. 7.15a illustrates other DIC measurements during compressive testing of masonry samples, where stone texture naturally acts as a speckle [3], and the analysis of deformation in an airless wheel [4], in Fig. 7.15b, where a painted pattern was necessary to obtain displacement fields.

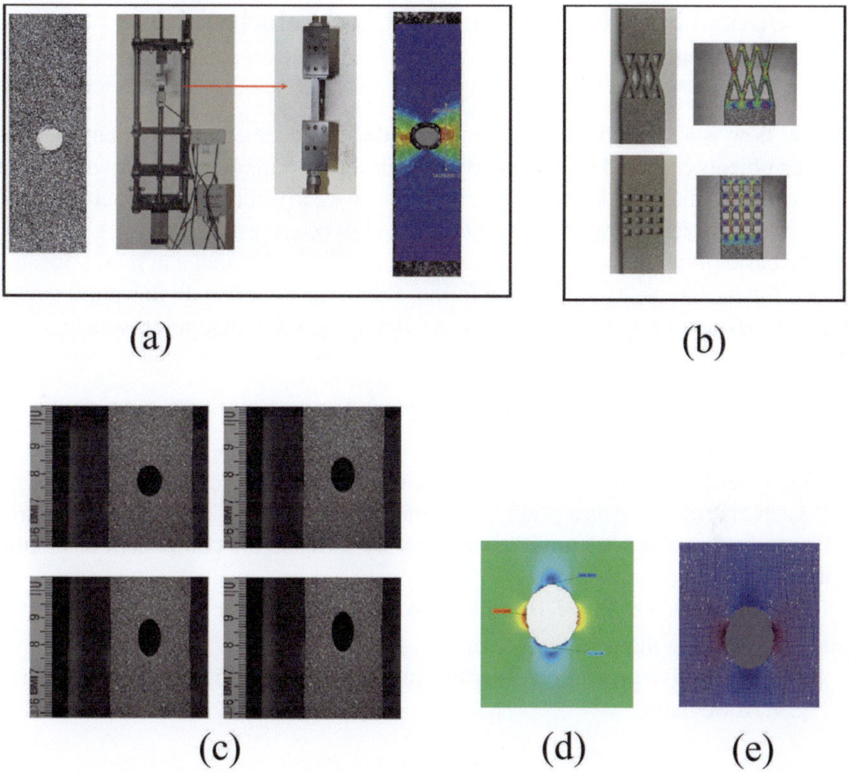

Fig. 7.14 Strain concentrations around specimen's holes and on trabecular 3D printed titanium specimens in (**c**), deformations calculated by GOM Correlate software in (**d**) and by DIC Engine freewere in (**e**)

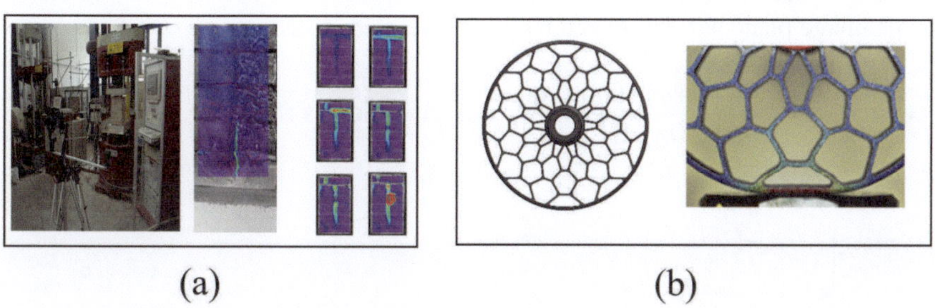

Fig. 7.15 Measurements on a brick wall samples during a compression test and on an airless wheel

7.3 Optical Flow for Displacement and Velocity Field Measurement

Optical flow is a technique used to estimate the displacement and velocity field of moving objects captured in a video. By measuring how image points move between consecutive frames, and knowing the frame rate, we can evaluate the velocity of each point as the displacement divided by the time between frames.

The basic principle is as follows: consider a point in the image at time t, located at coordinates $\mathrm{I(x,\ y,\ t)}$. In the next frame, after a time interval dt, this point moves to a new position $(\mathrm{x + dx,\ y + \ dy})$. The velocity vector at that point is then:

$$\mathrm{v} = \left(\frac{\mathrm{dx}}{\mathrm{dt}}, \frac{\mathrm{dy}}{\mathrm{dt}}\right) \tag{7.17}$$

This velocity is what we refer to as the optical flow. However, since a standard video camera (modeled as a pinhole camera, as described in Chap. 2) only captures a 2D projection of the 3D motion, we can only estimate the component of motion that lies within the image plane.

This is illustrated in Fig. 7.16, where the real motion vector $\boldsymbol{v}_0 = \frac{d\boldsymbol{r}_0}{dt}$ is projected onto the 2D image plane as the vector $\boldsymbol{v}_i = \frac{d\boldsymbol{r}_i}{dt}$.

From perspective projection described in Chap. 2 (Eq. 2.25) and vectors defined in Fig. 7.16 we have:

$$\frac{\boldsymbol{r}_i}{f} = \frac{\boldsymbol{r}_0}{\boldsymbol{r}_0 \cdot z}$$

But there are conditions, like in Fig 7.17b and c, where motion exists but no optical flow can be detected. Anyway, from the above equation, the optical flow velocity $\mathbf{v}_\mathrm{i}$ on the image plane will be related by the 3D velocity $\mathbf{v}_0$ as:

$$\boldsymbol{v}_i = \frac{d\boldsymbol{r}_i}{dt} = f\frac{(\boldsymbol{r}_0 \cdot z)\boldsymbol{v}_0 - (\boldsymbol{v}_0 \cdot z)\boldsymbol{r}_0}{(\boldsymbol{r}_0 \cdot z)^2} \quad \text{and simplified as :} \quad \boldsymbol{v}_i = f\frac{(\boldsymbol{r}_0 \times \boldsymbol{v}_0) \times z}{(\boldsymbol{r}_0 \cdot z)^2}$$

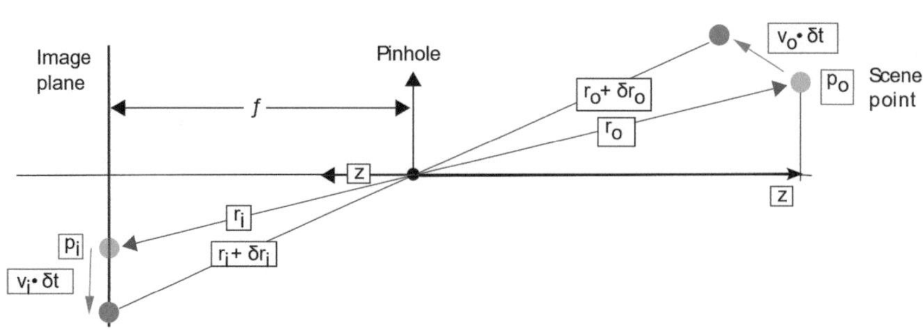

Fig. 7.16 Image-plane projection (optical flow) of real 3D motion of a point

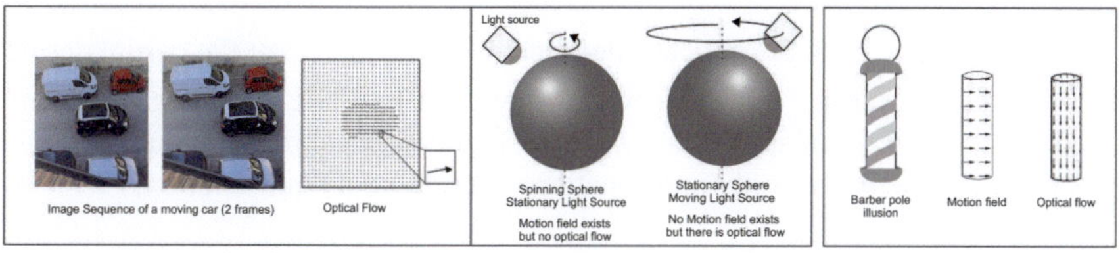

Fig. 7.17 Optical flow due to real motion and of only moving image pattern

7.3.1 Mathematical Basis of Optical Flow

To better understand the optical flow principle, consider Fig. 7.18, where the motion of an object is shown across two frames. The apparent movement of image patterns corresponds to the actual motion of the object in space.

The optical flow calculation relies on two fundamental assumptions:

1. Brightness constancy: The brightness of a point in the image remains constant between frames:

$$I(x, y, t) = I(x + dx, y + dy, t + dt) \tag{7.18}$$

2. Small motion approximation: The displacements dx, dy, and time step dt are small. This allows the use of a first-order Taylor expansion:

$$I(x + dx, y + dy, t + dt) \approx I(x, y, t) + \frac{\partial I}{\partial x}dx + \frac{\partial I}{\partial y}dy + \frac{\partial I}{\partial t}dt \tag{7.19}$$

Substituting Eqs. (7.18) into (7.19) and rearranging terms yields the optical flow constraint equation:

$$I_x u + I_y v + I_t = 0 \tag{7.20}$$

Fig. 7.18 Optical flow basic principle

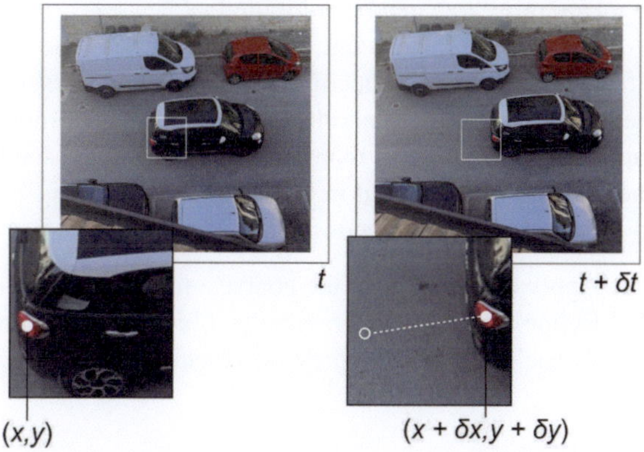

where:

$(u, v) = \left(\frac{dx}{dt}, \frac{dy}{dt}\right)$ is the optical flow vector, and I_x, I_y, I_t are the partial derivatives of image intensity with respect to x, y, and t. While the image gradients $\left(I_x, I_y, I_t\right)$ can be estimated from two consecutive frames, this equation contains two unknowns (u, v) but only one equation so it has, no unique solution.

7.3.2 *Lucas-Kanade Method*

To overcome this limitation, Lucas and Kanade [5] proposed a method that assumes the optical flow is constant within a small neighborhood (e.g., a square region of N × N pixels). For each pixel (i, j) in this sub-image, the optical flow constraint equation can be applied. This gives a system of N^2 equations, one for each pixel, which can be written in matrix form as:

$$\begin{bmatrix} I_x(1,1) & I_y(1,1) \\ \vdots & \vdots \\ I_x(i,j) & I_y(i,j) \\ \vdots & \vdots \\ I_x(N,N) & I_y(N,N) \end{bmatrix} \begin{bmatrix} u \\ v \end{bmatrix} = \begin{bmatrix} I_t(1,1) \\ \vdots \\ I_t(i,j) \\ \vdots \\ I_t(N,N) \end{bmatrix} \quad \text{or simply} \quad A \cdot \begin{bmatrix} u \\ v \end{bmatrix} = B \tag{7.21}$$

where A contains the image gradients I_x and I_y, and B contains the temporal gradients I_t. The least-squares solution to this system is:

$$\begin{bmatrix} u \\ v \end{bmatrix} = \left(A^T A\right)^{-1} A^T B \tag{7.22}$$

To obtain a reliable solution the matrix $\left(A^T A\right)$ must be invertible and well-conditioned. This requires both eigenvalues λ_1 and λ_2 of the matrix $(A^T A)$ to be sufficiently large and of similar magnitude. Only in such cases as shown in Fig. 7.19c a meaningful flow can be computed.

When the displacement between frames is too large, the linear approximation from the Taylor series becomes invalid. To address this, the image resolution can be reduced by averaging pixel blocks [6–8]. This produces a smoothed image where apparent motion is less than 1 pixel, allowing the small displacement assumption to hold again, as illustrated in Fig. 7.20.

The algorithm proceeds as follows:

1. Apply optical flow at the coarsest resolution (where displacement is small).
2. Refine the result progressively at higher resolutions (finer scales).
3. At each step, use the previously estimated flow as an initial guess.

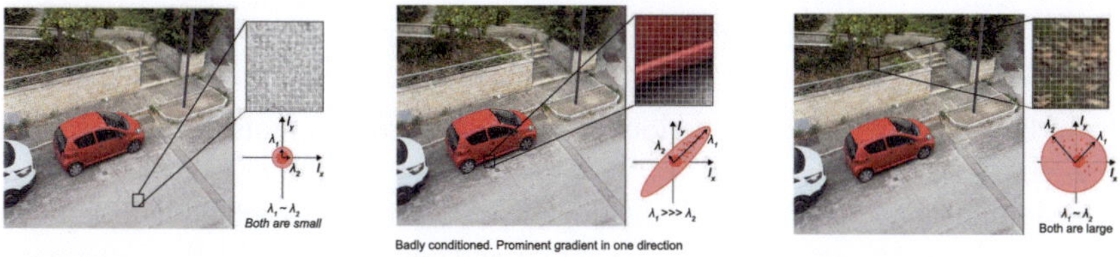

Fig. 7.19 Eigenvalues of three different subset windows matrices ($\mathbf{A}^T\mathbf{A}$)

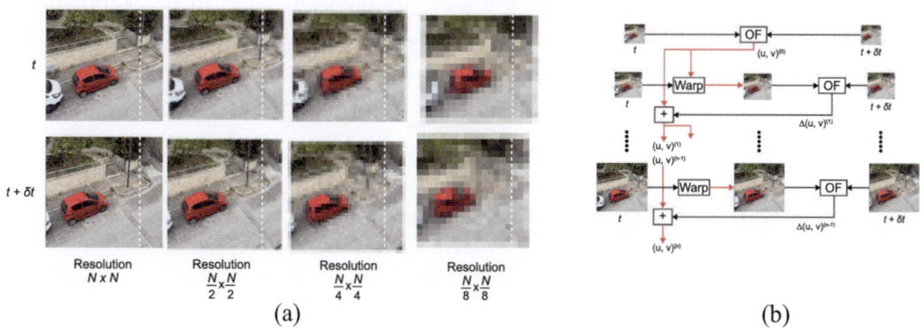

Fig. 7.20 Optical flow for large displacements: resolution reduction in (**a**) and coarse to fine processing step by step in (**b**)

7.3.3 Software Implementations

Several software libraries and platforms offer implementations of optical flow techniques:

- OpenCV (Python, C++) provides built-in functions like cv2.calcOpticalFlowPyrLK().
- MATLAB includes functions such as opticalFlowLK or estimateFlow.
- Other dedicated tools exist for more advanced or real-time applications.

Lucas-Kanade is generally faster and more robust than other methods for small search windows.

Another way to estimate motion is via template matching, discussed in Chap. 4. This method finds corresponding patterns between frames. However:

- It is computationally more expensive when search areas are large.
- It is more prone to mismatches compared to optical flow methods like Lucas-Kanade.

7.3.4 Applications of Optical Flow

Optical flow is widely used in various applications, from motion tracking to fluid dynamics.

One interesting use case is in evaluating the efficiency of a kitchen range hood, a measurement system developed by the author and collaborators for Elica company [9].

By illuminating vapor with a laser sheet and recording the motion with a camera, optical flow algorithms can compute the velocity field of the vapor particles. Integrating the velocity over control lines (A, B, and C in Fig. 7.21) gives the corresponding flow rates crossing the lines Q_A, Q_B, Q_C as:

$$Q_A = l_1 \cdot \sum v_y^- \quad Q_B = l_2 \cdot \sum v_x^- \quad Q_C = l_2 \cdot \sum v_x^+$$

The capture efficiency UPTI of the range hood can be defined as:

$$\mathrm{UPTI} = \frac{Q_A}{Q_A + Q_B + Q_C}$$

A motion sensor based on optical flow is something we often have in our hands. This sensor uses a low-resolution camera and a lighting LED to apply the optical flow technique for measuring displacements: is the optical mouse, as illustrated in Fig. 7.22a.

Another common application of optical flow is in traffic monitoring, as illustrated in Fig. 7.22b.

7.4 Marker Based Displacement and Velocity Measurement Techniques

Measuring displacements through image processing can be performed using clearly defined markers, such as the one shown in Fig. 7.23, instead of relying on natural surface textures or randomly applied speckle patterns. These markers can be easily

Fig. 7.21 Optical flow used to measure vapour flow captured by a kitchen range hood

Fig. 7.22 Optical flow applications in optical mouse and speed control

detected on the surface of a moving object using efficient and optimized template matching algorithms. Once the markers are decoded, each one can be uniquely identified by a number. Furthermore, if the physical dimensions of the markers are known in millimeters, they can also serve as a reference to calibrate the camera and convert image measurements from pixels to millimeters.

There are various types of markers used in image marker based measurement systems. Some kind of markers are shown in Fig. 7.23.

(a) Intersense circular markers

These are simple round markers commonly used in motion tracking systems.

(b) AprilTag markers

AprilTag is an open-source fiducial marker system developed by the University of Michigan. Each tag consists of a square with a black-and-white pattern inside a white border. The white border is essential for reliable detection. The tag encodes information in a binary matrix; for example, the 36h11 family uses a 6×6 grid of black (0) and white (1) squares.

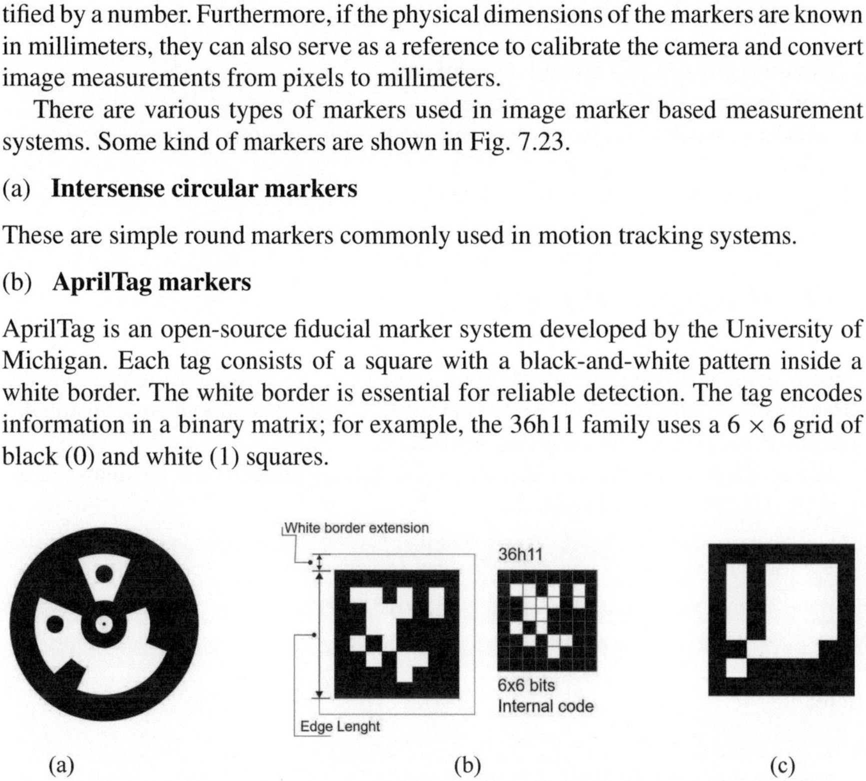

Fig. 7.23 Different kinds of markers: Intersense marker in (**a**), AprilTag marker in (**b**), ArUco marker in (**c**).

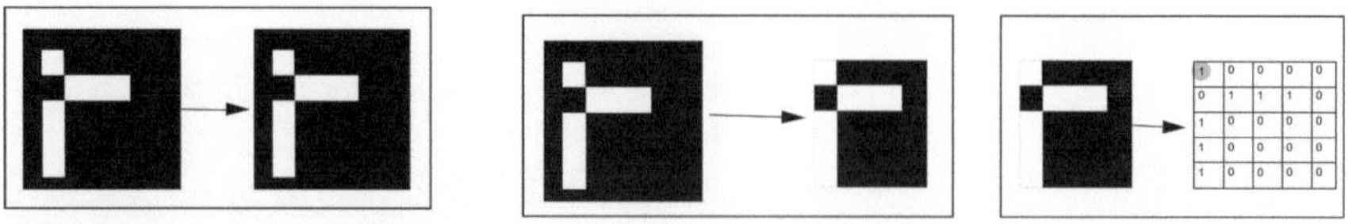

(a) (b) (c)

Fig. 7.24 ArUco marker identification: from initial detection to ID extraction via image normalization and binary matrix decoding.

(c) **ArUco markers**

ArUco is another marker system, developed by the University of Córdoba. Each ArUco marker is a square with a thick black border and an internal 5×5 binary matrix. This matrix encodes the marker ID. The columns of the matrix follow a specific structure: the first, third, and fifth columns contain parity bits, while the second and fourth columns encode the data. This structure allows for $2^{10} = 1024$ unique marker IDs.

The decoding process for an ArUco marker (illustrated in Fig. 7.24) includes the following steps:

1. Remove any padding around the marker.
2. Divide the marker into a 5×5 grid.
3. Read the second and fourth columns from top to bottom to extract the binary code.
4. Convert the binary code into a decimal number representing the marker's ID.

7.4.1 Marker Pose Detection

Once a marker is detected, its 3D pose (position and orientation) relative to the camera can be estimated as illustrated in Fig. 7.25. This is commonly done using the pinhole camera model described in Chap. 2.

For a 3D point on the marker, $P_i = (x_i, y_i, z_i)$, projected to an image point $p_i = (u_i, v_i)$, the projection equation is:

$$\lambda_i \begin{bmatrix} u_i \\ v_i \\ 1 \end{bmatrix} = K(RP_i + t) \tag{7.23}$$

Here:

- λ_i is a scale factor,
- K is the camera intrinsic matrix, defined as:

$$K = \begin{bmatrix} f_x & 0 & c_x \\ 0 & f_y & c_y \\ 0 & 0 & 1 \end{bmatrix} \tag{7.24}$$

where f_x, f_y are the focal lengths and $(c_x,\ c_y)$ is the principal point.

- R is a 3×3 rotation matrix, and
- t is a 3×1 translation vector.

For a square ArUco marker with side length s, it is common to define the marker's coordinate system such that the marker lies in the $Z = 0$ plane. A typical assignment of 3D coordinates for the marker's corners is:

$$P_1 = \left[-\frac{s}{2}, \frac{s}{2}, 0\right],\quad P_2 = \left[\frac{s}{2}, \frac{s}{2}, 0\right],\quad P_3 = \left[\frac{s}{2}, -\frac{s}{2}, 0\right],\quad P_4 = \left[-\frac{s}{2}, -\frac{s}{2}, 0\right] \tag{7.25}$$

Their corresponding 2D image coordinates are:

$$p_i = \begin{bmatrix} u_i \\ v_i \end{bmatrix}, \qquad i = 1, 2, 3, 4 \tag{7.26}$$

This formulation leads to the well-known Perspective-n-Point (PnP) problem: estimate R and t given 3D-2D correspondences. The equations are nonlinear due to the unknown scale factors λ_i. A typical approach is to minimize the reprojection error, defined for each point as:

$$e_i = p_i - \widehat{p_i} \tag{7.27}$$

where $\widehat{p_i}$ is the projected point given by:

$$\widehat{p_i} = \frac{1}{\lambda_i} K(RP_i + t), \quad \lambda_i = [K(RP_i + t)]^3 \tag{7.28}$$

The goal is to minimize the total error:

$$\min_{R,t} \sum_{i=1}^{4} \left\| p_i - \frac{1}{\lambda_i} K(RP_i + t)^2 \right\| \tag{7.29}$$

Since R must satisfy orthonormality constraints $(R\top R = I, \det(R) = 1)$, the solution usually involves:

- Direct Linear Transform (DLT),
- EPnP (Efficient PnP), or
- Iterative methods such as the Levenberg–Marquardt algorithm.

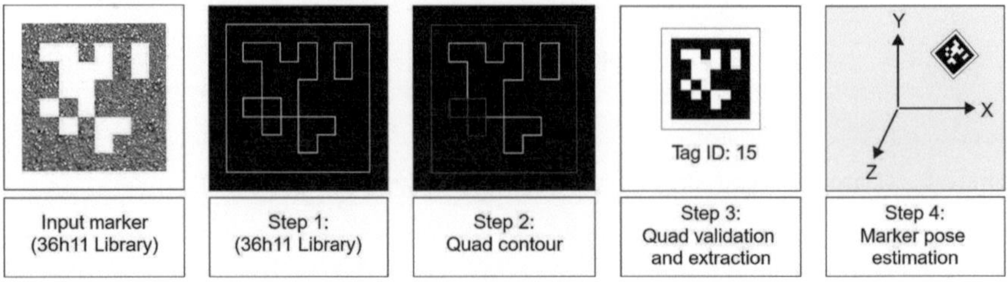

Fig. 7.25 ArUco marker pose detection: from raw image to 3D pose estimation relative to the camera

Libraries like OpenCV offer the solvePnP() function to compute *R* and *t*.

7.4.2 Marker Center Tracking and Calibration

Once the four marker corners positions are identified, the center of the marker in the image can be calculated frame by frame as:

$$C = (x_c, y_c) = \vec{G} \cdot \left(\frac{1}{4} \sum\nolimits_{r=1}^{4} |x_r|, \frac{1}{4} \sum\nolimits_{r=1}^{4} |y_r| \right) \tag{7.30}$$

Here:

- (x_r, y_r) are the pixel coordinates of the rth corner
- G is the spatial calibration factor, converting pixel distances to SI units (e.g., meters), defined as:

$$G = \frac{d_{SI}}{d_{px}} \tag{7.31}$$

where d_{SI} is the known side length of the marker in meters, and d_{px} is its measured size in pixels in the image.

This calibration factor is updated in every frame, allowing compensation for perspective distortions or marker deformations during motion.

7.4.3 Applications of Marker Detection and Marker Tracking

Marker detection and tracking systems, particularly those based on ArUco markers, are widely used in experimental mechanics for non-contact measurements of displacement, vibration, and dynamic behavior. These systems offer high spatial resolution and flexibility, making them suitable for both laboratory tests and field applications.

One application was carried out by the author and collaborators [10], who used ArUco markers to study the impulse response of the control robot of the new "San Giorgio" bridge in Genova. A total of 13 markers were placed along the main beam of the robot, and an accelerometer was mounted at its tip. The displacement of the markers was measured simultaneously and compared to the signal output of the accelerometer. With marker movements in the millimeter range, the system demonstrated an excellent spatial resolution, on the order of hundredths of a millimeter, validating its suitability for high-precision dynamic testing (Fig. 7.26).

ArUco marker tracking has also proven effective in wind tunnel studies for aerospace applications. In Fig. 7.27, fiducial markers were used to analyze the motion, dynamic response, and shape evolution of a sub-scale parachute subjected to wind forces [11, 12]. This approach enabled a detailed reconstruction of deformation patterns during flight-like conditions.

A similar methodology was adopted in the marine sector, as shown in Fig. 7.28, where markers and drone-mounted cameras were employed to investigate the shape changes and vibrations of a sail in motion. The use of UAVs allows the acquisition of dynamic 3D shape data from multiple angles, overcoming occlusion issues and enabling real-time monitoring in realistic operating conditions.

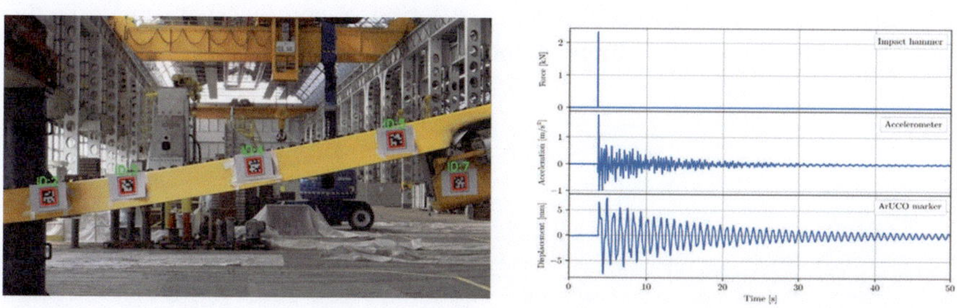

Fig. 7.26 Testing dynamic performances of a large beam by ArUco marker displacement measurements

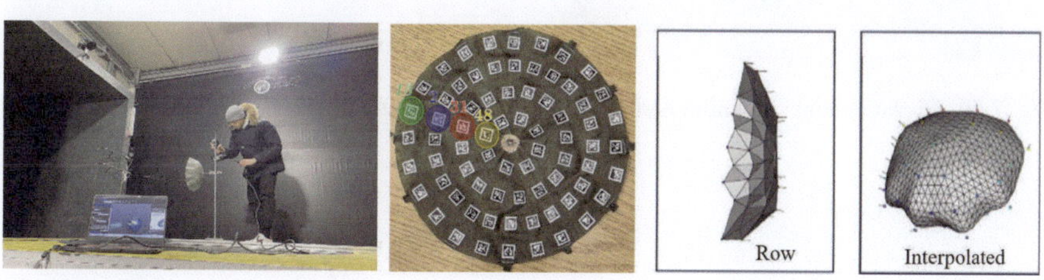

Fig. 7.27 Motion measurements, dynamics study and shape reconstruction of a sub-scale parachute for space veichle landing

(a) (b)

Fig. 7.28 Measurements by markers for dynamic shape reconstruction of a sail and its vibration detection from a drone video recording operated by Emanuele Rossi

In robotics, marker tracking supports vibration and motion studies of mechanical components. Fig. 7.29 illustrates the 3D motion tracking of a Cartesian robot manufactured by Campetella Italian company, using ArUco markers, by author and collaborators. The film was recorded inside the factory by Tommaso Tocci with a EOS 7D camera, providing sufficiently accurate film records for identifying structural dynamics and validating control strategies and robot performances.

These case studies demonstrate how marker-based systems can be useful to measure with sufficient accuracy displacements through marker-based video recording in various engineering fields, offering a non-invasive, cost-effective alternative to traditional measurement methods.

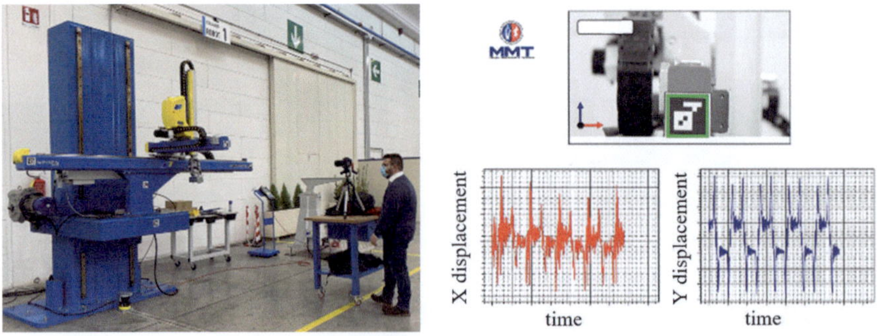

Fig. 7.29 Measurement of motion and vibration of a Cartesian robot by ArUco markers

7.5 Photoelasticity for Strain Field Measurements

Photoelasticity is a measurement technique used to evaluate stress and strain fields by exploiting the optical behavior of polarized light interacting with birefringent materials [13]. The fundamental principles and phenomena of polarized light, its generation and detection were discussed in Chap. 1.

When a beam of light passes through a birefringent material, it splits into two rays polarized in orthogonal directions:

- The ordinary ray, which follows Snell's law.
- The extraordinary ray, which does not follow Snell's law and oscillates in a plane perpendicular to that of the ordinary ray.

These two rays travel at different speeds due to the material's anisotropy, which causes different refractive indices n_1 and n_2 along the two polarization directions (see Fig. 7.30a).

7.5.1 Birefringence and Phase Delay

If a light beam of wavelength λ passes through a material of thickness h, the travel times t_1 and t_2 for each ray (with velocities $v_1 = \frac{c}{n_1}$ and $v_2 = \frac{c}{n_2}$) are:

$$t_1 = \frac{h}{v_1} = \frac{h}{c} n_1 \quad t_2 = \frac{h}{v_2} = \frac{h}{c} n_2 \tag{7.32}$$

The time delay Δt and optical path difference Δs between the two rays can be calculated as:

$$\Delta t = t_2 - t_1 = \frac{h}{c}(n_2 - n_1) \quad \Delta s = c\ \Delta t = h(n_2 - n_1) \tag{7.33}$$

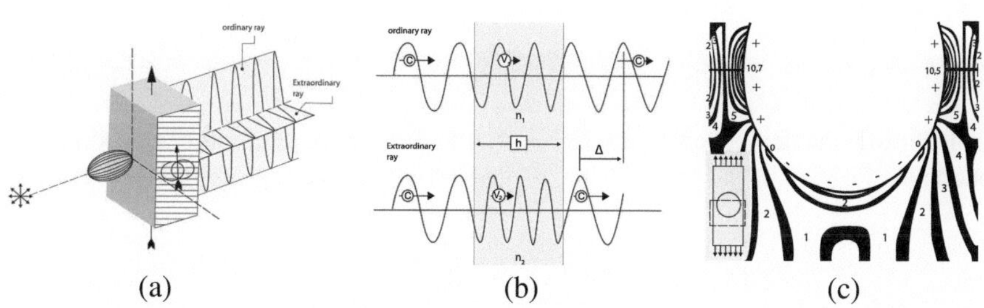

Fig. 7.30 Birefringence: light splits into ordinary and extraordinary rays with different refractive indices, causing a phase delay and distinct polarization states.

These differences cause interference when the rays exit the material, producing a visible fringe pattern on a screen as shown in Fig. 7.30c thanks to the exploitation of polariscopes apparatuses (see Sect. 7.5.3).

7.5.2 Photoelastic Materials

Some materials become birefringent only when subjected to mechanical stress. This phenomenon is known as stress-induced birefringence, and such materials are called photoelastic. In stressed regions, the polarization directions align with the principal stress difference, and the difference in refractive indices is directly related to the principal stresses. The relationships, known as the Maxwell-Neumann (named as stress-optic law) equations [13], are:

$$n_1 = n_0 + A\sigma_1 + B(\sigma_2 + \sigma_3) \tag{7.34}$$

$$n_2 = n_0 + A\sigma_2 + B(\sigma_1 + \sigma_3) \tag{7.35}$$

By subtracting the two equations we obtain:

$$n_2 - n_1 = (B - A)(\sigma_1 - \sigma_2) = C(\sigma_1 - \sigma_2) \tag{7.36}$$

where C is the photoelastic constant of the material, and the difference in refractive indices is proportional to the difference between the principal stresses $\sigma_1 - \sigma_2$, independent of the third stress σ_3.

Using the optical delay:

$$\Delta s = h\,(n_2 - n_1) \tag{7.37}$$

And the relationship between optical phase difference ϕ, fringe order N, and wavelength λ can be expressed by:

$$N = \frac{\Delta s}{\lambda} = \frac{Ch}{\lambda}(\sigma_1 - \sigma_2) \tag{7.38}$$

Defining the fringe constant $F = \frac{\lambda}{C}$, the fundamental photoelastic equation becomes:

$$\sigma_1 - \sigma_2 = \frac{F}{h}N \tag{7.39}$$

This equation allows us to determine the stress difference at any point by measuring the fringe order N in the interference pattern.

7.5.3 *Transmission Photoelasticity*

In the transmission photoelastic test bench (Fig. 7.31), the specimen is placed between a polarizer and an analyzer, which is a polarizer as well, but placed at the end of the optical path. The polarizer transforms natural light into linearly polarized light, which passes through the stressed birefringent material specimen. The analyzer, oriented orthogonally to the polarizer, modulates the transmitted light according to the polarization changes induced by stress.

The interference pattern formed reveals two families of fringes:

- **Isochromatic fringes**: are fringes due to the phase delay induced by the principal stress difference $\sigma_1 - \sigma_2$.
- **Isoclinic fringes**: are fringes due to the coincidence of the principal stress direction with the polarizer and analyzer axis.

Isochromatic Fringes

In the locations where the two principal stresses are equal, the two polarized rays remain in phase, canceling out after the analyzer, producing dark fringes. If the stress difference increases, a phase delay arises, proportional to N, resulting in colored fringes corresponding to the fringe order, if white light is used, or intensity modulated fringes if monochromatic light is used.

Isoclinic Fringes

If one of the principal stress directions aligns with the incident polarization, one of the rays is blocked entirely by the analyzer, resulting in a dark point, since no orthogonally polarized ray is generated by the specimen, thus no phase delay either. The collection of such points forms isoclinic lines. By rotating both the polarizer and analyzer together, one can trace multiple isoclinic fringes corresponding to different principal stress orientations (see Fig. 7.32).

To isolate the isochromatic fringes and suppress isoclinic ones, the setup may include quarter-wave ($\lambda/4$) filters to convert linear to circular polarization and back.

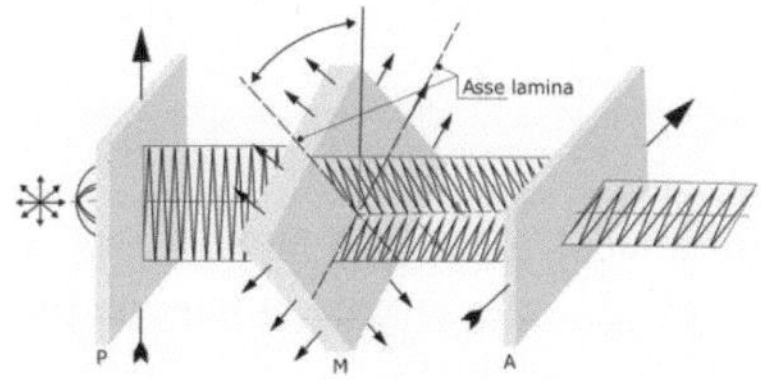

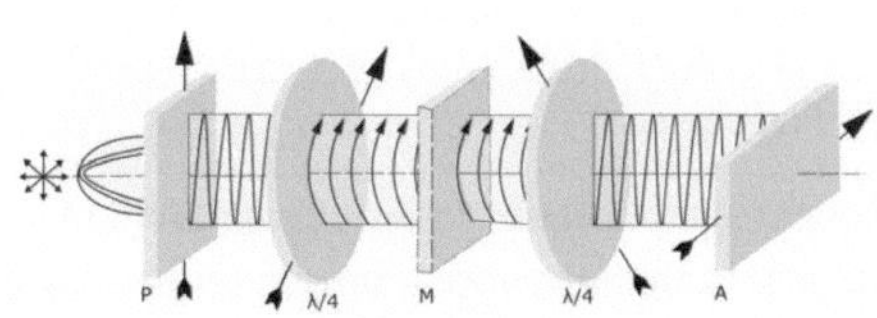

Fig. 7.31 Two transmission photoelastic benches: (left) standard setup with linear polarizers; (right) advanced setup with crossed polarizers and quarter-wave plates for circular polarization, enhancing stress fringe visualization.

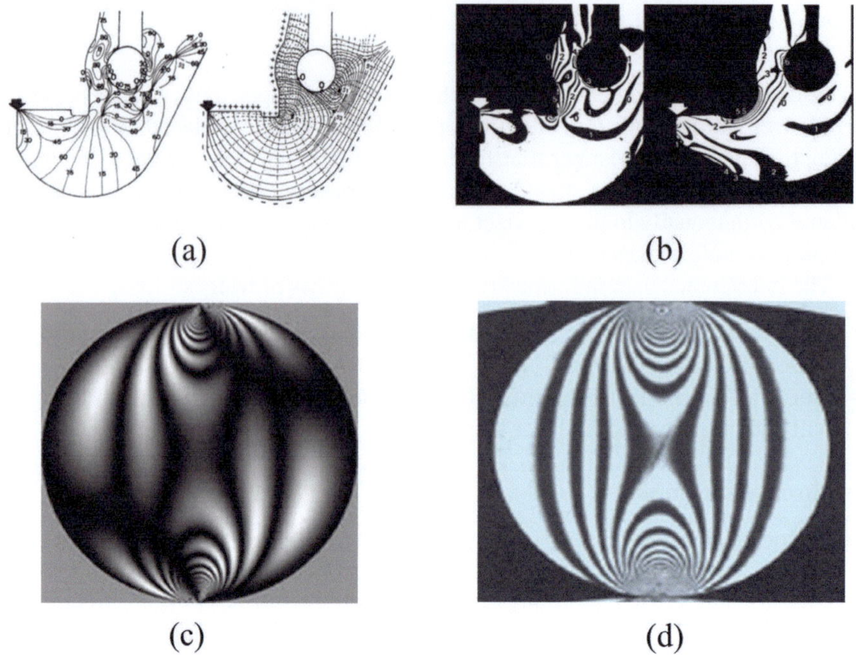

Fig. 7.32 Comparison of analytical (**a**) and experimental (**b**) photoelastic results in a loaded curved element, in (**c**) a loaded disc sample observed in a plane polariscope in which the isopclinics are clearly visible, in (**d**) the disc is observed via circular polarisciope setup, only the isochromates are observable

Circularly polarized light eliminates isoclinic fringes, making the isochromatic pattern easier to interpret, especially at low stress levels (Fig. 7.32).

Classically, plane polariscope set up is used to map the principal stress direction over the specimen surface; subsequently, the circular polariscope layout is used for achieving a precise measurement of the stress magnitude distribution.

7.5.4 Reflection Photoelasticity

Reflection photoelasticity is used when the component is opaque (e.g., made of metal). In this method, a birefringent coating is applied to the surface of the component. Polarized light illuminates the coating and reflects off the metal surface underneath. The reflected light passes through an analyzer, revealing a fringe pattern like the transmission method, as illustrated in Fig. 7.33.

To ensure accurate measurements:

- The coating must be thin enough so as not to alter the structural behavior.
- It must adhere perfectly to the component, so the strains in the coating match those of the substrate.

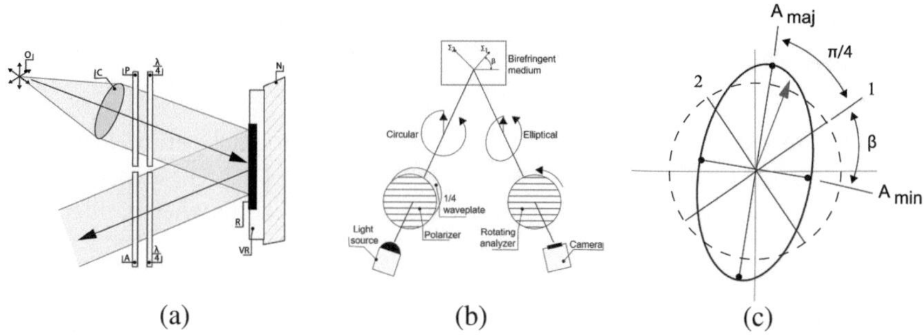

Fig. 7.33 Reflective photoelasticity: polarized light passes through a birefringent coating on an opaque specimen, reflects off the underlying layer and returns, producing stress-induced fringe patterns visible to the observer.

Typical coatings are made of epoxy resins with thicknesses ranging from 0.5 to 3 mm.

7.5.5 *Gray-Field Polariscopes*

Advanced measurements can be performed using gray-field polariscopes, which are optical systems designed to analyze the polarization state of reflected light. These devices use a kaleidoscopic camera setup that splits the image into four, each filtered through a different orientation analyzer, as illustrated in Fig. 7.34.

When circularly polarized light reflects off the coating, it becomes elliptically polarized due to differences in amplitude along the two principal directions. By measuring the ellipse parameters (major and minor axes), the refractive indices n_1 and n_2 can be determined. From this, the principal stress difference $\sigma_1 - \sigma_2$ is calculated. Each quadrant of the camera detects a different polarization component. By combining the four images, a full-field map of stress distribution can be reconstructed.

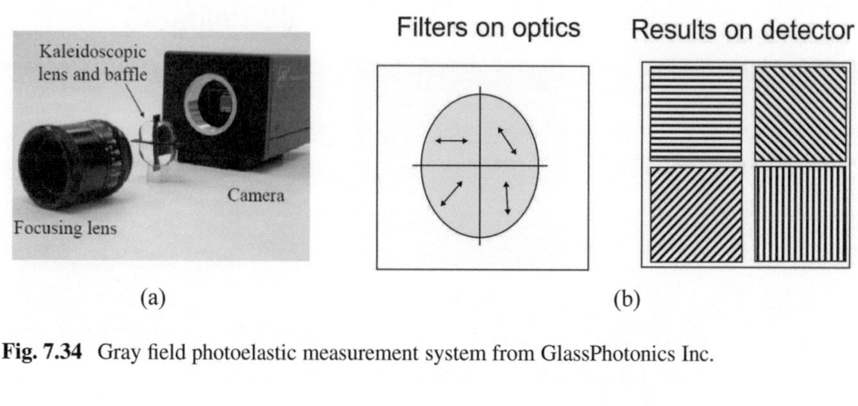

Fig. 7.34 Gray field photoelastic measurement system from GlassPhotonics Inc.

7.6 Electronic Speckle Pattern Interferometry (ESPI) and Sherography

Electronic Speckle Pattern Interferometry (ESPI) and shearography are measurement techniques based on the principle of laser light interference [14]. These methods use the unique property of laser light to produce interference patterns when it reflects off a surface.

7.6.1 Principle of Operation

In these techniques, the entire surface of the object being analyzed is illuminated with laser light. A video camera captures the reflected light from the surface. Due to the microscopic variations in the surface, the reflected light creates a random grainy image called a speckle pattern, which represents the fine surface structure of the object. This speckle pattern is then combined with a uniform reference laser beam. The interaction between the reference beam and the speckle pattern produces a new interference image, which highlights specific points on the object's surface. This image is stored as a reference in a computer. If the object moves or deforms, the speckle pattern changes. By comparing the new pattern with the reference image, it is possible to detect and measure even very small displacements or deformations. An example of this result is shown in Fig. 7.35.

By comparing the current speckle pattern with the reference pattern, a set of interference fringes, called correlation fringes, is generated. These fringes represent the displacement or deformation of the object's surface in the direction defined by the specific measurement setup and technique used [14].

Specialized data processing software then analyzes these correlation fringes. It automatically counts and interprets them, converting the visual fringe patterns into a precise set of numerical values that describe the deformation or strain on the object's surface.

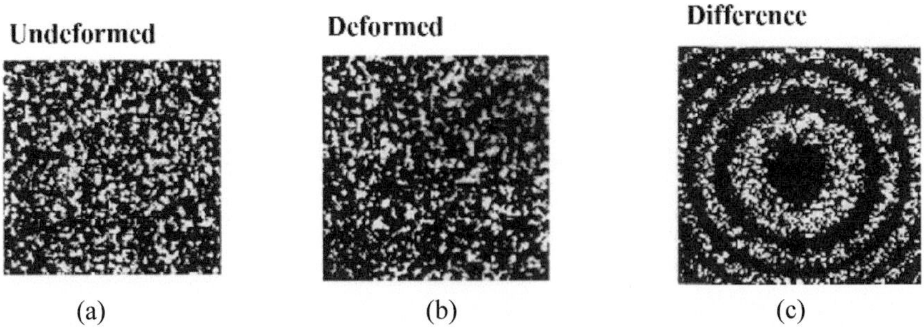

Fig. 7.35 ESPI speckle analysis: **a** speckle on undeformed surface, **b** speckle after deformation, **c** interference pattern of their difference highlighting surface displacement

7.6.2 *Electronic Speckle Pattern Interferometry (ESPI)*

Electronic Speckle Pattern Interferometry (ESPI) is an advanced optical measurement technique [14] that uses laser light to illuminate a surface, allowing for precise, full-field, non-contact measurements of surface displacements. ESPI can be applied to virtually any material or surface type and is especially useful for analyzing how an object deforms under mechanical or thermal loading. It provides three-dimensional maps of displacement, strain, and stress across the entire surface of the object under test. ESPI can be seen as a modern, electronic evolution of holographic interferometry. Instead of using photographic film to record interference patterns, ESPI uses a digital CCD camera to capture the images, and a computer to store and process them. When a rough surface is illuminated with coherent laser light, the scattered light produces a grainy pattern known as speckle. This speckle pattern is unique to the surface's microscopic texture, like a fingerprint. A portion of the laser light is split and directed as a reference beam, which is then combined (or superimposed) with the speckled light reflected from the surface. The interference between these two beams creates an interferogram, recorded by the camera as a speckle pattern. In ESPI, both the object beam and the reference beam travel along different optical paths from the laser source to the camera. The phase difference ϕ between these beams leads to the formation of interference patterns. If the object moves or deforms, the path of the reflected light changes, introducing a new phase $\phi + \Delta$. The difference Δ indicates how much the object's surface has moved during the measurement. This change results in visible correlation fringes, which are counted and analyzed.

By measuring the number N of fringes at each point on the object, the surface displacement in the measurement direction can be determined with sub-wavelength accuracy. This measurement direction lies midway between the illumination and observation angles.

The sensitivity can be expressed using the following formula [14]:

$$d = \frac{N\lambda}{2\cos\left(\frac{\alpha}{2}\right)} \tag{7.40}$$

where:

- d is the deformation component in the measuring direction
- N is the fringe order (number of fringes)
- λ is the wavelength of the laser light
- α is the angle between the illumination and observation directions.

While ESPI is a powerful technique, it requires stable environmental conditions, as the interference fringes are highly sensitive to external vibrations or disturbances. Despite this, ESPI is extremely effective for detecting very small deformations, whether on tiny components or large structures. A notable example is the analysis of vibrations in high-frequency systems such as a violin body (Fig. 7.37), where structural deformations can be detected at frequencies starting from around 140 Hz and above.

7.6.3 *Shearography: Working Principle and Setup*

Shearography is an optical, non-contact measurement technique that uses laser light to detect surface deformations [14]. Unlike traditional Electronic Speckle Pattern Interferometry (ESPI), shearography works by superimposing two slightly shifted (sheared) images of the same object surface, as shown in Fig. 7.36b.

This shearing means that points on the object that are a small distance apart are compared directly in the image. When these two points are illuminated by coherent laser light, the difference in the path lengths to the camera creates a phase shift (φ), which leads to the formation of an interference pattern known as a speckle pattern.

If the object deforms, for example, due to thermal, pressure, or mechanical loading, the light paths change differently for each point, generating a new phase difference (φ + Δ). The change in phase (Δ) between the two sheared images produces correlation fringes, which indicate the relative displacement between neighboring surface

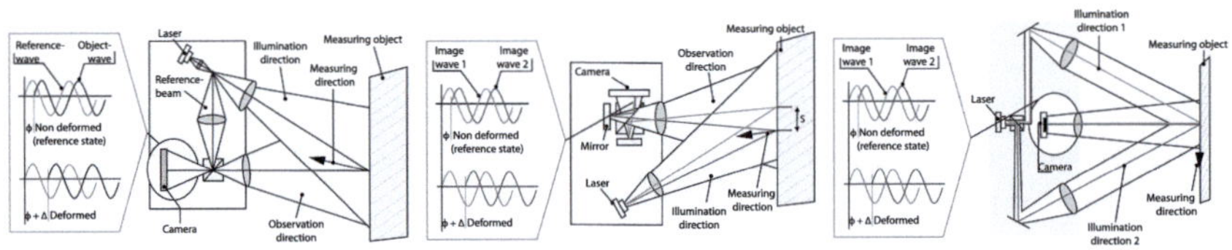

Fig. 7.36 ESPI optical setups: **a** shearography for strain via displacement gradients, **b** speckle interferometry for displacement, **c** ESPI system for full-field deformation measurement.

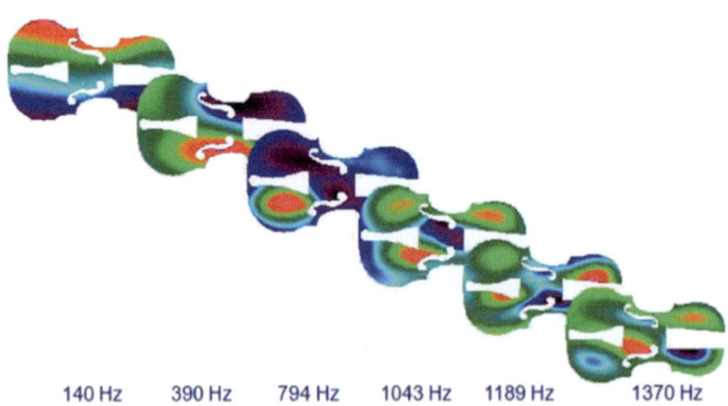

Fig. 7.37 ESPI vibration analysis: deformation patterns of a violin body under dynamic excitation from Dantec company

points. By counting the number of fringes (N) at each image point, one can determine the surface deformation gradient in the direction defined by the shear.

The sensitivity of this technique is determined by the shear distance (s), which is the physical distance between the compared points on the surface. The deformation gradient (d/s) can be calculated using the following formula [14]:

$$\frac{\mathrm{d}}{\mathrm{s}} = \frac{\mathrm{N}\,\lambda}{2\cos\left(\frac{\alpha}{2}\right)} \tag{7.41}$$

where:

- d is the deformation component along the measuring direction,
- s is the shear distance,
- N is the number of fringes observed,
- λ is the laser wavelength,
- α is the angle between the illumination and observation directions.

Shearography is widely used in industrial non-destructive testing (NDT) due to its high sensitivity and ability to detect very small deformations. It is less sensitive to environmental vibrations compared to ESPI, making it more robust for field use. Applications include testing composite structures in the aerospace, automotive, marine, and wind turbine industries.

For instance, in Fig. 7.38, shearography is used to inspect honeycomb composite panels commonly used in aircraft. When the component is slightly heated or placed under vacuum (typically 50–300 mbar), internal flaws like delaminations cause local surface deformations. These are detected by fringe patterns.

Fig. 7.38 Tyipical Shearography measurement systems and results, from Dantec company

7.6.4 Speckle Interferometry (Dual-Beam ESPI)

Speckle Interferometry (also known as dual-beam ESPI) is another optical method for measuring surface deformations. It uses two laser beams to illuminate the object from symmetric directions relative to the camera (see Fig. 7.36c). This setup generates a speckle pattern based on the difference in phase (φ) between the two light paths. The initial pattern serves as a reference.

When the object deforms, the phase difference changes ($\varphi + \Delta$), producing a new speckle pattern. The difference between the original and new patterns appears as correlation fringes, which can be analyzed to determine surface displacement at each point. The measuring direction is perpendicular to the camera view and lies in the plane defined by the two laser beams. If the beams are parallel, the measuring direction remains constant across the entire field.

The sensitivity [14] is given by the formula:

$$d = \frac{N\,\lambda}{2\sin\left(\frac{\alpha}{2}\right)} \tag{7.42}$$

where:

- d is the deformation in the measuring direction,
- N is the number of fringes observed,
- λ is the laser wavelength,
- α is the angle between the two laser beams.

In Fig. 7.39, two examples of deformation measurement using speckle interferometry are shown:

- Figure 7.39a: A ceramic plate with two holes is loaded with a force ranging from 5.2 kN to 62 kN. The resulting deformation is visualized using pseudo-color mapping, where blue indicates minimum deformation and red indicates maximum (horizontal elongation).

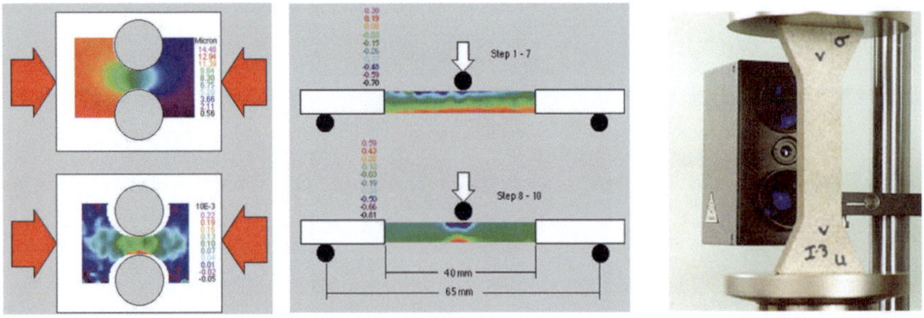

(a) (b) (c)

Fig. 7.39 Typical results of Speckle interferometry and Laser optical strain sensor from Dantec company

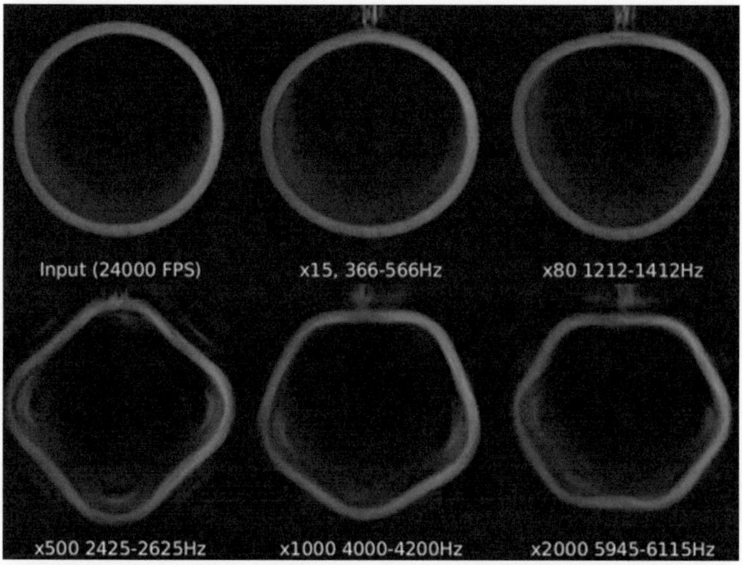

Fig. 7.40 Mode shapes of a vibrating tube section obtained by motion magnification

- Figure 7.39b: A bending specimen is subjected to incremental loading. At low loads (steps 1–7), the strain distribution is uniform: compression on the top surface (blue) and tension on the bottom (red). At higher loads (steps 8–10), localized deformation becomes apparent, highlighting the onset of material failure or stress concentration.

7.7 Motion Magnification

Motion Magnification is a computational technique that processes videos to amplify small movements of objects or surfaces, making them visible [15, 16]. The first algorithm developed to magnify motions in a video was published in 2005 by Liu et al. [17]. This video processing technique allows detection of vibrations or displacements that occur at micrometer or even nanometer scales using just recorded video data.

There are two main classes of motion magnification techniques:

(a) Eulerian Video Magnification (EVM) that amplifies changes in pixel intensity over time in specific frequency bands. Easier to implement and good for periodic motion like heartbeat or structural vibration, but can introduce artifacts for large or complex motion.
(b) Phase-Based Motion Magnification (PBMM) that extracts and manipulates phase information in the image's local frequency components and is more robust to noise, capable of subpixel accuracy but computationally more intensive.

After magnifying motion, the movement of specific points or regions in the video can be tracked and analyzed quantitatively. This is typically done by optical flow, feature tracking and by DIC processing. Tools such as OpenCV, MATLAB, and custom Python pipelines can be used to perform motion magnification.

An example of possible results obtained by application of these processing techniques to an high speed video (24000 fps) of a vibrating section of a tube is illustrated in Fig. 7.40, from [18].

An example of possible results obtained by the application of these processing techniques is shown in Fig. 7.40.

Some key application areas include structural health monitoring by detecting vibrations in bridges, buildings, and machinery; biomedical imaging by observing subtle movements in the human body (e.g., carotid pulse); mechanical systems analysis by identifying resonant frequencies, mode shapes or malfunctioning components.

References

1. G.W. Housner, T. Vreeland, The analysis of stress and deformation, in *Library of Congress Catalog Card Number: 65–22615* (1991)
2. M.A. Sutton, J.-J. Orteu, H.W. Schreier, *Image Correlation for Shape, Motion and Deformation Measurements Basic Concepts, Theory and Applications* (Springer, 2009)
3. M. Moretti, G.L. Rossi, E. Speranzini, Structural integrity detection of historical structures by dynamic digital image correlation, in *International Conference on Experimental Vibration Analysis for Civile Engineering Structures—EVACES 2011 Starrylink Editrice* (Varenna, Italy, 2011)
4. G. Allevi, F. Docchio, S. Pasinetti, G. Rossi, P. Castellini, R. Marsili, A. Quattrocchi, G. Sansoni, P. Chiariotti, R. Montanini, R. Rossetti, E.P. Tomasini, Qualification of additive manufactured trabecular structures using a multi-instrumental approach, in *IEEE I2MTC* (Auckland, New Zealand, 2019)
5. B.D. Lucas, T. Kanade, An iterative image registration technique with an application to stereo vision. Proc. DARPA Image Underst. Workshop **2**, 674–679 (1981)
6. B. K. P. Horn, B. G. Schunck, Determinig optical flow. Artif. Intell. (1981)
7. M. J. Black, P. Anandan, A framework for the robust estimation of optical flow. ICCV (1993)
8. J.L. Barron, D.J. Fleet, S. Beauchemin, Performance of optical flow techniques. IJCV (2005)
9. T. Tocci, L. Capponi, R. Marsili, F. Chiavarini, J. Pirisinu, G. Rossi, Optical-flow based analysis for range hoods captured flow measurement. J. Phys. Conf. Ser. 29th AiVeLa Conf. **2293** (2022)
10. L. Capponi, T. Tocci, M. D'Imperio, S.H.J. Abidi, M. Scaccia, F. Cannella, R. Marsili, G. Rossi, Thermoelasticity and ArUco marker-based model validation of polymer structure: application to the San Giorgio's bridge inspection robot. Acta Imeko **10**(4), 177–184 (2021)
11. T. Tocci, L. Villafane Roca, F. Panerai, R. Marsili, G. Rossi, Motion measurements, dynamics study and shape reconstruction of a sub-scale parachute in wind tunnel tests through fiducial markers, in *XXX AIVELA National Meeting* (Rome, 2022)
12. T. Tocci, L. Capponi, G. Rossi, ArUco marker-based displacement measurement technique: uncertainty analysis. IEEE Trans. Instrum. Meas.
13. K. Ramesh, *Digital Photoelasticity* (Springer, 2000)
14. W. Steinchen, L. Yang, *Digital Shearography—Theory and Application of Digital Speckle Pattern Shearing Interferometry* (SPIE Press, 2003)
15. H.-Y. Wu, M. Rubinstein, E. Shih, J. Guttag, F. Durand, W. Freeman, Eulerian video magnification for revealing subtle changes in the world. ACM Trans. Graph. (2012)

16. N. Wadhwa, M. Rubinstein, F. Durand, W. Freeman, Phase-based video motion processing. ACM Trans. Graph. (2013)
17. C. Liu, A. Torralba, W.T. Freeman, F. Durand, E.H. Adelson, Motion magnification. ACM Trans. Graph. **24**, 519–526 (2005)
18. N. Wadhwa, J.G. Chen, J.B. Sellon, D. Wei, M. Rubinstein, R. Ghaffari, S.H. Kang, Motion microscopy for visualizing and quantifying small motions. Proc. Natl. Acad. Sci. **114**(44), 11639–11644 (2017)

Chapter 8
Temperature and Stress Field Measurements by Thermographic Image Analysis

Abstract This chapter presents the fundamental concepts, devices and techniques underlying the measurement of temperature fields through infrared radiation detection. It introduces the primary physical principles governing infrared emission, including Planck's and Stefan–Boltzmann's laws, and discusses the role of emissivity and infrared optics, along with the key components of a thermal camera. The performance characteristics of thermographic systems are described, with guidance on selecting and using them appropriately for surface temperature distribution measurements. Dynamic thermography enables the detection not only of steady-state temperature fields but also of temporal fluctuations, capturing a "temperature video" that can reveal subsurface defects and surface stress fields. The chapter illustrates the main techniques for stress field measurement using dynamic and differential thermography and highlights a range of applications in experimental mechanics, structural component qualification, model validation, and beyond.

8.1 Introduction

Thermography, also known as thermal imaging, is a measurement technique that detects infrared radiation emitted by objects and converts it into images in which the grey levels correspond to temperature. Its development spans several centuries, with significant contributions from the fields of physics, military technology, medicine, and industry. The origins of thermography can be traced back to 1800, when astronomer William Herschel discovered infrared radiation while experimenting with sunlight and prism. By placing a thermometer just beyond the red end of the visible spectrum produced by the prism, he observed an increase in temperature, indicating the presence of an invisible form of radiation. This discovery marked the beginning of infrared science, leading to the development of numerous theories and instruments for detecting infrared radiation. However, modern thermographic imaging systems were developed much later, between the 1930 and 1940s, primarily for military applications such as night vision during World War II. In the 1960s, thermography found its first medical applications, being used to detect abnormal heat patterns associated

G. Rossi, *Mechanical and Thermal Measurements by Images and Waves*, Smart Sensors, Measurement and Instrumentation 51, https://doi.org/10.1007/978-3-032-03464-9_8

with diseases such as breast cancer and circulatory disorders. During the 1970s, the technique gained traction in the oil and gas industry, as well as in electrical engineering, for inspecting pipelines, power grids, and structural components. The 1980s saw major advancements in infrared detector technology, which led to the development of more compact and accurate thermal cameras. These improvements broadened the range of applications, including law enforcement, building diagnostics, and wildlife monitoring. The 1990s ushered in the integration of digital imaging and computer processing, significantly enhancing the accuracy, speed, and usability of thermal cameras. In the 2000s, infrared cameras became increasingly affordable and began to be incorporated into consumer electronics, such as smartphones and home security systems. In recent decades, artificial intelligence and machine learning have further advanced thermal analysis, enabling automated diagnostics in healthcare, firefighting, and predictive maintenance. The COVID-19 pandemic notably accelerated the global adoption of thermographic screening for fever detection in public spaces. Today, thermography is widely applied across medicine, industry, environmental monitoring, and security [1–4]. As technological innovation continues, thermal imaging is expected to become even more accessible, intelligent, and integrated into everyday life.

8.2 Infrared Radiation Emission, Transmission, Reflections, Absorption

Every object with a temperature above absolute zero emits thermal energy in the form of electromagnetic radiation. Objects at typical environmental temperatures primarily emit radiation in the infrared region of the electromagnetic spectrum, that is, at wavelengths longer than those of visible light, typically above 0.7 μm. When infrared radiation, which is an electromagnetic wave, reaches the surface of a material, it can be reflected, absorbed, or transmitted, depending on the optical and physical properties of the material, as illustrated in Fig. 8.1a and previously discussed in Chap. 1).

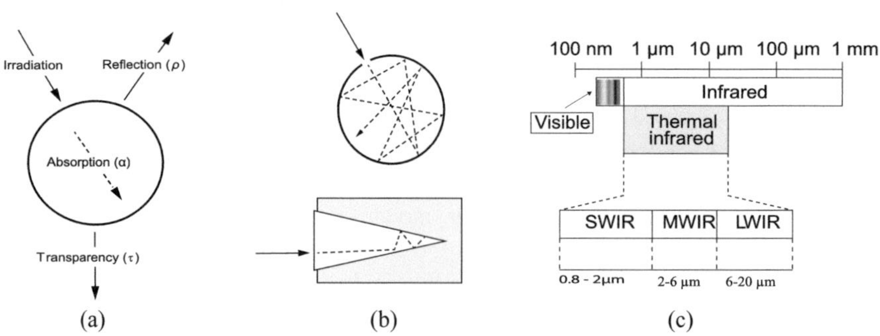

Fig. 8.1 **a** absorption, reflection, transmission; **b** the black body; **c** Infrared spectrum

If we define the coefficient α as the ratio of the absorbed infrared power to the incident power, ρ as the ratio of the reflected power to the incident power, and τ as the ratio of the transmitted power to the incident power, then, according to the principle of energy conservation:

$$\alpha + \rho + \tau = 1 \tag{8.1}$$

We define a black body, illustrated in Fig. 8.1b as a body for which $\alpha = 1$, $\rho = 0$, and $\tau = 0$. It absorbs all incident radiation, reflecting and transmitting none, and emits the maximum possible radiation at a given temperature T across all wavelengths. The term "black" arises from the observation that darker surfaces tend to absorb more radiation than lighter ones. However, a black body does not necessarily appear black to the human eye. A practical realization of a black body can be achieved, as illustrated in Fig. 8.1b, by using a cavity with a small hole; radiation entering the hole undergoes multiple reflections and is eventually absorbed, making the cavity effectively non-reflective and non-transmissive. Interestingly, a very effective natural approximation of a black body is a surface of fresh white snow, which, due to its multitude of microcavities and the strong absorption properties of ice, demonstrates excellent infrared absorption.

For most materials that are not transparent to infrared radiation, the transmitted component is negligible ($\tau = 0$), and Eq. (8.1) reduces to:

$$\alpha + \rho = 1 \tag{8.2}$$

This means that a non-transparent, non-black body reflects a portion of the incident infrared radiation. Consequently, observing its surface involves detecting both its own emitted radiation and the radiation reflected from other surrounding sources. There are also idealized materials with special properties. A perfectly infrared-transparent medium has $\alpha = 0$, $\rho = 0$, and $\tau = 1$, making it suitable for use in infrared optics. Conversely, a perfect infrared mirror has $\alpha = 0$, $\rho = 1$, and $\tau = 0$. The infrared radiation observed from a surface is therefore a combination of internally emitted radiation and external radiation reflected by the surface. The resulting spectral distribution of electromagnetic power, denoted by $W(\lambda, T)$, depends on the surface temperature T and the wavelength λ, and includes both emission and reflection components. In contrast, a black body emits only its own radiation, independent of external sources. Its spectral power distribution $W_b(\lambda, T)$ as a function of wavelength λ and temperature T is described by the fundamental physical principle [5] known as Planck's Law:

$$W_b(\lambda, T) = \frac{C_1}{\lambda^5\left(e^{\frac{C_2}{\lambda T}} + 1\right)} \tag{8.3}$$

where $C_1 = 3.742 \times 10^8$ [(W·μ m^4)/m^2] and $C_2 = 1.439 \times 10^4$ [μ m·K]. A plot of Planck's Law is shown in Fig. 8.2.

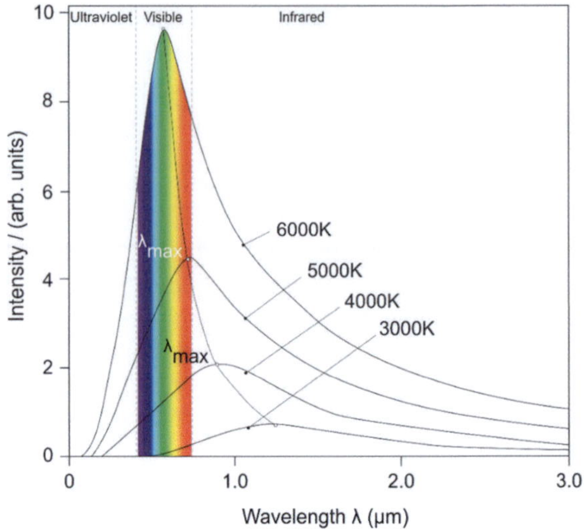

Fig. 8.2 Plot of the Plank's law

The infrared energy emitted by a body within a specific wavelength bandwidth, from λ_1 to λ_2, at temperature T, is given by:

$$W_{\lambda_1,\lambda_2}(T) = \int_{\lambda_1}^{\lambda_2} W_b(\lambda, T)\, d\lambda \tag{8.4}$$

This quantity represents the area under the Planck's law curve at temperature T, between the wavelengths λ_1 and λ_2. From this relationship, it follows that at low temperatures, significant radiation is emitted at wavelengths outside the visible range (not perceived by the human eye). As temperature increases, the peak of the emitted energy spectrum shifts toward shorter wavelengths, eventually entering the 0.4–0.7 μ m range associated with visible light. This explains why cold objects are visible only when illuminated, while incandescent bodies emit visible light due to their high temperatures. A significant fraction of solar radiation, given the sun's surface temperature of approximately $T = 5800$ K, falls within the visible spectrum. This aligns with the biological evolution of the human eye, which developed sensitivity under solar radiation. In contrast, at temperatures below 800 K, most of a black body's emission occurs in the infrared range. As shown in Fig. 8.2, increasing the temperature shifts the peak of the emission spectrum to shorter wavelengths. The position of this maximum, denoted by λ_m is found by differentiating Planck's law and solving for the condition [5]:

$$\frac{\partial W_b(\lambda, T)}{\partial \lambda} = 0 \tag{8.5}$$

This results in Wien's law:

$$\lambda_m \cdot T = 2898\ \mu m \cdot \mathrm{K} \tag{8.6}$$

This law states that the wavelength at which the emission is maximum is inversely proportional to the temperature. By integrating the spectral emission over the entire wavelength domain, we obtain the total radiated power of a black body. This leads to the Stefan–Boltzmann law:

$$W_{t,b}(T) = \int_0^\infty W_b(\lambda, T)\, d\lambda = \sigma \cdot T^4 \tag{8.7}$$

where $\sigma = 5.67 \times 10^{-12}$ W/(cm^2 K^4) is the Stefan–Boltzmann constant.

The Stefan–Boltzmann law tells us that the total power emitted by a black body increases rapidly with temperature, following a fourth-power relationship. This principle is also useful for indirectly measuring temperature by observing the total radiated power. However, real bodies emit less energy than black bodies. This discrepancy is accounted for by emissivity, defined as the ratio of the power emitted by a real body to that emitted by a black body at the same temperature. Emissivity generally depends on the wavelength λ, the temperature T, and the observation angle θ:

$$\varepsilon(\lambda, T, \theta) = \frac{W(\lambda, T, \theta)}{W_b(\lambda, T)} \tag{8.8}$$

where $W(\lambda, T, \theta)$ is the emitted power of the real body at a given angle θ, wavelength λ, and temperature T.

In many practical applications, the dependence of emissivity on temperature can be neglected, particularly if the temperature range of interest is narrow. The hemispherical total emissivity (averaged over all angles) can be obtained by integrating over the hemisphere of observation ($\gamma = \pi$), or over a given solid angle γ:

$$\varepsilon(\lambda) = \int_0^\gamma \varepsilon(\lambda, \theta)\, d\theta \tag{8.9}$$

The total emissivity across all wavelengths is:

$$\varepsilon = \int_0^\infty \varepsilon(\lambda, T)\, d\lambda \tag{8.10}$$

Values of total emissivity for many common materials and surface finishes can be found in scientific literature. Given a known emissivity ε, the total power radiated by a real body is:

$$W_{t,r} = \varepsilon \cdot \sigma \cdot T^4 \tag{8.11}$$

Therefore, if ε is known, the temperature of a real object can be determined from detection of its emitted thermal power.

It is important to note that two objects at the same temperature but with different emissivity will emit different amounts of thermal power, as shown in Fig. 8.3a. If their emissivity is assumed to be equal, temperature measurements based solely on emitted thermal power calculated by Eq. (8.11) will be incorrect. Furthermore, since emissivity can vary with the observation angle, detecting infrared radiation from different viewing directions may yield varying thermal power levels and thus by Eq. (8.11) different measured temperatures for same body, as illustrated in Fig. 8.3b. In Table 8.1 some typical emissivity values of common materials are given.

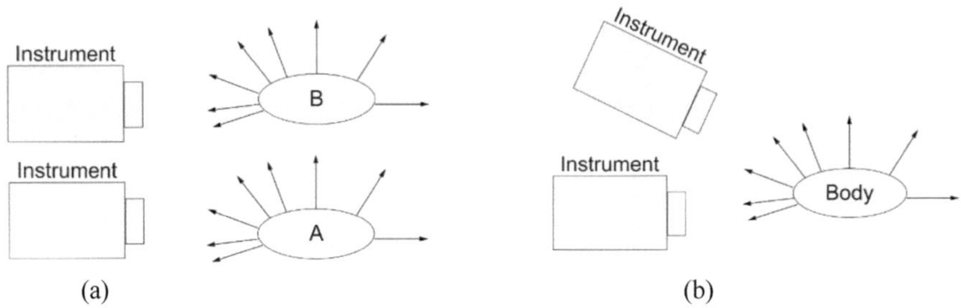

Fig. 8.3 Effects of emissivity due to its angular dependence

Table 8.1 Some emissivity values of common materials

Material	Emissivity
Aluminum (polished)	0.05–0.10
Aluminum (anodized)	0.77–0.95
Copper (polished)	0.02–0.05
Copper (oxidized)	0.60–0.80
Gold (polished)	0.02–0.05
Iron (polished)	0.10–0.20
Iron (rusted)	0.60–0.80
Stainless Steel (polished)	0.10–0.15
Stainless Steel (oxidized)	0.45–0.85
Brick	0.90–0.95
Concrete	0.85–0.95
Glass (smooth)	0.85–0.95
Ice	0.95–0.97
Snow	0.80–0.90
Water	0.95–0.98
Wood (varnished)	0.80–0.90
Human Skin	0.97
Paint (matte, dark)	0.90–0.98

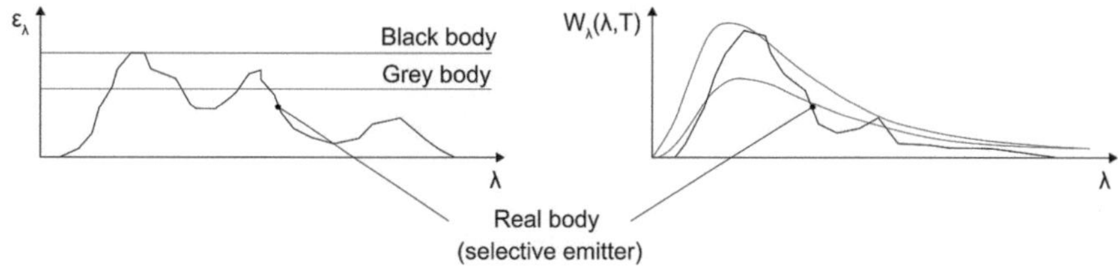

Fig. 8.4 Emissivity and emitted spectrum changes

Another model is that of a body with emissivity less than one but not changing with wavelength, at least in a certain wavelength range. Bodies that maintain constant emissivity with wavelength are called *gray bodies.* In Fig. 8.4, the emissivity plots for a black body, a gray body, and a real body are illustrated. The infrared emission spectrum for the black body follows the characteristic shape defined by Planck's law. The gray body exhibits the same spectral shape but with reduced intensity across all wavelengths, due to its constant emissivity being less than 1. In contrast, the real body may show a completely different spectral distribution, as its emissivity varies with wavelength, leading to a non-uniform deviation from the ideal Planck distribution.

To measure the temperature of an object, its thermal radiation must be detected. However, during the propagation path, the thermal radiation emitted by the object passes through the surrounding air and the infrared optics of the receiving instrument and attenuated as illustrated in Fig. 8.5.

Now, let us analyze how the atmosphere transmits thermal radiation. By considering a cubic meter of air through which thermal radiation passes at various wavelengths, it is possible to measure the thermal energy entering and exiting the volume to determine the absorption coefficient. It is observed that the absorption coefficient varies significantly depending on the wavelength.

Based on these absorption characteristics, three low-absorption transmission windows can be identified, primarily influenced by the absorption of water vapor and carbon dioxide present in the air. These windows are known as the near-infrared (1–2.5 μm–NIWR), mid-infrared (3.5–5.5 μm–MIWR), and far-infrared (7.5–14 μ m–LIWR) windows.

The following table shows the relative concentration of gases present in dry air and their absorption of radiation in the infrared frequency range (2–14 μm) [5] (Fig. 8.6).

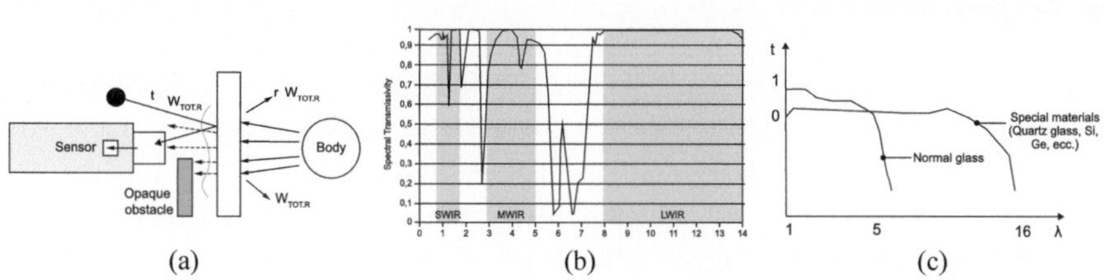

Fig. 8.5 Infrared radiation attenuation through air and infrared optics materials, **a** scheme of transmission, attenuation and reflection, **b** atmosphere transmission coefficient, **c** optical materials transmission coefficient

Constituent (major)	Chemical formula	Percent by volume	Absorbs between 2 and 15 μm
Nitrogen	N_2	78.084	No
Oxygen	O_2	20.946	No
Argon	A	0.934	No
Carbon Dioxide	CO_2	0.032	Yes
Neon	Ne	1.818X10^{-3}	No
Helium	He	5.24X10^{-4}	No
Methane	CH_4	2.0X10^{-4}	Yes
Krypton	Kr	1.14X10^{-4}	No
Nitrous oxide	N_2O	5.0X10^{-5}	Yes
Hydrogen	H_2	5.0X10^{-5}	No
Xenon	Xe	9.0X10^{-6}	No
Carbon monoxide	CO	7.5X10^{-6}	Yes

(a)

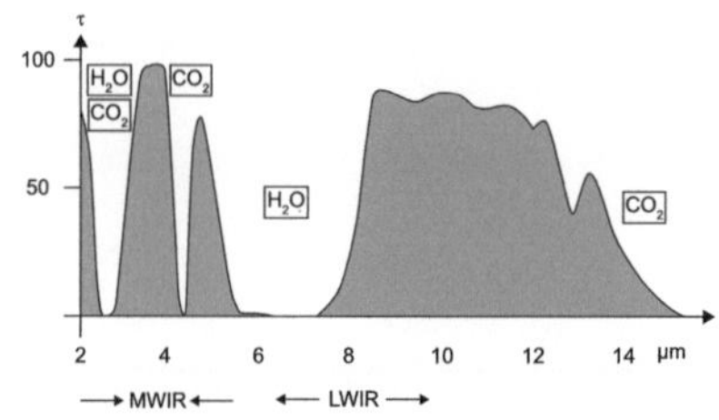

(b)

Fig. 8.6 Infrared transmission of atmospheric air components (**a**) and atmosphere transmission coefficient at different wavelenght

Approximately, the transmittance of the atmosphere [1] depends on gases such as CO_2 (carbon dioxide) and H_2O (water vapor). By observing how the transmittance varies with wavelength in the infrared region, four absorption "windows" can be noticed, which are wavelength bands where absorption is particularly high. Specifically, the first absorption window is influenced by both CO_2 and H_2O, while the second and fourth windows are primarily due to CO_2, and the third window is primarily due to H_2O. Therefore, we focus on the "clear" regions of the graph, where transmittance is higher. To evaluate the absorption of water in the atmosphere, the precipitable water vapor is considered. This represents the amount of water vapor present along the optical path of the radiant energy. It is defined as the thickness, in millimeters, of the layer of water that would be formed if all the water vapor along the path were condensed into a container with the same cross-sectional area as the energy beam. The amount of water vapor present in the air is measured using two parameters: absolute humidity (AH) and relative humidity (RH). Absolute humidity represents the mass of water vapor contained in a volume of air at temperature T. Relative humidity represents the mass of water vapor contained in the same volume of air when it is saturated at the same temperature T.

Therefore, the precipitable water vapor can be derived from the formula [5]:

$$p_W = 10^{-3} \cdot AH = 10^{-3} \cdot AH_{\text{sat}} = 5.072 \cdot e^{0.06T} \cdot RH \tag{8.12}$$

where AH_{sat} represents absolute humidity. For temperatures between 0 and 35 °C, it can be approximated as:

$$p_W = 5.072 \cdot e^{0.06T} \cdot RH \tag{8.13}$$

Alternatively, a chart can be used [5], like that illustrated in Fig. 8.7a. Given the measured temperature and relative humidity, a line is plotted on the chart, and the point of precipitable water vapor amount is found by intersection.

Once this value is determined, it can be used in the diagram of Fig. 8.7b. The diagram allows determining the air transmission index, considering the absorption of CO_2. By drawing a vertical line, it intersects with two lines, and the two obtained points represent the air transmission coefficient for the far-infrared (8–12 μm) and

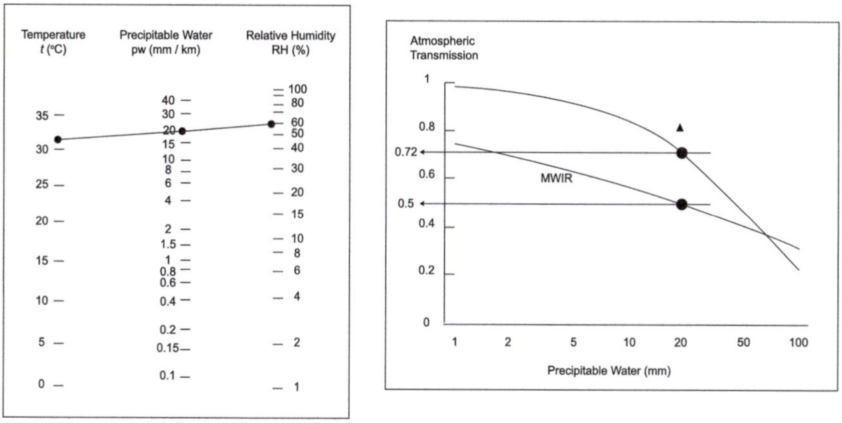

(a) (b)

Fig. 8.7 Defining transmissivity of air by nomogram and graph

Fig. 8.8 Typical infrared lenses and objectives

near-infrared (3–5 μm) ranges. It is observed that the transmissibility of atmosphere in the far-infrared range is better.

Regarding transmission through optical components such as lenses and mirrors, it is important to select materials with a high transmission coefficient at the specific wavelength used in the instrument, like germanium or zinc selenide (Fig. 8.8).

Choosing the appropriate optical material helps maximize the transmission of thermal radiation. Another aspect that affects transmission through lenses is cleanliness. A dirty lens can compromise the calibration effect of the instrument, so it is essential to keep the optical elements clean to ensure optimal performance.

8.3 Infrared Radiation Measurement by Detectors

Sensors used to detect infrared radiation from a point on a surface can be classified as follows:

- Broadband sensors (bolometric, uncooled sensors) detect the total thermal power emitted at all wavelengths

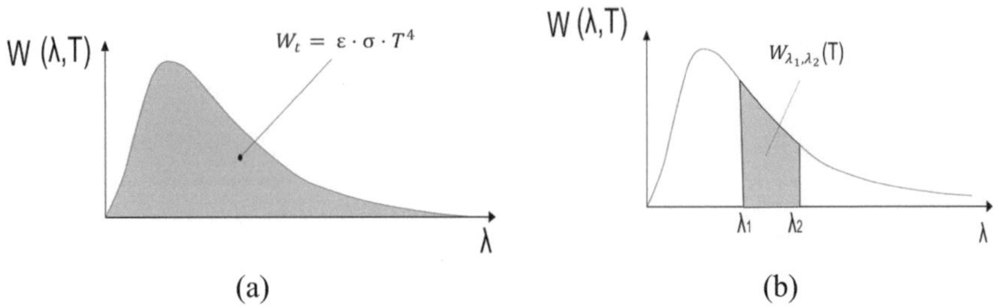

Fig. 8.9 Thermal infrared power captured by broadband detectors (**a**) and band-limited detectors (**b**)

- Monochromatic (band limited) sensors (photonic photodiodes normally cooled) detect a narrow bandwidth of the thermal radiation emitted by a body.

The two approaches are illustrated in Fig. 8.9a, b respectively, in terms of thermal energy detected, quantified by the area under the emission spectrum of a blackbody.

Broadband thermal sensors are designed to absorb thermal energy at all wavelengths, causing an increase in the sensor's temperature until it reaches equilibrium. The most used in array configurations are resistive materials whose resistance changes with temperature, named bolometers. As single detectors, other types of temperature sensors such as thermocouples and thermistors can also be used. They are considered broadband sensors as they capture all the wavelengths emitted by the measured object. However, they have a slow dynamic response because it is necessary to wait for a temperature change of the sensor itself due to infrared radiation heating.

Photodiodes, as described in Chap. 3, are semiconductor photodetectors sensitive only to specific narrow-band wavelengths and are called monochromatic sensors. They have a fast dynamic response as they do not require a change in temperature by heating. They collect less energy (lower area below Planck's low) but photodiodes offer high sensitivity and fast dynamic response, making them ideal for temperature measurement in specific spectral ranges. The thermal radiation to be measured must be focused on the sensor by lenses or mirrors as illustrated in Fig. 8.10. This radiation

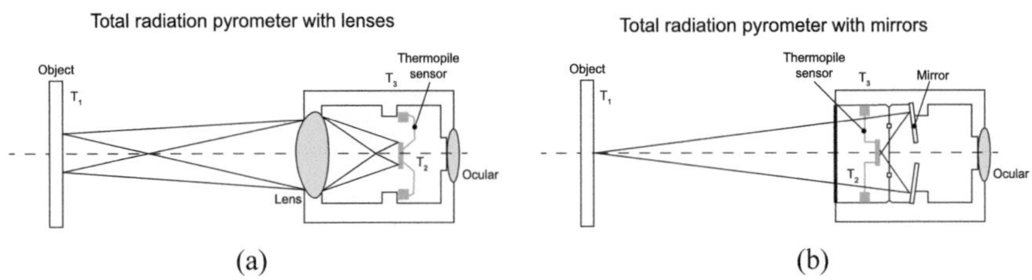

Fig. 8.10 Optical scheme of infrared radiation detection by single point pyrometers with lenses (**a**) and mirrors (**b**)

either heats the sensor (such as a thermocouple or thermistor) or is directly converted into an electric current by photodiodes.

Temperature measurement by total radiation sensors can be performed by reaching a thermal equilibrium between the object being measured at temperature T_1, the sensor at temperature T_2, and the instrument at temperature T_3. If the heat exchange between the instrument and the sensor always occurs through convection and between the instrument and the object occurs through radiation, we can write the following equation:

$$Q_{\mathrm{OUT}} = Q_{\mathrm{IN}} \quad \Rightarrow \quad K \cdot (T_2 - T_3) \propto \varepsilon \cdot \sigma \cdot T_1^4 \tag{8.14}$$

where K is the adductive heat transfer coefficient (convection + radiation). The thermal sensor generates an output E:

$$E \propto (T_2 - T_3) \propto T_1^4 \tag{8.15}$$

If the sensor is a photodiode, it allows the passage of electrons to different energy levels in response to incident light. Considering the total photon flux emitted by the object:

$$N_f = k \cdot \varepsilon \cdot T^3 \quad \left[\mathrm{photons/s \cdot m^2}\right] \tag{8.16}$$

The photodiode output will be:

$$E_0 = K \cdot \varepsilon \cdot T^3 \quad \left[\mathrm{photons/s \cdot m^2}\right] \tag{8.17}$$

From which

$$T = \left(\frac{E_0}{K \cdot \varepsilon}\right)^{\frac{1}{3}} \tag{8.18}$$

The uncertainty can be derived by differentiating and dividing by the temperature, yielding:

$$\frac{dT}{T} = -\frac{1}{3}\frac{d\varepsilon}{\varepsilon} \tag{8.19}$$

So even if:

$$\frac{d\varepsilon}{\varepsilon} = 15\% \quad \text{then} \quad \frac{dT}{T} = 5\% \tag{8.20}$$

A significant advantage is evident as the temperature uncertainty is one-third compared to the uncertainty on the emissivity.

It is also possible to perform radiation measurements at two wavelengths, i.e., two narrow wavelength bandwidths. The two-wavelength measurement technique allows

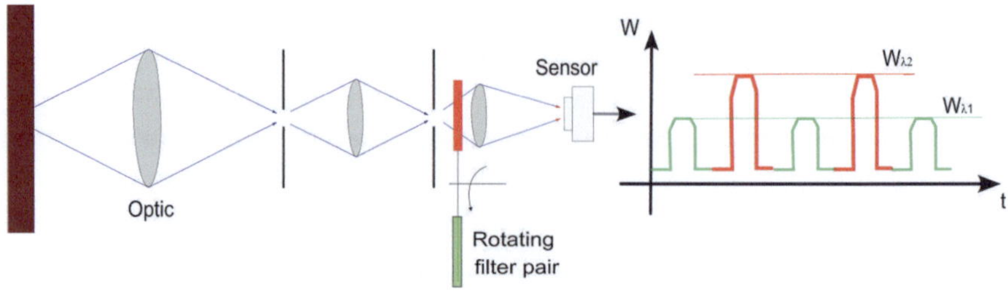

Fig. 8.11 The two-color temperature measurement technique measurement

solving the emissivity problem by calculating the temperature through the ratio of monochromatic emissions at two different wavelengths λ_1 and λ_2. For every λ_i, we have:

$$W_{\lambda_i} = \frac{\varepsilon_{\lambda_i} C_1}{\lambda_i^5 e^{\frac{C_2}{\lambda_i T_0}}} \tag{8.21}$$

Then:

$$\frac{W_{\lambda_1}}{W_{\lambda_2}} = \frac{\varepsilon_{\lambda_1}}{\varepsilon_{\lambda_2}} \cdot \left(\frac{\lambda_1}{\lambda_2}\right)^5 \cdot e^{\frac{C_2}{T_0}\left(\frac{1}{\lambda_2} - \frac{1}{\lambda_1}\right)} \tag{8.22}$$

If the object is a gray body ($\varepsilon_{\lambda_1} = \varepsilon_{\lambda_2}$), and T_0 can be determined by measuring $\frac{W_{\lambda_1}}{W_{\lambda_2}}$ and $\frac{W_{\lambda_2}}{W_{\lambda_1}}$, knowing the two wavelengths. This instrument can be implemented using two rotating color filters that alternate the arrival of light at two power levels, as illustrated in Fig. 8.11 or using two different detectors and an infrared beam splitter.

There are several advantages to using this methodology:

- The measurement of power ratios overcomes the issues faced by monochromatic pyrometers.
- It allows for accurate temperature measurement of objects even when the image is smaller than the sensor.
- This method is not affected by errors due to the absorption of radiation by obstacles such as windows, gases, or dirt.

Thermal sensors detect thermal radiation and translate it into a thermal image, where different colors or shades represent different temperatures of the detected areas.

The figure of merit for a thermal sensor is typically given by [5]:

$$\frac{D^*}{\sqrt{\Delta f}} \tag{8.23}$$

where D^* is the sensor detectivity parameter [5], proposed by Clark Jones in the 1950s, i.e. the signal-to-noise ratio when 1 W of radiation is incident on an area of 1 cm^2, and the output signal is viewed with a bandwidth of 1 Hz.

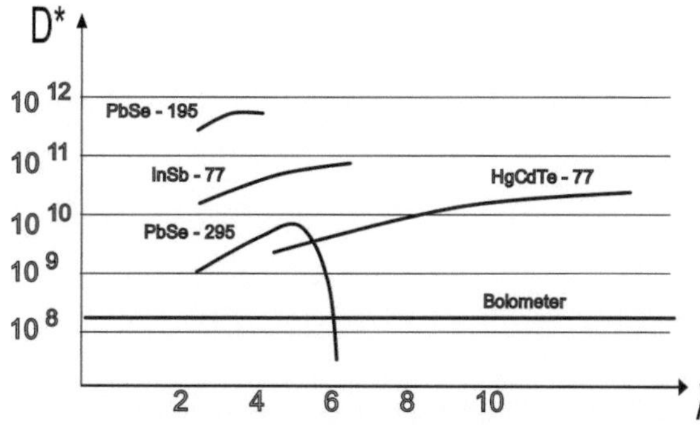

Fig. 8.12 Typical detectivity values of thermal detectors

From a graph of D^* values illustrated in Fig. 8.12 for lead Selenide sensors (PbSe) cooled at 195 K, indium antimonide sensors (InSb) cooled at 77 K, lead selenide sensor uncooled and uncooled bolometers it is possible to obtain the performance of different types of sensors. These sensors are made from semiconductor materials, and their responsivity varies depending on the wavelength, as illustrated in Chap. 3. Each type has a peak detectivity at a specific wavelength. Semiconductor sensors generally have a detectivity about two orders of magnitude higher than bolometers, as illustrated in Fig. 8.12. Although they operate within a narrow wavelength and receive less energy compared to broadband thermal sensors, their high responsivity makes them highly sensitive. However, a common issue with semiconductor sensors is their need for cooling to function properly. Cooling reduces thermal noise, thereby improving measurement accuracy by maintaining low operating temperatures. Bolometer sensor arrays are sensitive to all wavelengths. The key variation among individual sensitive elements in a bolometric array is the electrical resistance, which is monitored under direct current (DC). In an optical system, this means the focal plane array continuously observes the thermal scene, requiring a resistance reading system that is straightforward and easy to control.

The basic concepts of bolometer detectors are:

- A bolometer consists of an absorber connected to a heat reservoir (area of constant temperature) through a thermal insulation system that allows the passage of an electrical signal.
- Radiation absorbed by the absorber raises its temperature above that of the heat reservoir.
- A thermometer connected to the absorber (or the absorber itself if its resistance varies with temperature) measures this temperature increase.
- The absorbed power is calculated based on the temperature and resistance variation.

In detail, a bolometer consists of an absorptive element, such as a thin metal layer, connected to a thermal reservoir through a thermal link. Radiation impinging on the absorptive element raises its temperature above that of the reservoir—the greater the absorbed power, the higher the temperature. The intrinsic thermal time constant,

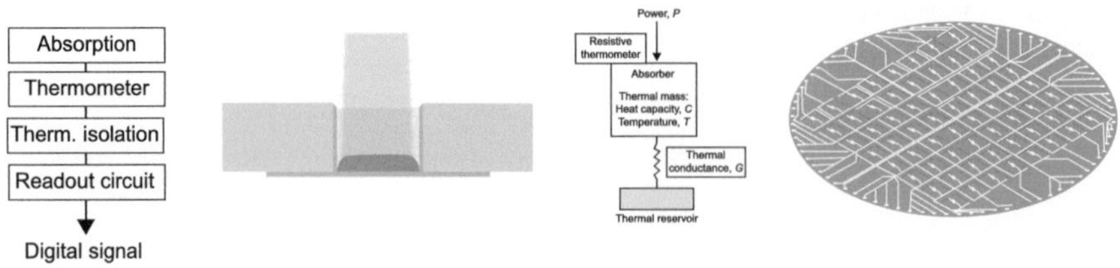

Fig. 8.13 Bolometric pixel structure and arrays

which determines the detector speed, is the ratio of the heat capacity of the absorptive element to the thermal conductance between it and the reservoir (Fig. 8.13).

If P is the thermal power absorbed from an incident infrared radiation, it heats a thermal mass with heat capacity C and temperature T. This thermal mass is connected to a reservoir of constant temperature through a thermal conductance G. The temperature increase is:

$$\Delta T = \frac{P}{G} \tag{8.24}$$

This ΔT is measured by a resistive thermometer, allowing determination of P. The intrinsic thermal time constant is:

$$\tau = \frac{C}{G} \tag{8.25}$$

A microbolometer is a specific type of bolometer used as a detector in thermal cameras. It consists of a grid of vanadium oxide (VOx) or amorphous silicon heat sensors atop a corresponding silicon substrate grid. Infrared radiation in a specific wavelength range strikes vanadium oxide or amorphous silicon, causing a change in electrical resistance. This resistance change is measured and processed into temperatures, which can be represented graphically. Microbolometer arrays commonly come in sizes such as 640 × 480, 320 × 240 (or 384 × 288 for amorphous silicon), or lower cost 160 × 120 arrays. The 640 × 512 VOx arrays are often used in static security cameras where low shock resistance is acceptable. Unlike thermal detectors, cooled detectors directly convert photons into an electric signal via quantum events such as electron band transitions (*interband* and *intraband*) in semiconductor materials. Since no energy absorption (heating) occurs, these detectors do not warm up and are often called quantum detectors. They are mainly used in the MWIR (Medium Wavelength Infrared) and LWIR (Long Wavelength Infrared) ranges [6].

In most cases, especially in the medium to far infrared range, these sensors require cooling to function properly. They are typically cooled to extremely low temperatures, sometimes as low as the temperature of liquid nitrogen (77 K). However, using liquid nitrogen cooling can be inconvenient because it requires periodic replenishment of the coolant within the camera. An alternative and more practical cooling method is provided by a Stirling engine-based cryocooler. A Stirling engine, when

Fig. 8.14 Thermal camera sensor arrays

operated in reverse as a heat pump, removes heat from the sensor by converting mechanical energy applied to its shaft. This mechanical action causes heat to flow from the expansion space (cold side) to the compression space (hot side), effectively cooling the sensor (Fig. 8.14).

8.4 Thermal Cameras

Early thermal cameras were based on mechanical scanning systems using oscillating or rotating mirrors to direct the field of view across a single thermal sensor, as illustrated in Fig. 8.15.

This approach required a long time to generate a full thermal image because the total acquisition time was the exposure time of the sensor multiplied by the total number of image points (n × m), and obviously a stationary thermal field in this time.

Modern thermal cameras, however, use focal plane arrays (FPAs) composed of many thermal sensor pixels, either photonic or bolometric elements, combined with infrared optics (lenses and objectives) and digitizing electronics like those in video cameras (see Chap. 3). These arrays capture the entire thermal image simultaneously,

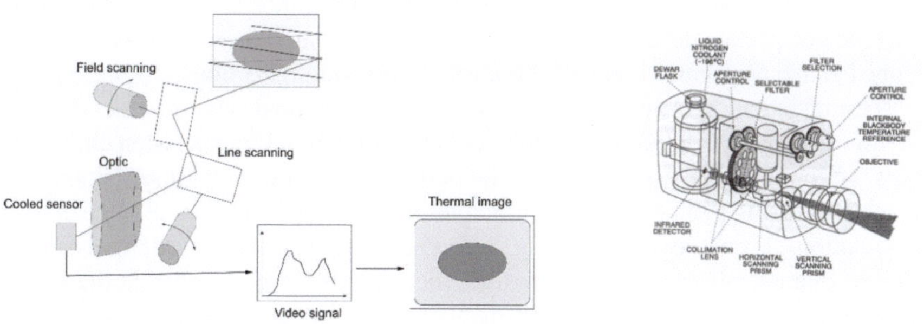

Fig. 8.15 Thermo-scanning cameras

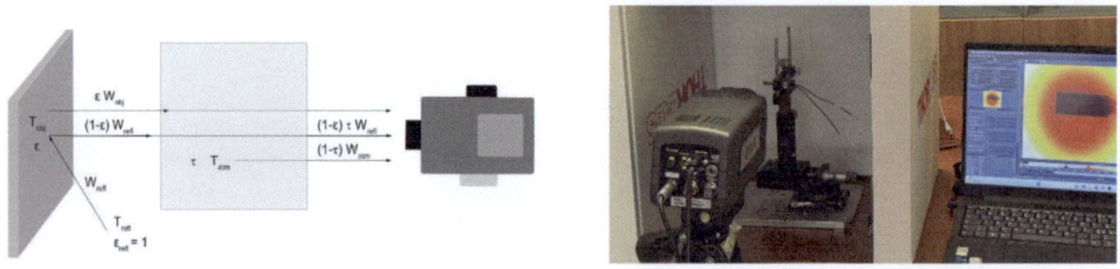

(a) (b)

Fig. 8.16 Focal plane array thermal camera model (**a**) and picture (**b**)

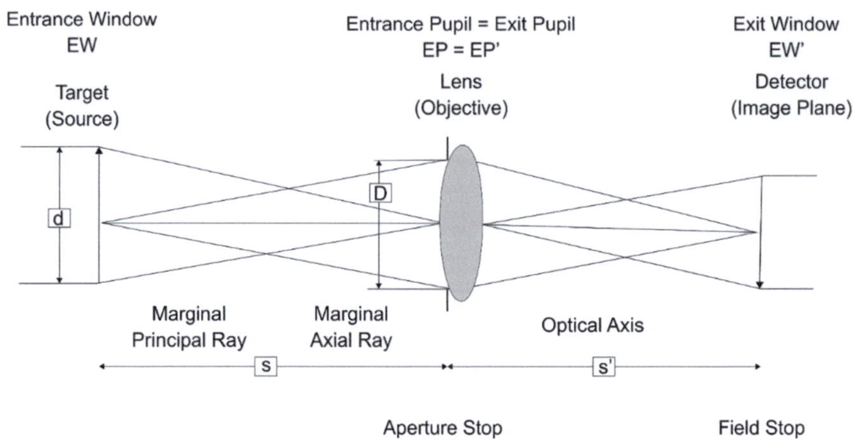

Fig. 8.17 Infrared optic model of a thermal camera

so the exposure time is equal for all pixels, greatly speeding up image acquisition. They are not anymore "thermal scanner". A thermal camera available at Department of Engineering of University of Perugia (Italy) is illustrated in Fig. 8.16b.

During thermal power detection by the thermographic sensor array the surface under examination emits thermal radiation based on its temperature and emissivity. At the same time, it reflects some incidents of thermal radiation from the environment. The reflected portion depends inversely on emissivity; materials with emissivity close to one behave nearly like ideal blackbodies, minimizing reflection.

Thermal radiation traveling from the surface to the thermal camera passes through the atmosphere, which has its own temperature and thermal transmission characteristics, affecting the measured signal. The optical system of the thermal camera can be analyzed using geometric optics principles described in Chap. 2 and same optical lens relations. From the scheme of Fig. 8.17, considering an entrance thermal camera pupil of diameter D acting as an aperture, focusing the image formed at the image plane located at distance s' from the lens, with the object at distance s.

Thermal cameras can be well characterized by their signal-to-noise ratio (SNR). Background noise is estimated by measuring signal fluctuations when the camera sensor do not receive any thermal radiation. The SNR can be expressed as:

$$SNR = \frac{P}{NEP} \tag{8.26}$$

where P is the radiant power incident on the sensor and NEP (Noise Equivalent Power) is the minimum incident power required to produce a signal equal to the background noise level of the sensor.

The radiant power reaching the sensor depends on the camera optics. It can be approximated by [5]:

$$P = \frac{E_P \times E_W}{s^2}, \qquad Po = \frac{E'_P \times E'_W}{s'^2} \tag{8.27}$$

where Po is the power emitted by the object, E_P and E_W represent the entrance pupil area and entrance window area, respectively, and s is the distance between them. The terms with primes correspond to the image side of the optics. For focal plane optics, the object is often assumed at infinity, i.e., on the focal plane of the lens. Hence, the output pupil area and distance simplify to [5]:

$$P = \frac{D^2 \pi d'^2}{4f^2} Po \tag{8.28}$$

where d' is the pixel dimension of the thermal array, D is the lens diameter, and f is the focal length. The factor multiplying Po in Eq. (8.28) represents the optical reduction or amplification of thermal power P reaching the detector.

Introducing the infrared optics f-number $N = \frac{f}{D}$, same for visible optics defined in Chap. 2, the equation for power reaching a single pixel becomes:

$$P = \frac{d'^2}{4(N)^2} W \tag{8.29}$$

This relationship highlights the importance of pixel size and optics: an $N = 1$ lens provides four times more thermal power to the sensor than an $N = 2$ lens. However, increasing lens diameter risks optical aberrations. The NEP depends on the sensor pixel size d', the bandwidth Δf of the electronics, and the sensor's figure of merit which represents the signal-to-noise ratio for 1 W incident on a 1 cm^2 sensor area with a 1 Hz bandwidth. Higher D^* means better sensitivity:

$$NEP = \frac{d' \cdot \sqrt{\Delta f}}{D^*} \tag{8.30}$$

The output of the sensor can be considered generated by many effects:

- The radiation emitted by the target surface, depending on temperature and emissivity,
- The transmission through the atmosphere, which may absorb or scatter part of the radiation,

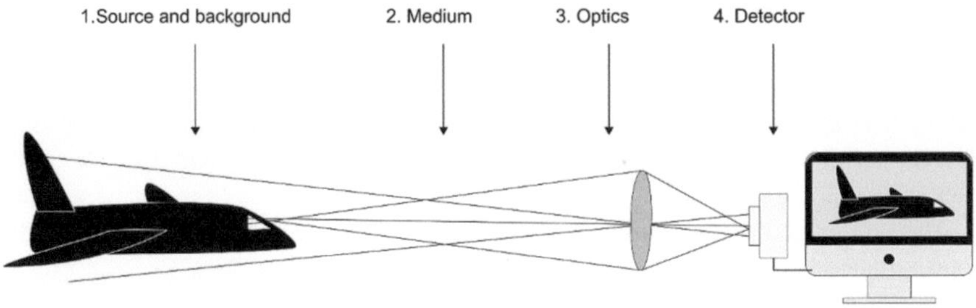

Fig. 8.18 Signal to noise model of a thermal camera

- The optical transmission efficiency of the camera optics,
- The intrinsic characteristics and sensitivity of the thermal sensor.

Considering all these, a comprehensive model can be constructed to predict the thermal sensor output and its relation to the thermal signal, as illustrated in Fig. 8.18.

The total thermal radiation power detected by a thermal camera is influenced by several important factors, each representing a different stage in the radiation's journey from the target to the sensor. The first and most fundamental component is the net radiation difference between the target object and its surrounding background. This difference can be expressed mathematically as the difference in emitted power weighted by the emissivity of the target and background, written as $W_T\varepsilon_T - W_B\varepsilon_B$. Here, W_T and W_B represent the spectral radiances of the target and background respectively, while ε_T and ε_B denote their respective emissivity. This quantity can be evaluated either over a specific spectral band or across the entire spectrum of thermal radiation, using Planck's law as the underlying physical principle. The higher the emissivity of the target, ideally close to that of a perfect blackbody, the stronger and more reliable the emitted thermal signal becomes.

The second important factor in the measurement chain is the transmittance of the atmosphere between the target and the camera. The atmosphere can absorb or scatter a portion of the thermal radiation, thereby reducing the amount of radiation that reaches the sensor. This atmospheric transmittance, denoted by τ, depends on environmental conditions such as humidity, temperature, and the presence of gases or particulates, and must be carefully estimated to accurately interpret the sensor readings.

Next, the thermal radiation that has passed through the atmosphere reaches the camera optics, which themselves contribute to attenuation. The optical transmissivity τ_O describes the fraction of incident radiation that successfully passes through the lens system and reaches the sensor pixels. This effect can be further quantified by considering the pixel size d′ and the lens f-number N, leading to the factor:

$$\frac{\tau_O d'^2}{4(N)^2} \tag{8.31}$$

This term highlights the importance of optical design in maximizing the thermal power incident on the sensor. For example, lenses with smaller f-numbers (such as N = 1) allow more radiation to reach the sensor compared to lenses with higher f-numbers (such as N = 2), potentially improving the system's sensitivity. However, increasing lens diameter or lowering the f-number can also introduce optical aberrations, which must be managed carefully in the design.

Finally, the sensor itself introduces constraints on the detection performance. A key parameter is the figure of merit D^*, which represents the sensor's intrinsic ability to distinguish a signal from noise. This value defines the signal-to-noise ratio (SNR) achieved when one watt of power is incident on a 1 cm^2 detector area with a bandwidth of 1 Hz. A higher D^* indicates better sensor sensitivity, meaning it can detect weaker signals above its own noise floor.

Finally, a model of signal to noise ratio of a thermal camera can be given by combining the above described three factors (atmosphere attenuation, infrared optics attenuation and sensor signal conversion performances) on the thermal contrast $(W_T\varepsilon_T - W_B\varepsilon_B)$ by the product:

$$SNR = \frac{P}{NEP} = (W_T\varepsilon_T - W_B\varepsilon_B)(\tau_A)\left(\frac{\tau_O d'^2}{4(N)^2}\right)\left(\frac{D^*}{\sqrt{\Delta f}}\right) \tag{8.32}$$

This model can be used to estimate uncertainty in temperature measurements or feasibility of thermal camera applications and for making a proper choice of main thermal camera optical and sensor parameters to assure a useful output thermal image.

In addition to these steady-state factors, the dynamic performance of a thermal camera is characterized by two related temporal parameters: the integration time and the response time. The integration time is like the exposure time in photography and represents how long the sensor collects thermal radiation during each frame. During this period, the sensor accumulates electrical charges generated by the incident radiation, acting much like a capacitor. The response time, on the other hand, is the sensor's speed in responding to changes in incoming radiation, influencing how quickly the sensor can adapt to variations in the scene. Choosing the integration time involves a trade-off. A shorter integration time enables capturing a greater number of frames per second, which is beneficial for monitoring fast-changing thermal events or dynamic scenes. However, because less radiation is collected during a shorter exposure, the resulting signal contains fewer generated charges, making the sensor's background noise more prominent and reducing the overall signal-to-noise ratio. Conversely, longer integration times allow the sensor to accumulate more charge per frame, increasing the signal strength and improving the signal-to-noise ratio, but at the cost of temporal resolution, since fewer frames can be acquired per unit time.

This inherent trade-off means that it is generally not possible to maximize both the imaging speed and the image quality simultaneously. Consequently, system designers and operators must carefully balance these parameters based on the specific requirements of their application, whether that is to capture rapid thermal fluctuations with moderate clarity or to obtain high-quality images with slower temporal sampling.

Figure 8.19 illustrates the dependencies between the key components involved in thermal radiation detection. Thermal radiation emitted by the target first passes through the camera lens, defined by its aperture and focal length, resulting in a flow

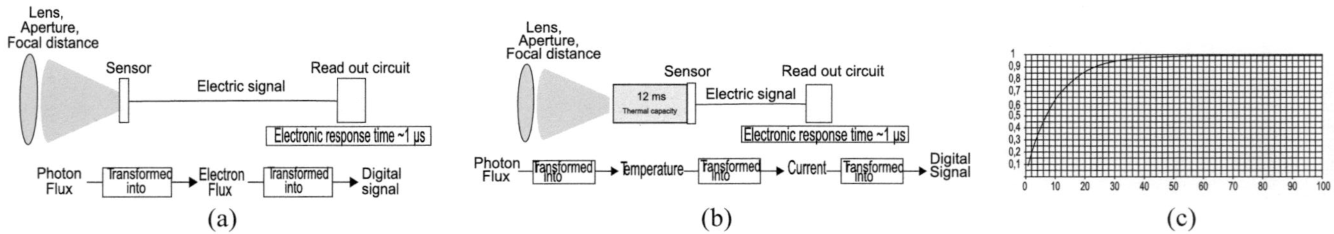

Fig. 8.19 Dynamic performances of thermal cameras: **a** photonic thermal camera, **b** bolometric, **c** thermal response in time of bolometric thermal camera

of photons that reach the sensor, as previously illustrated in Fig. 3.4 of Chap. 3 with the rain analogy filling by drops (the photons) the pixel basket. The sensor converts this incident radiation into an electrical output signal, which is then processed by a reading circuit. This reading circuit typically operates with very short response times, on the order of approximately 1 μs, thus defining part of the overall timing characteristics of the system. The response time of the sensor, however, depends on its underlying technology. For bolometric sensors, which detect changes in electrical resistance caused by temperature variations of a sensing element, the response time is an intrinsic physical property that cannot be altered externally. In contrast, photonic sensors convert infrared photons directly into an electrical signal nearly instantaneously, but the subsequent generation and electronic reading of the charge may require a waiting period ranging from 1 to 1000 μs. Bolometers, on the other hand, operate through a fundamentally different mechanism. The sensing element must first warm up to the equilibrium temperature determined by the absorbed radiation. This warm-up period means the initial reading differs from subsequent readings, necessitating a waiting time long enough to stabilize the resistance value that reflects the target's thermal radiation. Because of this, uncooled bolometric thermographic systems are generally unsuitable for measuring temperatures that vary rapidly over time. Typically, bolometric sensors have a response time around 8–12 ms, three orders of magnitude slower than the microsecond-scale electronic readout speed. This slower response time reduces the system's ability to track fast temperature fluctuations by nearly 1000 times. It is important to note that other system parameters, including the lens aperture, sensor size, and readout electronics, also influence the overall timing but usually operate much faster than the detector itself.

The figure below shows typical response times for different bolometric materials. For example, vanadium oxide (VOx) sensors commonly exhibit a response time around 12 ms, whereas amorphous silicon (a-Si) sensors respond faster, typically around 8 ms. From a dynamic perspective, each pixel or sensing element in a bolometer behaves approximately as a first-order dynamic system with low heat capacity.

Because of the step response nature of such a system, it is necessary to wait roughly 5–6 times the time constant of the first order dynamic system to obtain an accurate, stable reading of the sensor's electrical resistance. This behavior limits bolometers' ability to capture fast transient thermal events or rapid movements of the target. The key characteristics summarized in the table below highlight the trade-offs and limitations inherent in bolometric sensors compared to photonic sensors, especially in applications requiring fast temporal response.

As the temperature of the observed target increases, it becomes possible to use shorter integration times on photonic sensors. This is because higher temperatures naturally produce more radiant energy, which means that the sensor can accumulate sufficient signal in less time to generate an accurate temperature measurement. It is important to note that uncooled bolometric sensors typically have response times at least 100 times slower than those of cooled photonic sensors, limiting their ability to track rapid temperature changes, and from Table 8.2 it is possible to see that at higher temperatures response time of bolometric sensors do not decrease.

Table 8.2 Thermal camera typical dynamic performances

	Photonic detector in MWIR	Bolometer detector in LWIR
Room temperature 10–90°	Integration time needs for an N = 2,5 infrared optics: Typical 350 μs	Response time 8–12 ms: Typical measurement intervals about 50 ms
Mid temperature 35–150°	Integration time needs for an N = 2,5 infrared optics: Typical 120 μs	Response time 8–12 ms: Typical measurement intervals about 50 ms
Mid-high temperature 80–200°	Integration time needs for an N = 2,5 infrared optics: Typical 60 μs	Response time 8–12 ms: Typical measurement intervals about 50 ms

The process of generating a temperature image frame consists of two distinct phases: integration and data readout. During the integration phase, the sensor accumulates thermal radiation over a set exposure time. In bolometric sensors, this phase involves physically heating the sensitive element to register the temperature variation through changes in resistance. In photonic sensors, integration corresponds to the time required for electrical charges to be generated and stored in a capacitive element. After integration, the sensor enters the data readout phase, where the collected signal, whether electrical charge or resistance value is measured and digitized.

The relative timing between these two phases defines two main integration modes. The first is "Integrate Then Read" (ITR), where integration and data readout occur sequentially: the sensor completes its integration phase fully before starting to read out the data. The second is "Integrate While Read" (IWR), where integration of the current frame overlaps with the data readout of the previous frame. This overlapping allows for higher frame rates, enabling the camera to capture images more rapidly. However, this increased speed comes with a trade-off in terms of noise; the IWR mode generally exhibits more noise compared to ITR.

Frame synchronization in thermal imaging cameras relies on two main operational modes: Frame Start Integration (FSSI) and Frame Start Readout (FSSR). In FSSI mode, the integration phase begins immediately at the synchronization signal for the frame, meaning the sensor starts collecting thermal data "right now." This applies regardless of whether the camera is operating in ITR or IWR mode. In contrast, in FSSR mode, the synchronization signal triggers the start of the readout phase for the previous frame, and the integration of the current frame is scheduled accordingly by the camera's internal timing. Essentially, FSSR means "start reading out the image now."

When a camera is set to operate in trigger mode, streaming of image data only begins once an external trigger signal is received. Combining trigger mode with FSSI is the most intuitive approach since the sensor begins integrating new data precisely when the trigger occurs. On the other hand, if FSSR mode is selected, the first data streamed after the trigger corresponds to an image that began integration before the trigger signal, potentially introducing an unwanted delay depending on how much time had elapsed since the integration started prior to the trigger.

Because IWR mode allows integration and readout to occur simultaneously, it can support higher frame rates than ITR mode, which is strictly sequential. However, the drawback of this improved speed is increased noise in the output images. To reduce this noise, thermal cameras operating in IWR mode typically perform offset updates or calibration routines to mitigate fixed pattern noise artifacts.

As a result of the different integration and readout modes in thermal imaging cameras, a variety of strategies and timing intervals are employed, which directly influence the achievable frame acquisition frequencies. Understanding these reading strategies is key to optimizing frame capture rates while ensuring the integration time per pixel remains within acceptable limits to maintain image quality. During the sensor readout process, pixels are typically organized in lines. Within one line time, all pixels on that sensor line are biased simultaneously, meaning that the voltage across the detector elements in that line is sampled together via a shift register. Once sampled, the stored data in the shift register is sequentially shifted out during the following line time, enabling the orderly readout of pixel values one line at a time. In uncooled thermal cameras, this readout is generally performed line by line, starting from the first row and progressing sequentially to the last. On each line, all pixels can be read simultaneously by applying a potential difference across the sensor elements and measuring the corresponding resistances. This line-by-line approach balances the need for reasonably fast data acquisition with the constraints of the sensor's physical and electronic design.

To control the dynamic acquisition process, thermal cameras use specific signals and parameters related to synchronization, triggering, and image tagging. Synchronization refers to the activation of the frame generation process itself, typically, one synchronization pulse corresponds to the streaming of a single image frame, while the absence of a pulse halts frame generation. Triggering, on the other hand, initiates the start of the streaming sequence, which can be either internally or externally controlled, allowing the camera to begin capturing frames at precise, user-defined moments. Image tagging involves marking or labeling individual frames for recording or processing purposes, often combining hardware signals and software controls to manage timing and metadata.

Many modern thermal imaging cameras, both cooled and uncooled, provide the option to control this synchronization, triggering, and tagging signals externally. This allows for flexible integration into complex measurement setups where precise timing control is crucial, such as synchronized multi-camera systems or triggered acquisition tied to external events.

From an optical standpoint, infrared systems act much like optical filters, characterized by a specific spectral bandwidth, like lenses in the visible range. When imaging an ideal point source, the resulting image is not a perfect point but rather a diffraction-limited blur pattern known as the Airy disk. The size of this Airy disk, which represents the fundamental resolution limit imposed by diffraction, can be calculated as described in Chap. 1 by:

$$\text{Diameter of Airy disk} = 2.44 \times \lambda \times N \tag{8.33}$$

where λ is the wavelength of the infrared radiation and N is the infrared optic's f-number, defined as the ratio of the focal length f to the aperture diameter D, so (f/D), as already seen in Chap. 1. Because the Airy disk diameter scales with the wavelength, diffraction effects tend to be more pronounced in infrared imaging systems than in visible light systems, often becoming the dominant factor limiting spatial resolution. For a Focal Plane Array (FPA) sensor, the aperture corresponds to the opening through which incident rays pass and are focused onto the sensor array. In cooled cameras, this aperture is typically defined by the cold stop, a physical aperture designed to block unwanted lateral rays and reduce stray light, improving image contrast. In the case of lenses, the aperture is given by the diameter of the diaphragm inside the lens system and again is expressed as the f-number. Ensuring that the aperture of the lens is properly matched to the FPA aperture is essential. Proper coupling between the lens diaphragm and the sensor's cold stop or entrance pupil is necessary to optimize image quality and overall system performance. Mismatches can lead to reduced signal strength, increased noise, or optical aberrations, as illustrated in Fig. 8.20.

In thermal imaging optics, the f-number N plays a crucial role in determining how much infrared radiation reaches the sensor. A low number, such as N = 1 or N = 1.3, corresponds to a wide aperture, allowing more infrared energy to pass through the lens to the sensor. This characteristic is often described as "fast" optics, indicating the system can capture more infrared energy in a shorter time. Conversely, a high f-number, like N = 3 or N = 4, means the aperture is smaller, allowing less infrared radiation to reach the sensor. In such cases, the sensor requires a longer time to heat up (in bolometric detectors) or for charge to accumulate (in photonic detectors), leading to "slow" optics. Another fundamental factor influencing image quality in thermal cameras is the pixel size of the focal plane array (FPA). Ideally, the pixel size should be at least as large as the diameter of the Airy disk, the diffraction-limited spot size formed by the optical system. If pixels are smaller than the Airy disk, the energy from a single imaged point spreads across multiple adjacent pixels. This results in blurring and reduced sharpness in the image, effectively degrading spatial resolution despite the increased pixel count. In other words, simply having more pixels on an FPA does not guarantee better image detail if the optics cannot focus energy finely enough to match the pixel size. The image contours can become distorted or smeared, highlighting that true resolution depends heavily on the optical design, not just the pixel density. In practice, optimizing thermographic imaging systems involves a

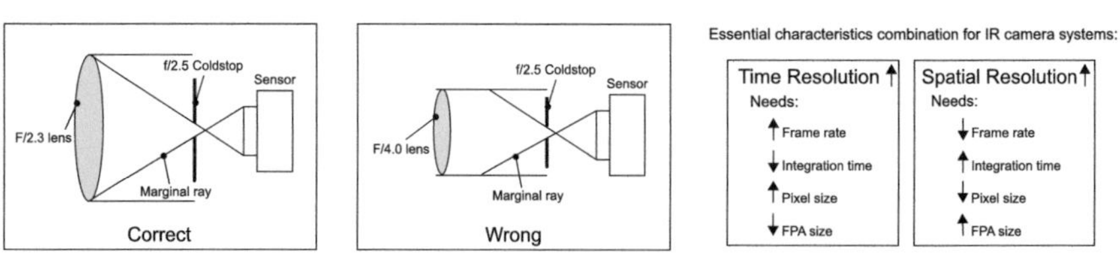

Fig. 8.20 Coupling infrared optics to camera stop and sensor

careful compromise. Fast optics are necessary to ensure adequate infrared energy reaches the sensor to maintain sensitivity and temporal resolution. Meanwhile, the industry trend is towards FPAs with more pixels and smaller pixel pitch to increase resolution while keeping sensor size manageable and lenses affordable. Table 8.3 gives ideal pixel size and some actual typical camera pixel dimensions.

This push for miniaturization and higher pixel density must be balanced with optical performance. Reducing pixel size and increasing the resolution enhances the thermographic array's capability, but only if the lens system can provide the necessary image quality to fully exploit these finer pixels.

Pixel dimensions are fundamental not only for spatial resolution but also for accurate digital image representation, as illustrated in Chap. 3. Ideally, each image detail or "dot" should be represented by an area of at least 3 by 3 pixels. While the Nyquist sampling theorem, as described in Chap. 3, states a minimum of 2x2 pixels is required to properly sample an image feature, opting for a 3x3-pixel grid often provides a better practical compromise between resolution and noise. However, the optimal pixel area depends on various factors including sensor configuration, optical quality, and the specific imaging system. This approach is generally effective for cooled photonic thermal cameras, where each pixel captures energy directly and spatial crosstalk is minimal.

For bolometric sensors, however, the situation is more complex. Bolometers detect temperature changes via thermal effects, and heat diffusion between adjacent pixels can cause spatial blurring, which reduces image sharpness. To counteract this, it's advisable to design systems with larger effective pixel areas, such as 5x5 or even 6x6 pixels, as determined by the optics and system requirements.

Thermographic imaging systems are often designed and evaluated according to established standards that define resolution, optics characteristics, and system capabilities. In bolometric systems, blurring is further complicated by the rolling shutter operation mode. In this mode, different lines of the sensor are exposed and read sequentially, causing temporal lag across the image.

Table 8.3 Optimal and typical pixel size of infrared cameras

Ideal pixel size in MWIR (central wavelength 4 μm)		Ideal pixel size in LWIR (central wavelength 10 μm)	
N = 2	19, 5 μm	N = 0.8	19.5 μm
N = 2.5	24, 4 μm	N = 1	24.4 μm
N = 3	29, 3 μm	N = 1.2	29.3 μm
N = 4	39 μm	N = 1.3	31.7 μm
Current pixel size in MWIR photonic cooled cameras		Current pixel size in LWIR bolometers	
320 × 256 pixels N = 3	25–30 μm	320 × 240 pixels N = 1.3	25 μm
640 × 512 pixels N = 2.5–3.4	15–25 μm	640 × 480 pixels N = 1	17 μm
1280 × 1024 pixels N = 2	15 μm	1024 × 768 pixels N = 1.2	17 μm

Additionally, the response time of the sensitive elements adds a memory effect, where a pixel's measured value reflects not only the current temperature but also residual effects from previous frames. This combination can cause spatial–temporal blur, especially if the observed object or scene moves before the sensor fully responds. To minimize this, the object ideally needs to remain stationary for at least six times the sensor's response time.

Cooled photonic systems, by contrast, typically operate in snapshot mode, capturing the entire frame simultaneously. In these systems, blurring is primarily governed by the integration time chosen, longer integration times improve signal-to-noise ratio but can reduce temporal resolution and introduce motion blur if the scene changes during exposure.

Another optical characteristic affecting thermographic imaging is the depth of field, which depends on the aperture or diaphragm of the system. Just as in visible photography, as illustrated in Chap. 2, the depth of field in thermal imaging is influenced by the aperture size and focusing distance, following a nonlinear relationship. Mathematically, the depth of field increases with the distance between the camera and the target. As illustrated in Chap. 2, it can be expressed as:

$$\text{Depth of field} = \frac{\text{Aperture} \times \text{Detector pitch} \times \text{Distance} \times (\text{Distance} - \text{Focal length})}{\text{Focal length}^2} \tag{8.34}$$

In practice, uncooled thermal detectors usually employ fast optics with low f-numbers, which allow more infrared radiation but reduce depth of field, making focus more critical. Cooled detectors, which often use slower optics with higher f-numbers, generally offer greater depth of field and easier focusing.

To perform thermography on little surfaces it is possible to get magnification using spacers or other optical elements, though the diffraction limit imposes fundamental constraints on achievable magnifications. Typical magnifications in thermal imaging systems range from $1\times$ to $4\times$. Spatial resolution in these systems typically spans from around 25 microns to over 100 microns, depending on the sensor type and system design.

Finally, diffraction, wavelength band (waveband), and f-number also limit the maximum useful magnification in thermal microscopes. For example, Mid-Wave Infrared (MWIR) microscopes can achieve smaller Instantaneous Field of View (IFOV) and finer spatial resolution than Long-Wave Infrared (LWIR) microscopes, due to their shorter operating wavelengths. A range of possible best values for spatial resolution (in terms of IFOV) is given in Table 8.4.

The relationship between image quality and noise levels in thermal imaging cameras, whether cooled or uncooled, is largely characterized by the signal-to-noise ratio (SNR). The SNR quantifies how clearly the useful signal, which is the thermal radiation captured by the camera, stands out against the background noise. This noise can come from various sources, including the sensor itself, the camera's electronics, and the inherent background noise present even when no radiation is detected.

Table 8.4 Best spatial resolution possible for thermographic microscopy

MWIR	IFOV for 15 μm pixel size	LWIR bolometer 320 × 240	IFOV	LWIR bolometer 640 × 480	IFOV
		4X (G 0,25)	100 μm	5.8X (G 0,172)	100 μm
		2X (G 0,5)	50 μm	2.9X (G 0,344)	50 μm
G1	15 μm	1X (G 1)	25 μm	1.5X (G 0,66)	25 μm
G1 at 300 mm	15 μm				
G4	3,77 μm				

When comparing cooled and uncooled thermal imaging cameras with a similar baseline noise level, a clear difference emerge: cooled cameras tend to produce images with significantly less noise. This advantage primarily stems from the design differences, cooled cameras typically have smaller pixels combined with faster optics, meaning a wider aperture or lower f-number.

These features enhance the amount of infrared energy collected by each pixel and improve energy transfer across pixels, which reduces noise and results in a clearer image.

On the other hand, uncooled cameras usually have larger pixels and slower optics, which means narrower apertures or higher f-numbers. This configuration limits the amount of infrared radiation reaching the sensor, reducing energy transfer between pixels and thus leading to comparatively higher noise levels in the images. It's important to note that the aperture of the optical diaphragm plays a key role in this difference.

To make a fair and direct comparison between cooled and uncooled systems, one should consider removing the diaphragms and focusing solely on the lenses. By doing so, the influence of aperture size on image noise is eliminated, isolating the fundamental sensor and optics characteristics responsible for noise performance in each system.

Informations about performances of the commercial available thermocameras can be founded on the website of the main producer, like Flir Systems, Fluke Corporation, Testo SE & Co. KGaA, Seek Thermal, Hikvision, InfraTec GmbH, ULIRVISION, etc.

Some classical applications of thermography are illustrated in Figs. 8.21, 8.22 and 8.23.

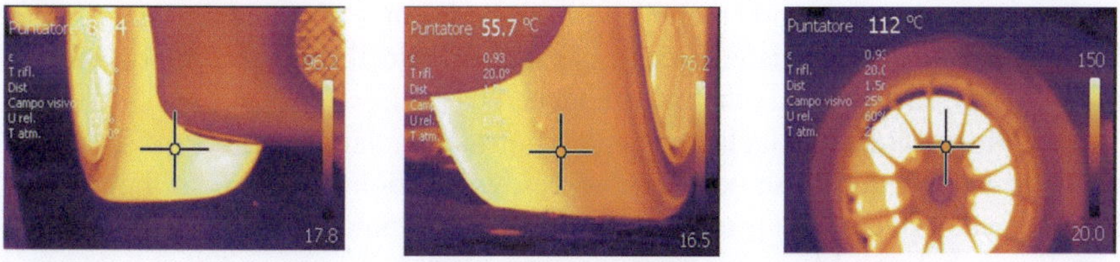

Fig. 8.21 Typical applications of thermography for racing car tire and brake control

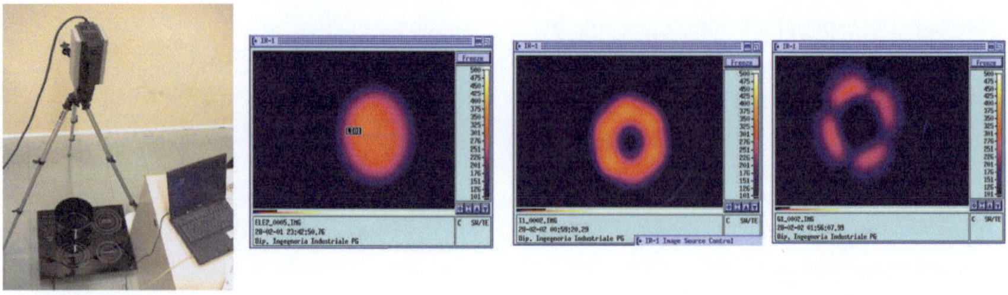

Fig. 8.22 Typical applications of thermography for cooking system characterization

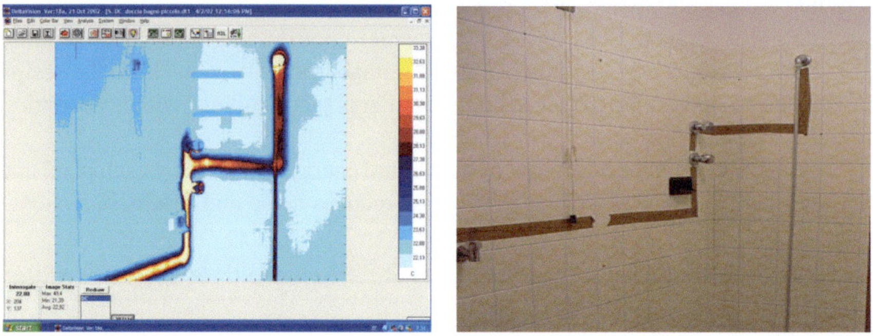

Fig. 8.23 Typical application of thermography for hot water tube identification inside walls

8.5 Dynamic Differential Active Thermography

Many measurement techniques are used to detect defects in materials by dynamically heating and recording the temperature field with an infrared camera. The principle is illustrated in Fig. 8.24.

Active thermography is based on the excitation of a tested material by an external energy source [7]. Halogen lamps, flash lamps, ultrasonic horns, or other devices can be used to deliver energy to the material. This excitation causes dynamic temperature changes and propagation of thermal waves, which are recorded by an infrared camera. Information about surface and sub-surface defects or material inhomogeneities can be obtained by choosing a suitable combination of excitation source, procedure, infrared camera, and evaluation method [8–13].

Thermographic inspection is applicable to a wide range of materials. It is regarded as a method of infrared defectoscopy capable of detecting imperfections such as cracks, voids, cavities, and other inhomogeneities [14].

Excitation sources can include visible or infrared radiation, microwaves, electromagnetic induction, ultrasonic elastic waves, convection heating (e.g., hot air), or thermoplastic effects during mechanical loading. Common sources include lasers, flash lamps, halogen lamps, electrical heating, ultrasonic horns, eddy currents, and microwaves. The object may be directly heated by an external source such as halogen

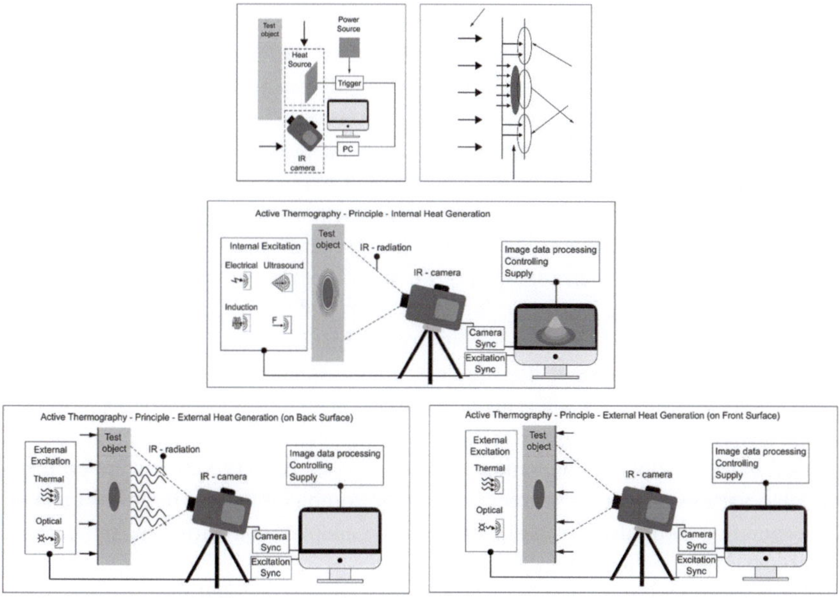

Fig. 8.24 Active thermography principle

lamps or hot air. Inhomogeneities in the material cause distortion in the temperature field, which is detected as surface temperature variations. Thermal image processing depends on the application, excitation source, and excitation type (pulse, periodic, or continuous). Advanced image analysis techniques are commonly used. The most popular methods include Lock-In, Pulse, and Transient (Step) thermography. Continuous excitation is used in special cases. An example of vibro-thermography is shown in Fig. 8.25 from [15], illustrating ultrasound-activated Lock-In vibro-thermography used for quantitative assessment of buried flaws in complex cast parts of automotive turbines manufactured by Fonderie Tacconi–Perugia-Italy. Amplitude-modulated ultrasonic heat generation allows selective detection, as the defect itself becomes a local thermal wave emitter.

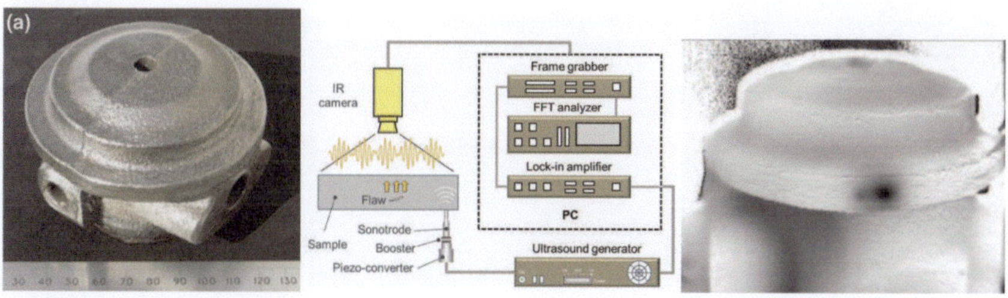

Fig. 8.25 Vibro-thermography identification of defects on cast iron components using ultrasound activated lock-in vibro-thermography

In Lock-In thermography [16] a modulated periodic energy source is used for excitation [17]. Phase and amplitude shift in the thermal signal are evaluated using various techniques. Suitable excitation sources include halogen or LED lamps, ultrasound, lasers, or electric currents. Pulse thermography uses a very short pulse, typically milliseconds in duration, to excite the material. A flash lamp is usually used, and the subsequent cooling is analyzed. The method is fast and enables estimation of defect depth. However, it has limitations in penetration depth, inspection area, and sensitivity to defect orientation. Transient (Step) thermography uses longer excitation pulses than Pulse thermography [18]. It requires less powerful sources (e.g., halogen lamps), making it suitable for larger areas and offering faster measurements compared to Lock-In. However, sensitivity is still limited by defective geometry [19]. Continuous excitation is the simplest method and is applicable only in specific scenarios. A high-speed, cooled infrared camera with high sensitivity is commonly used in IRNDT applications [20]. However, uncooled bolometric IR cameras may be used in some applications, reducing system costs. IRNDT systems are typically modular, various excitation sources that can be combined with different IR cameras and evaluation methods. This modularity allows adaptation to different applications, materials, time constraints, and test area sizes.

An application of this technique for detecting defects in "La Resurrezione" fresco painting of Piero della Francesca in Sansepolcro Civic Museum (Italy), as illustrated in Fig. 8.26. Wall heating was performed by infrared illuminators and cooling transient by natural convection was recorded and processed.

Dynamic active thermography is also used for tangential stress measurements in fluid flow, based on Reynolds' analogy [4]:

$$\frac{\mathrm{Nu_x}}{\mathrm{Re_x} \cdot \mathrm{Pr}} = \frac{C_{fx}}{2} \tag{8.35}$$

Here, Nu_x relates to the convection coefficient and is evaluated thermographically in two ways:

- Active technique by heating or cooling the solid body surface and evaluating the temperature evolution in time and space.

Fig. 8.26 Defect analysis by active thermography on "La Resurrezione" wall painting in Sansepolcro, Tuscany, Italy

- Passive technique by using the natural temperature difference between the fluid and body (e.g., a cold motorbike in a hot wind tunnel).

Once the heat transfer coefficient h_x is known, the surface shear stress coefficient C_{fx} can be determined. Assuming negligible conduction, the energy exchange by convection can be described as:

$$\Phi = h_x(T - T_\infty) \tag{8.36}$$

The convection coefficient h_x changes depending on whether the flow is laminar or turbulent [21]. When a body moves through a fluid, the temperature field observed thermographically can reveal flow characteristics. According to Reynolds' analogy [4]:

$$\frac{\mathrm{Nu_x}}{\mathrm{Re_x} \cdot \mathrm{Pr}} = \frac{C_{fx}}{2} \text{ and so : } C_{fx} = 2 \cdot \frac{\nu \cdot h_x}{c \cdot k} \tag{8.37}$$

This relation connects h_x with C_{fx}, enabling boundary layer pattern analysis [21]. The temperature distribution reflects both fluid and body properties. For example, if a cooler body moves through warmer air, temperature increases due to tangential conduction, followed by a dip in acceleration zones, then a rapid decrease from h_x variation, and finally a rise again. Temperature trend inversion often indicates flow transition (e.g., laminar to turbulent) or the presence of aerodynamic bubbles. The largest temperature gradient is usually near the profile antinode. Experimental visualizations using different materials and setups support this interpretation [22]. However, Reynolds' analogy alone is insufficient; internal energy of the flow (linked to Mach number) must also be considered. At high Mach numbers (transonic and above), direct energy measurements are possible. At low Mach numbers (subsonic flow), thermal gradients are weaker. This can be resolved by:

- Active technique by heating the body surface.
- Passive technique by using the temperature difference between the body and the airflow.

The heated-thin-foil technique [21] is the most used approach for active thermography [23]. It involves heating the body using metallic sheets or PCBs heater. By monitoring temperature and knowing the power dissipated by Joule's effect, the convection coefficient is calculated from:

$$h = \frac{\Phi_j - \Phi_{\mathrm{cond,lat}} - \Phi_{\mathrm{rad}}}{T - T_\infty} \tag{8.38}$$

Alternatively, in transient conditions with passive setups, the *thin skin* technique assumes low thermal inertia of the body [24]. Neglecting internal conduction, the energy balance becomes:

$$\Phi = c \cdot \rho \cdot e \cdot \frac{\partial T}{\partial t} \tag{8.39}$$

So, the convection coefficient can be expressed as:

$$h = \frac{c \cdot \rho \cdot e \cdot \frac{\partial T}{\partial t}}{T_\infty - T} \tag{8.40}$$

Another passive approach assumes a semi-infinite body, where the surface temperature from conduction can be modeled by:

$$T(x, t) = \frac{1}{2\sqrt{\pi a t}} \int_{-\infty}^{\infty} f\left(x'\right) \cdot e^{-\frac{(x-x')^2}{4at}} dx' \tag{8.41}$$

This also supports flow-heated passive techniques.

Some applications to wing profiles [4], car and motorcycle flow measurements in "Raffaele Balli" wind tunnel located at the Department of Engineering of the University of Perugia (Italy), illustrated in Fig. 8.27.

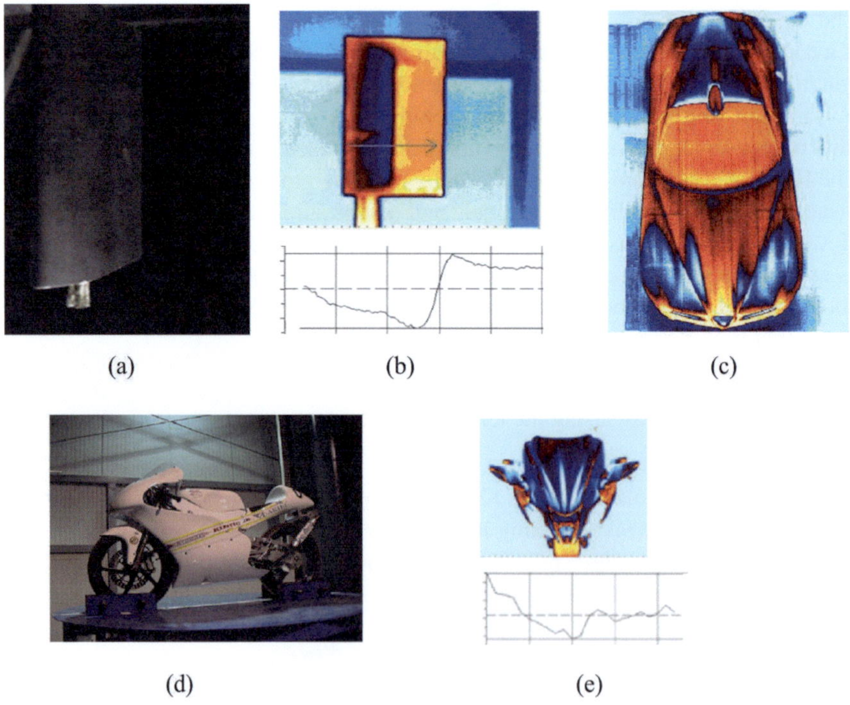

Fig. 8.27 **a** Temperature distribution for aluminium air foil black painted and Mylar® 12 um covered in open circuit wind tunnel

8.6 Thermoelasticity Stress Measurement Technique

The thermoelastic effect is the reversible temperature variation that occurs in a solid material when it is deformed within the elastic range [25]. The magnitude of this effect depends on the volume change accompanying the deformation of the solid in the elastic field. This phenomenon is analogous to the behavior of gases, which increase in temperature when compressed and decrease when expanded [26, 27]. While this effect is more pronounced in gases, in solids, the temperature variation is only in the order of hundredths, or at most tenths, of a Celsius degree [28]. In areas subjected to tensile stress (increase in specific volume), a temperature decrease is generally observed [29]. Conversely, in areas under compressive stress, the temperature increases. In pure shear conditions (no volume change), the temperature remains constant. Under quasi-static loading, the transformation is isothermal (temperature variation negligible), whereas under dynamic loading, the process tends to be adiabatic (no heat transfer) [30]. Here, the surface temperature variation reflects the thermal change in the infinitesimal material volume beneath it [31]. Beyond the thermoelastic effect, other irreversible thermomechanical exchanges, such as plastic deformations, also contribute to temperature variation [32].

Historically, the thermoelastic effect was observed by Gough (1805) on rubber and Weber (1830) on metals. Lord Kelvin developed the first theory for homogeneous, isotropic bodies under adiabatic conditions (1853). Biot [33] in 1956 extended this to anisotropic materials within thermodynamic frameworks. Early studies relied on point-contact techniques (e.g., thermocouples). In 1967, Belgen [34] proposed detecting infrared emission changes using a single-sensor thermal camera. The first system prototype, SPATE (Stress Pattern Analysis by Thermal Emission), illustrated in Fig. 8.28a appeared in 1978, and was commercialized in 1982 [35]. In 1994, the DeltaTherm system (by StressPhotonics Ltd.) marked hardware advancements, using sensor arrays for full-field acquisition (Fig. 8.28b). The initial TSA SPATE system had a single HgCdTe detector and scanning mirrors (choppers), requiring up to 45–60 min for full-field measurements. Modern focal plane systems use 320 × 256-pixel arrays with solid-state antimony sensors. These enable simultaneous full-field acquisition in ~30 s. The system includes a thermal camera and a reference signal for loading, like the CEDIP System illustrated in Fig. 8.28c.

TSA (Thermoelastic Stress Analysis) emerged formally with Belgen's work and the SPATE 9000 system, by OMETRON Company, illustrated in Fig. 8.28a.

The thermoelastic effect is quantified by the equation (from Biot [33]):

$$dT = \frac{T}{\rho \cdot c_v} \cdot \frac{\partial \sigma_{ij}}{\partial T} d\varepsilon_{ij} + \frac{\delta q}{c_v} \tag{8.42}$$

where:

- T: absolute temperature
- ρ: density
- c_v: specific heat at constant volume

(a) (b) (c)

Fig. 8.28 The first thermoelastic measurement systems

- δq: heat exchanged with environment
- σ_{ij}: stress component
- ε_{ij} : strain component.

Assuming isotropic linear-elastic material and temperature-independent properties, and using [26]:

$$\frac{\partial \sigma_{ij}}{\partial T} = -\gamma \delta_{ij} = -\frac{E}{1-2\nu}\alpha \delta_{ij} \tag{8.43}$$

Substituting into Eq. (8.42):

$$T = -\frac{T \cdot \gamma}{\rho \cdot c_v} \cdot d\left(\varepsilon_x + \varepsilon_y + \varepsilon_z\right) + \frac{\delta q}{c_v} \tag{8.44}$$

Under adiabatic conditions ($\delta q = 0$):

$$dT = -\frac{T \cdot \gamma}{\rho \cdot c_v} \cdot d\left(\varepsilon_x + \varepsilon_y + \varepsilon_z\right) \tag{8.45}$$

Which yields:

$$\Delta T = -\frac{\gamma}{\rho \cdot c_v} \cdot T_0 \cdot \Delta\left(\varepsilon_x + \varepsilon_y + \varepsilon_z\right) \tag{8.46}$$

Using constitutive laws:

$$\Delta\left(\varepsilon_x + \varepsilon_y + \varepsilon_z\right) = \Delta\left(\sigma_x + \sigma_y + \sigma_z\right)\frac{1-2\nu}{E} + 3\alpha \Delta T \tag{8.47}$$

Given:

$$c_p - c_v = \frac{3E\alpha^2 T}{\rho(1-2\nu)} \tag{8.48}$$

We derive:

$$\Delta T = -\frac{\alpha}{\rho \cdot c_p} \cdot T_0 \cdot \Delta\left(\sigma_x + \sigma_y + \sigma_z\right) \tag{8.49}$$

This is the classical thermoelastic law. For plane stress on the unloaded surface of an object ($\sigma_z = 0$)

$$\Delta T = -K_0 T_0 \Delta\left(\sigma_x + \sigma_y\right) \tag{8.50}$$

where:

$$K_0 = \frac{\alpha}{\rho_0 \cdot c_p} \tag{8.51}$$

K_0 is the thermoelastic constant, indicating material sensitivity. Since $\alpha > 0$, tension leads to cooling and compression to heating (Table 8.5).

The table shows temperature variations at the proportional limit and corresponding stress/strain variations that produce 10^{-3} K temperature changes (typical thermographic resolution). Resistance strain gauges have a resolution of ~1 μm/m.

If we consider an infinitesimal element of a solid body subjected to a state of stress, whether compressive, tensile, or general deformation, this element may undergo either compression or dilation. Such mechanical actions induce a corresponding variation in temperature, as illustrated in Fig. 8.29. Let us suppose that stress is applied over time, following the trend shown by the blue curve in Fig. 8.30. Initially, the stress increases, and the temperature rises accordingly.

However, once the stress becomes constant, the initial adiabatic condition begins to break down due to heat exchange processes. Specifically, thermal conduction within the material and convection with the surrounding environment cause the temperature to gradually decrease, as represented by the red curve, until thermal equilibrium is approached and the temperature distribution becomes uniform.

Table 8.5 Thermoelastic response of some common materials

			for $\Delta\sigma =$ σ_P (elastic limit)	for $\Delta T =$ 0.001 K	for $\Delta T =$ 0.001 K
Material	σ_P [N/mm^2]	K_0 [m^2/N]	ΔT [K]	$\Delta\sigma$ [N/mm^2]	$\Delta\varepsilon$ [μm/m]
Steel	400	$3.5 \cdot 10^{-12}$	0.4	1	5
Titanium	400	$3.5 \cdot 10^{-12}$	0.4	1	8
Alluminium	150	$8.8 \cdot 10^{-12}$	0.4	0.4	6
Magnesium	150	$14 \cdot 10^{-12}$	0.6	0.24	5
Epoxy	40	$62 \cdot 10^{-12}$	0.9	0.06	18

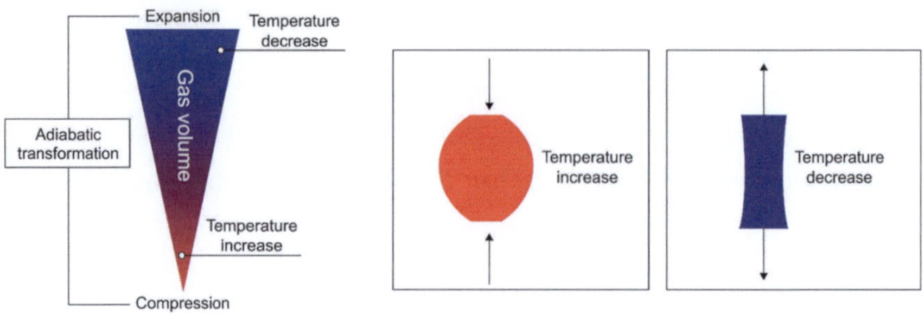

Fig. 8.29 Temperature variation vs. volume variation in gases and solids

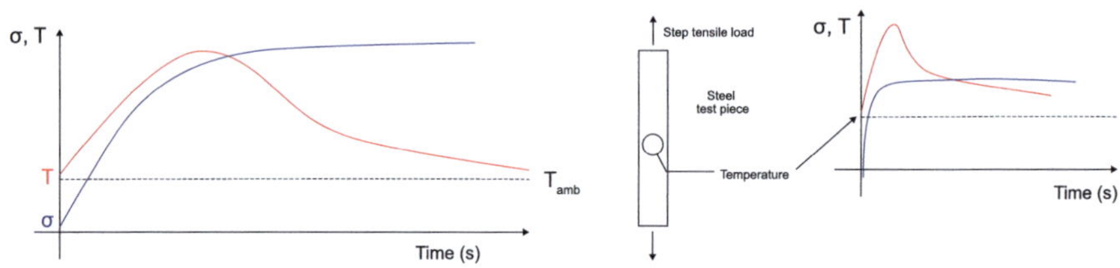

Fig. 8.30 Temperature (red) versus time during a step loading; stress (blue) versus time

Consider now a wedge-shaped beam with length L, width a, and thickness b. If a force or bending moment is applied at its end, the rod undergoes bending, as illustrated in Fig. 8.31. According to Saint–Venant's theory, the upper fibers experience tensile stress, while the lower fibers undergo compression, resulting in a butterfly-shaped stress distribution. One might expect a similar pattern in the temperature field. However, if the load or moment is held constant, the expected temperature distribution does not persist. This discrepancy is due to the internal redistribution of heat: thermal energy flows from warmer (compressed) regions to cooler (tensile) regions, progressively evening out the temperature. As a result, the distinct temperature gradients expected from the stress field are not sustained. Therefore, such measurements are impractical under static or slowly varying loads, where adiabatic conditions are not maintained.

To maintain adiabatic conditions and preserve the measurable thermoelastic response, a dynamically varying load must be applied. The load variation must occur at a frequency high enough to inhibit significant internal heat exchange. In other words, the thermal diffusion timescale must be longer than the mechanical loading cycle. By applying a sufficiently rapid cyclic load, there is no time for heat to flow from hotter to colder regions. In this case, the temperature response closely follows the load waveform (see Fig. 8.32), with only minor attenuation due to residual heat fluxes. These attenuation effects are generally incorporated into a system calibration factor [36, 37].

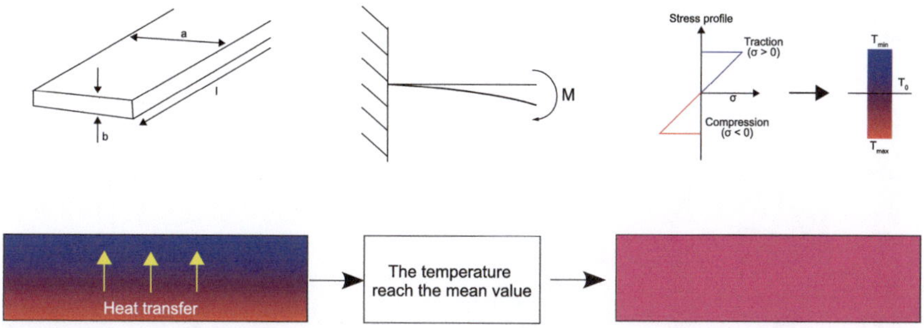

Fig. 8.31 Heat redistribution during static loading

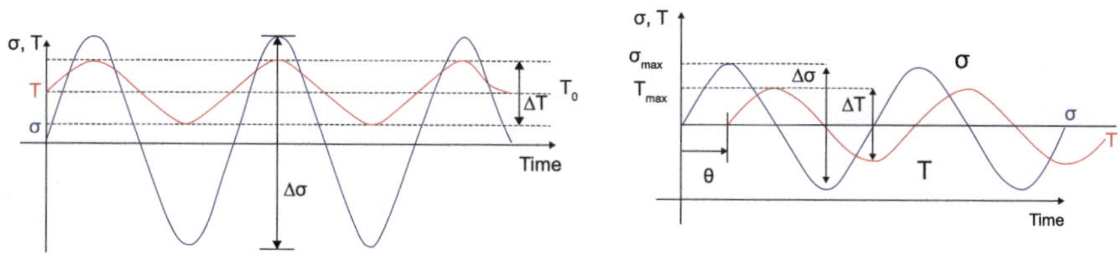

Fig. 8.32 Temperature (red) and stress (blue) over time under rapid cyclic loading

In thermoelastic measurements, a differential thermal imaging camera is typically used. This type of camera does not measure absolute temperatures but rather temperature variations over time. Conceptually, it continuously acquires thermal images and subtracts them from a reference frame, usually the first frame, thus highlighting only changes in temperature [38, 39]. To understand this, it is important to distinguish between DC and AC thermography:

- DC thermography (traditional thermal imaging) provides an absolute temperature field. For instance, as shown in Fig. 8.33, the fingers of a hand appear at a higher temperature than the surrounding environment, while a chain link appears cooler. However, this provides no information about the stress state of the chain, since constant or steady-state temperatures dominate the image.

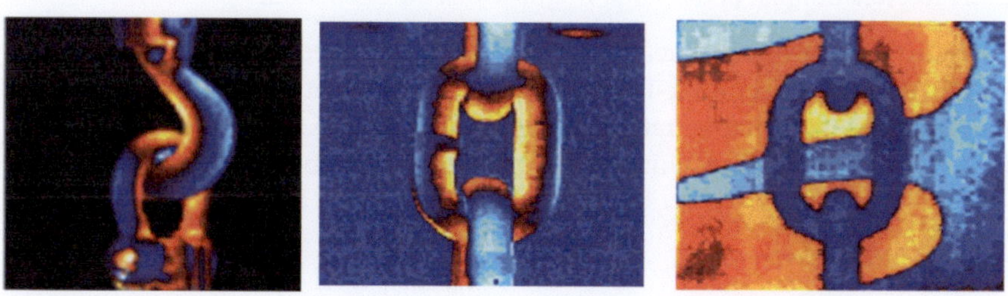

Fig. 8.33 A common thermal image of a cyclically loaded chain and a hook, from StressPhotonics

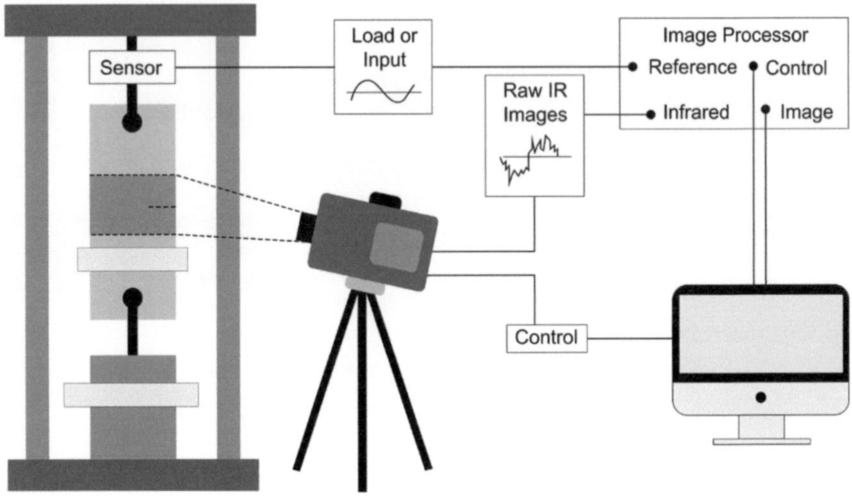

Fig. 8.34 The typical Thermoelastic stress analysis measurement chain

- In contrast, AC thermography reveals only those areas subjected to time-varying thermal fields. In the same figure, the fingers disappear because their temperature remains constant over time, whereas stressed areas of the chain undergoing tension or compression due to cyclic loading are visible. AC thermography is thus better suited for thermoelastic stress analysis, as it highlights regions experiencing stress-induced temperature fluctuations and filters out regions at thermal equilibrium.

Despite its advantages, a limitation exists: the temperature resolution of commercial thermal cameras rarely falls below 10 mK. Since the thermoelastic effect can generate even smaller temperature variations, this can hinder the detectability of stress-induced signals. To overcome this, Lock-In Signal Processing techniques, illustrated in Chap. 1, are employed to improve sensitivity and reduce noise, particularly when the excitation frequency is known, as illustrated in Fig. 8.34. This technique uses two input signals:

1. The measured temperature signal, which includes both signal and noise.
2. A reference signal at the same frequency as the excitation.

Even if the temperature fluctuation signal given by the thermoelastic effect is much smaller than the noise, Lock-In detection allows its isolation by demodulating the temperature signal with the reference signal. This process is applied independently at each pixel over time, resulting in two orthogonal components:

- I_x: in-phase component
- I_y: quadrature component

From these, the amplitude and phase of the temperature fluctuation can be computed as:

$$I = \sqrt{I_x^2 + I_y^2} \tag{8.52}$$

According to Kelvin's law, the temperature variation is proportional to the stress variation:

$$\Delta T \propto \Delta\sigma \tag{8.53}$$

Therefore, the amplitude I of the temperature signal is proportional to the local variation in stress. Additionally, the detection of the second harmonic component of the thermal response can provide information on constant (mean) stress levels and residual stresses [40]:

$$\Delta\sigma_{\text{residual}} \propto \text{Amplitude of 2nd harmonic of } \Delta T \tag{8.54}$$

This processing leads to a final image proportional to the first stress invariant, i.e., the sum of the principal stress variations [41–43]:

$$\Delta\sigma = \Delta\sigma_x + \Delta\sigma_y \tag{8.55}$$

as described in reference [44].

All that we have seen so far are temperature fluctuation maps, but if we want stress maps, we must perform a calibration. We use a thermal imaging camera and a lock-in technique to measure the amplitude of temperature fluctuation. The thermal imaging camera measures thermal radiation, and between radiation and temperature there is the emissivity of the surface.

We apply a dynamic load that generates stress fluctuations (in pascals), which in turn induce temperature variations (in kelvins). These temperature variations cause fluctuations in emitted thermal radiation (in watts), which the camera sensor transforms into an electrical signal (in volts), as illustrated in Fig. 8.35a. We now ask whether it is possible to directly relate the output electrical voltage to the stress on the surface. The calibration constant k relates the electrical output voltage V of the instrument to the surface stress σ, assuming a linear relationship. It is determined by comparing the TSA-derived signal to the actual stress measured in a known area of the specimen:

$$k = \frac{\sigma_x + \sigma_y}{S_{\text{avg}}} \tag{8.56}$$

A common method involves using electrical resistance strain gauges applied to a point on the surface to measure $\Delta\sigma_x$ and $\Delta\sigma_y$. To do this, we use a 90° strain gauge rosette connected to a Wheatstone bridge circuit, as illustrated in Fig. 8.35b. In the area where the strain gauges are applied, we also evaluate the average signal of the thermal camera pixels [45]. At the intersection of the two-gauge directions, the strain gauges measure strain components ε, which can be related to stress using the Lamé relations (7.7):

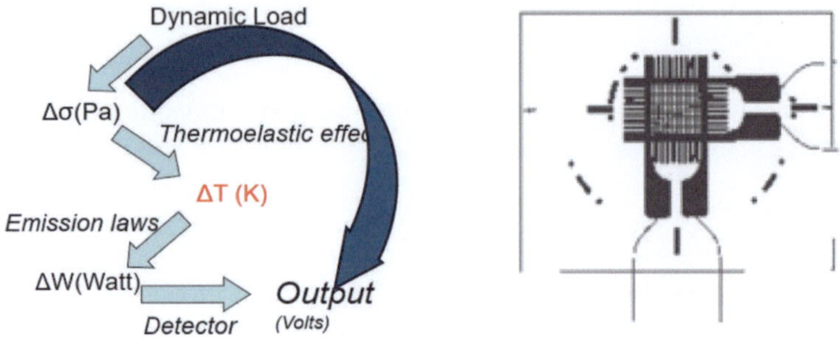

(a) (b)

Fig. 8.35 Physical variable conversion steps and strain gauge reference sensor

$$\varepsilon_x = \frac{\sigma_x}{E} - \frac{\nu\sigma_y}{E} \qquad \varepsilon_y = \frac{\sigma_y}{E} - \frac{\nu\sigma_x}{E} \tag{8.57}$$

The sensitivity k is computed as the ratio between the stress (calculated from the sum of strains) and the mean signal amplitude S_{avg}, corrected by Poisson's ratio:

$$k = \frac{E(\varepsilon_x + \varepsilon_y)}{S_{\text{avg}}(1 - \nu)} \tag{8.58}$$

Strain gauges should be placed in an area where the stress is sufficiently high and uniform, as illustrated in Fig. 8.36. Once calibrated in this region, the scaling is assumed valid across the surface—provided that the material and emissivity remain constant.

To perform TSA, it's also possible to use other kind of dynamic loading, for example impact or random excitation instead of a periodic one. This allows for very fast measurements and often does not require a dedicated test bench. A test case example is a tubular steel 50cc motorcycle frame illustrated in Fig. 8.37 [46]. The frame was impulsively loaded on the gray tube visible in the image by subsequent pulses. In this configuration, the reference signal was provided by an optical proximity sensor located at the lower part of the frame and average output correlation are performed.

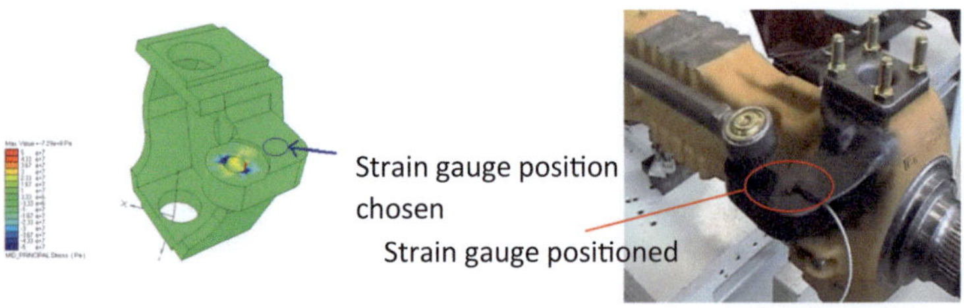

Fig. 8.36 Strain gauge calibration

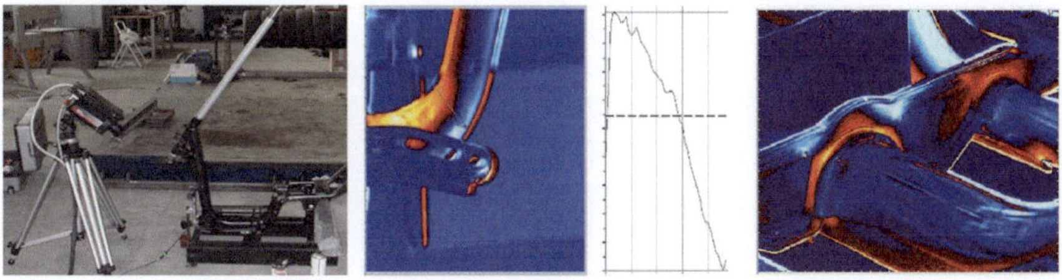

Fig. 8.37 Stress analysis by dynamic impact loading of motorcycle frames

Another possible loading is by dynamic random signals, as illustrated in Fig. 8.38. Also in this case averaged correlation amplitudes are detected on each point of the component surface. ΔT in this case will be the RMS value of random temperature signals on each pixel, correlated with a reference.

Whenever possible, it is advisable to spray the component with a highly emissive black paint [47]. This increases the surface emissivity and reduces its reflectivity, thereby minimizing disturbances caused by temperature fluctuations reflected from the environment. Uncertainty in the measurement can be reduced by repeating the acquisition process multiple times. When extracting the signal from a single pixel, a longer averaging time improves the estimation of the signal fluctuation amplitude. However, it is generally unnecessary to exceed a 60 s capture duration, as the additional benefits become negligible [36].

On the surface of a component, temperature variations may result from the thermoelastic effect, but they can also arise from other factors such as airflow, solar heating, or nearby thermal radiation sources (e.g., a person). Fortunately, lock-in detection techniques effectively eliminate disturbances that occur at frequencies different from that of the reference excitation signal. Modern focal plane array (FPA) thermal imaging cameras are now the most used due to their optical advantages: better depth of field, easier setup, and simpler focusing. In contrast, scanning thermal imaging cameras can offer advantages in specific applications, particularly because a single sensor element may exhibit uniform sensitivity across all measurement points.

Fig. 8.38 Random excitation differential thermography

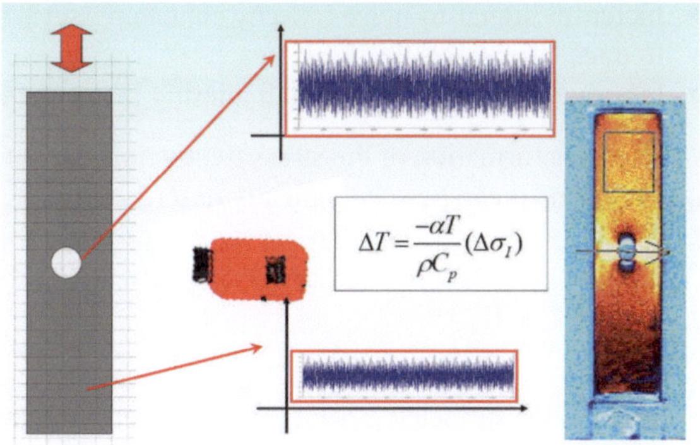

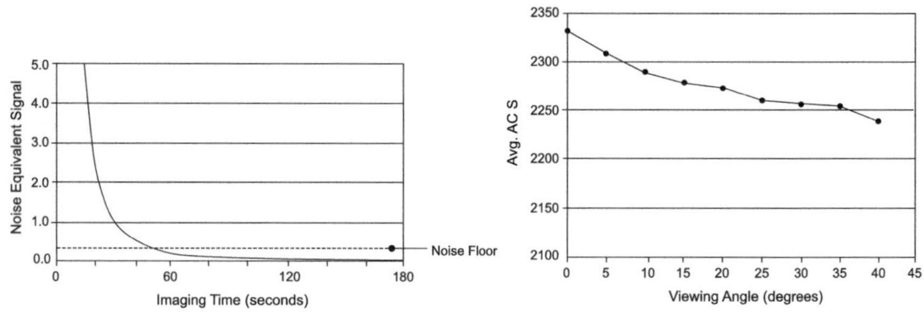

Fig. 8.39 Effects on signal-to-noise ratio due to averaging time and viewing angle

The maximum achievable spatial resolution depends on the wavelength of infrared radiation and the optical setup (lenses, focal length). Since infrared wavelengths are longer than those in the visible range, the resolving power is inherently lower. Using high-quality infrared optics and magnification systems, spatial resolutions down to approximately 10 microns are attainable. When analyzing a surface with complex geometry, the viewing angle significantly affects the measured amplitude of thermal radiation. This is due to the directional dependency of emitted radiation. Experimental observations show that signal degradation remains acceptable (with uncertainty $<4\%$) for viewing angles up to around 40° (Fig. 8.39).

The main sources of uncertainty in a thermoelastic system are [48]:

1. Disturbing infrared power and temperature changes not related to the thermoelastic effect that causes surface temperature variations.
2. Infrared power reflections on the surface.
3. Non-uniform or low emissivity.
4. Heat transfer through the paint layer and with the environment.

However, lock-in processing filters out all temperature variations except those occurring at the frequency of the mechanical load. To mitigate emissivity-related issues, a high-emissivity, infrared-opaque paint is applied to the specimen's surface. This:

- Increases signal-to-noise ratio by enhancing surface emissivity,
- Reduces thermal reflections,
- Introduces minimal load, albeit altering the system's thermoelastic response.

The transformation of the stress fluctuation $\Delta\sigma$ into the output signal S from the infrared sensor can be schematically described as:

$$\Delta\sigma \rightarrow \Delta T_{\text{sup}} \rightarrow \Delta T_{\text{paint}} \rightarrow \Delta J_{\text{paint}} \rightarrow \Delta J_{\text{det}} \rightarrow S \tag{8.59}$$

where:

- ΔT_{sup} is the surface temperature fluctuation (beneath the paint) due to mechanical loading.

- ΔT_{paint} is the temperature fluctuation on the paint surface.
- ΔJ_{paint} is the variation in radiative flux emitted by the paint (via Stefan-Boltzmann or Planck's law).
- ΔJ_{det} is the radiative flux reaching the detector.
- S is the digitized output level (grayscale intensity).

The transformation:

$$\Delta\sigma \to \Delta T_{\text{sup}} \to \Delta T_{\text{paint}} \tag{8.60}$$

is particularly affected by heat conduction between the substrate and the paint layer, which depends on:

- Load frequency f,
- Stress gradient $\nabla\sigma$,
- Paint thickness d,
- Thermal conductivity k,
- Specific heat c_p,
- Thermal exchange with the environment.

According to McKelvie (1987), the attenuation of the thermal signal due to heat transfer can be significant. If the substrate is poorly conductive, the heat transferred to the paint is only from a shallow subsurface region. Consequently:

$$\Delta T^*_{\text{sup}} > \Delta T^*_{\text{paint}} \tag{8.61}$$

As frequency f increases, thermal exchange reduces, leading to:

- Higher signal attenuation at high frequencies due to limited heat transfer from the substrate.
- Higher signal attenuation at low frequencies due to heat loss to the environment.

Experimental data for steel indicates that:

- At $f < 10$ Hz: Attenuation due to environmental convection dominates.
- At $f > 10$ Hz: Attenuation due to conduction within the paint becomes relevant.

This gives rise to a frequency window (typically 10–20 Hz for the material tested, aluminium in this case) where thermal signal loss is minimized and the calibration constant k remains reliable.

Let ΔT_{ad} be the adiabatic temperature change resulting purely from the thermoelastic effect (*i.e.*, no heat exchange). The ratio:

$$\frac{\Delta T}{\Delta T_{\text{ad}}} \tag{8.62}$$

depends on:

- The load frequency,

- The paint properties,
- The local stress gradient $\nabla\sigma$.

Thus, the calibration factor k, which maps $\Delta\sigma \rightarrow S$, is spatially and frequency dependent.

A proper surface coating for thermoelastic stress analysis should:

- Have high and uniform emissivity (e.g., 0.9–0.95),
- Be opaque in the IR range,
- Exhibit low thermoelastic sensitivity,
- Be thin and uniform to reduce heat capacity and allow faster thermal response,
- Have good thermal conductivity to minimize thermal lag.

Painting also standardizes the emissivity across the surface and reduces angular dependence of the measured signal, which is particularly beneficial for complex geometries or curved surfaces.

To experimentally determine whether the surface is operating under adiabatic conditions:

1. Apply a sinusoidal load at varying frequencies.
2. Monitor the temperature fluctuations ΔT.
3. Compute stress concentration factor K_t from thermal data.
4. Identify the frequency range where K_t matches theoretical values (i.e., signal is no longer frequency-dependent).

This defines the adiabatic range for measurements (Figs. 8.40 and 8.41).

Main Applications of thermoelastic measurement technique are in the following described.

Design Comparison

A differential gear of a heavy-duty machine produced by Carraro SpA in Campodarsego—Italy was analyzed before and after a design modification [49]. The first activity was to develop a test bench capable of applying realistic operational loads to the component. The goal was to generate an adjustable cyclical torque with torque amplitude from 0 to 800 Nm and loading frequency: from 0 to 10 Hz. This was

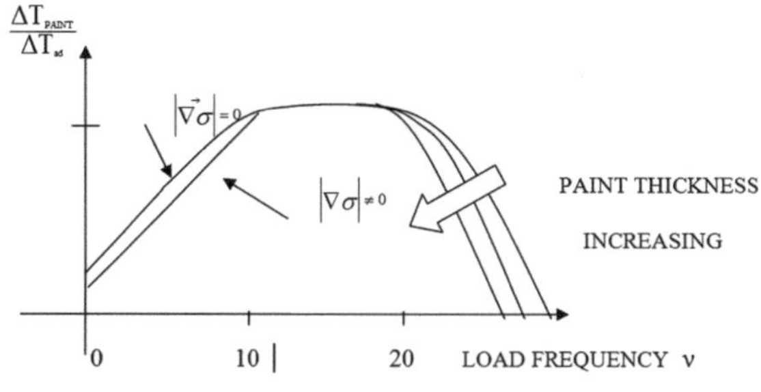

Fig. 8.40 Theoretical effects of load frequency on stress concentration measurements

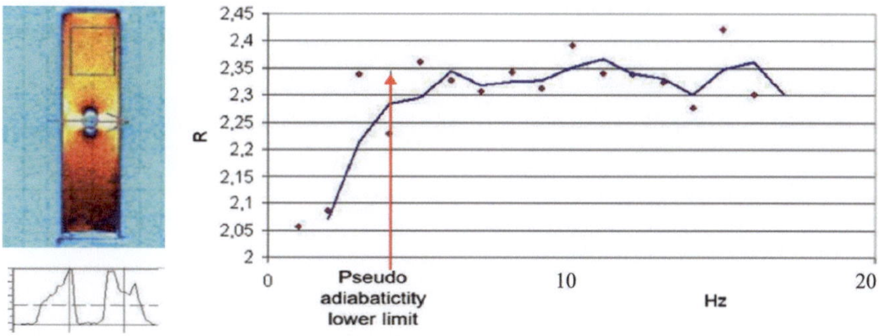

Fig. 8.41 Experimental effects of the loading frequency

achieved by welding a beam to the pinion of the differential gear and applying the load via a pneumatic actuator. The torque could be adjusted by changing the connection point and piston stroke, while the loading frequency could be controlled by varying the air pressure. A high-precision load cell was used for amplitude control, and its signal was also used to synchronize the thermographic acquisition with the periodic surface temperature variations induced by the external load.

Design A focused on saving weight and reducing the number of components in the differential. However, Thermoelastic Stress Analysis (TSA) revealed a large concentration of stress at a critical point. Design B appears to have addressed and solved the issues observed in Design A (Fig. 8.42).

FEM Validation

The steering joint of a heavy-duty machine produced by Carraro SpA in Campodarsego (Italy) was investigated. Finite Element Method (FEM) simulations predicted specific zones of high stress, and TSA was used to validate those predictions by measuring the actual stress distribution. The comparison confirmed that TSA is a valuable tool for experimental validation of numerical models (Fig. 8.43).

Plastic Radiator Fan

A plastic radiator fan, typically mounted in cars, manufactured by Rosati Fratelli of Leini in Turin (Italy) was studied. TSA clearly showed stress concentrations at the

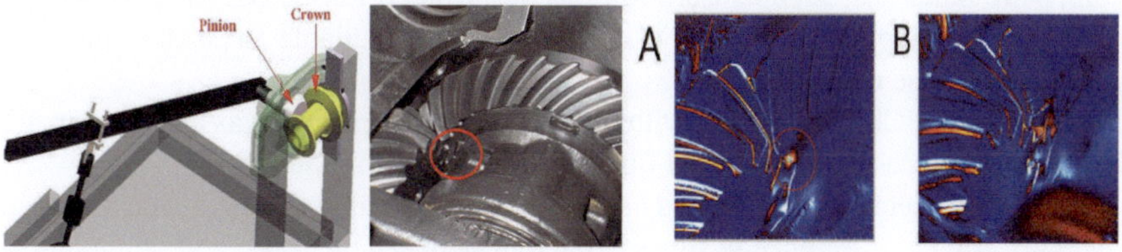

Fig. 8.42 Comparison between the two designs using TSA results

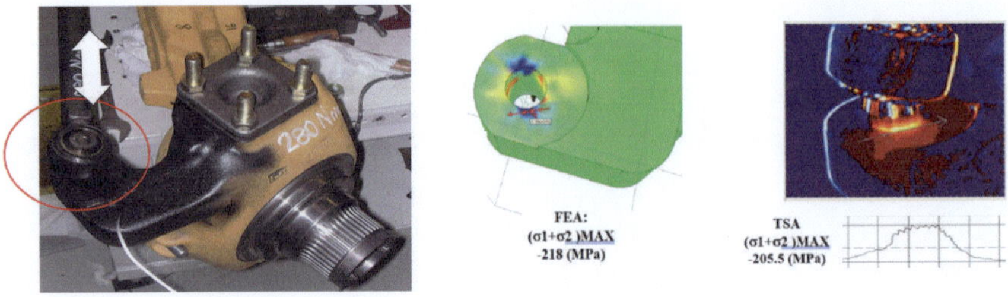

Fig. 8.43 Comparison between TSA measurements and FEM predictions

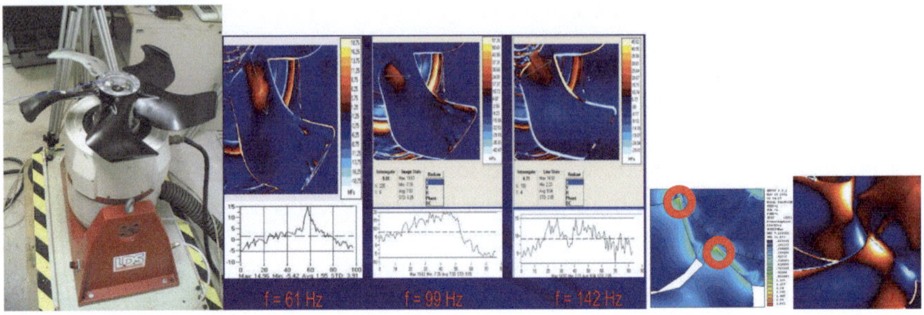

Fig. 8.44 Stress concentration on the fan blade-hub connection

junction between the blades and the hub. These areas were highlighted in the results and are visible as high-stress regions (Fig. 8.44).

Axle Testing Under Endurance Loads

In another test, TSA was used to analyze a component, manufactured by Carraro SpA in Campodarsego (Italy), under endurance loading conditions. The component was tested on the company hydraulic bench, and the applied force direction was indicated with a red arrow. A maximum working load was applied to the steering rod, and TSA allowed visualization of the surface stress field under critical conditions [50], as illustrated in Fig. 8.45.

Thermoelasticity for Composite Materials

Composite materials, such as Carbon Fiber Reinforced Plastic (CFRP), are widely used in modern applications. It is essential to develop reliable stress measurement methods for these materials [51–53].

In one test, a CFRP laminate with a twill 2/2 weave pattern was analyzed using TSA, which successfully provided full-field stress distribution [54–57].

The CFRP specimen was loaded using an electrodynamic shaker, as illustrated in Fig. 8.46. A clear difference in stress behaviour between the transverse and longitudinal fibres can be observed. Figure 8.46 shows the stress profile along the yellow line plotted over the surface. This represents one of the latest applications of TSA in

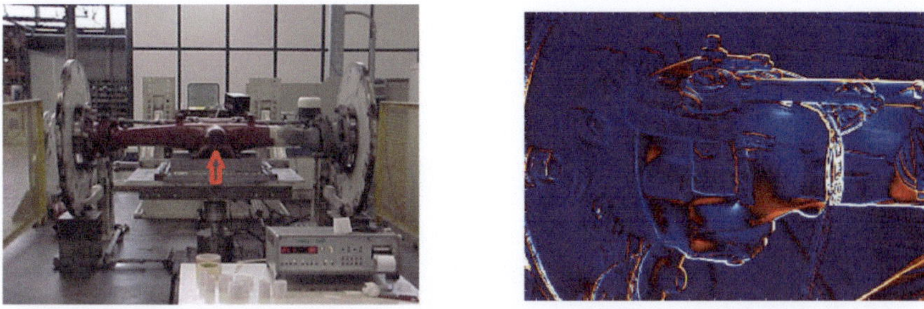

Fig. 8.45 Stress concentration on an axle during endurance testing

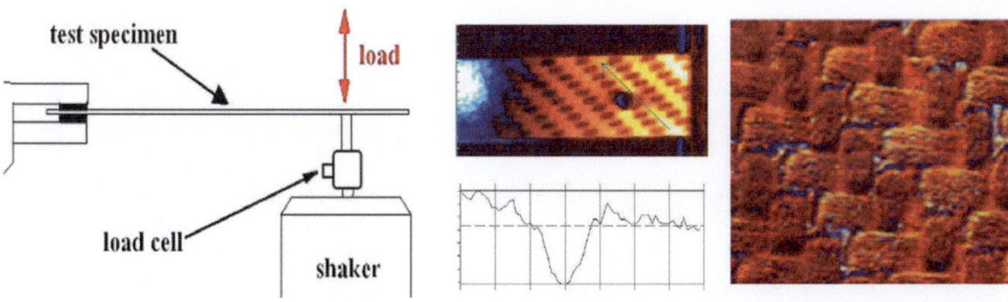

Fig. 8.46 TSA measurements on CFRP with twill 2/2 weave

the field of composite materials, where several open problems remain. The primary issue lies in the different emissivity values among the various composite components, which affect the thermal signal.

Modal Stress Analysis and Fatigue Life Prediction

Dynamic stress distribution at a given frequency gives us a mode shape of a component in terms of stress, as illustrated in Fig. 8.47 from [58].

By detecting the surface temperature fluctuations during dynamic loading, TSA can also be used for fatigue life estimation, as proposed in [58–69].

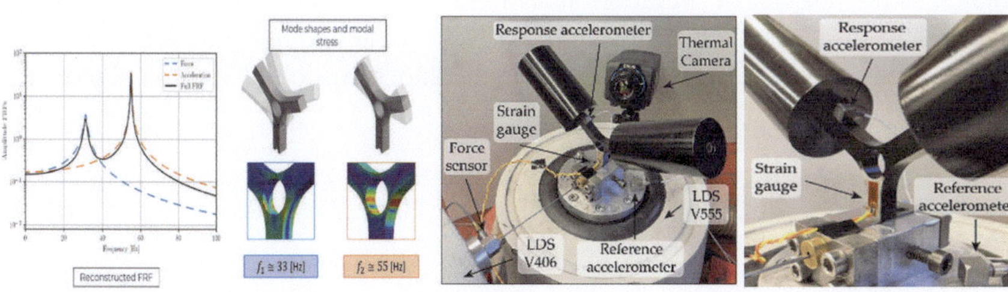

Fig. 8.47 Stress modal analysis of a Y-shaped specimen

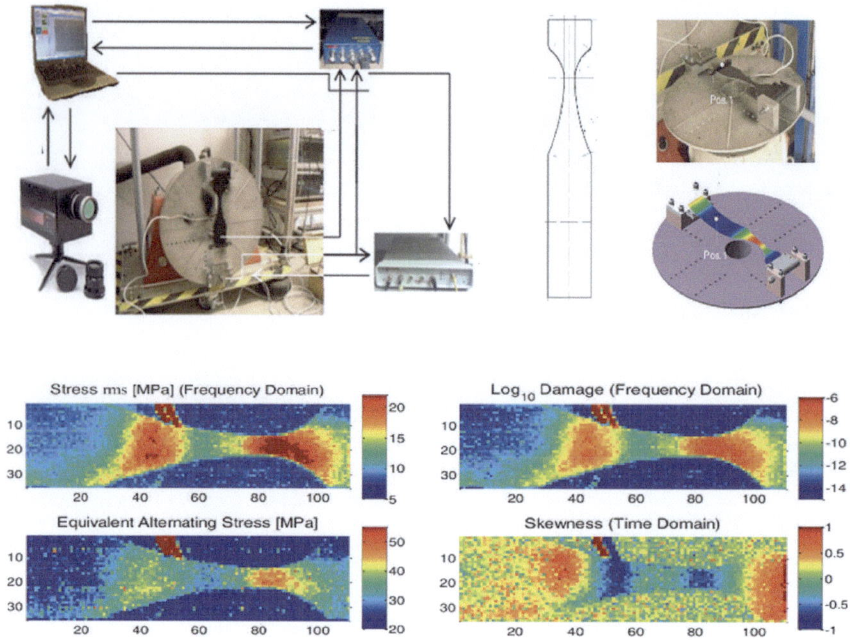

Fig. 8.48 TSA-based fatigue analysis

The dynamic behaviour of mechanical systems is governed by their resonance frequencies, which depend on the number and nature of degrees of freedom used to describe the system (Fig. 8.48).

The instantaneous strain ε at a point P in a vibrating system can be represented as the sum of modal contributions:

$$\varepsilon_{i,j}(P,t) = \sum_{k=1}^{n} A_k \cdot \varepsilon_{i,j,k}(P,t) \quad \text{for } i,j = 1,2,3 \tag{8.63}$$

where n is the number of vibration modes, A_k is the modal amplitude for mode k, independent of the point P, and $\varepsilon_{i,j,k}(P,t)$ is the strain component due to mode k [70–72]. Similarly, the stress at point P is the superposition of stresses from each mode:

$$\sigma_{i,j}(P,t) = \sum_{k=1}^{n} A_k \cdot \sigma_{i,j,k}(P,t) \quad \text{for } i,j = 1,2,3 \tag{8.64}$$

Each modal stress component can be written as a sinusoidal function:

$$\sigma_{i,j,k}(P,t) = \left|\sigma_{i,j,k}(P)\right| \cdot \sin(\omega_k t + \varphi_k) \quad \text{for } i,j = 1,2,3 \tag{8.65}$$

So, the total stress tensor becomes:

$$\sigma_{i,j}(P,t) = \sum\nolimits_{k=1}^{n} A_k \cdot \left|\sigma_{i,j,k}(P)\right| \cdot \sin(\omega_k t + \varphi_k) \quad \text{for } i,j = 1,2,3 \tag{8.66}$$

TSA measures the first stress invariant at the surface by detecting thermal fluctuations, defined as:

$$\sigma_I(P,t) = \sum_{k=1}^{n} A_k \cdot |\sigma_I(P)| \cdot \sin(\omega_k t + \varphi_k) \tag{8.67}$$

According to Kelvin's thermoelastic equation:

$$\delta T = \frac{-\alpha T \cdot \delta\sigma}{\rho C_p} \tag{8.68}$$

where $\delta\sigma$ is the stress variation (first invariant), α is the coefficient of thermal expansion, T is the absolute temperature, ρ is the material density, and C_p is the specific heat at constant pressure.

Combining equations and evaluating the temperature variation between two instants t and $t + \Delta t$:

$$\delta T(P, t+\Delta t, t) = -\frac{\alpha T}{\rho C_p} \sum_{k=1}^{n} A_k \cdot |\sigma_I(P)| \cdot [\sin(\omega_k(t+\Delta t) + \varphi_k) - \sin(\omega_k t + \varphi_k)] \tag{8.69}$$

This shows that the temperature signal is a linear combination of the modal stress contributions, each oscillating at its natural frequency. Hence, modal contributions can be separated and analysed in the frequency domain. A new function called *stressance* $S_t(\omega)$ is proposed analogous to mechanical compliance or mobility [63, 73]:

$$St(\omega) = \frac{\sigma_I(\omega)}{F(\omega)} \left[\frac{\mathrm{N/mm^2}}{\mathrm{N}}\right] \tag{8.70}$$

Rotating Components

This method was also applied to a 1:100 scale wind turbine model, tested in a subsonic closed-loop wind tunnel. A thermal film was recorded during rotation, and thermal markers were placed on the blades to aid in motion tracking and stress analysis (Fig. 8.49).

The results demonstrated the effectiveness of TSA in evaluating stress distribution on rotating composite structures in aerodynamic environments [74, 75].

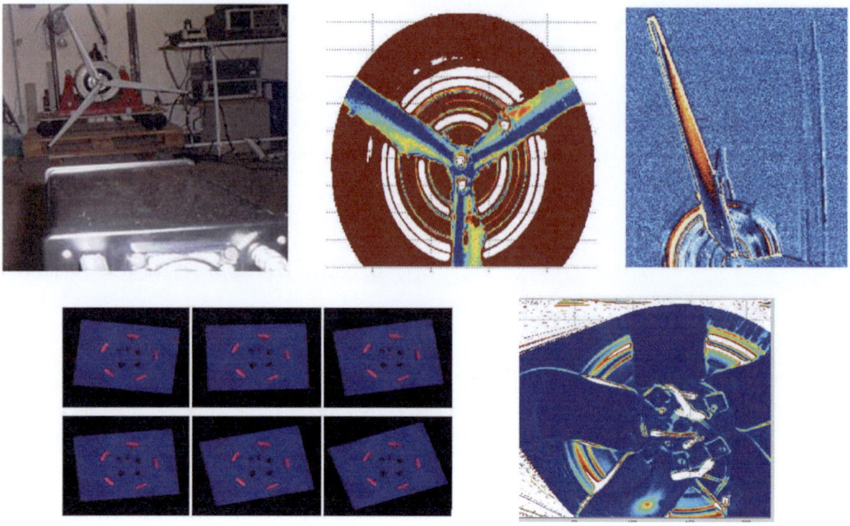

Fig. 8.49 TSA on rotating components

References

1. J.A. Wright, S.B. Palmer, Infrared thermography: an emerging NDT technique. Insight **41** (1999)
2. V.P. Vavilov, Active thermography methods for nondestructive testing, in *Review of Progress in Quantitative Nondestructive Evaluation*, vol. 21 (2002)
3. H. Sohn, C.R. Farrar, F.M. Hemez, J.J. Czarnecki, Review of structural health monitoring literature: 1996–2001. Los Alamos National Laboratory Report (2003)
4. A. Kumar, D.H. Kim, K.W. Lee, B.S. Kang, Non-destructive testing by active thermography. NDT E Int. **34**(3), 209–217 (2001)
5. M.J. Riedl, *Optical Design Fundamentals for Infrared Systems* (SPIE Press, 2001)
6. Bonanno, Rivelatori Bidimensionali per la Radiazione dall'X all'Infrarosso (Osservatorio Astrofisico di Catania)
7. H. Budzier, G. Gerlach, Active thermography, in *Handbook of Advanced Non-Destructive Evaluation*. ed. by H. Budzier, G. Gerlach (Springer International Publishing, Cham, 2018), pp.1–19
8. O.W. Dillon, T.R. Taucher, The experimental technique for observing the temperatures to the coupled thermoelastic effect. Int. J. Solid Struct. **2**, 385 (1966)
9. N. Ida, N. Meyendorf, *Handbook of Advanced Nondestructive Evaluation* (Springer, 2019)
10. H. Matsunaga, Thermoelastic stress analysis using an infrared thermographic technique. Exp. Mech. **44**(2) (2004)
11. T. Nagayama, T. Yamamoto, A. Asakura, New approach to thermal wave imaging for nondestructive evaluation. IEEE Trans. Ultrason. Ferroelectr. Freq. Control **45**(3) (1998)
12. V.P. Pavlovic, Experimental thermoelasticity for engineering applications. Exp. Tech. **24**(1) (2000)
13. V.P. Pavlovic, D.J. Tucker, An experimental investigation of thermoelastic stress analysis. Exp. Mech. **24**(1) (1984)
14. M. Choi, K. Kang, J. Park, W. Kim, K. Kim, Quantitative determination of a subsurface defect of reference specimen by lock-in infrared thermography. NDT E Int. **41**(2), 119–124 (2008)

15. R. Montanini, F. Freni, G.L. Rossi, Quantitative evaluation of hidden defects in cast iron components using ultrasound activated lock-in vibrothermography. Rev. Sci. Instrum. **83** (2012)
16. O. Breitenstein, W. Warta, M.C. Schubert, *Lock-In Thermography* (Springer, 2018)
17. J.R. Lesniak, R.E. Rowlands, Lock-in thermoelastic stress analysis. Exp. Mech. **26**(2) (1986)
18. M. Sacks, R. Owen, Infrared thermography for material characterization. NDT E Int. (2013)
19. H. Ogi, K. Maruyama, H. Tsuda, Damage detection in composites by lock-in thermography. Compos. Struct. **94**(7) (2012)
20. C. Pradere, D. Théron, S. Huguet, J.-P. Larpin, Comparison of active thermography techniques for the detection of hidden corrosion. NDT E Int. **39**(7) (2006)
21. M. Malerba, M. Argento, S. Salviuolo, G.L. Rossi, A boundary layer inspection on a wing profile through high resolution thermography and numerical methods. WSEAS Trans. Fluid Mech. (2008)
22. J. Balas, Some Applications of experimental analysis of models and structures. Exp. Mech. 127–139 (1967)
23. I.O. Kotovshchikov, Active thermography as a method of NDT. NDT E Int **13**(2), 95–101 (1980)
24. Y.T. Pei, X. Zhang, J.T.M. de Hosson, Infrared thermography in non-destructive evaluation of coatings and thin films. Surf. Coat. Technol. **202** (2008)
25. N. Harwood, W.M. Cummings, *Thermoelastic Stress Analysis* (IOP Publishing Ltd, Bristol, 1991)
26. E.A. Patterson, R.J. Greene, Thermoelastic stress analysis. Exp. Mech. **29**(4) (1989)
27. R.E. Rowlands, R.J. Greene, E.A. Patterson, Thermoelastic stress analysis–a review. Exp. Mech. (1992)
28. R.E. Rowlands, E.A. Patterson, R.J. Greene, Thermoelastic stress analysis. Exp. Mech. (1986)
29. K.T. Compton, D.B. Webster, Temperature changes accompanying the adiabatic compression of steel. verification of W. Thomson's theory to a very high accuracy. Phys. Rev. **5**, 159 (1915)
30. X. Wang, Y. He, X. Xiao, Thermoelastic stress analysis: principles and applications. Meas. Sci. Technol. (2019)
31. R.J. Greene, E.A. Patterson, R.E. Rowlands, *Thermoelastic Stress Analysis–Part C–26* (Springer)
32. E.O. Doebelin, *Strumenti e metodi di misura* (McGraw-Hill, 2001)
33. M.A. Biot, Thermoelasticity and irreversible thermodynamics. J. Appl. Phys. **27**(3), 240–253 (1956)
34. M.H. Belgen, *Infrared Radiometric Stress Instrumentation Application Range Study* (NASA Contractor Report, USA, 1968)
35. J.M. Dulieu-Barton, P. Stanley, Reproducibility and reliability of the response from four SPATE systems. Exp. Mech. **37**(4), 440–444 (1997)
36. S.M. Dulieu-Smith, Alternative calibration techniques for quantitative thermoelastic stress analysis. Strain 9–16
37. J.M. Dulieu-Smith, P. Stanley, On the interpretation and significance of the Gruneisen parameter in thermoelastic stress analysis. J. Mater. Process. Technol. **78**, 75–83 (1998)
38. G. Allevi et al., Investigating additive manufactured lattice structures: a multi-instrument approach. IEEE Trans. Instrum. Meas. **69**(5), 2459–2467 (2020)
39. G. Allevi, M. Cibeca, R. Fioretti, R. Marsili, R. Montanini, G. Rossi, Qualification of additively manufactured aerospace brackets: a comparison between thermoelastic stress analysis and theoretical results. Measurement **126**, 252–258 (2018)
40. A. Garinei, M. Becchetti, E. Pucci, G. Rossi, Constant stress field measurements through thermoelasticity. J. Thermal Stresses **36**(7) (2013)
41. S. Barone, E.A. Patterson, Full-field separation of principal stresses by combined thermo- and photoelasticity. Exp. Mech. **36**(4), 318–324 (1996)
42. S. Barone, G. Pitarresi, The development of a hybrid technique employing the boundary element method for thermoelastic stress separation. Strain **36**(4), 187–196 (2000)
43. Y.M. Huang, R.E. Rowlands, J.R. Lesniak, Simultaneous stress separation, smoothing of measured thermoelastic isopachic information and enhanced boundary data. Exp. Mech. **30**(4), 398–404 (1990)

44. M.H. Belgen, Structural Stress Measurements with an Infrared Radiometer. ISA Trans. **6** (1967)
45. M. Lacava, G. Agazzi, G. Rossi, R. Marsili, Comparison of thermoelasticity and strain gauges in fatigue load tests. Strain (2008)
46. R. Marsili, M. Moretti, J. Pirisinu, G.L. Rossi, A new measurement and testing procedure based on thermoelasticity for motorcycle frame prototype development. Appl. Mech. Mater. **3**, 411–416 (2005)
47. S. Barone, E.A. Patterson, Polymer coatings as strain witness in thermoelasticity. J. Strain Anal. **33**(3), 223–232 (1998)
48. R. Marsili, G. Rossi, E. Speranzini, Causes of uncertainty in thermoelasticity measurements of structural elements. Smart Struct. Syst. **20**(5), 539–548 (2017)
49. S. Geremia, R. Marsili, M. Moretti, J. Pirisinu, G. Rossi, Measurements of stress pattern on differential gear by thermoelasticity, in *Proceedings of ICEM*, vol. 12 (2004)
50. P. Lazzarin, F. Cannella, R. Marsili, G. Rossi, Experimental thermoelastic analysis for the vibration fatigue assessment of automotive components. Int. J. Fatigue **92** (2016)
51. F. Cannella, A. Garinei, M.P. D'Imperio, G. Rossi, A novel method for the design of prostheses based on thermoelastic stress analysis and finite element analysis. J. Mech. Med. Biol. **14**(5) (2014)
52. B.B. Lahiri, S. Bagavathiappan, T. Jayakumar, J. Philip, Medical applications of infrared thermography: a review. Infrared Phys. Technol. **55**, 221–235 (2012)
53. Z. Su, L. Ye, Y. Lu, Guided Lamb waves for identification of damage in composite structures: a review. J. Sound Vib. 295(3) (2006)
54. J. Pirisinu, G.L. Rossi, R. Marsili, Measurements of stress pattern on carbon fiber specimen by thermoelasticity, in *Icem12–12th International Conference on Experimantal Mechanics*
55. J. Shao, H. Li, X. Zhou, D. Wang, Lock-in thermography for detection of defects in composite materials. NDT E Int. **47** (2012)
56. T. Tocci, L. Capponi, G. Rossi, Thermoelastic model updating for fatigue damage detection. Struct. Health Monit. (2021)
57. L. Capponi et al., Thermoelasticity and ArUco marker-based model validation of polymer structure: application to the San Giorgio's bridge inspection robot. Acta Imeko **10**(4), 177–184 (2021)
58. L. Capponi, J. Slavič, G. Rossi, M. Boltežar, Thermoelasticity-based modal damage identification. Int. J. Fatigue **137** (2020)
59. C. Braccesi, M. Moretti, F. Cianetti, G. Rossi, Random load fatigue: experimental approach through thermoelasticity. Procedia Eng. (2015)
60. S. Saito, M. Watanabe, Y. Haruyama, Infrared thermography for fatigue crack detection. NDT E Int. 36(5) (2003)
61. P. Lazzarin, F. Cannella, R. Marsili, G. Rossi, Thermoelastic analysis of automotive components for vibration fatigue evaluation, in *Procedia Structural Integrity* (2016)
62. R. Marsili, G. Rossi, M. Becchetti, R. Flori, M. Moretti, Full-field modal parameter identification by thermoelasticity. Exp. Tech. **34**(5) (2010)
63. R. Marsili, G.L. Rossi, Thermoelasticity: a tool for experimental modal analysis, in *Proceedings of IMAC XXVIII* (2010)
64. M. Moretti, G. Rossi, R. Marsili, Thermoelasticity and strain field reconstruction. Exp. Mech. (2013)
65. M. Moretti, G. Rossi, R. Marsili, Applications of thermoelasticity in fatigue analysis. Strain **50** (2014)
66. M. Moretti, G. Rossi, R. Marsili, Full-field fatigue assessment of advanced materials by thermoelasticity, in *Procedia Engineering* (2013)
67. P. Cavaliere, G.L. Rossi, R. Di Sante, M. Moretti, Thermoelasticity for the evaluation of fatigue behavior of 7005/Al2O3/10p metal matrix composite sheets joined by FSW. Int. J. Fatigue **30**, 198–206 (2008)
68. R. Flori, R. Marsili, G.L. Rossi, "Operational modal analysis by thermoelasticity, in *International Operational Modal Analysis Conference—IOMAC 09* (Ancona, Italy)

69. T. Nguyen, M.C. Rojo, C. Petitjean, S. Reynaud, Advances in infrared thermography for fatigue monitoring. NDT E Int. (2019)
70. M. Becchetti, R. Flori, R. Marsili, G.L. Rossi, Measurement of stress and strain by a thermal camera, in *SEM Annual Conference and Exposition on Experimental and Applied Mechanics* (Albuquerque, NM, USA)
71. M. Becchetti, R. Flori, R. Marsili, G.L. Rossi, Stress and strain measurements by image correlation and thermoelasticity, in *SEM Annual Conference and Exposition on Experimental and Applied Mechanics* (Albuquerque, NM, USA)
72. M. Becchetti, R. Marsili, M. Moretti, R. Flori, Comparison between digital image correlation and thermoelasticity for strain field analysis, in *9th International Conference on Vibration Measurements by Laser and Noncontact Techniques* (Ancona, Italy)
73. N.F. Enke, B.I. Sandor, Cyclic plasticity analysis by differential infra-red thermography, in *Proceedings of VIth International Congress on Experimental Mechanics* (Portland, Oregon, 1988), pp. 836–842
74. G. Brustenga, R. Marsili, M. Moretti, J. Pirisinu, G.L. Rossi, Measurement on rotating mechanical component by thermoelasticity, in *International Conference of BSSM* (Southampton)
75. F. Castellani, M. Malerba, R. Marsili, A. Salviuolo, An innovative method to measure stress distribution of a wind turbine model by thermoelasticity, in *European Wind Energy Conference and Exhibition* (EWEA, 2008)

Chapter 9
Laser Doppler Anemometry, Vibrometry and Particle Image Velocimetry

Abstract In this chapter, we describe an interferometric measurement technique based on the use of laser beams and the Doppler effect, along with its two primary applications for measuring velocity in fluids and on solid surfaces. These two techniques are known as Laser Doppler Anemometry (LDA) and Laser Doppler Vibrometry (LDV). Complementary to LDA, the technique used to measure fluid velocity fields based on image analysis is called Particle Image Velocimetry (PIV). Its measurement principle is similar to that of Digital Image Correlation (DIC), as introduced in Chapter 7. However, PIV employs an optical and experimental setup specifically tailored for fluid flow applications, using tracer particles, light sheet lighting and high-speed imaging to capture and analyze the motion of fluids.

9.1 Laser Doppler Measurement Technique Basis

Laser Doppler Vibrometers (LDV) and Laser Doppler Anemometers (LDA) are velocity measurement systems used to determine the motion of solid surfaces or particles suspended in fluids. These systems operate based on the Doppler effect, a physical phenomenon introduced in Chap. 11, and employ coherent laser beams as the measurement medium. Sometimes are called "measurement systems by light".

To understand the fundamental principle of these techniques, consider a laser beam emitted by a stationary source S, illuminating a moving target located at point P, which could be a solid surface or a fluid-borne particle, moving with an arbitrary velocity vector u, as shown in Fig. 9.1a. Point P reflects back the incident laser beam if it corresponds to a solid surface, whereas it scatters the light if it is a particle with dimensions comparable to the laser wavelength. In both cases, the interaction leads to the diffusion of light in multiple directions as a result of scattering phenomena [1, 2].

We assume that the laser beam incident at point P can be described by a planar wavefront. Over a unit time interval, point P, moving with velocity u, travels a distance $u \cdot \cos\theta$ along the direction of wave propagation, where θ is the angle between the velocity vector and the laser beam's propagation direction. As a result, point P

G. Rossi, *Mechanical and Thermal Measurements by Images and Waves*, Smart Sensors, Measurement and Instrumentation 51, https://doi.org/10.1007/978-3-032-03464-9_9

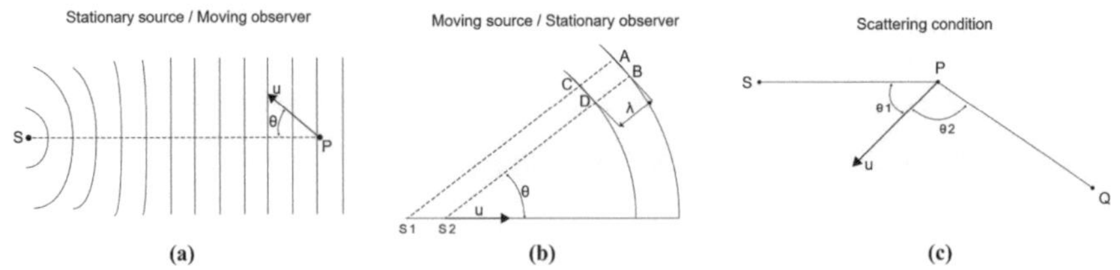

Fig. 9.1 Laser beam source and its reflections from a point P of a moving particle or surface

encounters (u·cos θ)/λ more wavefronts than it would if it were stationary, where λ is the laser wavelength, i.e. the distance between wavefronts.

Since the frequency of a wave is defined as the number of wavefronts observed per unit time, the moving observer at P perceives an increased frequency compared to that emitted by the source. The resulting frequency increase is therefore given by:

$$\Delta\nu = \frac{u \cdot \cos\theta}{\lambda} \tag{9.1}$$

Let us consider, at time t, two small portions of the wavefront, AB and CD, emitted at times t_1 and t_2 respectively [2], when the source was at positions S_1 and S_2 as in Fig. 9.1b. We can determine the lengths:

$$S_1A = c\,(t - t_1)\ S_2D = c\,(t - t_2) \tag{9.2}$$

where c is the speed of light. The interval between the two wave fronts is given by the period τ of the wave itself, given by:

$$\tau = t_1 - t_2 = 1/\nu \tag{9.3}$$

during this interval, the source has moved by $S_1S_2 = u\tau$, so the wavelength detected by the observer is:

$$\lambda' = AC = S_1A - S_2D - S_1S_2 \cdot \cos\theta \tag{9.4}$$

where θ is the angle between the wave propagation direction and the velocity vector. Based on the considerations above, we obtain:

$$\lambda' = c(t - t_1) - c(t - t_2) - u \cdot \cos\theta = (c - u \cdot \cos\theta)/\nu \tag{9.5}$$

but $\lambda' = c/\nu\prime$, where ν' is the reflected back frequency of the laser light, so [2]:

$$\frac{\Delta\nu}{\nu} = \frac{\nu'}{\nu} - 1 = \frac{c}{c - u \cdot \cos\theta} - 1 = \frac{u/c \cdot \cos\theta}{1 - u/c \cdot \cos\theta} \tag{9.6}$$

This expression is different from the one obtained for the case of a stationary source and a moving observer. However, if the velocity u is small compared to the speed of light c, by expanding Eq. (9.6) in a Taylor series [2], we obtain:

$$\frac{\Delta\nu}{\nu}=\frac{u\cdot\cos\theta}{c}+\frac{u^2\cdot\cos\theta}{c^2}+\cdots \tag{9.7}$$

It can be observed that this first-order approximation coincides with the expression derived in the previous case. The Doppler effect resulting from the scattering or reflection of a laser beam by a moving particle or surface can be interpreted as a combination of the two previously discussed frequency shift generated one by stationary source and moving observer, the other by moving source and stationary observer [1, 2]. Let ν be the frequency of the wave emitted by the stationary source S, and let P denote the point on the surface or the particle that reflects the beam. According to Eq. (9.1), the frequency of the backscattered signal, as perceived at the source, will be:

$$\nu'=\nu\left(1+\frac{u}{c}\cos\theta_1\right) \tag{9.8}$$

In point O of Fig. 9.1c will arrive the frequecy ν'' equal to [2]:

$$\nu''=\nu'\frac{1}{1-\frac{u}{c}\cos\theta_2}=\nu\frac{1+\frac{u}{c}\cos\theta_1}{1-\frac{u}{c}\cos\theta_2} \tag{9.9}$$

From above equation we get:

$$\frac{\nu''}{\nu}=\frac{1+\frac{u}{c}\cos\theta_1}{1-\frac{u}{c}\cos\theta_2} \tag{9.10}$$

If u is much smaller than the speed of light c the previous expression can be expanded in a Taylor series, stopping at the first term, obtaining:

$$\frac{\nu''}{\nu}=1+(\cos\theta_1+\cos\theta_2)\frac{u}{c}. \tag{9.11}$$

And also, multiplying both terms of the above expression by ν:

$$\nu''-\nu=\nu(\cos\theta_1+\cos\theta_2)\frac{u}{c} \tag{9.12}$$

By applying the trigonometric identity already seen in Chap. 1, Eq. 1.66, considering that c = λf, from Eq. 9.12 the following expression for the Doppler frequency shift $\Delta f_D=\left(\nu''-\nu\right)$ can be derived:

$$\Delta f_D=\frac{2\nu}{\lambda}\cos\left(\frac{\vartheta_1+\vartheta_2}{2}\right)\cos\left(\frac{\vartheta_1-\vartheta_2}{2}\right) \tag{9.13}$$

That is the most used basic formula of the Doppler effect in laser Doppler anemometry and vibrometry.

9.2 Laser Doppler Anemometry

Laser Doppler Anemometry (LDA) is a non-intrusive optical technique used to measure the velocity of particles suspended in a fluid, whether a gas or a liquid. The underlying assumption is that these tracer particles faithfully follow the motion of the fluid, allowing their velocity to represent the local fluid velocity [3].

Its development spans several decades and involves key milestones in both theoretical, tecnological and experimental advancements. The idea of using light to measure the velocity of particles in a fluid originates from the development of Doppler shift theory, which was initially formulated in the context of sound waves. The Doppler effect, first described by Christian Doppler in 1842, explains how the frequency of a wave changes due to the relative motion between the source and the observer. This fundamental principle was introduced in Chap. 1, and further explored in Chap. 6, where we examined its application to ultrasound-based techniques for measuring velocity profiles in fluids.

The idea of using light to measure fluid flow velocity was proposed by P. M. S. Black and C. L. H. T. Hamilton in the 1950s. The invention of the laser in 1960 by Theodore Maiman was a crucial development for LDA. Lasers provide coherent light, a wave having a single frequency, which is ideal for measurements of the Doppler shift. In the early 1960s, researchers like S. G. McNaughton and R. B. Combes explored the use of laser light in fluid dynamics experiments, laying the groundwork for LDA.

In 1971, the first Laser Doppler Anemometer was developed by B. J. W. Shurcliff at the Massachusetts Institute of Technology (MIT). This early system was based on a single laser beam, which was split into two beams, one scattered by particles in a fluid. Is the nowadays called the reference beam configuration. The Doppler shift in the scattered light was used to measure the velocity of the particles.

The use of dual-beam differential LDA systems was later introduced, where two laser beams intersected in the flow, improving performances and allowing measurements of multi-dimensional velocity components. This development further enhanced the application of LDA in both academic research and industry. By the 1990s, LDA systems were commercially available by companies like Dantec, Aerometrics, LaVision, etc., and the technique gained widespread use in industries like aerospace, automotive, and fluid dynamics research.

Laser-Doppler Anemometry (LDA) is an advanced, non-intrusive optical technique used to measure the velocity field of fluid flows. It is widely employed across various domains, including industry (for evaluating technical specifications and optimizing performance), engineering (to identify critical parameters and support model development), and theoretical fluid dynamics.

Unlike traditional intrusive methods, such as Pitot tubes or hot-wire anemometry, LDA does not require the insertion of physical probes into the fluid stream. This non-intrusive nature prevents disturbance of the flow field, making LDA particularly valuable for studying complex or sensitive flow environments. LDA is especially useful in the investigation of turbulent flows, where the governing Navier–Stokes equations, introduced in Chap. 1, cannot be solved analytically due to their nonlinear and complex nature.

Although Computational Fluid Dynamics (CFD) provides a numerical approach to solving these equations by discretizing them in space and time, such simulations often demand substantial computational resources and can become highly complex.

In this context, LDA represents a powerful and efficient alternative, enabling accurate, real-time velocity measurements without interfering with the flow. This is particularly advantageous in experimental validation of CFD models, flow diagnostics, and high-resolution turbulence studies.

The technique involves seeding the fluid with microscopic tracer particles, which, due to their small size and low inertia, are assumed to faithfully follow the motion of the surrounding fluid. Thus, measuring the velocity of the particles yields an accurate representation of the local fluid velocity.

The measurement process in LDA is based on the interaction of a laser beam with these moving particles. When illuminated by the laser, the particles scatter light; due to the Doppler effect, the frequency of the scattered light shifts in proportion to the particle's velocity. This Doppler frequency shift is then captured and analyzed by a photodetector.

To illustrate the principle, consider a laser source S emitting light at a frequency f, and a particle P moving through the fluid at velocity v. The moving particle scatters the incident laser light, resulting in a frequency shift, such that an observer at point O receives the scattered light at a frequency $f + \Delta f_D$, where Δf_D represents the Doppler shift. Simultaneously, a reference beam, identical in frequency f, but not scattered, also reaches the photodetector. The superposition of these two laser beam at the detector, as shown in Fig. 9.2c, enables the detection of the Doppler frequency and thus the computation of the particle velocity [3].

For the velocity component v_t, which is perpendicular to the bisector of the angle α (as illustrated in Fig. 9.2b), and considering the geometric relationship between the angles θ_1, θ_2, and α, specifically, $\theta_1 + \theta_2 + \alpha = \pi$, as shown in Fig. 9.2b, we can derive the following expression from Eq. (9.13):

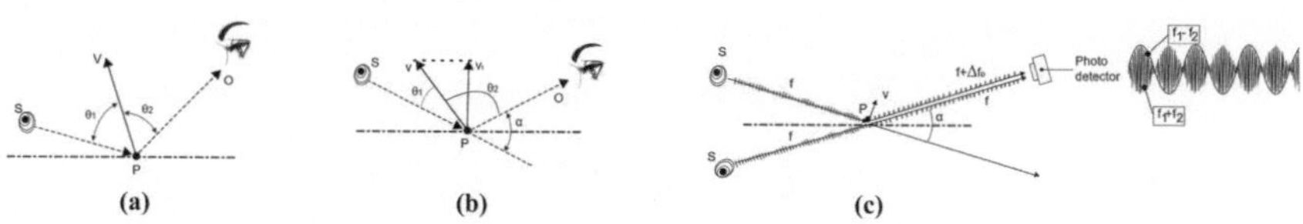

Fig. 9.2 Doppler effect generated by light scattering particles

$$\Delta f_D = \frac{2v_t}{\lambda}\sin\left(\frac{\alpha}{2}\right) \tag{9.14}$$

The two laser beams converge at the photodetector (observer O) and produce a beat phenomenon, as introduced in Chap. 1. This interaction results in an amplitude modulation of the photodetector's output signal at a frequency equal to the difference between the two optical frequencies, so at the Doppler frequency Δf_D. By measuring this modulation frequency, the Doppler shift can be accurately determined. Given the known values of the angle α and the laser wavelength λ, Eq. (9.14) can then be used to calculate the transverse velocity component v_t of the moving particle.

It is important to note that the Doppler frequency shift resulting from the particle's motion is extremely small compared to the original frequency of the laser light. For instance, a Helium–Neon (He–Ne) laser emitting at a wavelength $\lambda = 0.632$ μm corresponds to an optical frequency of approximately 4.7×10^{14} Hz. A particle moving at 10 m/s would induce a Doppler shift of only 3.2×10^{7} Hz, resulting in a relative frequency change on the order of 7×10^{-8}.

Detecting such a minute frequency variation directly would require an optical spectrum analyzer with extremely high resolution of 7×10^{-8}, well beyond the capabilities of conventional direct frequency measurement instruments. For this reason, an interferometric setup or other techniques are employed in LDA systems to convert the Doppler shift into a measurable beat frequency in the electrical domain, by removing the optical high frequency of the light, enabling precise and practical detection of particle velocity.

The first of these techniques is illustrated in Fig. 9.3. As discussed in Chap. 1, the two laser beams converging on the same photodetector can be represented by their respective electric field waveforms, described as follows:

$$E_A(t) = E_0\cos(2\pi f_1 t)\ E_B(t) = E_0\cos(2\pi f_2 t)$$

where $\Delta f_D = f_2 - f_1$, and the total field at the detector will be is the sum:

$$E_T(t) = E_A(t) + E_B(t) = E_0\left[\cos(2\pi f_1 t) + \cos(2\pi f_2 t)\right]$$

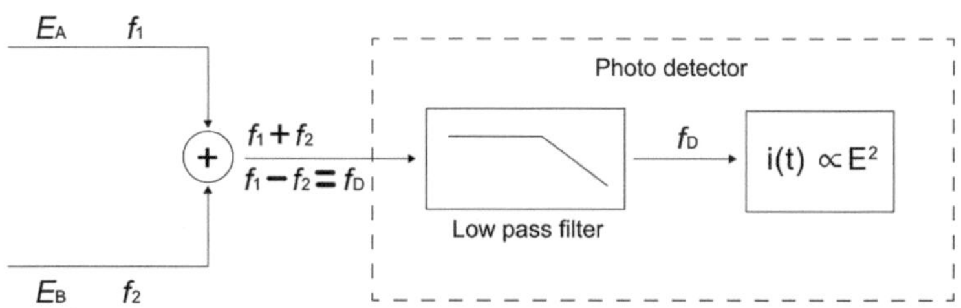

Fig. 9.3 Optical heterodyne on LDA systems and filtering effect of photodetector

This superposition is referred to as optical heterodyning, as it involves the combination of two waves with different frequencies (in contrast to homodyning, which involves the combination of two waves having the same frequency). The same principle is illustrated in Fig. 1.25 of Chap. 1. Using once again the trigonometric identity (1.64) in the above equation we get:

$$E_T(t) = 2E_0 \cos\left(2\pi \cdot \frac{f_1 + f_2}{2} \cdot t\right) \cdot \cos\left(2\pi \cdot \frac{f_1 - f_2}{2} \cdot t\right)$$

The photodiode is inherently band-limited, effectively working as a low-pass filter, as discussed in Chap. 3. As such, it suppresses the high-frequency component at $(f_1 + f_2)/2$, and outputs only the lower-frequency signal i(t) at $(f_1 - f_2)/2 = \Delta f_D$**,** as illustrated in Chap. 1. This output signal is proportional to the intensity of the incident light, which in turn is proportional to the square of the electric field amplitude. Consequently, the photodiode output is a signal i(t) described by:

$$i(t) = K \cdot [E_1(t) + E_2(t)]^2 = K' \cdot \cos^2\left(2\pi \cdot \frac{f_1 - f_2}{2} \cdot t\right) = K' \cdot \cos^2\left(2\pi \cdot \frac{\Delta f_D}{2} \cdot t\right) \tag{9.15}$$

Here, K and K′ are constants that also account for the sensitivity of the photodetector. The resulting signal oscillates at the Doppler frequency Δf_D, which, according to Eq. (9.15), is proportional to the transverse velocity component v_t of the particle. By applying a demodulation process, such as a frequency-to-voltage conversion, this signal can be processed to extract a direct measurement of v_t. The complete setup of a reference beam LDA system, the first one proposed by B. J. W. Shurcliff at the Massachusetts Institute of Technology in 1971, is shown in Fig. 9.4a. To ensure that the two beams have the same optical frequency, a single laser source is used in combination with a beam splitter.

Using two independent laser sources is typically impractical, as their frequency differences would far exceed the Doppler frequency range, preventing accurate detection.

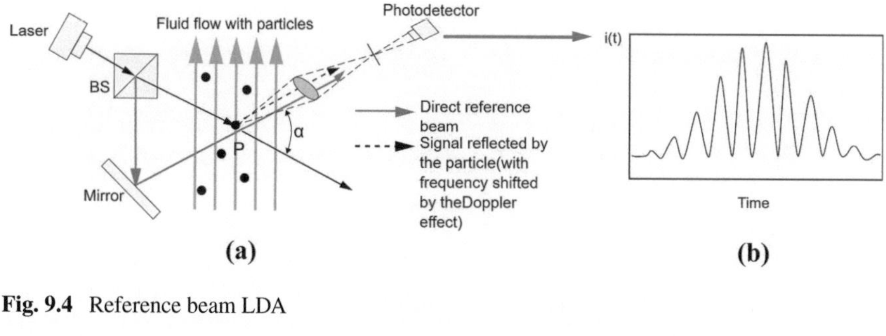

Fig. 9.4 Reference beam LDA

In this configuration, one beam is directed toward the measurement volume (focused by the receiving optics), where it interacts with particles in the fluid, while the reference beam is directed straight to the photodetector.

A modulated signal at the Doppler frequency is generated only when a particle enters the laser beam and scatters the light. This signal is transient, as illustrated in Fig. 9.4b, and appears only during the particle's passage through the measurement volume. It exhibits oscillations at the Doppler frequency Δf_D, with the amplitude peaking when the particle is at the center of the beam, where the laser intensity profile is at its maximum.

This measurement system configuration is useful for illustrating the fundamental physical principles underlying the Laser Doppler Anemometry (LDA) technique. However, it presents several practical limitations that hinder its effective application:

- Optical heterodyne detection requires that both laser beams have approximately the same intensity. In practice, the reference beam, delivered directly to the photodetector, has a much higher intensity than the light scattered by the moving particle, which is typically very weak and collected through receiving optics. To balance the intensities, attenuation of the reference beam is necessary (e.g., using neutral density filters). However, this approach significantly reduces the signal-to-noise ratio, compromising measurement accuracy.
- The angle α, which is critical for determining the velocity component via Eq. (9.13), is difficult to measure with high precision. This angle depends on the precise alignment of the optical components around the flow field. Even small misalignments can lead to significant uncertainty in the velocity estimation.

These two challenges, intensity imbalance and angle measurement sensitivity, combine to produce high uncertainty in the measurement results.

To overcome these limitations, a more robust configuration known as the Differential Laser Doppler system is employed. Its optical layout is shown in Fig. 9.5a.

Once again, a laser beam is split into two separate beams using a beam splitter as illustrated in Fig. 9.5. These two beams are then directed and focused to the measurement point P through the use of a mirror and a lens. At point P, the two

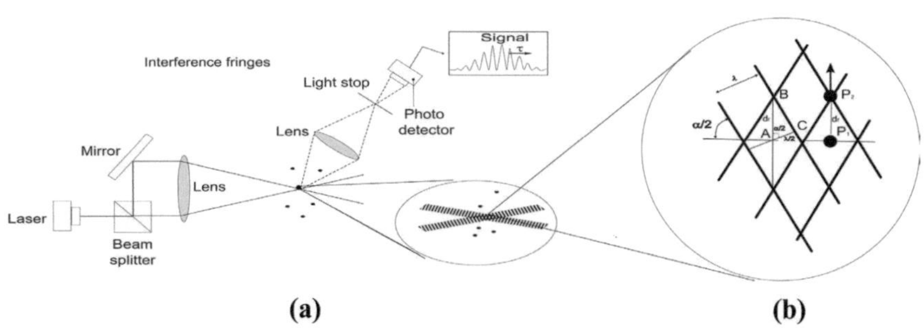

(a) (b)

Fig. 9.5 Differential LDA

laser beams, each represented by planar waves of the same wavelength λ, interfere and generate a spatial interference pattern, consisting of alternating bright and dark fringes, as previously illustrated in Fig. 1.26 of Chap. 1. These interference fringes effectively form a "virtual grid" in space.

Referring to triangle ABC in Fig. 9.5b, where two wavefronts, spaced by a wavelength λ and intersecting at an angle α, are geometrically expanded, we can derive the following relationship:

$$\lambda/2 = d_f \sin\left(\frac{\alpha}{2}\right) \text{ and therefore : } d_f = \frac{\lambda}{2 \cdot \sin\left(\frac{\alpha}{2}\right)} \tag{9.16}$$

A particle moving from point P_1 to point P_2 in Fig. 9.5b, and crossing two consecutive interference fringes with a transverse velocity v_t, requires a time $t = d_f/v_t$, where d_f is the fringe spacing. Consequently, the frequency Δf at which the photodiode detects the passage of the particle, corresponding to the rate at which fringes are crossed, is given by:

$$\Delta f = \frac{1}{t} = \frac{v_t}{d_f} = \frac{2v_t}{\lambda}\sin\left(\frac{\alpha}{2}\right) \text{ and also } v_t = \frac{\lambda \Delta f}{2\sin\left(\frac{\alpha}{2}\right)} \tag{9.17}$$

The previously described quantity is the "flashing frequency" Δf generated by the moving particle as it crosses the interference fringes. This frequency is identical to the frequency shift due to the Doppler effect, as previously derived through the analysis of interferometric phenomena in the time domain; hence, $\Delta f = \Delta f_D$.

By detecting this flashing frequency in the output signals, similar to the reference arm configuration, the transverse velocity component of the particle (i.e. the component perpendicular to the fringes) can be measured, as expressed by Eq. 9.17.

The region where the two laser beams intersect defines the measurement volume, which approximates a three-dimensional Gaussian ellipsoid. This shape results from the Gaussian intensity profiles of the intersecting laser beams, as illustrated in Fig. 9.6. Within this volume, the interference fringes exhibit lower intensity near the edges and peak intensity at the center. This intensity variation explains the characteristic shape of the backscattered signal.

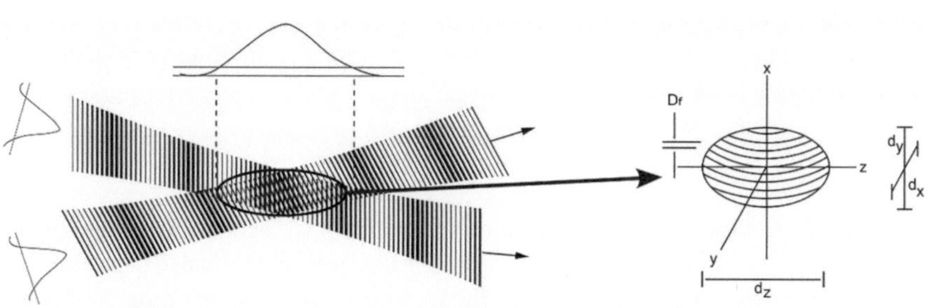

Fig. 9.6 Measurement volume of differential LDA

Models and formulas for estimating the size of the measurement volume can be found in [3]. Typically, for He–Ne or argon ion laser sources used in commercial LDA systems, the measurement volume dimensions are on the order of a few millimeters or less.

Compared to LDA systems using a reference beam, the differential laser Doppler configuration offers several key advantages, mainly:

- Stable and Precisely Defined Angle (α): The interference angle α between the two beams is fixed and accurately known, as it is determined solely by the optical setup and the geometry of the lens system, both typically manufactured with high precision.
- Balanced Beam Intensities: The intensities of the two laser beams (A and B) can be made equal. This results in a more efficient geometric heterodyning, yielding significantly brighter received optical signals.

As a result of these improvements, the signal-to-noise ratio achieved with this configuration is considerably higher than that of the reference beam LDA systems. For this reason, nearly all modern commercial LDA systems adopt the differential configuration.

LDA systems optical setup previously seen has a directional ambiguity limitation, as shown in Fig. 9.7: is its inability to determine the direction of particle motion. Specifically, a particle crossing the measurement volume with the same speed and trajectory, but in opposite directions, will generate the same output frequency. Thus, the system cannot distinguish between forward and reverse motion along the same line. To overcome this limitation, a Bragg cell, also known as an acousto-optic modulator, is introduced into the optical path. The Bragg cell shifts the frequency of the laser beam using the acousto-optic effect, whereby sound waves modulate the refractive index of a crystal medium, creating a moving diffraction grating.

In practical terms, a radio-frequency (RF) signal is applied to a piezoelectric transducer bonded to a crystal (e.g., quartz or tellurium dioxide). The transducer generates ultrasonic acoustic waves that propagate through the crystal, producing alternating regions of compression and rarefaction. These periodic variations form a traveling refractive index grating within the crystal.

When a laser beam enters the Bragg cell, it interacts with this dynamic grating. The light is diffracted by the periodic structure, analogous to X-ray diffraction by atomic planes in a crystal, hence the term Bragg diffraction. For optimal diffraction efficiency, the incident laser beam must strike the grating at the Bragg angle, defined by the following condition:

$$\lambda = 2n\Lambda\sin(\theta_B)$$

where λ is the light wavelength; n the refractive index; Λ the acoustic wavelength and θ_B the Bragg angle. The diffracted light gains or loses frequency equal to the frequency of the acoustic wave f_a. This is because the moving grating imparts momentum and energy to the photons. So if the laser light has frequency f_0, the first-order diffracted beam has frequency:

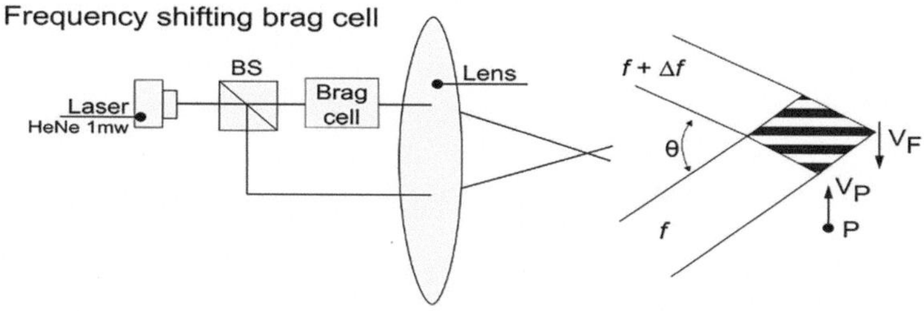

Fig. 9.7 Resolving directional ambiguity with a Bragg cell

$$f = f_0 \pm f_a$$

The sign of the frequency shift depends on the direction of diffraction relative to the acoustic wave. As a result, Bragg cells produce a frequency shift analogous to a moving particle crossing interference fringes. This effect generates a pattern of moving fringes with a velocity v_f, as defined by Eq. 9.17 and illustrated in Fig. 9.7.

If the velocity of the particle crossing the interference fringes (v_p) is in the same direction and versus as the fringe motion velocity (v_f), the photodiode output "flashing" signal will correspond to a particle with an effective velocity of $v_F - v_p$. As a result, the detected "flashing" frequency is reduced, and the Doppler shift becomes $\Delta f_D - f_b$, where Δf_D is the Doppler shift calculated using Eq. 9.17, corresponding to the particle velocity v_p.

Conversely, if the particle velocity is in the opposite versus to the fringe motion, the particle crosses more fringes per unit time. This results in an increased output signal "flashing" frequency equal to $\Delta f_D + f_b$. However, frequencies cannot be negative, negative frequencies are physically meaningless in this context. Therefore, the Bragg frequency f_b defines the maximum measurable velocity and sets the limits of the velocity measurement range.

A typical Bragg frequency is 40 MHz. According to Eq. 9.17, this value corresponds to a velocity of approximately 30 m/s for a laser beam with a wavelength of 633 nm and a beam crossing angle of 15°. So only particle velocity in the range ±30 m/s can be measured.

When light interacts with particles whose dimensions are comparable to the wavelength of the laser it is partially reflected backward, refracted, and diffracted forward, all components carrying the Doppler shift. As illustrated in Fig. 9.8a, the forward-diffracted light is the most intense, as shown in the polar diagram.

For this reason, collecting the forward-scattered light is generally more efficient. However, this approach requires precise optical alignment between the transmitting and receiving optics on the test bench, as shown in Fig. 9.8c. In contrast, backward-scattered light collection, illustrated in Fig. 9.8b, enables integration of all optical components into a single device with automatic alignment. This configuration is therefore the most commonly used in LDA systems.

Flow velocity sampling occurs only when a particle crosses the measurement volume, generating a signal such as the one shown in Fig. 9.9a. This signal typically contains both a low-frequency pedestal and a high-frequency oscillation. Signal

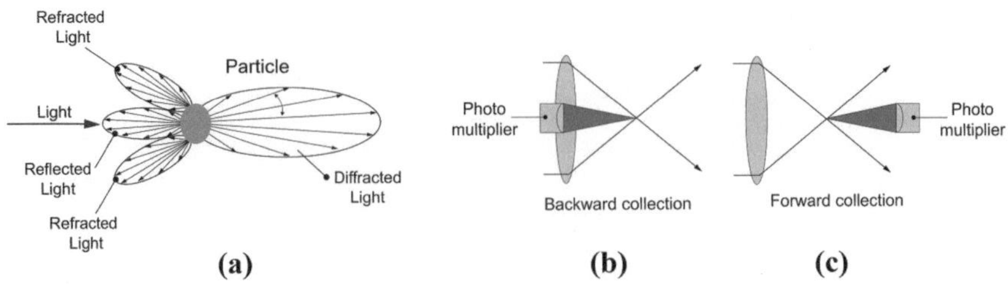

Fig. 9.8 Forward or backward scattering optical configuration of differential LDA systems

processing involves removing the pedestal and analyzing the high-frequency component, commonly using methods such as Fourier Transform or zero-crossing techniques. Once the signal is cleaned, the known Bragg cell frequency is subtracted to isolate the Doppler component. The direction of particle motion (versus) is determined, and Eq. 9.17 is applied to compute the particle's velocity from each valid ("good") burst signal.

The quality of a burst signal depends on the particle's size relative to the fringe spacing in the measurement volume. Ideally, the particle should be large enough to reflect sufficient light but small enough to interact with only one fringe at a time. Particles that are too large or too small result in weak or ambiguous signals.

Since velocity data is obtained only when a particle crosses the fringes, the sampling is inherently random. To reliably capture fast velocity fluctuations in turbulent flows, a high average sampling rate is necessary. This requires a sufficient concentration of appropriately sized seeding particles in the flow, as illustrated in Fig. 9.9b.

The number of particles that produce a valid burst signal from which frequency can be reliably estimated is referred to as the data rate. In some cases, multiple particles may simultaneously cross the measurement volume, resulting in a composite signal with overlapping frequencies. These signals cannot be reliably processed and are therefore discarded.

If the particle population contains too many excessively large or small particles, the proportion of unusable signals increases. The data valid parameter is defined as the ratio of valid bursts to total detected bursts (including those discarded). Achieving

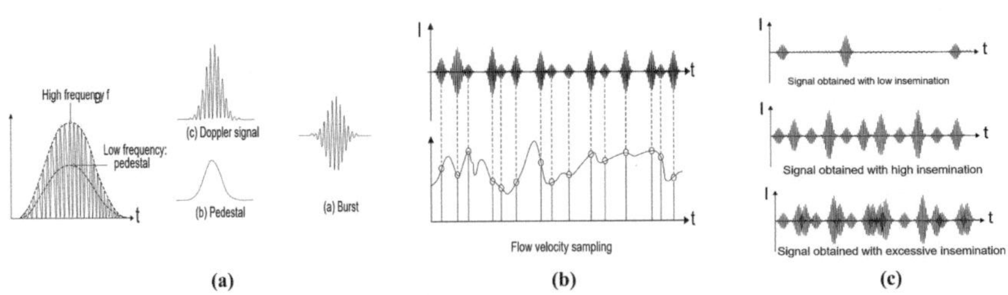

Fig. 9.9 Signal processing on LDA differential systems

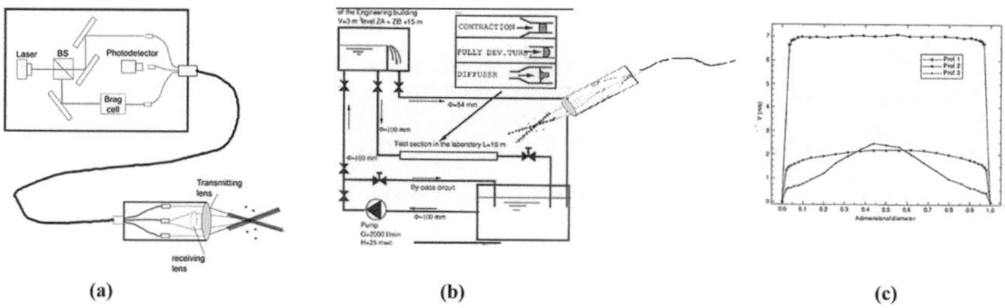

Fig. 9.10 Analysis of typical velocity profiles inside tubes by an LDA measurement system

optimal measurement conditions thus requires a careful balance between data rate, data valid, and the proper selection of particle size relative to fringe spacing.

In practical applications, particle seeding is one of the most critical parameters to define. Extensive guidance is available in the literature and on the websites of leading LDV system manufacturers, such as Dantec Dynamics, LaVision, Measurement Science Enterprise, ILA R&D.

Figure 9.10a shows a classical commercial LDA system from Dantec Dynamics. The system features a probe equipped with lenses connected via optical fibers to the laser source, a beam splitter, and a Bragg cell, which generates the two laser beams. These components, along with a photodetector, are used to measure velocity profiles of water flow inside a transparent Plexiglas tube with a diameter of 100 mm. The experiment was conducted using a water test rig at the University of Ancona, now Politecnica delle Marche, Italy [4]. By traversing the probe, and thus the measurement volume, across the diameter of the tube, three typical flow configurations were analyzed:

1. Post-contraction flow
2. Fully developed turbulent flow (measured after a straight tube length equal to 100 times the tube diameter)
3. Flow downstream of a diffuser.

The layout of the test rig is shown in Fig. 9.10b, and the corresponding measurement results are presented in Fig. 9.10c.

In the same laboratory and using the same LDA measurement system, a two-phase turbulent jet was analyzed [5]. The jet was generated within a transparent plexiglas reservoir by mixing air and water flows, as illustrated in Fig. 9.11a.

A photograph of the resulting jet is shown in Fig. 9.11b. As depicted in Figs. 9.11c and 9.12d, velocity profiles were successfully measured. These measurements revealed the anular structure characteristic of this type of two-phase jet near the outlet section. This anular configuration is attributed to the specific air–water mixing geometry upstream of the jet outlet, as illustrated by measured velocity profile in Fig. 9.11c.

As previously explained in the theoretical discussion of the measurement principle, an LDA system measures the velocity component that lies in the plane of the

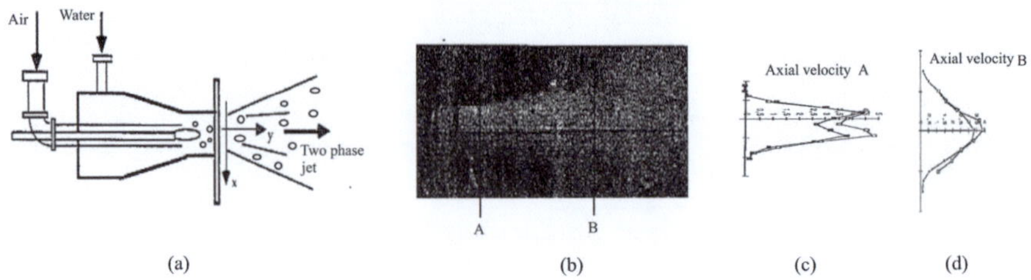

Fig. 9.11 Analysis of a two phase flow by LDA

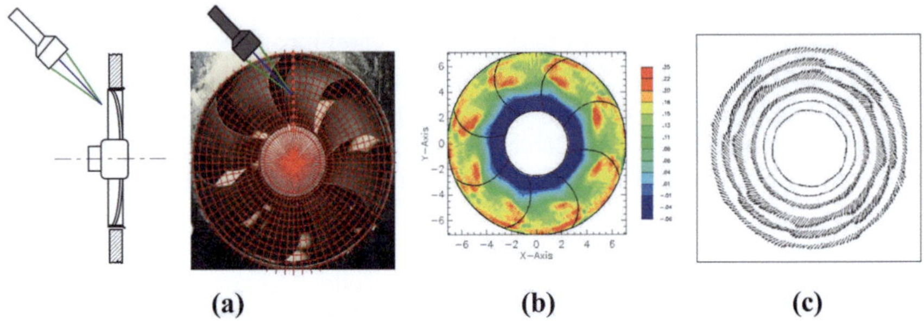

Fig. 9.12 An application to two component velocity field measurenemt on a fan

two intersecting laser beams, specifically, the component perpendicular to the optical axis of the probe, i.e., perpendicular to the bisector of the beam crossing angle.

To measure two velocity components simultaneously, it is possible to use a dual-wavelength laser source, such as an argon-ion laser, that emits both blue and red laser beams. With this setup, a probe can be designed to generate four laser beams arranged in two orthogonal planes. These beams are focused by a single lens into two superimposed measurement volumes at the same spatial location.

The backscattered light is collected and directed to two separate photodetectors, each equipped with a narrow-band optical filter corresponding to one of the laser wavelengths. This configuration enables the simultaneous measurement of two orthogonal components of particle velocity.

An application of this technique is illustrated in Fig. 9.12, where axial and radial components of the airflow generated by an automotive fan are measured by scanning measurement point on a circular grid [6].

A two-wavelength (two-color) LDA probe featuring a dual optical configuration in two perpendicular planes, similar to the one shown in Fig. 9.10a, was employed to perform these measurements. The system uses two pairs of laser beams: two blue and two green, on two perpendicular planes, which converge at the same measurement point to capture two orthogonal components of the velocity vector.

Typical results are presented in Fig. 9.12b for the axial component of the airflow (i.e., the air “pumped” by the fan), and in Fig. 9.12c for the combined radial-tangential component, representing the air flow lost laterally by the fan.

To obtain all three components of the velocity vector field, a third LDA probe, configured for single-component measurement, can be aligned and focused on the same point. In this way, the complete three-dimensional velocity vector at a specific location can be measured.

To map the velocity field across a volume or surface, measurements can be taken at multiple spatial positions by scanning, i.e., moving the probe using a traversing mechanism or a robotic system. However, this scanning approach is only valid for stationary (steady-state) flows, as each measurement point is acquired at a different time. The technique assumes that the velocity field varies spatially but remains constant temporally during the measurement process in different points.

9.3 Laser Doppler Vibrometry

Like laser Doppler anemometer the origins of this measurement technology is on laser technology recent development and in Doppler effect. In 1960s the early prototypes of laser-based vibrometers were developed, primarily for laboratory research. In 1970s they became more practical with the advent of He–Ne (helium–neon) lasers and high performances photodetectors. Researchers began applying them to measure vibration in mechanical structures without the need for physical contact. In 1980–1990s start commercialization from companies like Polytec and Ometron and industrial use. They were increasingly adopted in industries like aerospace for non-invasive testing, improving aircraft components, modal analysis; automotive for testing car frames, engine vibrations, component fatigue, acoustics analysis, quality control; consumer electronics to measure speaker membrane motion and hard disk vibrations; in biomedical applications and for meny others. Recently integration with digital signal processing improved accuracy and usability and last systems 3D Scanning LDVs enabled multi-dimensional vibration measurement for complex shapes; long range systems to measure vibration on far point of civil structures like buildings and bridges. Relevant also micro and nano-scale applications used in MEMS (micro-electromechanical systems) experimental development, medical devices, and semiconductor components. Modern LDVs can measure displacement, velocity, and acceleration across wide frequency ranges (from Hz to MHz). Surfaces with little or no reflectivity, using advanced signal conditioning. Measurements in harsh environments, such as rotating machinery or high-temperature components [1].

9.3.1 Single Point Laser Doppler Vibrometer

In vibrometers, the observer of the Doppler effect O coincides with the source S as illustrated in Fig. 9.13, which results in $\theta_1 = -\theta_2 = \theta$, thus simplifying the expression of Δf_D seen in Eq. 9.13 as follows:

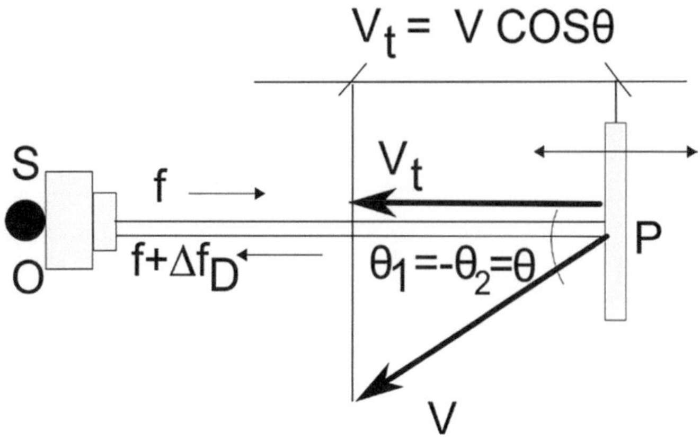

Fig. 9.13 Optical setup of source S and detector O of a laser Doppler vibrometer

$$\Delta f_D = \frac{2vf}{c}\cos\theta = \frac{2v}{\lambda}\cos\theta = \frac{2v_t}{\lambda} \tag{9.18}$$

The Doppler frequency shift, denoted as Δf_D, is directly proportional to the component of the target's velocity that lies along the direction of the incident laser beam. This component is expressed as $v_t = v\cos(\theta)$, where v is the magnitude of the target's velocity and θ is the angle between the velocity vector and the direction of the laser beam. As a result, a LDV is capable of measuring only this longitudinal component v_t by detecting the corresponding Doppler frequency shift Δf_D. A commonly used source in LDVs is a Helium–Neon (He–Ne) laser, which emits red light with a wavelength $\lambda = 0.632\ \mu m = 0.632 \times 10^{-6}$ m. Using the relationship between the speed of light, frequency, and wavelength $c = f \cdot \lambda \Rightarrow f = c/\lambda$ and substituting the values we have:

$$f = (3 \times 10\,\mathrm{m/s})/\left(0.632 \times 10^{-6}\,\mathrm{m}\right) \approx 4.75 \times 10^{14}\,\mathrm{Hz}$$

Thus, the frequency of the emitted radiation by an He–Ne (red) laser is approximately 475 THz.

The Doppler frequency shift Δf_D resulting from a surface moving with velocity $v_t = 1$ m/s toward or away from the laser can be calculated using the classical Doppler formula (assuming back-scattering) as $\Delta f_D = (2 v_t/\lambda)$. For $v_t = 1$ m/s and $\lambda = 0.632 \times 10^{-6}$ m we obtain $\Delta f_D = 3.16 \times 10^6$ Hz = 3.16 MHz. Therefore, a target moving at 1 m/s along the beam axis will cause a Doppler shift of approximately 3.16 MHz.

This means that to measure the vibration velocity of an object illuminating a point with a laser beam and detecting the back reflected light, it is necessary to measure very small frequency variations, requiring instruments (optical power spectrum analyser) with a resolution of $\Delta f_D/f \approx 10^{-8}$. Since instruments with this resolution do not exist, interferometry is nowadays used to measure the Doppler shift Δf_D. A first kind

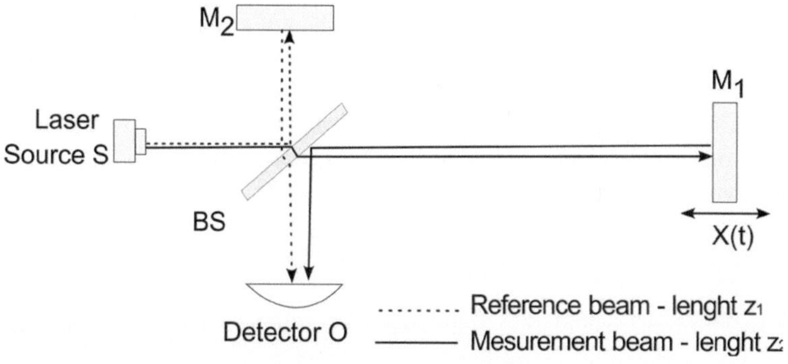

Fig. 9.14 Basic scheme of the Michelson interferometer

of interferometer used in laser Doppler vibrometers to detect Doppler shift is the classical Michelson inetrferomenter [1, 2], illustrated in Fig. 9.14.

In this optical setup, the radiation emitted by the laser source first reaches a beam splitter, which divides the incident light into two beams of equal amplitude. One beam is transmitted straight through toward the fixed mirror M2, while the other is reflected toward the moving mirror M1. After reflecting off their respective mirrors, both beams return to the beam splitter, where they are recombined and directed toward the photodetector O.

To analyze the interference at the detector, we represent the electric field of the laser beam using a complex exponential function:

$$E(t) = Ae^{-j\omega t}$$

where:

- A is the amplitude of the electric field,
- ω is the angular frequency ($\omega = 2\pi$f),
- j is the imaginary unit $\left(\sqrt{} - 1\right)$.

According to Euler's formula:

$$e^{J\theta} = \cos(\theta) + \mathrm{j}\sin(\theta) \tag{9.19}$$

This means that, in physical applications such as this, we typically consider only the real part of the complex exponential, namely, the cosine term, as it corresponds to the electric field.

This formalism allows us to define the electric field expressions for the reference beam and the measurement beam, respectively, as follows [1, 2]:

$$E_r(t) = \frac{1}{2}Ae^{j(\omega t - \theta_1)}\ E_m(t) = \frac{1}{2}Ae^{j(\omega t - \theta_2)} \tag{9.20}$$

Were θ_1 and θ_1 are the phase angles of the two beams arriving to the photodetector, following the two different optical path z_1 and z_2, where $\theta_1 = 2\pi z_1/\lambda$ and $\theta_2 = 2\pi z_2/\lambda$.

As described in Chap. 1 the intensity of the beam combining to photodetector will be proportional to the square of the electric field, so the photodetector signal output i, proportional to this optical intensity, will be:

$$i \propto E^2 \propto (E_r + E_m)(E_r + E_m)^* = |E_r|^2 + |E_m|^2 + 2\mathrm{Re}\left[E_r E_m^*\right]$$

That performing calculation with exponential functions, using trigonometric identities and performing semplifications [1, 2] reduce to:

$$i \propto A^2\left[1 + \cos(\theta_2 - \theta_1)\right] = A^2\left[1 + \cos(\Delta\theta)\right] \tag{9.21}$$

This result tells us that the output signal i from the photodetector depends only on the phase difference between the two beams, denoted as $\Delta\theta = \theta_1 - \theta_2$, and not on the angular frequency ω of the laser light.

If the moving mirror undergoes a displacement x(t), this motion induces a corresponding time-dependent phase shift Δθ(t) in the photodetector signal. According to Eq. 9.21, this relationship can be expressed as:

$$i(t) = A^2\left[1 + \cos(\Delta\theta(t))\right] = A^2\left[1 + \cos(2\pi x(t)/\lambda)\right] \tag{9.22}$$

So removing the signal constant component A^2, considering only its fluctuation, we get:

$$i(t) = A^2 \cos(2\pi x(t)/\lambda) \tag{9.23}$$

The resulting signal of Eq. 9.23 is periodic in time, reaching its maximum values when $|\cos(2\pi \mathrm{x(t)}/\lambda)| = 1$, i.e., when $2\pi \mathrm{x(t)}/\lambda = 0, \pi, 2\pi, \ldots$. These correspond to the occurrence of interference fringes, which are spaced by λ/2, since a mirror displacement of x results in a change of the optical path length of 2x (due to the round-trip of the beam).

Therefore, mirror displacement can be measured by counting the number of fringes, using the known wavelength λ of the light as a reference, and even by interpolating between fringes to achieve sub-wavelength resolution.

However, due to the mathematical nature of the cosine function (i.e. cos(x) = cos(–x)), a displacement of +x produces the same signal intensity output *i* as a displacement of −x. That is, the photodetector output is insensitive to the direction of motion, as can be confirmed by Eq. (9.23). This introduces an ambiguity in the simple Michelson interferometric configuration: the system cannot distinguish whether the mirror is moving forward or backward based on signal amplitude alone.

To resolve this versus ambiguity, more advanced optical configurations, such as heterodyne interferometers or quadrature detection schemes, are typically employed.

If the moving mirror has a velocity v(t), then for uniform motion or over small time intervals, we can express the displacement as $x(t) \approx v(t)\,t$. Substituting this into Eq. (9.23), the photodetector output becomes

$$i(t) = A^2 \cos(2\pi \mathrm{v(t)\ t}/\lambda)$$

But from (9.19) $2\mathrm{v(t)}/\lambda = \Delta \mathrm{f_D}$, so:

$$i(t) = A^2 \cos(2\pi t \Delta f_D(\mathrm{t})) \tag{9.24}$$

The photodetector output signal i(t) is therefore modulated by the Doppler frequency. By applying demodulation electronics, such as frequency-to-voltage conversion or other electronic signal processing methods, it is possible to track the time evolution of the Doppler frequency. This, in turn, allows for the determination of the proportional instantaneous velocity of the moving mirror by Eq. (9.18).

However, due to the symmetrical nature of the cosine function, the system remains insensitive to the versus of motion, meaning it cannot distinguish between positive and negative velocities.

To overcome this directional ambiguity and enhance the signal-to-noise ratio, the Mach–Zehnder interferometric configuration is widely used in laser Doppler vibrometers. This configuration enables unambiguous velocity detection, including the sign (direction) of the motion.

A typical optical layout of the Mach–Zehnder interferometer, as implemented within a vibrometer head, is illustrated in Fig. 9.15 [1, 2]

Once again, the laser source beam is split into two beams of equal intensity by the beam splitter BS1 (as shown in Fig. 9.15). The reference beam travels through a mirror, passes through a Bragg cell (which introduces a frequency shift), and is then

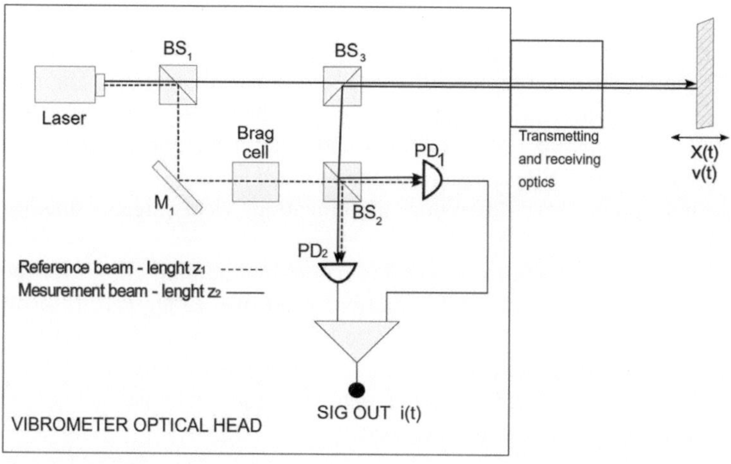

Fig. 9.15 Mach Zender interferometer

directed via a second beam splitter (BS2) toward the two photodetectors PD1 and PD2.

The measurement beam is directed to a third beam splitter (BS3), which sends it toward the moving target. The light reflected from the target returns along the same path through BS3 and is subsequently directed to BS2. At BS2, the returning measurement beam is recombined with the reference beam, and the resulting optical signals are directed to both photodetectors.

As a result, each photodetector receives a superposition of the reference and measurement beams, producing an interference signal analogous to that described by Eq. 9.21. However, due to the multiple beam-splitting stages, the intensity of each beam is reduced, resulting in half the optical power compared to the simpler Michelson configuration.

Thus, each photodetector (PD1 and PD2) records an interference signal similar to that of the Michelson interferometer, with the first photodetector signal being:

$$i_1(t) \propto \frac{A^2}{2}[1 + \cos(\theta_2 - \theta_1)] = \frac{A^2}{2}[1 + \cos(\Delta\theta)]$$

In the Mach–Zehnder interferometer, the two pairs of reference and measurement beams reach photodetectors PD1 and PD2 after undergoing an odd or even number of reflections on mirrors and beam splitters. And because of each reflection change wave phase from positive to negative this results in a relative phase shift between the two protodetector signals: the signal detected at PD2 is out of phase with respect to that at PD1 [5, 6].

Consequently, the second signal (at PD2) is phase-shifted relative to the first signal (at PD1), such that:

$$i_2(t) \propto \frac{A^2}{2}[1 - \cos(\theta_2 - \theta_1)] = \frac{A^2}{2}[1 - \cos(\Delta\theta)]$$

By using a differential amplifier to process the outputs from photodetectors PD1 and PD2, the system amplifies the difference between the two signals. This operation effectively doubles the useful signal amplitude (as the two signals are out of phase) while simultaneously suppressing common-mode noise components. As a result, the signal-to-noise ratio (SNR) is significantly improved [1, 2].

The final output, after differential amplification, yields signals analogous to Eq. 9.23.

Moreover, the Mach–Zehnder interferometer configuration allows for the integration on the reference arm of the interferometer of a Bragg cell to resolve the directional ambiguity. The Bragg cell introduces an additional angular frequency shift ω_B, as previously described in the context of the laser anemometer. This modification, depicted in Fig. 9.15, alters the reference beam such that the reference and measurement beams can now be described by:

$$E_r(t) = \frac{1}{2}Ae^{j((\omega+\omega_B)t-\theta_1)}\ E_m(t) = \frac{1}{2}Ae^{j(\omega t-\theta_2)} \tag{9.25}$$

And the intensity of the combined (heterodyne) beam to the photodetector will be:

$$\begin{aligned} I(t) &\propto E^2 \propto (E_r + E_m)\cdot(E_r + E_m)^* = |E_r|^2 + |E_m|^2 + 2\mathrm{Re}\big[E_rE_m^*\big] = \\ &= \frac{1}{2}A + \frac{1}{2}A\mathrm{Re}\big[e^{-j\omega_B t + j\omega_B T - j\omega T_m}\big] = \frac{1}{2}A + \frac{1}{2}A\cos[(\omega_B + \omega_D)t - \phi_B] \end{aligned} \tag{9.26}$$

where $\omega_D t = \omega T_m$ and $\phi_B = \omega_B T_r$ is a constant phase, not relevant. So, the resulting output signal of the photodetector will be given by:

$$i(t) \propto \frac{1}{2}A + \frac{1}{2}A\cos[(\omega_B + \omega_D)t] \tag{9.27}$$

This allow to solve the ambiguity versus problem.

The measurement point of a laser Doppler vibrometer (LDV) corresponds to the illuminated spot on the surface, similar to the dot produced by a laser pointer. If the surface is a perfect mirror-like reflector, as shown in Fig. 9.16a, the incident laser beam is reflected away at a single angle according to the law of reflection discussed in Chap. 2, resulting in no backscattered light returning to the interferometer. In such cases, no measurement is possible due to the absence of a detectable return signal. In practical applications, however, measurement surfaces are typically diffusive, exhibiting Lambertian or other types of scattering behavior, as discussed in Chap. 5, and illustrated in Fig. 5.31. For such surfaces, the scattered light is distributed over a range of angles. To enhance the collection of backscattered light, a focusing lens is employed in the optical path of the LDV, as illustrated in Fig. 9.16d. This setup significantly improves the system's ability to capture the scattered signal and perform accurate measurements. For some surface, if possible, it can be useful to apply on the surface diffusive paints, or use reflecting target on the measurement point.

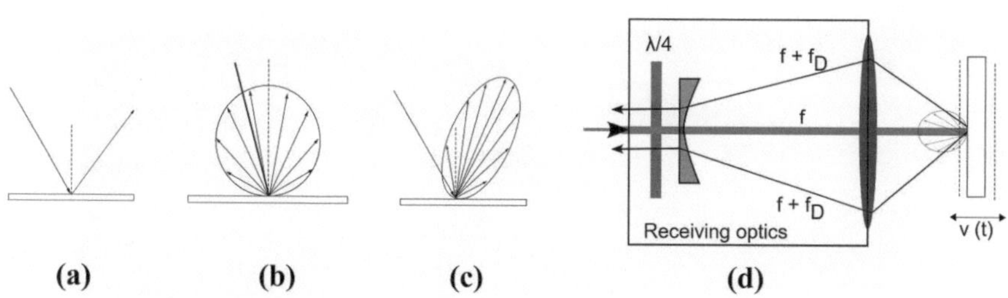

Fig. 9.16 Back reflections and receiving optics of a vibrometer

A $\lambda/4$ waveplate is also integrated into the receiving optics, and is crossed twice by the returning beam. This setup serves to compensate for the $\lambda/2$ phase shift introduced by the reflection of the laser light on the target surface, ensuring correct phase alignment for accurate interference detection.

To extract velocity information from the modulated Doppler signal, different types of frequency modulation (FM) demodulators can be employed. In laser Doppler vibrometers, the two most commonly used techniques are:

- Phase-Locked Loop (PLL) decoding
- Pulse-density decoding.

A typical laser Doppler vibrometer, such as the one available at the Department of Engineering, University of Perugia, is illustrated in Fig. 9.17. It consists of:

- an optical head, which includes the laser source and interferometric system,
- and demodulation electronics, which process the Doppler signal received from the photodetector housed in the optical head.

The demodulation electronics typically provide two output signals:

- one proportional to the velocity of the vibrating surface,
- and the other proportional to its displacement.

For additional technical details and specifications, extensive information on commercial vibrometers can be found on the manufacturer's website such as www.polytec.com.

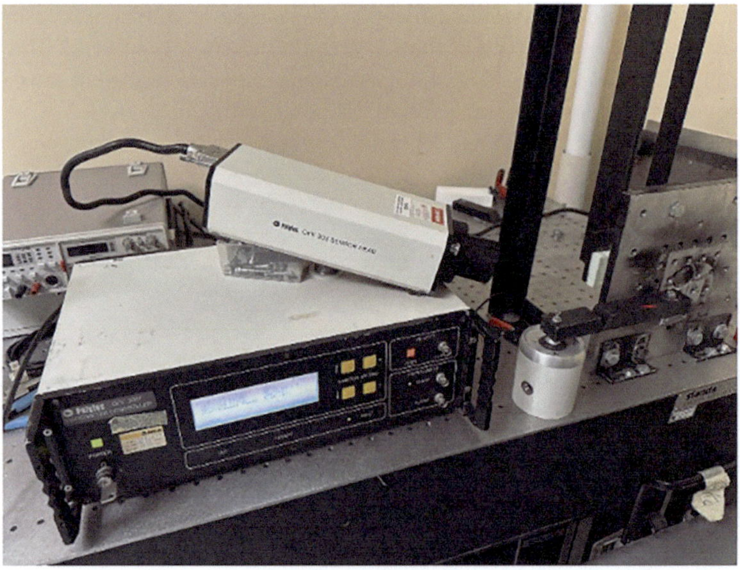

Fig. 9.17 A single point laser Doppler Vibrometer

Another optical setup to realize a laser Doppler vibrometer is that given by self-mixing or feedback interferometry. Is a compact and cost-effective alternative to traditional laser Doppler vibrometry. In a self-mixing laser Doppler vibrometer (SM-LDV), the interference between the reference and Doppler-shifted light takes place inside the laser cavity itself, rather than in an external interferometric setup. This technique simplifies the optical system while still enabling accurate velocity and displacement measurements. In SM-LDV a portion of the laser light is emitted from a semiconductor laser diode and directed toward the target surface. The backscattered light from the moving surface re-enters the laser cavity, where it interferes with the light inside the cavity. This optical feedback modulates the laser's output power, frequency, and phase, a phenomenon known as the self-mixing effect. These modulations are detected by an integrated photodiode within the laser diode package or via the laser's driving current. The resulting signal encodes information about velocity (via Doppler frequency shift) and displacement (via phase changes).

Main advantages are that no external interferometer is required: Optical mixing happens within the laser cavity, reducing size and complexity. Limitations are due to lower signal quality and shorter measurement range. More informations can be founded in www.julight.it.

9.3.2 Scanning Laser Doppler Vibrometer

By incorporating two mirrors into the optical head of the vibrometer, similarly to the mechanisms used in thermographic scanners (see Chap. 8) or LIDAR systems (see Chap. 7), it becomes possible to change the laser beam direction and scan multiple points across a surface. This enables the measurement of the complete vibration velocity field over the surface. At each scanned point, the time history of the vibration velocity is recorded by an integrated signal digital acquisition system.

Using a Fourier transform (see Eq. 3.21), the frequency spectrum of the vibration at each point can be computed and stored. If the vibration field is stationary, the spatial distribution of vibration amplitudes at individual frequencies can be used to derive the operational mode shapes of the vibrating structure.

An example application is shown in Fig. 9.18, where this method is employed to measure the vibration amplitudes at different frequencies on the membrane of a loudspeaker [7].

Another example is presented in Fig. 9.19, where a laser Doppler scanning vibrometer is used to measure the operational vibration modes of the same fan whose stress distribution had previously been analyzed using thermoelasticity (see Fig. 8.44 in Chap. 8) [8].

Additional full-field measurements using a laser scanning vibrometer are shown in Figs. 9.20 and 9.21, illustrating studies on how vibrations generated by handheld devices are transmitted to the human hand [9–11].

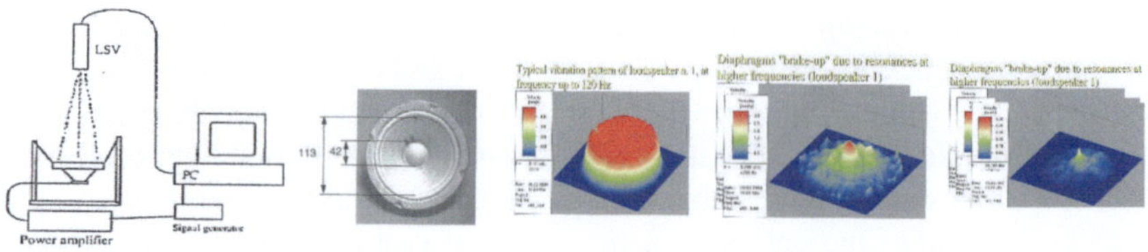

Fig. 9.18 Loudspeaker membrane vibrations at increasing frequencies

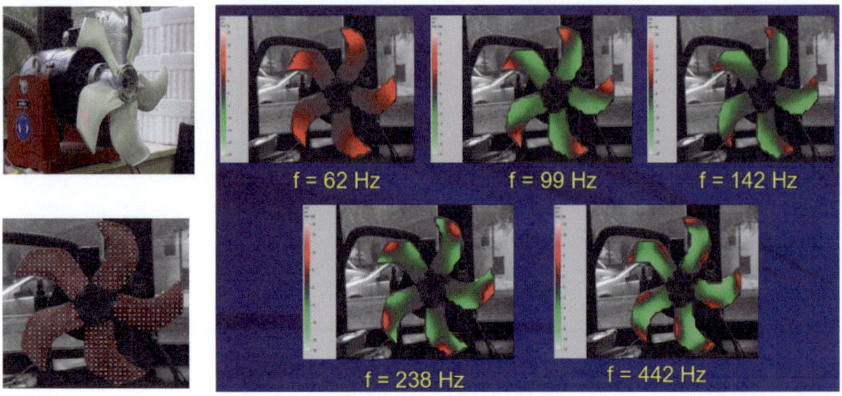

Fig. 9.19 Measurment of fan blades vibration modes by a laser scanning vibrometer

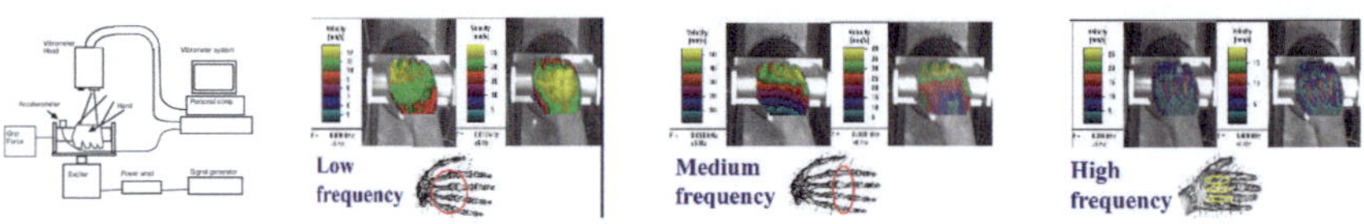

Fig. 9.20 Hand vibrations analysis on a reference vibrating (by an exciter) handle

Another application [12] is illustrated in Fig. 9.22, where operational mode shapes of a ropeway seat cover are measured in order to reduce vibrations and enhance acoustic comfort. The cover is excited using an electrodynamic shaker.

By synchronizing and precisely controlling the movements of the two scanning mirrors, it is possible to "track" a specific measurement point, for example, a rotating point on a propeller blade used in marine applications, as shown in Fig. 6.24 [13]. A typical result of this type of measurement, performed on propeller blades rotating inside a water tank with transparent walls, is illustrated in Fig. 9.23c [13].

An alternative approach for measuring vibration maps on translating or rotating surfaces of vibrating objects is known as continuous scanning, illustrated in Fig. 9.24. This method is thoroughly described in [6]. The core idea is that, by continuously moving the measurement point across a surface vibrating in stationary conditions, i.e. exhibiting standing waves as shown in Fig. 1.21 of Chap. 1 for a guitar string, the

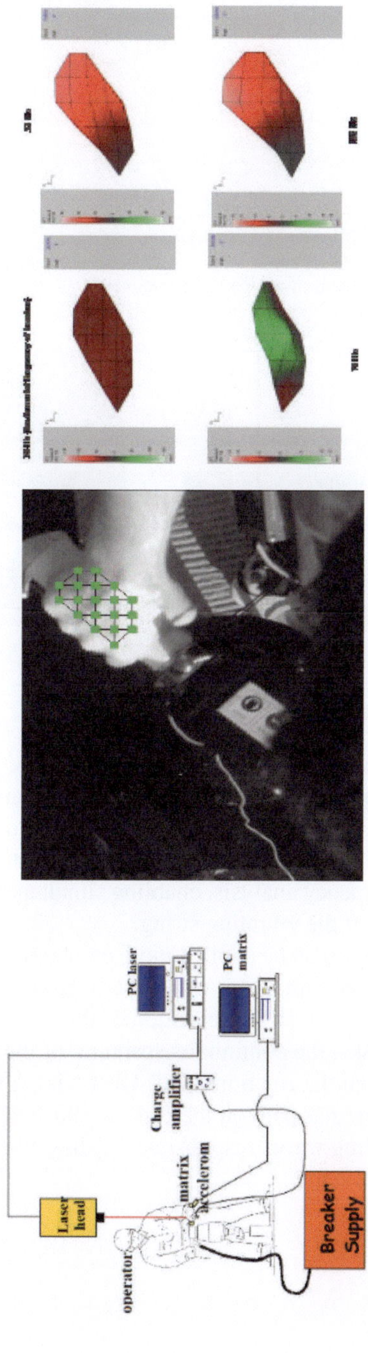

Fig. 9.21 Hand vibration measurements on working tools

Fig. 9.22 LDV vibration measurement of a ropeway sits cover

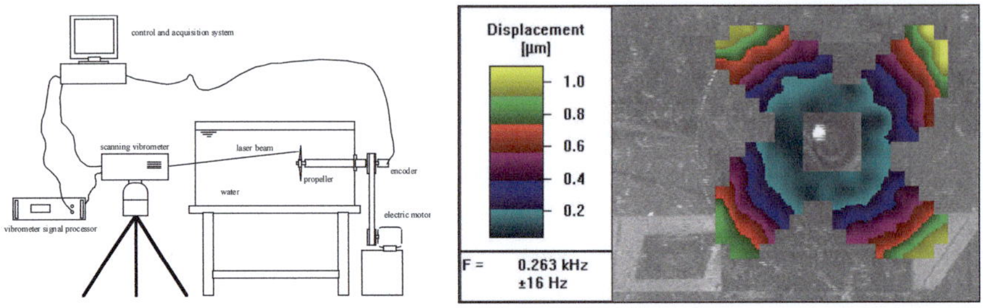

a) b)

Fig. 9.23 Rotational lased Doppler vibrometry

amplitude of the velocity signal captured by the vibrometer varies according to the shape of the standing wave. Consequently, the amplitude of the output velocity signal provides a representation of the vibration mode shapes of the surface. This technique allows for frequency-by-frequency analysis, enabling simultaneous measurement of all operational mode shapes of the vibrating string.

Naturally, the same principle can be applied more efficiently by using the mirrors of a scanning laser Doppler vibrometer to move the measurement point along the string, eliminating the need for physical translation of the guitar.

In the case of rotating disks, the continuous scanning of the measurement point is inherently achieved through the rotation itself. Observing a fixed point in space corresponds to scanning a circular path on the disk, as illustrated in Fig. 9.25. This allows the detection of standing waves responsible for disk vibrations, based on the same continuous scanning principle.

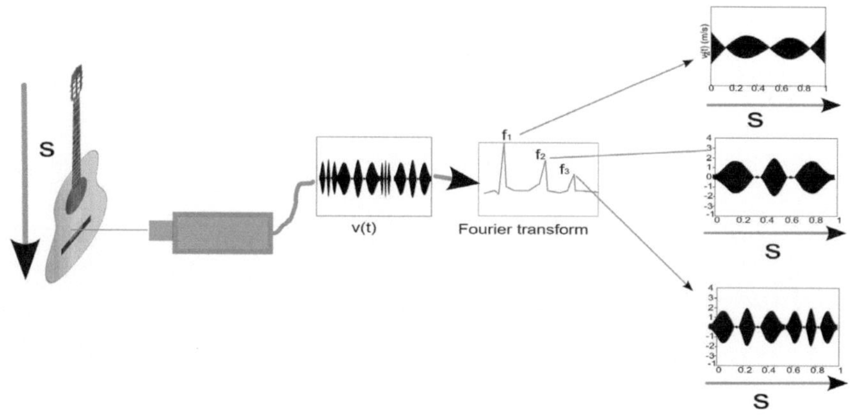

Fig. 9.24 Continous scanning concept

To extend the measurement across the entire disk surface, the vibrometer's mirrors can be used to reposition the measurement point at various radii, enabling full-field scanning of the rotating disk surface.

9.3.3 Tangential Laser Doppler Vibrometer

By adopting a configuration similar to that of a laser Doppler anemometer, as illustrated in Fig. 9.26, it is also possible to measure the tangential in plane component of a surface's velocity. In this setup, the scattered light originates from the surface's natural roughness, which acts as the scattering medium. As with particle-based Doppler systems, the scattered light produces a signal at the Doppler frequency, which is directly proportional to the tangential velocity of the surface.

An alternative optical configuration can also be employed, in which two focused laser spots on the surface generate scattered light detected by two separate photodetectors. By performing cross-correlation of the two resulting signals, a time delay Δt can be measured. Given the known spacing Δs between the two focal points, the tangential velocity of the surface can then be calculated as $v = \Delta s/\Delta t$.

Laser Doppler vibrometers based on this principle are commonly referred to as "in-plane" vibrometers, as they measure velocity components within the plane of the surface.

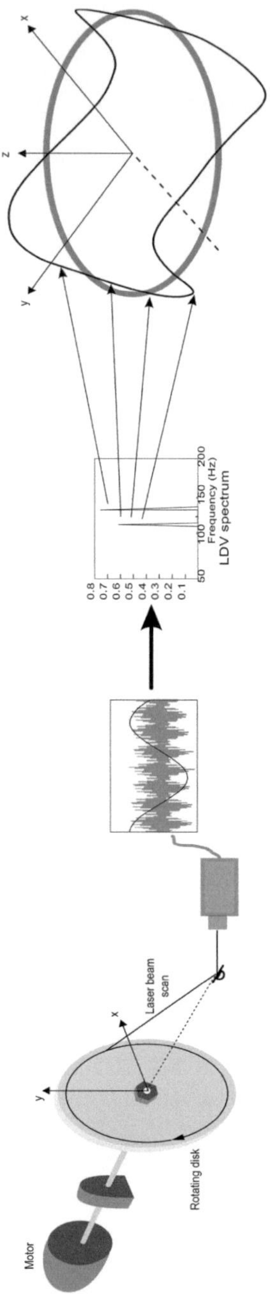

Fig. 9.25 Continous scanning on vibrating and rotating disks

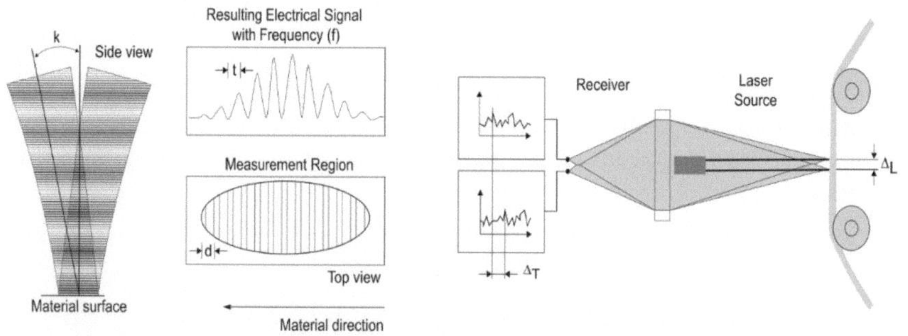

Fig. 9.26 Tangential laser Doppler and cross-correlation vibrometer

9.3.4 *Torsional Laser Doppler Vibrometer*

Torsional lased Doppler vibrometers allow to measure the angular instantaneous rotational speed ω(t) of a rotating shaft, disk or other kind of rotating mechanical component.

As previously explained a single point vibrometer measure the out of plane velocity component along the incident (and received) direction of the laser beam, as in Fig. 9.27a.

Therefore, by implementing an optical head equipped with two single-point laser Doppler vibrometers, configured as shown in Fig. 9.27b having two parallel laser beams, each measuring at points P_1 and P_2 on a rotating surface, it is possible to obtain two Doppler frequency signals, at frequencies f_{D1} and f_{D2}, both proportional to the velocity components along the respective incident (and backscattered) laser beams. Based on the geometric configuration illustrated in Fig. 9.27b, we obtain the following expressions for the two Doppler frequencies:

$$f_{D1} = 2v_1/\lambda f_{D2} = 2v_2/\lambda \quad (9.28)$$

But if ω is the instantaneous rotational velocity of the body, from the classical cinematic laws of the rotating body as illustrated in Fig. 9.27b we have also:

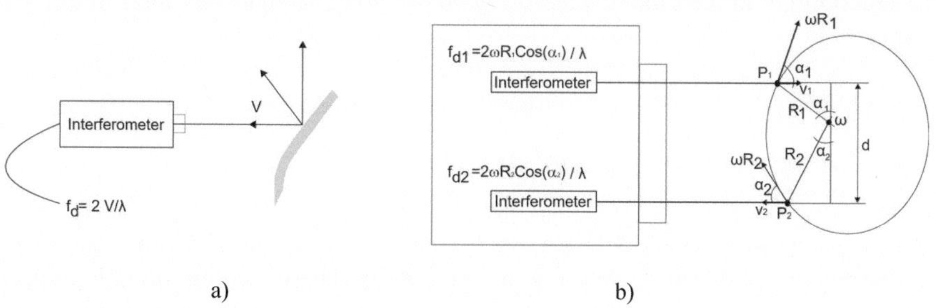

Fig. 9.27 Torsional vibrometer measurement principle

$$v_1 = \omega R_1 \cos\alpha_1 \; v_2 = \omega R_2 \cos\alpha_2 \tag{9.29}$$

Multiplying the output signals from the two Doppler photodetectors effectively results in the mixing of the two signals, as explained in Chap. 1, producing a new signal whose frequency corresponds to the sum of the two frequencies. Therefore, we obtain:

$$\begin{aligned} f_{D1} &= \frac{2\omega R_1 \cos\alpha_1}{\lambda} + \\ f_{D2} &= \frac{2\omega R_2 \cos\alpha_2}{\lambda} = \\ f_D = [R_1\cos\alpha_1 + R_2\cos\alpha_2] * \frac{2\omega}{\lambda} &= \frac{2d\omega}{\lambda} \end{aligned} \tag{9.30}$$

Here, d represents the distance between the two parallel laser beams, a constant given by the expression ($R_1\cos\alpha_1+R_2\cos\alpha_2$), as derived from the geometrical configuration shown in Fig. 9.27b. As a result, the mixed signal of the two interferometers exhibits a frequency f_D that is proportional to the angular velocity ω, following the same expression as those for the single-point laser Doppler vibrometry, but with the angular velocity ω replacing the linear velocity v.

Any translational velocity component u, due for example to flexural shaft vibrations, as illustrated in Fig. 9.28b, does not affect the sum of the Doppler frequencies. This because:

- The transverse component u_t of u, being perpendicular to the laser beams, has no projection along the beam direction and thus generates no Doppler shift.
- The parallel component u_p of u, aligned with the beam direction, produces Doppler shifts of equal magnitude but opposite sign on the two interferometers (since $\alpha_1 = -\alpha_2 = \pi$), resulting in a net cancellation in the frequency sum.

Therefore, the frequency of the mixed signal, obtained by multiplying the outputs of the two interferometers, is determined solely by the rotational velocity ω, multiplied by a constant.

By applying demodulation electronics, similar to those used in single-point vibrometers, the frequency-modulated Doppler signal can be converted into a voltage signal proportional to the Doppler frequency, and thus, according to Eq. 9.30, to the instantaneous angular velocity ω. This enables the measurement of the time evolution of angular velocity ω(t), allowing torsional vibrations in shafts and disks to be measured either in the time domain or by computing and analysing their frequency spectrum.

9.4 Particle Image Velocimetry

Particle image velocimetry (PIV) is a technique that allows the global measurement of velocity fields of entire flow regions of a fluid. The technique literally means velocimetry via particle image analysis. The concept of visualizing and measuring fluid motion dates back to the nineteenth century, with pioneers such as Ludwig

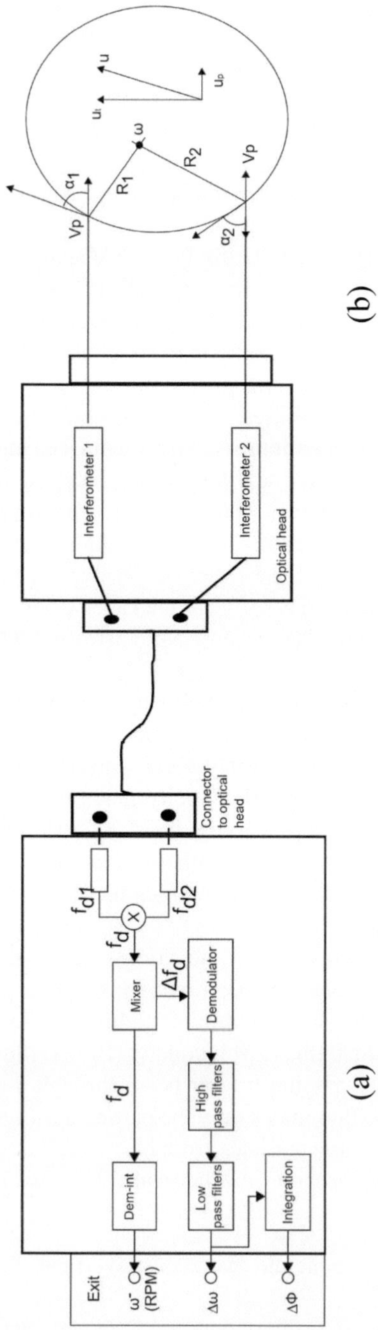

Fig. 9.28 Torsional laser Doppler vibrometer components

Prandtl and Henri Bénard using smoke or dye streaks to observe flow patterns. However, these early techniques were mostly qualitative. Anyway, also today used in wind tunnel tests for flow analysis.

The quantitative roots of PIV trace back to the 1980s, when advances in laser technology and digital imaging enabled researchers to simultaneously illuminate and record multiple particles suspended in a fluid. The idea was simple yet revolutionary: track the movement of tiny particles across sequential images to determine velocity vectors in a fluid flow. Is the same concept illustrated in Chap. 7 for speckle particles on a solid surface by DIC.

The introduction of pulsed lasers, particularly Nd:YAG lasers, allowed researchers to capture high-speed images at precise time intervals. Laser sheet optics realized by cylindrical lenses helped restrict illumination to a 2D plane, increasing measurement accuracy. The shift from film-based cameras to CCD and CMOS digital cameras dramatically improved temporal and spatial resolution.

The development of correlation-based algorithms, same already seen for DIC in Chap. 7, enabled accurate computation of particle displacements between frames. Interrogation window techniques refined this further, breaking images into smaller regions to track local flow features. The rise of digital image processing and computational power made it feasible to analyze large datasets in reasonable timeframes.

Open-source and commercial PIV software packages emerged, streamlining the process. From the 1990s onward, PIV became a mainstream tool in fluid mechanics research.

Its applications include: aerospace by analyzing airflow over wings and in jet engines; biomedical engineering for studying blood flow in vessels or air flow in lungs; environmental sciences for tracking ocean currents or atmospheric turbulence; combustion research for mapping velocity fields in flames and engines, to reduce pollution. PIV recent developments are the stereo PIV providing 3D velocity information by using at least two cameras; time-resolved PIV capturing sequences at high frequency to study unsteady flows; tomographic PIV reconstructing 3D flow fields in volumes, not just planes.

Modern PIV continues to evolve with innovations such as machine learning for better feature detection and displacement estimation; microscopic-PIV for flows in microfluidic devices; volumetric PIV combining multiple angles and perspectives.

The basic principle of Particle Image Velocimetry (PIV) is illustrated in Fig. 9.29. A laser sheet is generated by passing a laser beam through a cylindrical lens (as described in Chap. 2), which illuminates only the particles within a thin plane of the flow. To capture the motion of these particles, a double image is acquired, representing their positions at two closely spaced time intervals. This can be achieved in two primary ways:

1. By using a pulsed laser to generate two successive light flashes recorded on a single image frame.
2. By using a continuous laser to illuminate the flow and capturing two separate images with a high-speed camera.

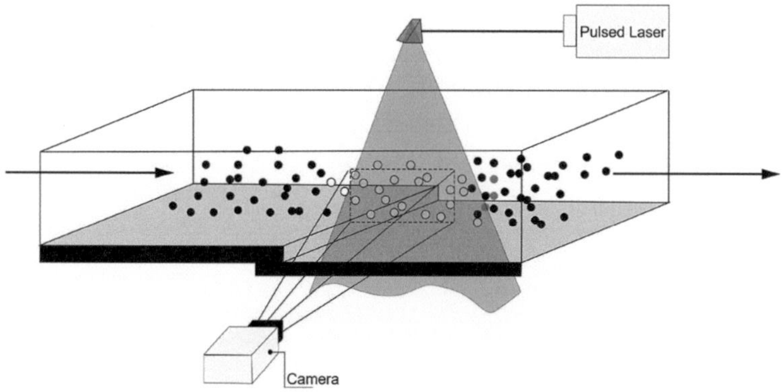

Fig. 9.29 Particle image velocimetry optical setup

These images are then used to compute particle displacements and determine the velocity field of the flow within the illuminated plane.

Different image processing techniques are used depending on how the particle images are acquired:

(1) On single double exposed image:

   (a) In the case of a double-exposed single digital image acquisition, autocorrelation is performed on small sub-regions (interrogation windows) of the image. By identifying the location of the autocorrelation peak, the displacement vector **Δs** is determined, similar to the approach used in DIC in Chap. 7. Knowing the time delay Δt between the two laser pulses, the local velocity is then calculated as $\mathbf{v} = \mathbf{\Delta s}/\Delta t$. This technique is illustrated in Fig. 9.30a using digital image analysis.
   (b) An older analog approach is shown in Fig. 9.30b, where a laser beam is passed through a transparent photographic film containing the double-exposed image of the moving particles. As the laser crosses the film, it generates an interference fringe pattern on a screen. The orientation of the fringes indicates the direction of particle displacement, while the fringe spacing corresponds to the magnitude of the displacement. Although this analog method is now outdated, it remains an elegant and historically significant solution, particularly from the time when images were captured on photographic film.

(2) By processing the two sequential images acquired by a camera through cross-correlation of sub-images, as done in DIC in Chapter 7, the displacement vectors **Δs** of the tracer particles can be determined by locating the correlation peak. Knowing the time interval Δt between the two image acquisitions, the local velocity can then be calculated as $\mathbf{v} = \mathbf{\Delta s}/\Delta t$. This process is illustrated in Fig. 9.31.

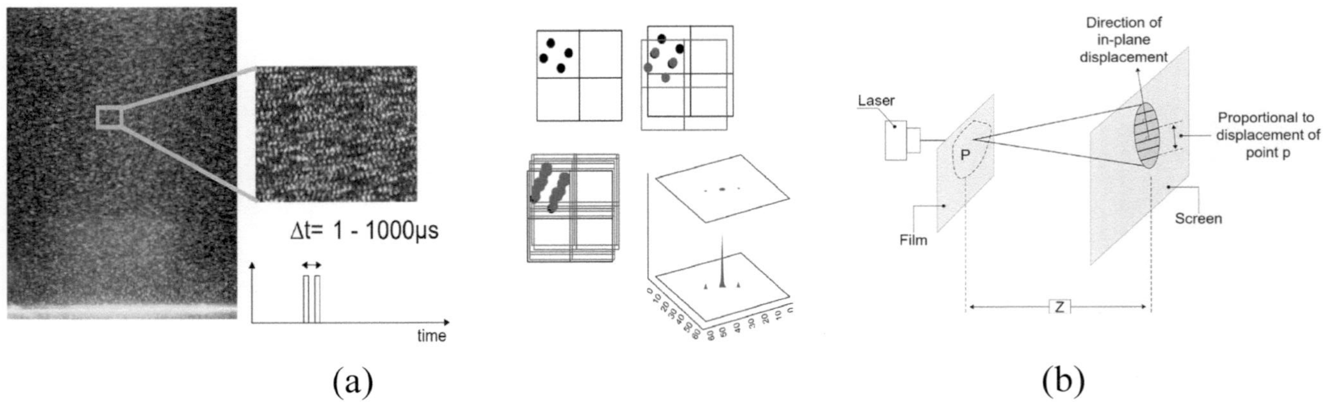

Fig. 9.30 Data processing by autocorrelation: **a** by digital image processin and **b** by a laser beam

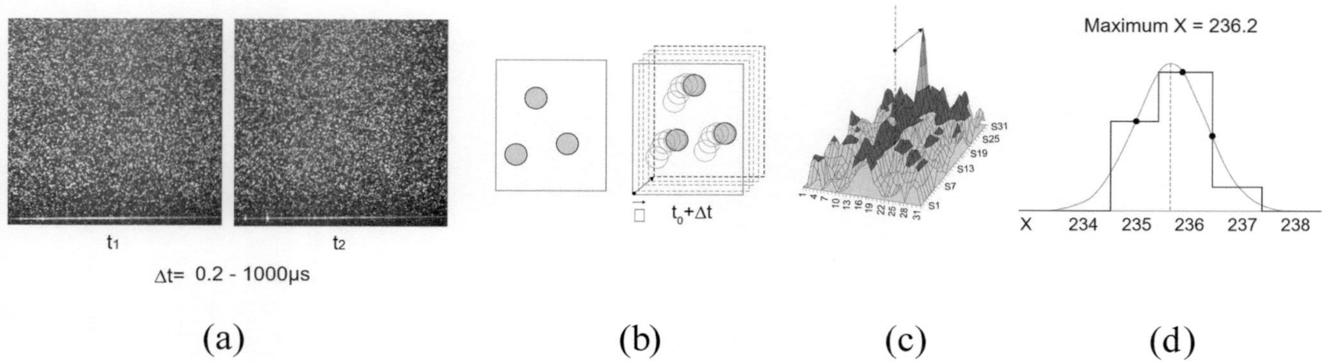

Fig. 9.31 Data processing by crosscorrelation

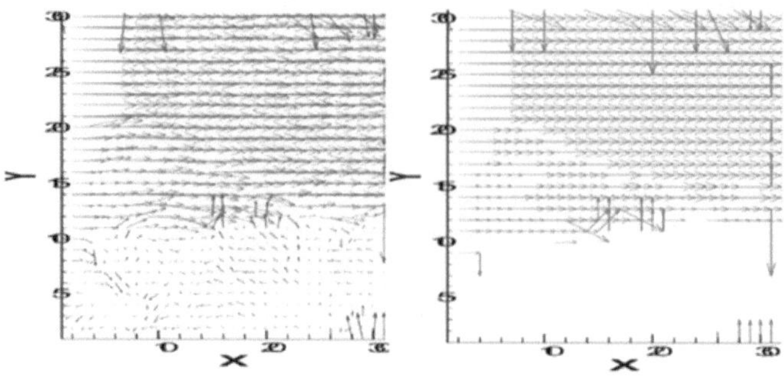

(a) (b)

Fig. 9.32 Image processing same results with a) and without b) subpixel interpolation

This is the most widely used data processing technique today, largely due to the high performance and resolution of modern digital cameras. By capturing a sequence of images (a video), it becomes possible to measure velocity fields in both space and time across multiple frames, making this one of the most comprehensive and powerful measurement methods in experimental fluid dynamics.

Moreover, subpixel interpolation techniques, as discussed in Chapter 7 for DIC and illustrated in Figure 9.31d, can be applied to enhance the resolution of velocity field measurements. A comparison of processing results with and without subpixel interpolation is shown in Fig. 9.32, highlighting the significant improvement achievable through this approach.

(3) Particle tracking, whether on single or double images, can also be employed, similar to the techniques described for DIC in Chap. 7 and illustrated in Fig. 9.33. This approach can be applied to both double-exposed images and to pairs of sequential images captured by a camera.

Particle identification and tracking can be performed using advanced image processing algorithms such as SIFT (Scale-Invariant Feature Transform), SURF (Speeded-Up Robust Features), template matching, and other methods detailed in Chap. 4. These techniques enable robust detection and correspondence of particle features across frames, supporting accurate velocity field reconstruction.

As previously discussed for DIC, particle tracking provides a random spatial sampling of the velocity field. In flows with high levels of turbulence, this can pose limitations: particles may be dispersed by vortical structures, resulting in sparse or even missing data in the core regions of vortices. This effect is illustrated in Fig. 9.33b.

More informations about commercially available PIV systems, characteristics and details can be founded in the website of the companies like Dantec Dynamics and LaVision.

Using PIV to measure the flow field enables the acquisition of the complete velocity field, as shown in Fig. 9.34a. By computing the spatially averaged flow vector

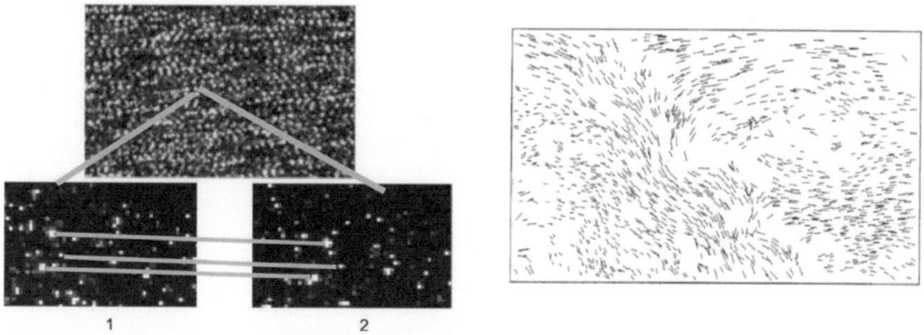

(a) (b)

Fig. 9.33 Data processing by particle traking

and subtracting it from the instantaneous velocity vectors, the turbulent (fluctuating) component of the flow can be isolated, as illustrated in Fig. 9.34b. This demonstrates the powerful capability of PIV to analyze complex and unsteady flow phenomena, offering detailed insights into both mean and turbulent flow structures.

A comparison between LDA and PIV for analyzing a turbulent air jet is presented in Fig. 9.35 [15]. In this study, air is ejected through a circular nozzle with a diameter of 25 mm, at a maximum Reynolds number of 6.8×10^4, corresponding to an axial velocity of 42 m/s.

Both techniques are capable of identifying turbulent fluctuations and estimating the integral length scale of the flow. Measurements show a difference of less than 8% in the length scale values obtained using PIV and LDA. LDA, being a single-point measurement technique, is particularly well-suited for time scale analysis due to its high temporal resolution. However, PIV offers more comprehensive spatial data

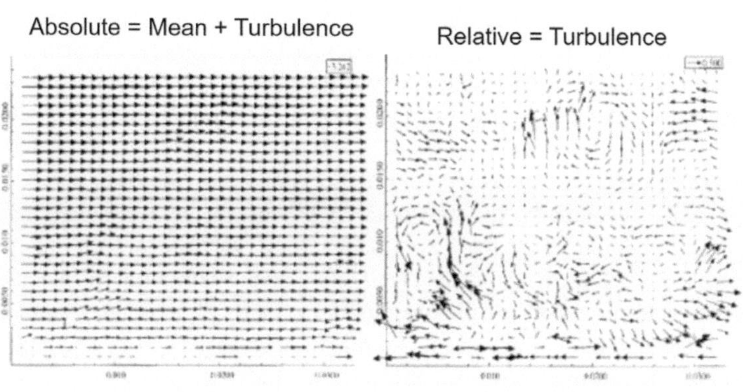

(a) (b)

Fig. 9.34 Turbulence structure detection by PIV field measurement processing

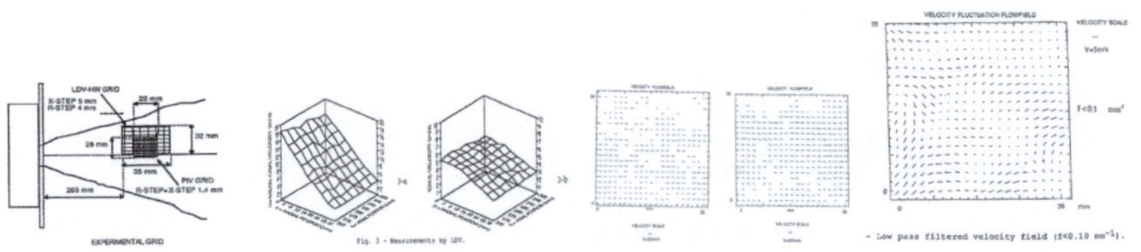

Fig. 9.35 Analysis of a turbulent jet by LDA and PIV

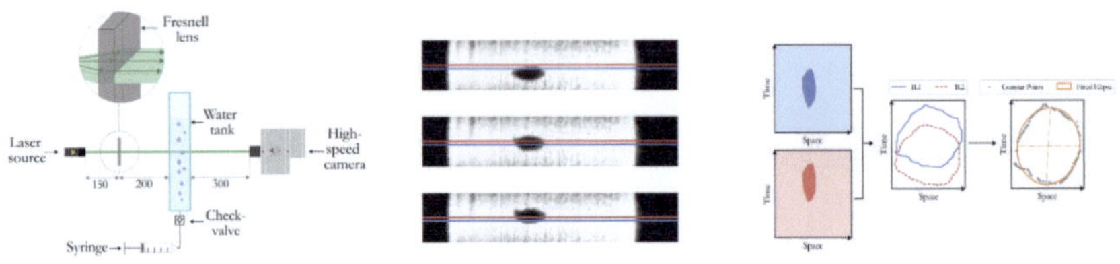

Fig. 9.36 Velocity and shape measurements of air bubbles by laser sheets line images

and, through appropriate post-processing, can provide more accurate and detailed estimates of the integral length scale, especially across the full flow field.

PIV and Particle Tracking Velocimetry (PTV) [16, 17] are well-established full-field techniques widely used to measure velocity fields in single-phase and two-phase flows in fluid mechanics. Recently, a novel image-based technique has been developed that is complementary to PIV and particle tracking, enabling the simultaneous measurement of bubble velocity, shape, and volume in bubbly flows along a densely sampled linear region [18].

In this method, a high-speed camera records the variation in light intensity as bubbles pass through a laser sheet generated by a Fresnel lens. Unlike conventional PIV setups where the laser sheet is perpendicular to the camera's optical axis, this configuration aligns the laser in the plane of the optical axis, directly illuminating the image sensor and producing a bright line on the camera image, as shown in Fig. 9.36a.

Two lines of pixels are analyzed from the region where the laser sheet lightens the image plane; eventually, the two lines of pixels can be substituted with two laser sheets. As bubbles cross these two planes, they cast shadows, as illustrated in Fig. 9.36b. These shadows can be analyzed using spatio-temporal cross-correlation techniques to determine both the dimensions and velocity of individual bubbles. This allows for the calculation of important parameters such as the volumetric flow rate of air (the bubbles), offering a powerful new tool for studying bubbly two-phase flows.

References

1. E.P. Tomasini, P. Castellini, *Laser Doppler Vibrometry—A Multimedia Guide to its Features and Usage* (Springer, 2020)
2. M. Martarelli, Exploiting the Laser Scanning Facility for Vibration Measurements, Ph.D. Thesis (University of London, 2001)
3. H.E. Albrecht, M. Borys, N. Damaschke, C. Tropea, *Laser Doppler and Phase Doppler Measurement Techniques* (Springer, 2003)
4. G. Rossi, N. Paone, Uncertainty analysis of flow rate measurement by integration of experimental velocity profiles, in *Quaderni del Dip di Ing Meccanica* (Univ di Ancona, 1996)
5. O. Massi, G. Rossi, Confronti tra misure di velocità in un getto bifase aria acqua con sensori a fibra ottica doppi e tecnica LDA, in *I Convegno nazionale AiVeLa* (Capri, 1993)
6. M.Gasparetti, E.P. Tomasini Misure di velocita' nel flusso generato da un ventilatore per uso automobilistico, in *3° Congr. Naz. di Misure Meccaniche e Termiche* (1997)
7. G.L. Rossi, E.P. Tomasini, Vibration measurements of loudspeaker diaphragms by a laser scanning vibrometer, in *International IMAC—SEM Conference* (Nashville, 1995). SEM, Bethel, Connecticut, USA
8. G. Rossi, R. Marsili, A. Di Renzo, M. Martarelli, G. Rosati Simultaneous application of laser scanning vibrometry and thermoelasticity on mechanical components, in *Aivela Conference* (2006)
9. G.L. Rossi, E.P. Tomasini, Hand-arm vibration measurement by a laser scanning vibrometer. Meas. J. **16**, 113–124 (1995). Elsevier Science, Amsterdam, Olanda
10. G. Di Giulio, N. Paone, G. Rossi, G. Miccoli, P. Christ, M. Geuder, A. Cristalli, R. Deboli, M. Valentino, V. Rapisarda, Measurement methods for the improvement of hand-arm vibration testing: polymeric capacitive matrices and laser vibrometry applied to an hydraulic breaker, in *9th International Conference on Hand-Arm Vibration* (Nancy, France, 2001)
11. R. Deboli, G. Miccoli, G.L. Rossi, Human hand transmitted vibration measurements on a pedestian controlled tractor operators by a laser scanning vibrometer. Ergon. J. **42**(6), 880–888 (1999). Taylor & Francis Ltd., London, UK.
12. P. Evangelisti, M. Rinaldi, G.L. Rossi, C. Santolini, E.P. Tomasini, Application of laser vibrometers for dynamic characterization of ropeways components, in *VIII OITAF International Congress* (S. Francisco, USA, 1999)
13. P.Castellini, C. Santolini, Vibration measurements on blades of a naval propeller rotating in water with tracking laser vibrometer. Measurement **24**(1), 43–54 (1998). Elsevier
14. M. Raffel, C.E. Willert, F. Scarano, C.J. Kähler, S.T. Wereley, J. Kompenhans, *Particle Image Velocimetry A Practical Guide*, 3rd edn. (Springer Nature, 2018)
15. P. Diodati, N. Paone, G. Rossi, E.P. Tomasini, Comparison of laser doppler velocimetry, hotwire anemometry and particle image velocimetry for the investifation of a turbulent jet, laser anemometry advances and applications, in *Vth International Conference* (Koningshof, Veldhoven, The Netherlands, 1993), SPIE Vol 2052
16. A. Neogi, H.K. Mohanta, P.C. Sande, Particle image velocimetry investigations on multiphase flow in fluidized beds: a review, in *311 Flow Measurement and Instrumentation*, vol. 89 (2023), p. 102309
17. R. Cerqueira, E. Paladino, B. Ynumaru, C. Maliska, Image processing techniques for the measurement of two-phase bubbly 329 pipe flows using particle image and tracking velocimetry (PIV/PTV). Chem. Eng. Sci. **189**, 1–23 (2018)
18. G. Tribbiani, L. Capponi, T. Tocci, M. Mengoni, M. Marrazzo, G. Rossi, An image-based technique for measuring velocity and shape of air bubbles in two-phase vertical bubbly flows. MDPI, Fluids **10**(3), 69 (2025)

Chapter 10
Pressure Field Measurements

Abstract This chapter presents the most widely used techniques for measuring pressure fields through image and wave-based processing methods, highlighting how they contribute to the evaluation and enhancement of performance in dynamic fluid environments. The first category focuses on acoustic pressure field measurements, which utilize single microphones, microphone pairs, or microphone arrays. These systems capture the fluctuating pressure generated in the air by propagating sound waves. Such techniques are extensively employed in the analysis of noise fields, sound intensity, and sound power. They are essential tools in the study, control, and mitigation of noise in various conditions, including built environments, industrial products, and vehicles. Practical applications of these techniques, based on the author's hands-on experience, are described here and in the previous chapter, particularly in the context of solving industrial acoustic problems. Another important application of pressure field measurement involves the interface between solid bodies, such as the contact point between a tire and the road, a foot and the ground, or a finger and an object held by a human or a robotic system. This section illustrates the main techniques used to measure such contact pressures, with a focus on methods involving visible and infrared image acquisition and analysis. Finally, pressure and shear stress measurements are critical for studying fluid–solid interactions, such as those encountered in optimizing the aerodynamic performance of vehicles (e.g., cars, aircraft, racing machines, and ships). The chapter describes the principal image-based measurement techniques used in this domain.

10.1 Acoustic Fields Measurements

Acoustic and ultrasonic fields consist of pressure fluctuations that occur at one or more frequencies, either individually or in combination. At any given point in space, these fields can be represented in different ways depending on their characteristics. In the simplest case, the pressure variation may follow a sinusoidal waveform, fully described by its amplitude, frequency, and phase. More generally, the waveform may

G. Rossi, *Mechanical and Thermal Measurements by Images and Waves*, Smart Sensors, Measurement and Instrumentation 51, https://doi.org/10.1007/978-3-032-03464-9_10

be complex, varying irregularly over time. Wave characteristics and phenomena have been described in Chap. 1.

To analyze such complex pressure fluctuations, it is often useful to represent them in the frequency domain using Fourier analysis, which decomposes the time-domain waveform into its spectral components. This process, also discussed in Chap. 1, provides insight into the distribution of energy across frequencies of sound and ultrasound and is fundamental for understanding and characterizing acoustic phenomena.

10.1.1 Acoustic Fields

In general, acoustic fields, whether sonic or ultrasonic, consist of propagating pressure waves in air, liquids, or other solid elastic media [1] , as introduced in Chap. 1. In the case of air, these waves manifest as sound or noise, depending on the listener's opinion. Sound is defined as any periodic fluctuation in air pressure that can be detected by the human ear. These pressure variations span a wide dynamic amplitude range, from the faintest sounds just above the threshold of hearing to intense levels capable of causing auditory damage. The distinction between "sound" and "noise" is inherently subjective, often depending on how the listener perceives the stimulus, whether as pleasant (such as enjoyable music), neutral, or disturbing (for example, the same music played while trying to sleep).

Acoustic wave fluctuations typically range in amplitude from just a few microPascals (μPa) to several hundred Pascals (Pa). These values are significantly lower than the atmospheric pressure, which averages around 100,000 Pa, and also lower than the natural pressure variations caused by weather conditions or altitude differences across locations and cities, as illustrated in Fig. 10.1.

These acoustic fluctuations occur within a frequency range of approximately 2 Hz to 20 kHz, which corresponds to the average frequency range of human hearing sensitivity. To characterize an acoustic or ultrasonic field, three fundamental physical parameters are commonly used:

- Pressure Amplitude: The magnitude of pressure fluctuations at each point in the medium.

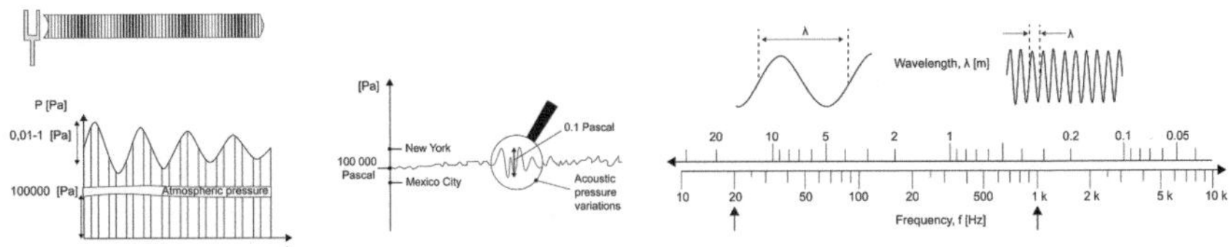

Fig. 10.1 Sound waves in air

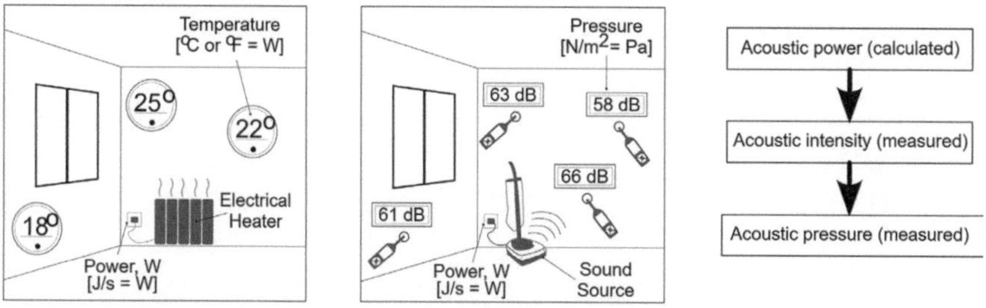

(a) (b) (c)

Fig. 10.2 Acoustic field parameters and analogy with thermal field

- Acoustic Intensity: The amount of acoustic energy transmitted through a unit area per unit time at each point.
- Acoustic Power: The total acoustic energy emitted by a source per unit time that generates the field.

These parameters can be effectively explained using a thermal analogy. Imagine a room where an electric heater emits energy at a constant rate, measured in Joules per second (J/s), which corresponds to a heating power expressed in Watts (1 Watt = 1 Joule/second). This heating power represents the intrinsic ability of the heater to generate heat, an absolute measure that does not depend on the surrounding environment.

However, the temperature that develops at different points within the room, as illustrated in Fig. 10.2a, is heavily influenced by external factors, such as the quality of the thermal insulation and the room's geometry. As a result, even if two rooms are heated by identical devices with the same power output, the resulting temperature distribution can differ significantly. This highlights how the same source energy can lead to varied thermal outcomes depending on environmental conditions.

A similar principle applies when an acoustic source is placed in a room, as illustrated in Fig. 10.2b. The source emits a certain amount of wave energy per unit time, measured in Joules per second (J/s), which corresponds to its acoustic power, expressed in Watts (W). This value represents the inherent capability of the source to generate sound energy and, like thermal power, is independent of the surrounding environment.

However, the acoustic pressure experienced at various points in the room is strongly influenced by the room's acoustic characteristics, particularly the reflection, absorption, and transmission coefficients of its surfaces. Just as thermal insulation affects the distribution of temperature, these acoustic properties determine how sound propagates, reflects, and dissipates within the space.

The amplitude of the acoustic pressure field is measured in Pascals (Pa), where 1 Pa = 1 N/m^2. Yet, because the range of audible pressure levels spans several orders of magnitude, from 20 μPa (the threshold of hearing) up to 100 Pa (the threshold of pain), a linear scale is impractical. To manage this vast dynamic range, logarithmic scales are used to define levels for the key acoustic parameters.

These logarithmic representations, known as the three decibel (dB) scales, are defined using standard reference values [1]:

- Sound pressure level (SPL) uses 20 μPa as the reference pressure (the threshold of hearing),
- Sound power level (SWL) uses 1 picowatt (1 pW) as the reference power,
- Sound intensity level (SIL) uses 1 pW/m^2 as the reference intensity.

These decibel scales provide a more practical and intuitive way to express variations in acoustic fields and are widely used in acoustics to describe sound environments and system performance.

The definitions of sound field values in terms of the so called three deciBel logaritmic scales, from [1], are as follows:

$$\begin{aligned} L_p &= 10\log\left(\frac{p_{rms}}{p_0}\right)^2 dB = 20\log\left(\frac{p_{rms}}{p_0}\right) dB \quad (p_0 = 20\mu Pa) \\ L_w &= 10\log\left(\frac{w}{w_0}\right) \quad (W_0 = 1pW) \\ L_1 &= 10\log\left(\frac{I}{I_0}\right) \quad \left(I_0 = \frac{1pW}{m^2}\right) \end{aligned} \tag{10.1}$$

Using the definition above, the sound pressure level (SPL) of an acoustic field ranges from 0 dB, corresponding to the threshold of hearing, up to 120 dB, which marks the threshold of pain.

To quantify the pressure amplitude of a general fluctuating signal, the root mean square (RMS) value is commonly used. This value is calculated from the time-domain waveform over a duration T, and is defined as:

$$\mathrm{p}_{rms} = \sqrt{\frac{1}{T}\int_0^T p(t)^2 dt} \tag{10.2}$$

The relationship between acoustic power, intensity, and pressure depends on the propagation conditions of the sound waves, which are determined by both the acoustic characteristics of the source and the environment in which the sound propagates. Two idealized types of acoustic fields are particularly useful for analysis and acoustic measurement and are illustrated in Fig. 10.3:

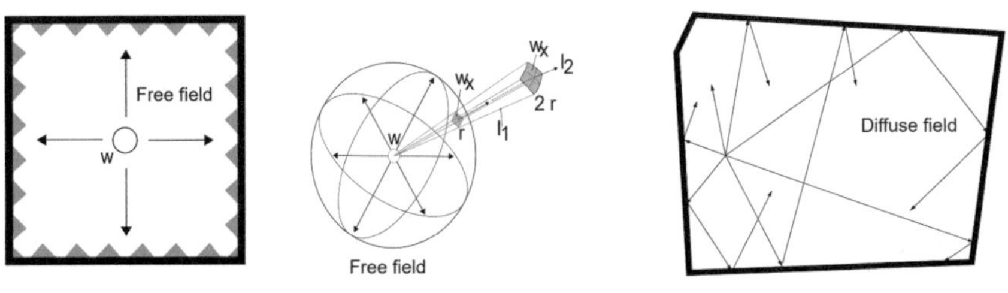

Fig. 10.3 Free and reverberating acoustic fields

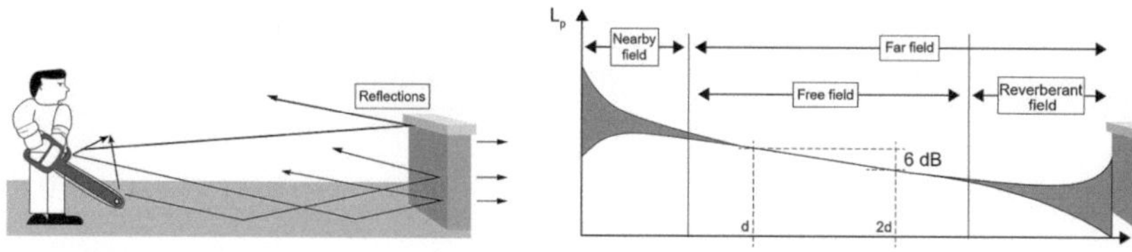

Fig. 10.4 Real complex acoustic fields

- Free-field conditions: These occur in environments such as anechoic or semi-anechoic chambers, or in open free space where there are no reflecting surfaces. In this case, sound waves emitted by a point source propagate spherically and symmetrically in all directions, without encountering any obstacles. Any interaction with surfaces, if present, is either negligible or entirely absorbed by the chamber walls.
- Diffuse-field conditions: These are created in a reverberation chamber, where the walls are non-parallel to avoid acoustic resonances, reflect in all directions and have a reflection coefficient close to one, meaning they reflect nearly all incident sound energy. In such an environment, sound waves undergo countless reflections, eventually arriving at each point in the room from all possible directions with statistically equal amplitude. The result is a field where the energy is uniformly distributed in space, and the directional characteristics of the waves are randomized.

In practice, perfect anechoic or perfect reverberant chambers do not exist. Any real acoustic field is typically a combination of these two idealized conditions, as illustrated in Fig. 10.4. In a free-field condition, however, it is possible to derive a simple and highly useful relationship between the three fundamental acoustic parameters: power, intensity, and pressure. Referring to Fig. 10.3c, we consider a point source radiating sound uniformly in all directions. The acoustic power W of the source is spread over the surface of an expanding sphere with radius r, whose area is given by $4\pi r^2$

Therefore, the sound intensity I, defined as the power per unit area, is given by the ratio:

$$I = \frac{W}{4\pi r^2} \tag{10.3}$$

As already described in Chap. 1 with Eq. (1.58).

Acoustic intensity vector I is given in general by Eq. (1.60) as the averaged product of pressure and velocity, so for a simple plane progressive acoustic wave propagating in free space, from the definition of acoustic impedance Z = p(t)/v(t) = ρc [1] we have:

$$p(t) = \rho c v(t) and also v(t) = \frac{p(t)}{\rho c} \tag{10.4}$$

where ρ and c are respectively air density and speed of sound wave propagation, and its product ρc the medium acoustic impedance. Substituting (10.4) in (1.60) give us:

$$I = \frac{1}{T}\int_{0}^{T} \frac{p^2(t)}{\rho c} dt = \frac{1}{\rho c}\frac{1}{T}\int_{0}^{T} p^2(t)dt \tag{10.5}$$

By applying the RMS definition in Eq. (10.2) and the relationship expressed in Eq. (10.3), we arrive at the fundamental equation that links the key parameters of sound wave propagation in free space:

$$I = \frac{p_{rms}^2}{\rho c} = \frac{W}{4\pi r^2} \tag{10.6}$$

The above equation gives us the modulus of intensity vector. Direction will be that of the radius of the spere of Fig. 10.3 and versus from the source to the point on the sphere, as illustrated on same figure.

An anechoic chamber is designed with walls that completely absorb incident sound waves, minimizing reflections. This is typically achieved using wedge-shaped foam or cone structures on the walls, which create multiple internal reflections within the material, effectively dissipating the wave energy. As a result, the amount of acoustic energy reflected back toward the source is extremely low, simulating free-field conditions.

In contrast, a reverberant chamber is constructed with highly reflective, rigid walls that reflect nearly all incoming sound waves. To ensure a uniform distribution of sound energy, the chamber's walls are intentionally non-parallel, promoting reflections in all directions and avoiding the formation of standing waves. This setup creates a diffuse acoustic field, ideal for measurements requiring a statistically uniform sound environment.

In a real acoustic field, it is possible to identify several distinct regions, each characterized by different wave behaviors:

- Near field
- Far field, which can be further divided into:
    - o Free field
    - o Reverberant field.

The near field is the region closest to the sound source, where sound pressure levels can vary significantly from point to point. This variation arises due to the complex interaction of multiple vibrating surfaces, wave interference, and non-uniform propagation phenomena. The acoustic behavior in this zone is highly dependent on the geometry and mechanics of the source.

The far field begins at a distance of at least one wavelength or one characteristic dimension of the source, whichever is greater. In the free field the source can be approximated as a point source, and the sound intensity decreases with distance according to the inverse square law, as described in Eq. (10.3). In decibel scale this corresponds to a 6 dB reduction of sound pressure level for each doubling of the distance from the source.

By dividing both sides of Eq. (10.7) by $I_0 p_0^2$, we obtain the normalized form of the relationship among the key acoustic field parameters:

$$\frac{I}{I_0 p_0^2} = \frac{p_{rms}^2}{I_0 p_0^2 \rho_0 c}$$

Using the definition of pressure and intensity levels (10.1) to the above equation we have:

$$10\log_{10}\left(\frac{I}{I_0}\right) = 10\log_{10}\left(\frac{p_{eff}^2}{p_0^2}\right) + 10\log_{10}\left(\frac{p_0^2}{I_0\rho_0 c}\right)$$

And so:

$$L_I = L_P + 10\log_{10}\left(\frac{p_0^2}{I_0\rho_0 c}\right)$$

In standard conditions the ratio $\frac{p_0^2}{I_0\rho_0 c} \approx 1$ therefore:

$$L_I = L_P \tag{10.7}$$

So, in the far free field and inside a perfect anechoic room the intensity level coincides with the pressure level SPL and attenuates both of 6 dB for each doubling of the distance r from the source.

Other common acoustic sources can be considered the line source or the plane wave source, both illustrated in Fig. 10.5.

For these kinds of sources, the fundamental relationship between pressure, power and intensity are different. For line sources wavefront are cylindrical, therefore surface area of power distribution is $2\pi rL$, therefore:

$$\begin{aligned}
&I = \frac{W}{2\pi rL} so \frac{I}{I_0} = \frac{WS_0}{W_0 2\pi rL} \\
&10\log_{10}\left(\frac{I}{I_0}\right) = 10\log_{10}\left(\frac{W}{W_0 L}\right) + 10\log_{10}\left(\frac{S_0}{2\pi r}\right) \\
&L_I = L_{W,L} - 10\log_{10} r - 8
\end{aligned} \tag{10.8}$$

where $L_{W,L}$ is the acoustic power per unit length W/L. So the sound pressure level produced by a linear source of known acoustic power level attenuates with distance, with an attenuation equal to 3 dB for each doubling of the distance.

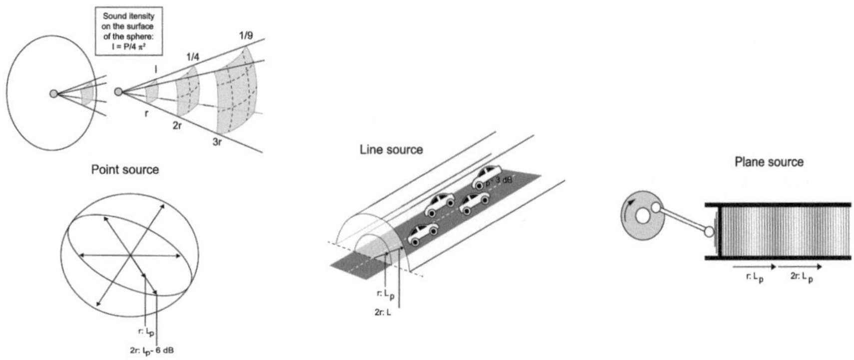

Fig. 10.5 Acoustic fields generated by different sources

A plane source consist of a piston from which energy is radiated into a tube which creates a plane wave inside the tube. Assuming that there is no energy loss through the walls of the tube, the intensity, i.e. the acoustic energy flowing in the tube, is independent of the distance from the source. Since the intensity is the same everywhere in the tube, the sound pressure level will not decrease with an increase in distance from the piston. This because in general the value is $I = \frac{W}{S}$, expressing the reference intensity as $I_0 = \frac{W_0(1pW)}{S_0(1m^2)}$ we have:

$$\frac{I}{I_0} = \frac{W}{I_0 S} = \frac{W S_0}{W_0 S} \text{and } 10\log_{10}\left(\frac{I}{I_0}\right) = 10\log_{10}\left(\frac{W}{W_0}\right) + 10\log_{10}\left(\frac{S_0}{S}\right)$$

and:

$$L_I = L_W + 10\log_{10}\left(\frac{S_0}{S}\right) = L_W - 10\log_{10}(S) \tag{10.9}$$

Remembering that in the far field $L_I = L_P$ we have that the sound pressure level produced by a flat source of known acoustic power level is independent of the distance, obviously assuming no loss through the tube walls.

In general, acoustic fields [1] can be estimated by defining boundary conditions of the source, shape and acoustical properties of the surroundings and performing numerical solution of the wave d'Alembert Eq. (1.34), discussed in Chap. 1.

Simplified approach can be done by simple beam tracing, similar to ray tracing using the principle of geometrical optics described in Chap. 2, but taking into account for acoustic "rays" the limitation given in acoustic field by the ratio wavelengh/(object dimension).

Many software are available to make acoustic field calulation like Sysnoise, Raynoise, etc.

10.1.2 Measurement of Sound Parameters

The microphone is the fundamental sensor for measuring sound [1]. It functions as a transducer, converting fluctuations in sound pressure at a specific point into a proportional electrical signal.

Microphones can be designed based on various operating principles. One common type relies on electromagnetic induction: a diaphragm is attached to a coil suspended within a magnetic field. When sound waves move the diaphragm, the coil oscillates, inducing an electrical current corresponding to the sound signal.

Another widely used type is the electret microphone, a variant of the condenser (capacitor) microphone. It incorporates a permanently charged material, known as an electret, placed between the condenser plates, eliminating the need for an external polarizing voltage.

To measure sound or ultrasound in water piezoelectric hydrophones are largely used, they use piezoelectric materials (such as quartz or lead zirconate titanate (PZT) that generate an electric charge when subjected to pressure changes from acoustic waves, as illustrated in Chap. 6 about ultrasonic scanners.

For precision acoustic measurements in air, the most commonly used microphones are condenser microphones that operate based on the capacitive principle and require an external polarizing voltage. These microphones offer high sensitivity, broad frequency response, and excellent linearity, making them ideal for laboratory and field measurements where accuracy is essential.

In this type of microphone, a thin diaphragm (membrane) is deformed by sound pressure fluctuations, causing a change in the distance between the diaphragm and a fixed backplate, which together form a capacitor. As the distance varies, the capacitance of the system changes accordingly.

These capacitance variations are converted into oscillating electrical signals, current and voltage, within an RC (resistor–capacitor) circuit. To enhance sensitivity, a high external polarizing voltage is applied across the capacitor. The resulting signal is then amplified for further processing or measurement, as illustrated in Fig. 10.6.

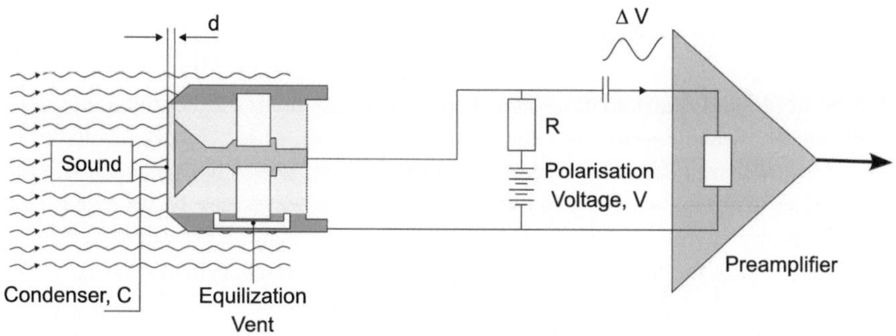

Fig. 10.6 The condenser microphone structure and conditioning circuit

The measurement principle can be illustrated considering that in the capacitor structure of the microphone we have (for a condenser of capacity C, charge Q, voltage V, dielectric constant ε, distance between the condenser plates d):

$$V = \frac{Q}{C} \text{ and } C = \frac{\varepsilon A}{d} \text{ therefore } V = \frac{Qd}{\varepsilon A} \tag{10.10}$$

So the fluctuating acoustic pressure p(t) generates on the microphone membrane a fluctuating force F(t) = p(t)A = p(t) π $D^2/4$ generating a displacement d(t) = k F(t) = k p(t)A where k is the microphone membrane stiffness, so we will have an output V(t) signal given by:

$$V(t) = \frac{Qd(t)}{\varepsilon A} = \frac{Qkp(t)}{\varepsilon} \tag{10.11}$$

Therefore, the electrical output V(t) of the microphone is linearly related to the sound pressure fluctuation p(t). Although this relationship appears to be independent of the diaphragm area A, in reality, the diaphragm area has a significant impact on its mechanical stiffness.

Microphones with a larger diaphragm diameter exhibit higher sensitivity, as the force acting on the diaphragm is proportional to the product of pressure and area.

Specifically, the force is given by F = p A = p π $D^2/4$ where D is the diaphragm diameter. This means that, all else being equal, sensitivity increases at least with the square of the diaphragm diameter.

However, increasing the diaphragm size also increases its mass and generally reduces its mechanical stiffness, both of which lower the resonance frequency. As a result, the microphone's usable frequency response range may be reduced. Therefore, in microphone selection, there is always a trade-off between sensitivity and frequency range, and the diaphragm size must be carefully selected to achieve the desired performance balance.

Microphones are standardized according to diameter, they are produced in different sizes with diameters of 1 inch, 1/2 inch, 1/4 inch and 1/8 inch. The most common is the 1/2 inch microphone which can be used for most of the applications in common acoustic field measurements.

When a microphone is placed in a sound field, as shown in Fig. 10.8a, it may alter the field itself, particularly when its physical dimensions are comparable to or larger than the wavelength of the incoming sound wave, as previously discussed in Chap. 1 and illustrated by Fig. 1.31 about interactions of waves with obstacles. If the microphone is not acoustically transparent, part of the wave may be reflected back by the diaphragm. In such cases, the microphone may register a pressure fluctuation up to twice that of the undisturbed field, due to wave complete reflection (Fig. 10.7a and b).

However, this effect is generally negligible at low and medium frequencies, where the wavelengths are on the order of meters or decimeters, and the microphone diameter is relatively small (typically 1–2 cm or less). At high frequencies, where the wavelength becomes comparable to the microphone size, the microphone's presence

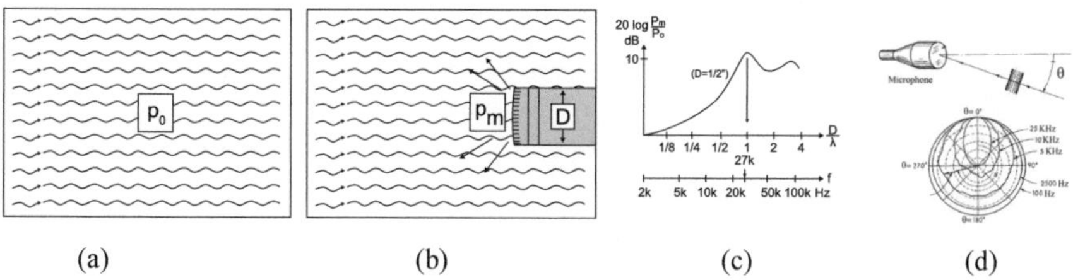

Fig. 10.7 The microphone directional response

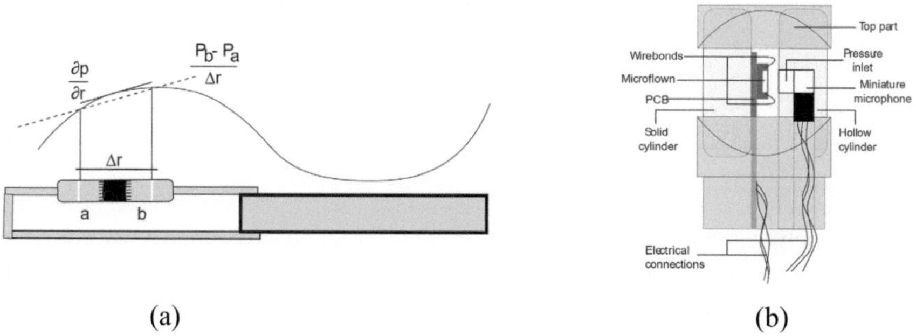

Fig. 10.8 The intensity probe with two microphones in (**a**) and with one microphone and a velocity sensor in (**b**)

can significantly disturb the sound field. In such situations, it is essential to consider the microphone's polar sensitivity pattern, as illustrated in Fig. 10.7b, to properly interpret the measurements and account for directional response effects.

The sound intensity in an acoustic field is measured using specialized instruments called intensity probes [1]. As previously introduced in Chap. 1 and derived from Eq. (10.5), intensity is a vector quantity, defined as the time-averaged product of acoustic pressure and particle velocity at a point in the sound field.

There are two main types of intensity probes, illustrated in Fig. 10.8:

1. p–v Probe: This type combines a microphone (to measure pressure fluctuations) with a hot-wire or hot-film anemometer (to measure particle velocity). It directly implements Eq. (10.5) by multiplying the instantaneous pressure and velocity signals and then averaging the product over time to compute the intensity vector.
2. Two-Microphone Probe: This design uses two closely spaced microphones, mimicking the way human hearing with two ear detects both amplitude and direction of incoming sound. This configuration allows the probe to estimate particle velocity based on the pressure gradient between the microphones, thereby determining the intensity vector. The principle relies on the Euler equation for fluids, which is conceptually similar to Newton's second law of motion ($F = ma$), relating force to mass and acceleration.

In the direction defined by the line connecting the two microphones and aligned with the intended intensity measurement, the Euler equation can be expressed as:

$$a = \frac{\partial v}{\partial t} = -\frac{1}{\rho}\frac{\partial p}{\partial r} \tag{10.12}$$

So, we can measure the wave fluctuation velocity from the pressure gradient as:

$$v(t) = -\int \frac{1}{\rho}\frac{\partial p}{\partial r} dt \tag{10.13}$$

Using two microphones in the position A and B and approximating the gradient $\frac{\partial p}{\partial r}$ with the finite difference (p_B-p_A) divided by the microphones spacing Δr and taking as pressure p(t) the mean value ($p_A + p_B$)/2 we have for the instantaneous intensity, from (10.13) the expression:

$$I(t) = -\frac{(p_A(\mathrm{t}) + p_B(\mathrm{t}))}{2} \int \frac{1}{\rho}\frac{(p_B(\mathrm{t}) - p_A(\mathrm{t}))}{\Delta r} \tag{10.14}$$

To measure the average sound intensity along the r-direction, that is, along the line connecting the two microphones, the data processing system must compute the following time-averaged product of the two pressure fluctuation signals, in accordance with Eq. (10.14):

$$I_r = \frac{1}{T}\int_0^T -\frac{(p_A(t) + p_B(t))}{2} \int \frac{1}{\rho}\frac{(p_B(t) - p_A(t))}{\Delta r} dt \tag{10.15}$$

This give us a component of the intensity vector, that along the line of the two microphones. Rotating the probe, it is possible to measure the maximum intensity amplitude and also more component of it, or using a probe with 3 microphone couples it is possible to measure all the tree components of the intensity vector.

As illustrated in Fig. 10.9, the particle velocity is estimated by approximating the pressure gradient of the acoustic wave using the finite difference between the pressure values measured at the positions of the two microphones. This corresponds to estimating the phase difference of the pressure wave between the two points where microphones are placed.

The accuracy of this gradient approximation by finite difference depends on the spacing between the microphones relative to the wavelength of the sound wave. When the spacing is small compared to the wavelength, the finite difference closely approximates the true gradient, making this method particularly suitable for low-frequency measurements, where wavelengths are long.

However, if the phase difference between the two microphone signals becomes too small, it may fall within the phase uncertainty of the microphone channels, leading to an inaccurate estimation of the pressure gradient.

This implies two important limitations in intensity probes:

- A smaller microphone spacing improves the estimation accuracy at higher frequencies, but introduces an upper frequency limit beyond which the phase resolution is insufficient.

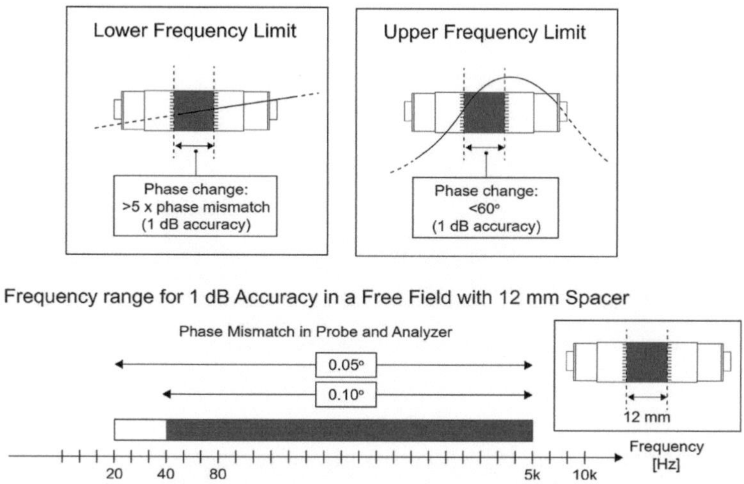

Fig. 10.9 Intensity probe frequency limitations due to phase uncertainties

- Conversely, a larger spacing increases the ability to resolve low-frequency components, but may reduce the accuracy at high frequencies and introduces lower frequency limitations due to phase mismatch between the two microphone channels and to noise.

Therefore, the choice of microphone spacing in an intensity probe involves a trade-off between frequency range and measurement accuracy, and must be carefully optimized based on the intended application and required uncertainty.

In Fig. 10.9 the horizontal bar at the bottom shows the frequency range for 1 dB accuracy in a free field using a 12 mm spacer. I also shows the effect of phase mismatch. In this example, with a 12 mm spacer, a phase mismatch improvement from 0.10 to 0.05° gives a frequency range improvement from 40 to 20 Hz.

The sound power emitted by a source generating an acoustic field can be determined using various indirect measurement techniques, primarily by measuring either sound pressure (with microphones) or sound intensity (with intensity probes).In free-field conditions or within an anechoic chamber, the sound pressure is directly related to the emitted sound power through Eq. (10.7). This method is standardized in numerous international norms, which typically require measurements at multiple spatial points around the source to verify field uniformity and conditions. Common measurement geometries, such as the hemisphere and shoebox shown in Fig. 10.10, are used to sample the sound field around the source. The floor is assumed as a perfect sound reflecting surface and the chamber in this case is named semi-anechoic. In a reverberant field, such as that created inside a reverberation chamber, other simplified expressions can be derived to relate the spatial and temporal average of sound pressure to the total sound power of the source. Measurements in this context follow different

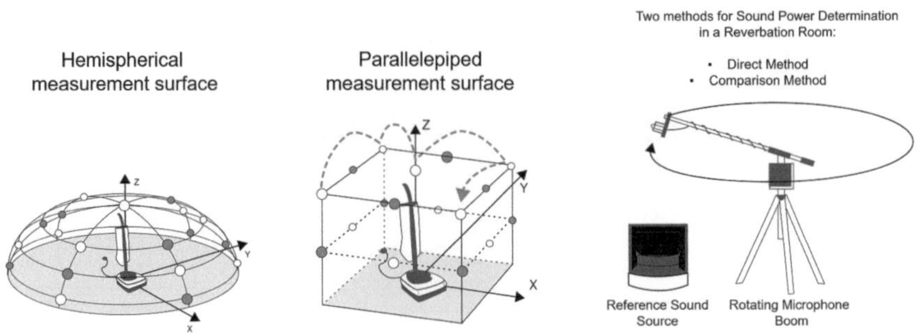

Fig. 10.10 Measurement of sound power in anechoic and reverberating room

standardized procedures and use specific geometric setups, such as the rotating microphone, as illustrated in Fig. 10.10, to verify the uniform acoustic energy distribution throughout the chamber.

According to the definition of sound intensity, the sound power emitted by a source is given by the surface integral of the sound intensity vector over a closed surface surrounding the source. Mathematically, it can be expressed as the surface flux integral of the intensity field:

$$\mathrm{W} = \int \boldsymbol{I} \cdot \boldsymbol{dA} \tag{10.16}$$

If we measure the component of sound intensity normal to the surface (I_n) at multiple points on a closed surface surrounding the source, we can approximate the surface integral using a finite summation. This involves multiplying each measured normal intensity value by the corresponding surface area element around the measurement point. The approximation is given by:

$$\mathrm{W} = \sum I_n \Delta A \tag{10.17}$$

Like illustrated in Fig. 10.11a. An alternative is to measure directly the average given by Eq. (10.18) by continously moving the intensity probe on each face of area A_1, A_2,... around the sound source.

Above sound power measurements can be performed in any real environment, not using an anechoic or reverberating room or other hypothesis of simple relationship between acoustic field parameters.

Intensity measurements on many points of a surface in front of an emitting source, also in near field, can be very useful to identify the main part of it emitting sound, as illustrated in Fig. 10.11c for a washing machine emitting the most of the noise from the bottom.

Another effective method for identifying the primary sources of sound emission through vibrating surfaces and estimating the resulting acoustic field is by directly measuring the surface vibrations of the sound source. This can be achieved using a scanning laser Doppler vibrometer (SLDV) as illustrated in Chap. 9, which records

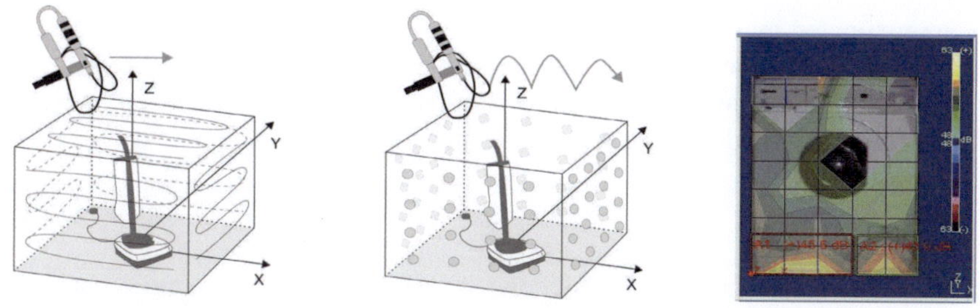

(a) Sweep measurement method (b) point measurement method (c) a typical result

Fig. 10.11 Measurement of sound power and source identification using an intensity probe

the vibration time history point by point across the surface. The collected data can then be analyzed using a Fourier transform to obtain the vibration spectrum frequency by frequency.

Amplitude and phase velocity data of each frequency, measured by using the SLDV on each point of the vibrating surface emitting sound, can be employed to compute the acoustic field in front for example of a vibrating belt as performed in [2, 3].

In general acoustic emission generated by a vibrating surface can be estimated by the superposition principle [12]. The idea is that the vibrating surface is composed by an array of simple sources; each source corresponds to the portion of area where vibration measurement is performed. Each acoustic wave component at frequency ω can be described by a complex exponential function $e^{j\omega t}$, so the acoustic pressure $p_i(t)$ radiated by each element A_i of this vibrating surface with amplitude v_i and area A_i in a generic point B_i at a distance r_i (as illustrated in Fig. 10.12c), can be estimated by [12]:

$$p_i(t) = S_i(i\,\rho\, c\, k\, v_i/2\pi\, r_i)e^{j}(\omega t\; - \; kr_i - \phi i\cdot) \tag{10.18}$$

Following an approach similar to light ray tracing and wave superimposition in space and time all these contribution can be added, as described in Chap. 2 for generic waves. The total pressure in each B_i grid points due to the whole surface vibration can

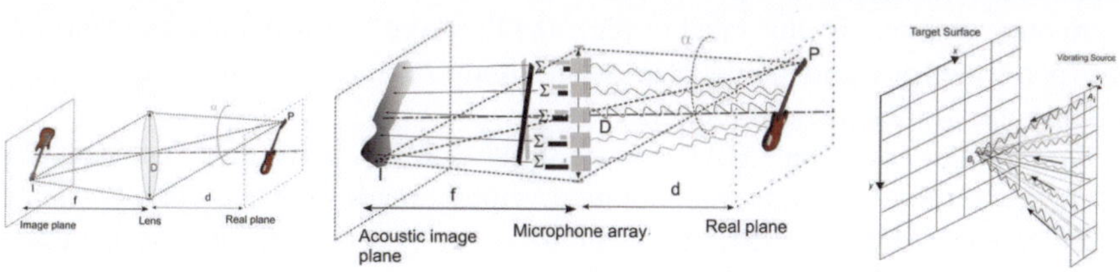

(a) (b) (c)

Fig. 10.12 Acoustic beam forming and optical imaging: scheme of pressure calculation on a grid of B_i points by the superposition principle

be approximated by summing the contributions of all the A_i vibrating elements on each B_i point of the acoustic field. Clearly, for each points of the measurement grid the phase difference must be considered. This phase difference has two components: the first is due to the difference of path length

$$\Delta_{ri} = r'_i - r_i \tag{10.19}$$

which depends on geometry; the second is due to the different initial phases ϕ_i generated by starting waves from each A_i vibrating elements, given by:

$$\Delta\phi_i = \phi_{i'} - \phi_i \tag{10.20}$$

A laser scanning vibrometer allows to measure $\Delta\phi_i$ phase differences of the vibrating surface if a reference vibration signal for the subsequent acquisitions on the different A_i vibrating elements is provided. Once the vibration spectrum on a grid of points of the surface has been measured, the total sound pressure in each point B_i of the acoustic field grid can be calculated [14] by summing all the waves on each point of the acoustic field as:

$$p_{tot}(t) = (i\,\rho\,c\,k/2\pi)\sum_i S_i(v_i/r_I)e^i(\omega t - kr\,\varphi_i) \tag{10.21}$$

Then the RMS value of this expression p_{RMS} can be calculated by Eq. 10.2. Considering no reflections (free field conitions) the acoustic intensity on each point B_i of the grid can be estimated by:

$$I_i = \left(p_{RMS\,i}\right)^2/\rho c \tag{10.22}$$

The total acoustic power flowing across the acoustic field surface grid can be calculated by summing the products of the sound intensity values I_i and the corresponding surface elements area S_i over the entire grid.

Intensity distribution can be also estimated by more complex Boundary Element Models, realized by Sysnoise or other commercial software for acoustic field modelling, and then compared with direct intensity measurements by using a probe positioned on the same grid points, as done in [4] and illustrated in Fig. 10.13.

Another example is illustrated in Fig. 10.14b where intensity measurements has been performed on a grid in front of an engine timing belt transmission cover, working inside a classical engine test room [5].

For the same test setup, a 3D acoustic simulation was performed once again by Sysnoise. In this simulation, the vibration velocity maps obtained via a Scanning Laser Doppler Vibrometer (SLDV), as shown in Fig. 10.14a, were used as input data for Sysnoise model. The resulting sound field calculated is presented in Fig. 10.14c.

To validate the model, this result was compared with direct acoustic intensity measurements taken using an intensity probe, illustrated in Fig. 10.14b. This approach enabled both the identification and quantification of the dominant noise sources. In

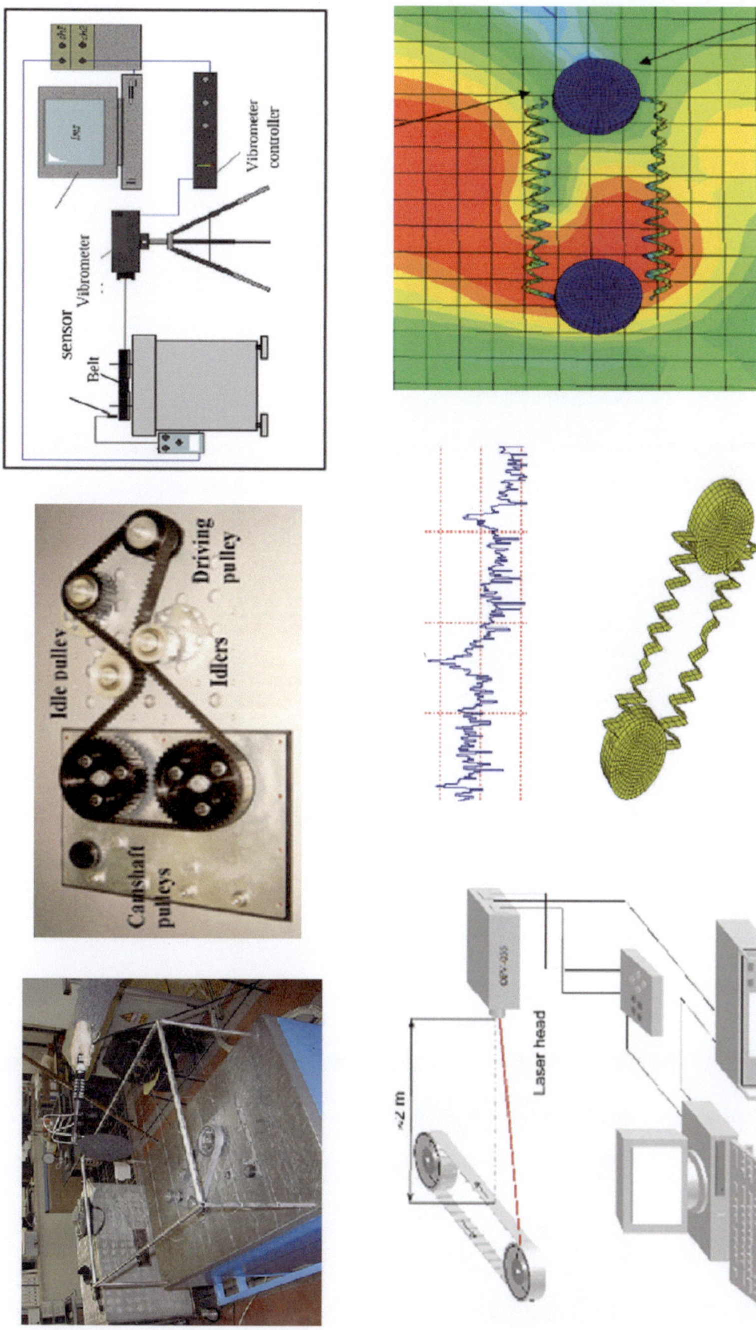

Fig. 10.13 Acoustic emission intensity measurement of timing belt transmission and calculation by a Sysnoise model from vibration measurements by LSDV

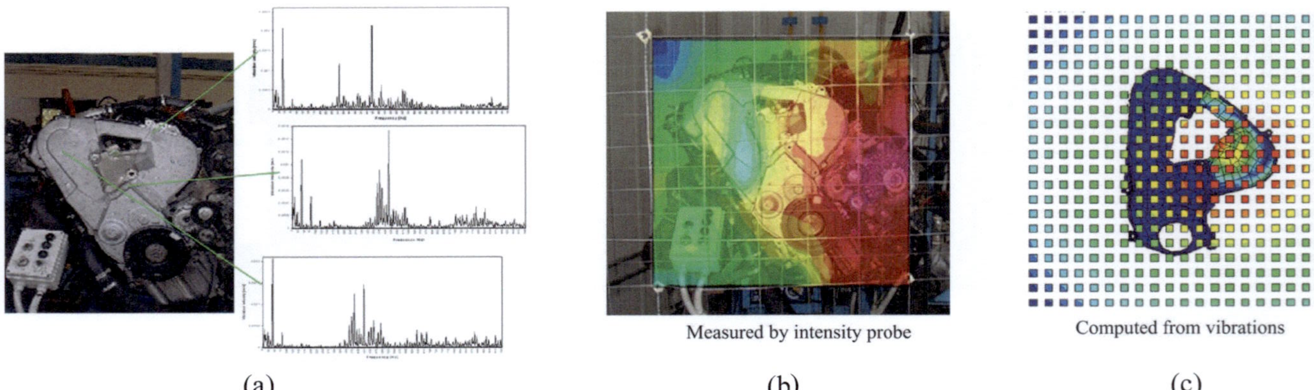

Fig. 10.14 intensity field measurement on an internal combustion engine operating in a testing room.

the analyzed case, the main source of noise was traced to a resonance occurring in the right section of the transmission's plastic cover [5]. The validated model thus serves as a predictive tool for evaluating by simulation the effectiveness of potential noise mitigation strategies.

To measure sound fields, systems based on microphone arrays can also be employed. A commonly used configuration involves positioning microphones along a circular array, with a central camera capturing images of the sound source. The sound field is then reconstructed from the pressure signals measured by the microphones. This setup is depicted in Fig. 10.15a, where the circular array of microphones was used to analyze the acoustic field generated by vortex shedding around car side mirrors, inside a wind tunnel.

Alternative configurations are also possible. For example, Fig. 10.15b shows an arrangement where microphones are distributed along three radial lines. This configuration was used to measure the sound field produced by the vibrating tip of a wind turbine blade.

Each microphone captures the time-domain waveform of the pressure fluctuations at its specific location. Through advanced signal processing techniques, the entire propagating sound field can then be reconstructed. This reconstructed field is subsequently "projected" onto the surface of the source image, captured by the central camera, using superimposition and wave propagation methods and models.

To identify better the noise sources in front of an emitting vibrating surface an arrays of microphones, following the shape of the object, can be placed in the very near field, as illustrated in Fig. 10.16. The measurement technique is named conformal mapping.

It is a mathematical method used in sound field analysis and modeling, particularly when dealing with complex geometries where conventional methods become computationally intensive or less accurate. Conformal mapping is a mathematical transformation that preserves angles but not necessarily lengths or areas. In simpler terms, it transforms a complex-shaped domain into a simpler one (often a circle or rectangle), while maintaining the local geometrical properties (like angles between curves). This makes it easier to solve physical problems (like wave propagation) that are hard to tackle directly in the original shape. In acoustics, conformal mapping techniques are used to simplify the analysis of sound propagation, vibration behaviour, and acoustic radiation from irregular surfaces by geometry simplification and solving the d'Alembert wave equations, described in Chap. 1. Once the solution is obtained in the simplified domain, it can be mapped back to the original geometry.

10.2 Measurements of Contact Pressure Fields Between Solid Bodies

Pressure distribution fields are particularly important in the analysis of contact areas between two elastic bodies. Typical applications include tire-road interactions, finger-object contact in robotic grasping and manipulation, as well as contact scenarios

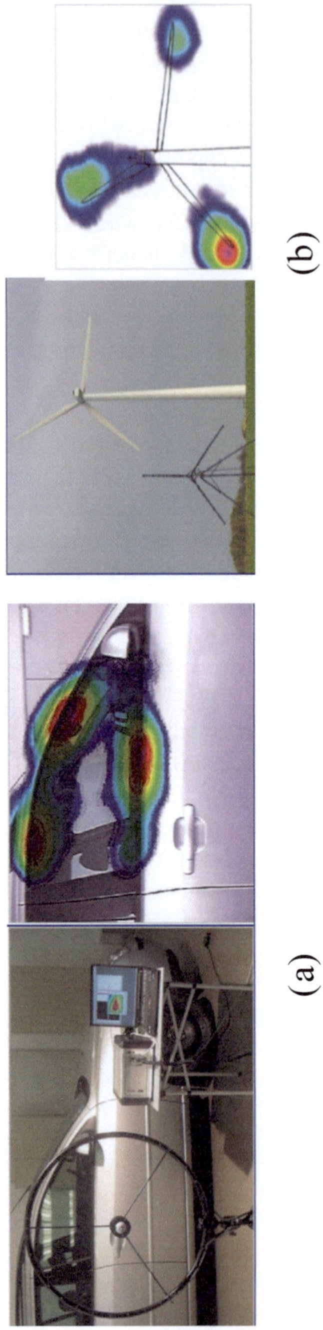

Fig. 10.15 Acoustic camera measurement systems from HBK company

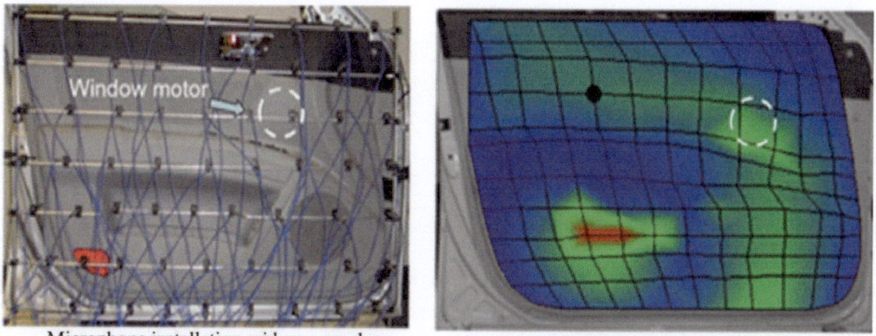

Microphone installation grid on a car door

Acoustic field measured

Fig. 10.16 Conformal mapping

involving machines, humans, and automated systems. Understanding these pressure fields is essential for optimizing performance, control, and safety across a wide range of engineering and biomechanical applications.

10.2.1 Images of Pressure Sensitive Paints and Films

The simplest pressure measurement techniques rely on pressure-sensitive paints, films, or polymers, materials that change their color intensity in response to the applied pressure, as illustrated in Fig. 10.17. Among the most widely used are pressure-sensitive films produced by Fujifilm. While these systems offer ease of use and immediate visual feedback, their measurement accuracy is relatively limited, typically within 10% under optimal application conditions. Other producer of pressure sensitive film are McMaster-Carr, Pressure Sensitive Label Films by Taghleef Industries, etc.

Pressure-sensitive films contain microcapsules embedded within a color-forming layer. When pressure is applied, these microcapsules rupture, releasing a colorless dye that is absorbed by an underlying developer layer. This triggers a chemical reaction that produces a red coloration, with the intensity of the color directly proportional to the applied pressure. The microcapsules are engineered with varying sizes and rupture thresholds, and are uniformly distributed across the film. As a result, the film generates a color density that accurately reflects the pressure magnitude, according to a predefined color scale.

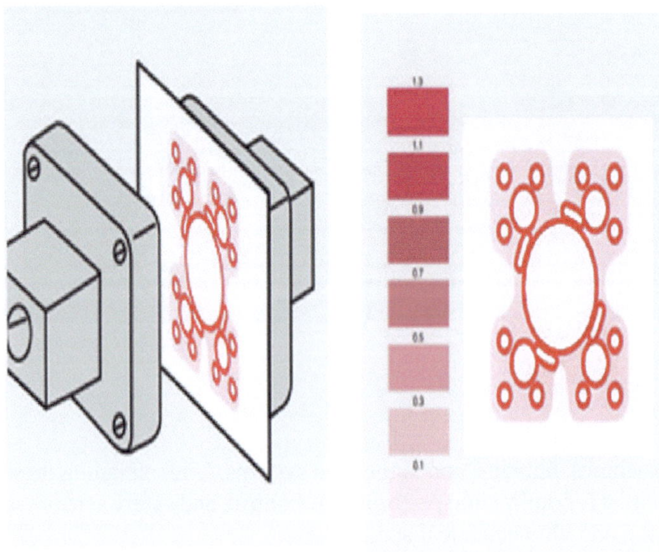

Fig. 10.17 Barosensitive film

10.2.2 Images of Scattered Light at the Contact Surface

Contact pressure distributions between two bodies can be measured by analyzing the increase in light scattering in the direction normal to the interface between rough surfaces. As the contact pressure rises, the micro-asperities of the rough surfaces become statistically more aligned parallel to the surface plane and more densely distributed.

This enhances surface reflections and increases the intensity of scattered light observed in the normal direction, as illustrated in Fig. 10.18.

By using a glass or plexiglass plate illuminated along its edges with an LED strip, it is possible to generate diffuse lighting within the plate.

When viewed perpendicularly to the plate at the contact interface, the intensity of scattered light visibly increases with the rise in contact force and pressure. This optical phenomenon provides a practical method to qualitatively assess contact pressure distribution. The technique is described in detail in reference [6].

Various techniques and material layers can be employed to enhance the linearity and sensitivity of this optical pressure measurement method, as detailed in [6]. Additionally, calibration procedures can be applied to define measurement uncertainty.

Figure 10.19a, b illustrate an experimental setup consisting of a black box enclosed by a plexiglass layer, with a recording camera positioned at the bottom, inside the box. Once the box is sealed with the plexiglass cover, a small tire model is placed on top of the plate. This configuration allows for the visualization and analysis of the pressure distribution at the tire–flat plate interface.

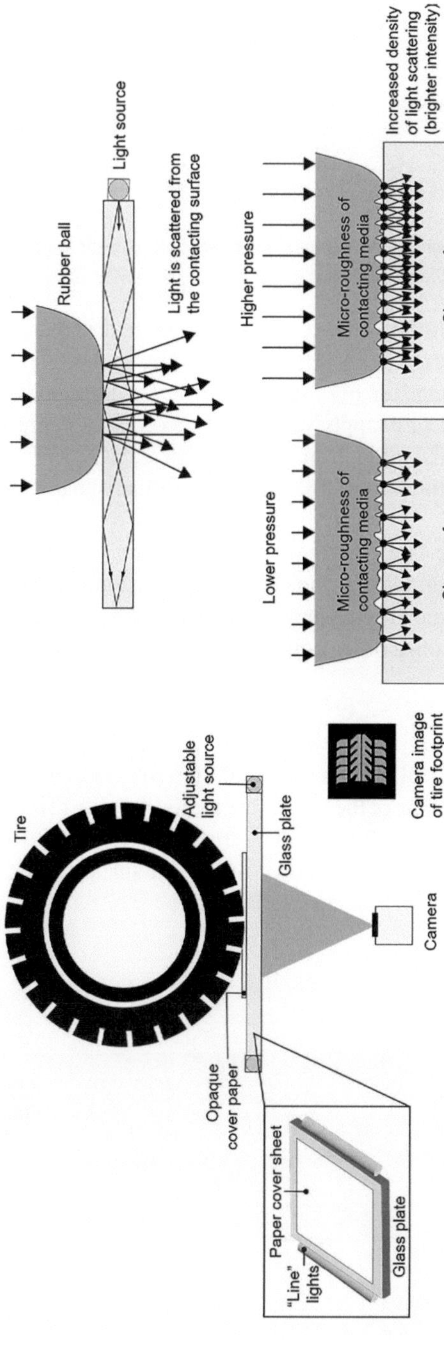

Fig. 10.18 Contact pressure measurements by light scattering images

Fig. 10.19 Test bench setup for contact pressure measurement by light scattering

Before performing measurements, a calibration procedure must be carried out to determine the system's sensitivity, defined as the pressure-to-light intensity ratio. This test bench is illustrated in Fig. 10.20a and b, where a rubber membrane is inflated using compressed air to apply varying levels of uniform pressure to the plate surface. The results of this calibration, showing the relationship between applied uniform pressure and light scattering intensity uniform levels, are presented in Fig. 10.20c.

From the calibration tests a non linear curve for each pixel is obtained, so four coefficient matrix **A**, **B**, **C** and **D** of the interpolating polynomial describing this curve were obtained. To determine the pressure matrix **P** of each pixel from the grey levels matrix **I** of the image the third order polinomial matrix relationship is therefore:

$$\boldsymbol{P} = \boldsymbol{A}\boldsymbol{I}^3 + \boldsymbol{B}\boldsymbol{I}^2 + \boldsymbol{C}\boldsymbol{I} + \boldsymbol{D} \tag{10.23}$$

After completing the calibration, placing the tire model on the plate and performing the data processing of the image, accordingly to Eq. (10.23), yields typical results such as those shown in Fig. 10.21. Other applications has been performed to measure pressure distributions at human finger-flat surface interface, to validate FEM haptic models [7] for robotic applications.

a) b) c)

Fig. 10.20 Calibration of light scattering contact pressure measurement technique

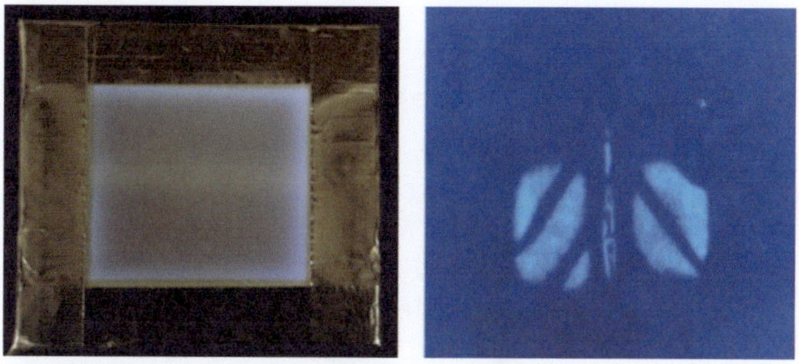

Fig. 10.21 Examples of results

10.2.3 Thermoelastic Techniques for Contact Pressure Measurements by Infrared Transparent Materials

Contact pressure can also be measured using the thermoelastic effect in combination with infrared (IR) transparent materials. Among the most effective IR-transparent materials is sapphire, though common polycarbonate plastics can also be used. Naturally, materials typically employed in the optics of infrared cameras possess good infrared transparency. By placing an IR-transparent material in contact with an object—such as the tire model mentioned in the previous section, and observing it with a thermal camera, one can visualize only the infrared radiation emitted by the contact surface, as seen through the transparent layer. Due to the thermoelastic effect described in Chap. 8, variations in temperature correspond to changes in the first stress invariant at the contact area. These variations are influenced by the full three-dimensional stress state, including the normal contact stress. Since the first stress invariant corresponds to the contact pressure p, the relationship can be quantitatively described by Eq. (8.40), so for the contact pressure changes Δp we can write:

$$\Delta T = -\frac{\alpha}{\rho \cdot c_p} \cdot T_0 \cdot \Delta(\sigma_x + \sigma_y + \sigma_z) = -\frac{\alpha}{\rho \cdot c_p} \cdot T_0 \cdot \Delta p \tag{10.24}$$

Therefore, by applying dynamic cyclic loading, from temperature fluctuation ΔT it is possible to measure the pressure amplitude Δp and thus determine its distribution across the contact surface. Figure 10.22 presents results for two configurations: the contact between a tire model and a flat surface, and the contact between a sphere and a flat surface.

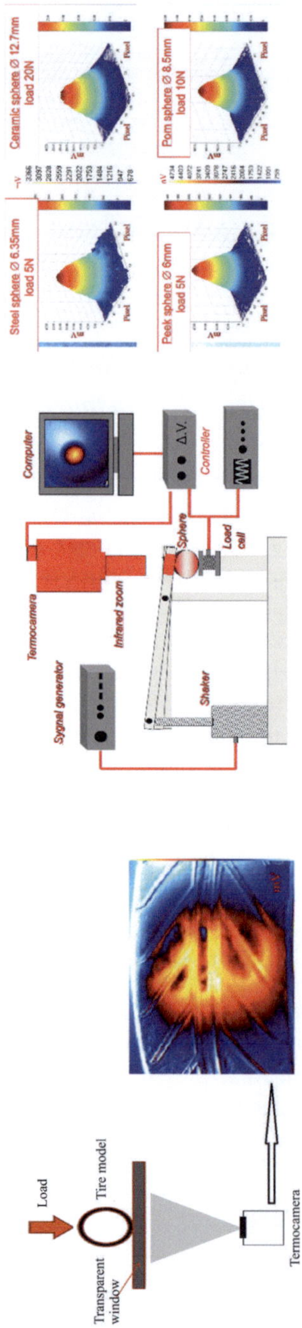

Fig. 10.22 Contact pressure measurements by thermographic techniques

10.3 Stress Distribution Measurements at Fluid–Solid Interfaces

Measuring pressure distribution on solid surfaces exposed to fluid flow is essential for experimentally optimizing the aerodynamic and hydrodynamic performance of objects such as aircraft, cars, ships, buildings, bridges, turbine blades, pumps, and more. To achieve this, scale models of these objects are typically constructed and tested in controlled environments such as wind tunnels or water flow channels.

The classical method for measuring pressure distribution does not rely on image analysis. Instead, it involves drilling small holes, named pressure taps, into the surface of the model. These pressure taps are connected via tubing to a pressure measurement system either to a single pressure transducer through a mechanical pressure scanner, or to multiple transducers connected to an electronic scanning unit for digital signal acquisition. This measurement technique is illustrated in Fig. 10.23.

In fluid dynamics applications, pressure distribution measurements using image analysis are primarily based on the application of barosensitive (pressure-sensitive) paints to the surface of the model [8]. These techniques involve capturing images of the painted surface and processing the data to extract pressure information, as illustrated in Fig. 10.24.

Typical barosensitive paints [9] consist of oxygen-sensitive fluorescent molecules embedded in an oxygen-permeable binder.

The method exploits the property of certain luminescent molecules that are quenched by the presence of oxygen. Once applied, the paint is illuminated using a high-intensity light source, typically a 400 nm ultraviolet (UV) LED. When the luminescent molecules absorb photons from the LED, they enter an excited singlet energy state.

The return to the ground state typically occurs via the emission of a longer-wavelength photon (red-shifted fluorescence).

However, when oxygen interacts with the excited molecules, the return to the ground state occurs through a non-radiative process—this is known as oxygen quenching. The intensity of emitted light is thus reduced in regions with higher oxygen concentration.

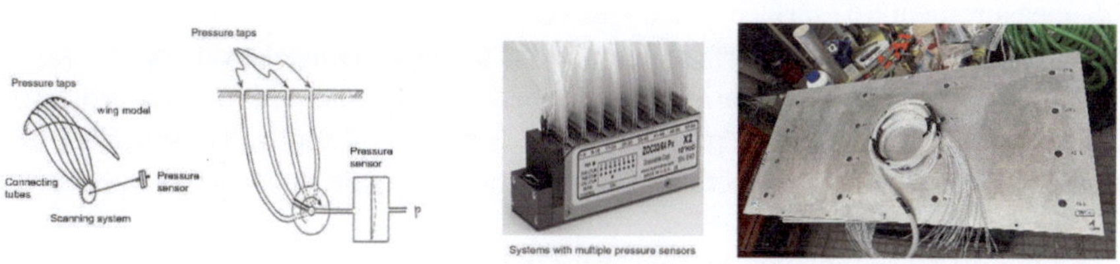

(a) (b) (c)

Fig. 10.23 Pressure scanners (**a**), sensor arrays (**b**) and a plate with holes (**c**) for pressure distribution measurements

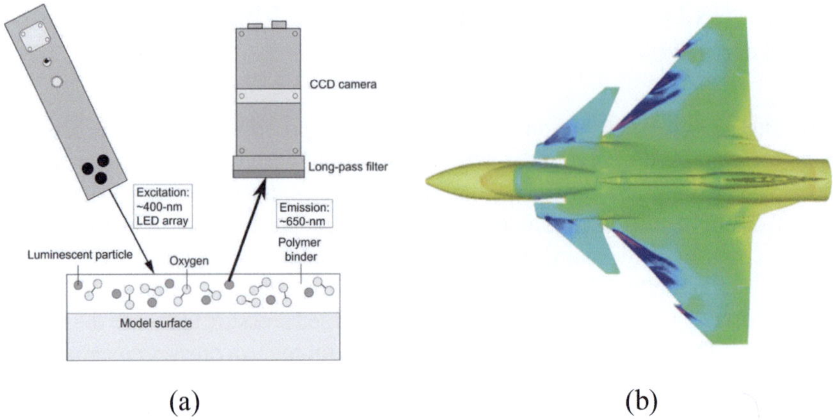

Fig. 10.24 Barosensitive paints technique for pressure distribution measurements

A camera, equipped with a long-pass optical filter, records the resulting fluorescence distribution. Since the oxygen concentration near the surface is directly influenced by the local barometric pressure (due to Henry's law), the captured intensity pattern can be quantitatively correlated with the surface pressure distribution.

To establish the relationship between image intensity and pressure, a sample of the pressure-sensitive paint is calibrated by subjecting it to a range of known pressures and temperatures. This process defines the correlation between gray levels in the recorded images and the actual pressure values. A typical example of this application on an aircraft model is shown in Fig. 10.24b. Through comparison with conventional pressure tap measurements, the technique has demonstrated an uncertainty of approximately 5% [10].

In addition to normal pressure, tangential (shear) stresses on the surface of a body interacting with a fluid are of significant interest. These stresses play a major role in generating drag forces, which in turn impact energy efficiency, fuel consumption, and overall performance in vehicles such as cars, aircraft, and ships.

An image-based technique for measuring tangential stresses, Differential Dynamic Thermography, was previously introduced in Chap. 8. This method enables the measurement of surface shear stress distribution by detecting temperature variations caused by fluid–surface interactions under dynamic conditions.

Another method for measuring surface shear stress is the oil film interferometry. This technique is based on the principle that a droplet of oil applied to the surface of a model will spread under the action of skin friction—i.e., the shear stress τ generated by the boundary layer flow.

As the oil layer thins due to this shear-induced motion, it eventually reaches a thickness at which interference effects become visible when the surface is illuminated with monochromatic visible light. These interference patterns, or fringes, arise from variations in the oil film thickness and can be captured with an optical system.

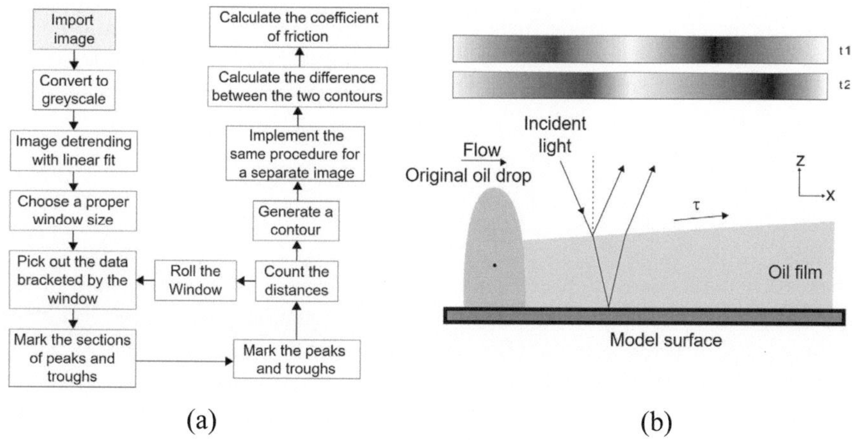

Fig. 10.25 Oil film interferometry principle and processing basics

The spacing and evolution of the fringes provide a direct visualization of the local shear stress distribution over time. An example of this phenomenon is shown in Fig. 10.25.

The basic equation form of the relationship between oil film thickness h_0 and wall shear stress is the Brown and Naughton equation given by [11]:

$$\frac{\partial h_o}{\partial t} + \frac{\partial}{\partial x}\left(\frac{\tau_{w,x} h_o^2}{2\mu}\right) + \frac{\partial}{\partial z}\left(\frac{\tau_{w,z} h_o^2}{2\mu}\right) = 0$$

where $\tau_{w,x}$ and $\tau_{w,z}$ are the wall shear stress components on the surface coordinates x and z respectively, and t is the time. For the one-dimensional situation, changes in the wall shear stress on the z coordinate can be ignored. Then, the equation can be rewritten as

$$\frac{\partial h_o}{\partial t} + \frac{\partial}{\partial x}\left(\frac{\tau_{w,x} h_o^2}{2\mu}\right) = 0$$

Under the flow and effect of shear stress, the oil becomes thin. Furthermore, when illuminated by a monochromatic light source, the fringe pattern of the oil film can be visible on a reflective surface. Then, the thickness of the oil film can be determined by.

$$h_o = \frac{\lambda\phi}{4\pi}\left(\frac{1}{\sqrt{n_0^2 - \sin^2\theta_i}}\right) and \frac{\partial h_o}{\partial t} + \frac{\partial}{\partial x}\left(\frac{\tau_{w,x} h_o^2}{2\mu}\right) = 0.$$

where ϕ is the phase difference between the portion of the beam reflected from the top of the oil and that transmitted through the oil, n_0 is the reflectance index of the oil, and θ_i is the local illumination angle. Skin friction coefficient can be determined as [11]:

$$C_f = \frac{2n_0 \cos\theta_r \Delta x}{N\lambda \int_{t_1}^{t_2} \frac{q_\infty(t)}{\mu(t)} d}$$

where θ_r is the angle of light refracted through the oil, λ is the wavelength of the light source, N is the number of fringes used in the equation, Δx is the total width of N fringes, q_∞ is the pressure free flow dynamics, μ is the viscosity of the oil, and t_1 and t_2 are the start time and end time for acquiring images, respectively.

Tangential (shear) stresses at solid–liquid or solid–gas interfaces can also be measured using specialized paints or liquid crystal (LC) mixtures that change their optical properties in response to applied shear stress. Liquid crystals are sensitive to mechanical stimuli, including shear stress, which induces changes in their molecular alignment. This reorientation alters their optical characteristics, such as birefringence or reflected color, especially when viewed under polarized light.

Typically, a thin film of cholesteric or nematic liquid crystal is applied to the surface under investigation. When a fluid (air or liquid) flows over the surface, the shear stress causes the LC molecules to reorient, with the greatest effect occurring near the boundary layer where tangential stress is dominant. This molecular reordering influences the way light is reflected or transmitted through the layer, leading to visible changes in brightness or interference color patterns. These optical changes correlate with the magnitude and direction of the local shear stress. Through calibration and image analysis software, the optical response can be quantitatively mapped to tangential stress values.

In Fig. 10.26a, based on data from [9], a liquid crystal-coated plate is calibrated in a wind tunnel. Illumination is provided by a tungsten–halogen lamp, and varying the flow velocity changes the tangential stress, which in turn modifies the color and intensity of the light reflected by the LC layer. Once calibration is complete, the same plate is used to measure the tangential stress distribution generated by an impinging air jet. The results are shown in Fig. 10.26b, c.

Liquid crystals are also employed under similar testing conditions to measure temperature distributions on vehicle models during aerodynamic experiments in wind tunnels, as illustrated in [13].

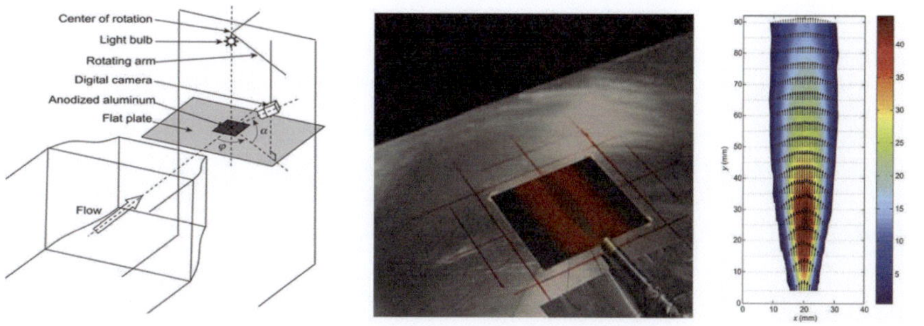

(a) (b) (c)

Fig. 10.26 Liquid crystals applications to measure tangential stresses

References

1. T.D. Rossing, *Springer Handbook of Acoustics* (Springer Hanbooks, 2014)
2. G.M. Revel, G.L. Rossi, Sound power estimation by laser Doppler vibration measurement techniques. J. Shock Vib. **5**, 297–305 (1998). IOS press, Amsterdam, Olanda
3. R. Di Sante, G.M. Revel, G.L. Rossi, Measurement techniques for the acoustic analysis of syncronous belts. Meas. Sci. Technol. J.11, 1463–1472 (2000). IOP Publishing Ltd, UK
4. R. Di Sante, G.L. Rossi, A new approach to the measurements of transverse vibration and acoustic radiation of automotive belts using laser Doppler vibrometry and acoustic intensity techniques. Meas. Sci. Technol. J. **12**, 525–533 (2001)
5. D. Sante, C. Santolini, G.L. Rossi, S. Lolli, D. Guyon, Vibro-acoustic design of covers for belt drives, in *Proceedings of the IMAC XIX Conference* (Kissimmee, Florida, USA). SEM, Bethel, Connecticut, USA.
6. V. Roth, B. Moore, J. Turner, An optical method for normal contact stress measurement with applications to tire-road interaction, Bridgestone/Firestone North American Tire, LLC Tire Development and Technology 1200 Firestone Parkway Akron, OH 44317
7. M.L. D'Angelo, F. Cannella, M.o Bianchi, M. D'Imperio, E. Battaglia, M. Poggiani, G. Rossi, A. Bicchi, D.G. Caldwell, An integrated approach to characterize the behavior of a human fingertip in contact with a silica window. IEEE Trans. Haptics **10**(1) (2017)
8. M. Jahanmiri, Pressure Sensitive Paints: The Basics and Applications, Research report 2011:07—Division of Fluid Dynamics (Department of Applied Mechanics, Chalmers University Of Technology, Göteborg, Sweden 2011)
9. Z. JiSong, S. Peter, G. Liang Xian, Color change characteristics of two shear-sensitive liquid crystal mixtures (BCN/192, BN/R50C) and their application in surface shear stress measurements. Chinese Sci. Bull. Fluid Mech. **56**(27), 2897–2905. https://doi.org/10.1007/s11434-011-4673-y
10. T. Nakano, N. Fujisawa, Wind tunnel testing of shear-stress measurement by liquid-crystal coating. J. Vis. **9**(2), 135–136 (2006)
11. J. Lunte, E. Schülein, Wall shear stress measurements by white-light oil-film interferometry. Exp. Fluids **61**, 84 (2020)
12. ISO/TR 7849—Estimation of airborne noise emitted by machinery using vibration measurement
13. F. Coletti, Isothermal lines detection on liquid crystals coated flat and concave surfaces, master degree thesis, University of Perugia, Italy, 2005.

MIX
Papier aus verantwortungsvollen Quellen
Paper from responsible sources
FSC® C105338

If you have any concerns about our products,
you can contact us on
ProductSafety@springernature.com

In case Publisher is established outside the EU,
the EU authorized representative is:
Springer Nature Customer Service Center GmbH
Europaplatz 3, 69115 Heidelberg, Germany

Printed by Libri Plureos GmbH
in Hamburg, Germany